Granular Computing

Studies in Fuzziness and Soft Computing

Editor-in-chief
Prof. Janusz Kacprzyk
Systems Research Institute
Polish Academy of Sciences
ul. Newelska 6
01-447 Warsaw, Poland
E-mail: kacprzyk@ibspan.waw.pl
http://www.springer.de/cgi-bin/search_book.pl?series=2941

Further volumes of this series can be found at our homepage.

Vol. 47. E. Czogała† and J. Łęski
Fuzzy and Neuro-Fuzzy Intelligent Systems, 2000
ISBN 3-7908-1289-7

Vol. 48. M. Sakawa
Large Scale Interactive Fuzzy Multiobjective Programming, 2000
ISBN 3-7908-1293-5

Vol. 49. L.I. Kuncheva
Fuzzy Classifier Design, 2000
ISBN 3-7908-1298-6

Vol. 50. F. Crestani and G. Pasi (Eds.)
Soft Computing in Information Retrieval, 2000
ISBN 3-7908-1299-4

Vol. 51. J. Fodor, B. De Baets and P. Perny (Eds.)
Preferences and Decisions under Incomplete Knowledge, 2000
ISBN 3-7908-1303-6

Vol. 52. E.E. Kerre and M. Nachtegael (Eds.)
Fuzzy Techniques in Image Processing, 2000
ISBN 3-7908-1304-4

Vol. 53. G. Bordogna and G. Pasi (Eds.)
Recent Issues on Fuzzy Databases, 2000
ISBN 3-7908-1319-2

Vol. 54. P. Sinčák and J. Vaščák (Eds.)
Quo Vadis Computational Intelligence?, 2000
ISBN 3-7908-1324-9

Vol. 55. J.N. Mordeson, D.S. Malik and S.-C. Cheng
Fuzzy Mathematics in Medicine, 2000
ISBN 3-7908-1325-7

Vol. 56. L. Polkowski, S. Tsumoto and T.Y. Lin (Eds.)
Rough Set Methods and Applications, 2000
ISBN 3-7908-1328-1

Vol. 57. V. Novák and I. Perfilieva (Eds.)
Discovering the World with Fuzzy Logic, 2001
ISBN 3-7908-1330-3

Vol. 58. D.S. Malik and J.N. Mordeson
Fuzzy Discrete Structures, 2000
ISBN 3-7908-1335-4

Vol. 59. T. Furuhashi, S. Tano and H.-A. Jacobsen (Eds.)
Deep Fusion of Computational and Symbolic Processing, 2001
ISBN 3-7908-1339-7

Vol. 60. K.J. Cios (Ed.)
Medical Data Mining and Knowledge Discovery, 2001
ISBN 3-7908-1340-0

Vol. 61. D. Driankov and A. Saffiotti (Eds.)
Fuzzy Logic Techniques for Autonomous Vehicle Navigation, 2001
ISBN 3-7908-1341-9

Vol. 62. N. Baba and L.C. Jain (Eds.)
Computational Intelligence in Games, 2001
ISBN 3-7908-1348-6

Vol. 63. O. Castillo and P. Melin
Soft Computing for Control of Non-Linear Dynamical Systems, 2001
ISBN 3-7908-1349-4

Vol. 64. I. Nishizaki and M. Sakawa
Fuzzy and Multiobjective Games for Conflict Resolution, 2001
ISBN 3-7908-1341-9

Vol. 65. E. Orłowska and A. Szalas (Eds.)
Relational Methods for Computer Science Applications, 2001
ISBN 3-7908-1365-6

Vol. 66. R.J. Howlett and L.C. Jain (Eds.)
Radial Basis Function Networks 1, 2001
ISBN 3-7908-1367-2

Vol. 67. R.J. Howlett and L.C. Jain (Eds.)
Radial Basis Function Networks 2, 2001
ISBN 3-7908-1368-0

Vol. 68. A. Kandel, M. Last and H. Bunke (Eds.)
Data Minining and Computational Intelligence, 2001
ISBN 3-7908-1371-0

Vol. 69. A. Piegat
Fuzzy Modeling and Control, 2001
ISBN 3-7908-1385-0

Witold Pedrycz
Editor

Granular Computing

An Emerging Paradigm

With 122 Figures
and 47 Tables

Physica-Verlag
A Springer-Verlag Company

Prof. Dr. Witold Pedrycz
University of Alberta
Department of Electrical and Computer Engineering
Edmonton, Alberta
Canada T6R 2G7
pedrycz@ee.ualberta.ca

and

Systems Research Institute
Polish Academy of Sciences
ul. Newelska 6
01-447 Warsaw
Poland

ISSN 1434-9922
ISBN 3-7908-1387-7 Physica-Verlag Heidelberg New York

Cataloging-in-Publication Data applied for
Die Deutsche Bibliothek – CIP-Einheitsaufnahme
Granular computing: an emerging paradigm; with 47 tables / Witold Pedrycz Ed. – Heidelberg; New York: Physica-Verl., 2001
(Studies in fuzziness and soft computing; Vol. 70)
ISBN 3-7908-1387-7

Physica-Verlag Heidelberg New York
a member of BertelsmannSpringer Science+Business Media GmbH

Printed in Germany

Hardcover Design: Erich Kirchner, Heidelberg

SPIN 10797154 88/2202-5 4 3 2 1 0 – Printed on acid-free paper

Contents

Interval Arithmetic and Interval Analysis: An Introduction

Jon G. Rokne *

Department of Computer Science, The University of Calgary
Abstract This article commences with a brief historical introduction to interval analysis and some applications in engineering. This is followed by a simple example motivating the use of interval analysis. A more detailed definition of interval analysis and some properties are then given followed by a discussion of the application of interval analysis to the problem of computing inclusions for the range of functions. A brief discussion of new types of interval based algorithms (mainly the interval Newton algorithm) is provided. The article is completed with a discussion of some application areas for interval analysis without claiming to be a complete survey.

Keywords: interval arithmetic, inclusion functions, range computations, interval analysis applications, interval filter

1 Introduction

The idea of performing calculations with sets of numbers and to define algebras over such sets is not new (see [76] for an early paper). It was only when electronic computers had been introduced that such calculations both became practical and also of interest for error control purposes. Computers employ a finite set of fixed precision floating point numbers, called *machine numbers* in the sequel. Similarly the computer arithmetic executed using machine numbers is called *machine arithmetic.* The requirement for error control in numerical calculations stem from the limited possibilities for representation from the finite set of machine numbers. Whereas hand calculations can control precision-related errors by representing results by increasing the number of digits dynamically it is much harder to do this on a modern computer automatically since extending the possible representations require multiple precision or adaptive precision arithmetic. This has been implemented in software, but the overhead is large which means that it is rarely used in practice. A partial solution to the representation problem is provided by interval arithmetic where a range of real numbers is enclosed between two machine numbers. In its simplest form the interval result of a numerical computation executed using machine arithmetic is an interval guaranteed to contain the result of the exact computation (the interval being represented by upper and lower machine number bounds).

* Author's address: Department of Computer Science, The University of Calgary, Calgary, Alberta, Canada, T2N 1N4.

This led to the concept of an interval arithmetic and an interval algebra over the reals since analysis over the set of machine numbers is intractable. For a given problem the analysis is therefore done over the set of real intervals and then the computation is done using machine numbers. This is a reasonable approach which is valid due to the *inclusion isotony* of interval arithmetic (see 9).

The modern history of interval arithmetic can be traced back to the thesis of R. E. Moore [36], and the publication of his book *Interval Analysis* [37]. Following this a number of conferences devoted to interval analysis were held with accompanying proceedings [44–46]. Further books on the subject have appeared [2,58,28] and bibliographies have been published [12–14]. A journal *Reliable Computing* [42], founded in 1995, serving the interests of the interval community, is "devoted to reliable mathematical computations based on finite representations and guaranteed accuracy".

No subject of current interest can exist without one or more web pages devoted to the subject. Interval analysis and interval application areas are therefore represented at the web address [21] where many links to interval researchers and to a variety of interval topics are found.

Recently there has been a growing interest in interval arithmetic and interval analysis in a number of application areas. Here we mention a few engineering applications.

In civil engineering the paper of Chen and Ward [5] provides a tutorial on the use of interval arithmetic in design. Rao and Berke [56] discusses the uncertainty in engineering analysis/design problems and they mention a number of situations where an uncertain parameter can be modeled as an interval number. In design and manufacture a geometric parameter can be modeled as an interval number if it is subject to a tolerance. Environmental parameters such as wind and snow load can be modeled as an interval parameter if subject to uncertainty (see also [41]). If a sensitivity analysis of a structure is required then this can be effectuated by interval analysis techniques. Similarly, truncation errors when truncating expansions of engineering formulations can be captured by intervals. Aging, wear, creep and changes in operating conditions can also be captured by suitably introducing intervals for the relevant values. In the paper by Muhanna and Mullen [40] we also find interval analysis applied as a tool for finite element error analysis and inclusion.

In mechanical engineering interval analysis is applied by Kulpa et. al [32] to the analysis of linear mechanical structures with uncertainties. In manufacturing the inspection of machine parts can be subject to variations leading naturally to an interval treatment as is discussed in [18]. Interval analysis applied to vibrating systems is considered by Dimargonoas [7].

Electrical engineering is normally the most mathematized of the engineering disciplines. This is reflected in the early applications of interval analysis in this discipline. In a series of papers, Oppenheimer and Michel [47–49] provide a basis for interval analysis of linear electrical systems. The first paper discusses fundamentals of interval analysis from the electrical engineering view point. In the second paper the interval matrix exponential function is treated and it is shown how to include this function using partial interval sums that can be made as narrow as required. The third paper applies these results to three examples: an RLC circuit, an instrument servomechanism and the design of a minimum plant sensitivity optimal linear regulator. This work is carried further in Leenaerts [33] where a top-down circuit design is considered from the interval point of view. A monograph in this area is *Interval Methods for Circuit Analysis* by Kolev [30]. Interval methods were also applied successfully to the solution of a previously unsolved transistor circuit analysis problem in [60].

The process of finding bounds on the parameters or state of a dynamical system is discussed in a series of papers in [35] where interval analytic methods form a central focus.

In control systems and robotics interval analysis has been applied to stability analysis [51,34], model conversion [8], and robot motion planning [52,53].

Recent research directions currently pursued relating to interval analysis include

- developing new theoretical results as for example in [71],
- developing new algorithms as is done in [75],
- developing reliable software, see for example *Numerica* [74], the software provided by the Sun Fortran 95 compiler [73] and PASCAL-SC [4],
- applying interval analysis to a number of application areas some of which are described in [31,28] and some of which are described later in this article.

In the remainder of this article section 2 provides a simple motivating example for the use of interval arithmetic. In section 3 the basic definitions and properties of interval arithmetic are given. Section 4 is concerned with the problem of inclusions for the range of functions. In section 5 we consider the interval Newton method as an interesting example of an interval algorithm. In the remaining sections several further application areas are considered.

2 A simple example

In this section we first consider an example of the use of interval arithmetic in a simple numerical problem of computing the intersection point of two lines in the plane taken from [10]. The lines are

$$0.000100x + 1.00y = 1.00, \tag{1}$$

$$1.00x + 1.00y = 2.00. \tag{2}$$

The intersection point is $(1.0001, 0.99990)$ computed using exact arithmetic rounded to 5 digits.

If the computation is performed in 3-digit floating point arithmetic with rounding and Gaussian elimination without interchanges then the result is $(0.00, 1.00)$ which is "awful" as stated in [10].

From the point of view of error control we now assume that the system is written as

$$[0.000100, 0.000100]x + [1.00, 1.00] = [1.00, 1.00] \tag{3}$$

$$[1.00, 1.00]x + [1.00, 1.00]y = [2.00, 2.00] \tag{4}$$

where each of the coefficient are now interpreted as *point intervals*, i.e. intervals of zero width.

Using Cramers rule we get the formulas

$$x = \frac{\begin{vmatrix} [1.00, 1.00] & [1.00, 1.00] \\ [2.00, 2.00] & [1.00, 1.00] \end{vmatrix}}{\begin{vmatrix} [0.000100, 0.000100] & [1.00, 1.00] \\ [1.00, 1.00] & [1.00, 1.00] \end{vmatrix}}, \tag{5}$$

and

$$y = \frac{\begin{vmatrix} [0.000100, 0.000100] & [1.00, 1.00] \\ [1.00, 1.00] & [2.00, 2.00] \end{vmatrix}}{\begin{vmatrix} [0.000100, 0.000100] & [1.00, 1.00] \\ [1.00, 1.00] & [1.00, 1.00] \end{vmatrix}}. \tag{6}$$

Evaluating (5) and (6) simulating three digit machine interval arithmetic with outward rounding we obtain

$$x \in [0.879,\ 1.03]$$
$$y \in [0.990,\ 1.02],$$

were the intervals computed are guaranteed to contain the correct result (but, note that if the computations were implemented as in the "awful" version in [10], then $x = 1$ and $y = 0$ would have had to be members of the respective resulting intervals).

Considerations such as these led to an interval arithmetic from which interval analysis was abstracted. Further, as noted in the introduction, we distinguish between real interval arithmetic and real interval analysis which we use in the analysis of interval algorithms and machine interval arithmetic which was simulated in the above calculations and normally implemented on a computer such that a real interval calculation containing the exact result is enclosed by the corresponding machine interval calculations.

3 Interval arithmetic

In the following the set of real numbers is denoted by **R**. The set of real compact intervals is $\mathbf{I} = \{A : A = [a,\ b], a, b \in R\}$ and we identify the set of point intervals $[a, a] \in \mathbf{I}$ with **R**. Let $A, B \in \mathbf{I}$. Then the interval arithmetic operations are defined by

$$A * B = \{\alpha * \beta : \alpha \in A, \beta \in B\}$$

where $* \in \{+, -, \cdot, /\}$ (note that / is undefined when $0 \in B$), that is, the interval result of $A * B$ contains all possible real point results $\alpha * \beta$ where α and β are real numbers such that $\alpha \in A$ and $\beta \in B$ and $*$ is one of the basic arithmetic operations.

This definition is motivated by the following argument. We are given two intervals A and B and we know that they contain exact values x and y respectively. Then the definition guarantees that $x * y \in A * B$ for any of the operations given above *even though we do not know the exact values of* x *and* y.

This definition is not very convenient in practical calculations. Letting $A = [a,\ b]$ and $B = [c,\ d]$ it can be shown that it is equivalent to

$$\begin{aligned}
[a,\ b] + [c,\ d] &= [a + c,\ b + d], \\
[a,\ b] - [c,\ d] &= [a - d,\ b - c], \\
[a,\ b] \cdot [c,\ d] &= [\min(ac, ad, bc, bd),\ \ \max(ac, ad, bc, bd)], \\
[a,\ b]\ /\ [c,\ d] &= [a,\ b] \cdot [1/d,\ 1/c] \text{ if } 0 \notin [c,\ d]
\end{aligned} \tag{7}$$

which means that each interval operation $* \in \{+, -, \cdot, /\}$ is reduced to real operations and comparisons. Note that we define

$$\begin{aligned}
A < B &\ \text{ iff } b < c, \\
A = B &\ \text{ iff } a = c \text{ and } b = d, \\
A > B &\ \text{ iff } b > c
\end{aligned} \tag{8}$$

and that this does not exhaust the possible relationships between intervals in contrast to comparisons between reals.

One very important property of interval arithmetic is that

$$A, B, C, D \in \mathbf{I}, A \subseteq B, C \subseteq D, \Rightarrow A * C \subseteq B * D \text{ for } * \in \{+, -, \cdot, /\} \tag{9}$$

if the operations are defined. In other words, if A and C are subsets of B and D respectively then $A * C$ is a subset of $B * D$ for any of the basic arithmetic operations. Therefore, errors introduced at any stage of the computations such as floating-point errors or input errors can be accounted for. The property given by (9) is been called the *inclusion isotony* of interval operations and it is recognized as the fundamental principle of interval analysis.

One consequence of this is that any function $f(x)$ described by an expression in the variable x which can be evaluated by a programmable real calculation can be embedded in interval calculations using the natural correspondence between operations so that if $x \in X \in \mathbf{I}$ then $f(x) \in f(X)$ where $f(X)$ is interpreted as the calculation of $f(x)$ with x replaced by X and the operations replaced by interval operations. The evaluation $f(X)$ is called the *natural interval extension* of the expression $f(x)$.

A number of operations on intervals $A = [a, b], B = [c, d] \in \mathbf{I}$ can be defined. Here we mention the *width* of an interval

$$w(A) = b - a, \tag{10}$$

the *midpoint* of an interval

$$m(A) = (a + b)/2, \tag{11}$$

the *absolute value* of an interval

$$|A| = \max\{|a|, |b|\}, \tag{12}$$

the distance between two intervals

$$q(A, B) = \max\{|a - c|, |b - d|\}, \tag{13}$$

which is known as the Hausdorff metric and the functional χ defined in [57] as $\chi : \mathbf{I} \to [-1, 1]$ with $\chi[0, 0] = -1$ and if $[a, b] \neq 0$, with

$$\chi[a, b] = a/b \text{ if } |a| \leq |b| \text{ and } b/a \text{ otherwise.} \tag{14}$$

These definitions satisfy some properties and are related in a number of ways. For example

$$q(AB, AC) \leq |A| q(B, C), \tag{15}$$
$$q(A + B, A + C) = q(B, C), \tag{16}$$
$$|AB| = |A||B|, \tag{17}$$
$$w(AB) \leq w(A)|B| + w(B)|A|, \tag{18}$$
$$m(A \pm B) = m(A) \pm m(B). \tag{19}$$

We also need the notation

$$x \vee y = \begin{cases} [x, y] & \text{if } x \leq y \\ [y, x] & \text{if } y < x. \end{cases} \tag{20}$$

Further properties and connections between these and other definitions for interval tools are found in [2,65]. The functional (14) was defined in [57] and used to characterize the solution of linear interval equations in [64] and the width of interval products in [62].

Let $\mathbf{I}^n$ be the set of n-dimensional interval column vectors. Then we can define the interval vector operations

$$AU = (aU_i) \text{ for } A \in \mathbf{I}, U \in \mathbf{I}^n, \tag{21}$$
$$U \pm V = (U_i \pm V_i) \text{ for } U, V \in \mathbf{I}^n. \tag{22}$$

Similarly, if $\mathbf{I}^{n\times m}$ is the set of interval matrices then interval matrix operations can be defined as

$$AP = (AP_{ij}) \text{ for } A \in \mathbf{I}, P \in \mathbf{I}^{n\times m}, \tag{23}$$

$$P \pm Q = (P_{ij} \pm Q_{ij}) \text{ for } P, Q \in \mathbf{I}^{n\times m}, \tag{24}$$

$$PQ = (\sum_{k=1}^{l} P_{ik}Q_{kj}) \text{ for } P \in \mathbf{I}^{n\times l}, Q \in \mathbf{I}^{l\times m}. \tag{25}$$

Matrix-vector operations are similarly defined.

The notions of width, midpoint and absolute value are also extended elementwise.

The notion of a n-dimensional linear equation can also be extended to the interval domain by defining it as

$$PX = B, P \in \mathbf{I}^{n\times n},\ B \in \mathbf{I}^{n}. \tag{26}$$

The solution of (26) is the set

$$\Box(PX - B) = \{x \in \mathbf{R}^n | px = b, p \in P, b \in B\}. \tag{27}$$

of solutions to the pointwise linear equations. It should be noted that this set is in general not an interval. By solving an interval linear equation we therefore mean the computation of an interval vector that contains all the solutions to the pointwise linear equations. It should be noted that even the solution of one linear interval equation is in general non-trivial [64].

Extensions of interval concepts to the complex plane has been done by a number of authors.

(A) An extension to rectangular complex arithmetic was defined in [1]. Here the multiplication of two rectangular intervals pose a problem, whereas the addition and subtraction is straightforward. The subdivision of a rectangular complex interval along the coordinate axes is also simple.
(B) A circular complex interval arithmetic was defined in [11]. Addition and subtraction is simple, whereas multiplication and division is more complicated. Also, the subdivision of a circle into subcircles is not possible and instead covering circles have to be used if the circular complex interval has to be subdivided (see [55] for an example).
(C) A complex sector arithmetic was discussed in [29]. More recently it was applied by [9] to qualitative reasoning, a subfield of artificial intelligence.

A recent monograph by Petković and Petković [50] deals with complex interval arithmetics and their applications in more detail.

Interval polynomials are polynomials where the coefficients are intervals. These polynomials occur with interval methods for differential equations where the solutions are guaranteed to lie in the function strip defined by the interval polynomial. A survey of some of the research can be found in [6].

Another important feature of interval arithmetic is that it can be implemented on a floating-point computer such that the resulting interval contains the result of the real interval computations using equations (7) and directed rounding. Several software systems are available for this such as PASCAL-SC [4] and the system under development at Sun Microsystems [73]. The implementations only has to take care that each calculation of interval endpoints are rounded outwards from the interior of the intervals. This is known as *machine interval arithmetic.* In practice the process of performing an interval analysis and implementation of a corresponding interval calculation is therefore handled by performing the analysis in real interval arithmetic and then implementing the calculation in machine interval arithmetic. The inclusion principle given above therefore guarantees that the resulting interval contains the interval obtained by a real interval calculation.

Interval arithmetic has some drawbacks as well.

- Subtraction and division are not the inverse operations of addition and multiplication.
- The distributive law does not hold. Only a subdistributive law

$$A(B+C) \subseteq AB + AC,\ A, B, C \in \mathbf{I} \tag{28}$$

  is valid.
- The interval arithmetic operations are more time-consuming than the corresponding real operations roughly by a factor of 3 (although interval arithmetic implementations of some problems may run faster than the corresponding real versions, see [72]). Interval arithmetic is often too conservative. That is, although the interval evaluation of a given expression will contain the result of the expression evaluated over the reals the width of the interval result might be so large that no information is obtained from the interval computation.

4 Inclusions for the range of functions

The range of a function $f(x)$ over an interval X is defined as

$$\square f(X) = \{f(x) | x \in X\}. \tag{29}$$

From equation (9) it follows for the natural interval extension that $\square f(X) \subseteq f(X)$.

It was noted in the previous section that the order of evaluation was assumed given in order to define the natural interval extension. The reason for this is the failure of the distributive law in general and that only the subdistributive law (28) holds.

As an example consider the function $f(x) = x - x^2$ which is defined over the interval $[0, 2]$.

It might be written as

$$f(x) = x - x^2, \text{ i.e. the original power sum definition} \tag{30}$$
$$f(x) = x(1-x), \text{ bracketted} \tag{31}$$
$$f(x) = (x-1) + (1-x)(1+x), \text{ another rearrangement} \tag{32}$$
$$f(x) = s - s^2 + (1-2s)(x-s) - (x-s)^2, \quad \text{as a Taylor expansion around } s. \tag{33}$$

Evaluating these expressions over $X = [0, 2]$ as interval expressions using the standard interval arithmetic (not interval arithmetic with the extended power evaluation mentioned in [58]) we get

$$I_1 = X - X^2 = [-4, 2] \tag{34}$$
$$I_2 = X(1-X) = [-2, 2] \tag{35}$$
$$I_3 = (X-1) + (1-X)(1+X) = [-4, 4] \tag{36}$$
$$I_4 = 1 - 1^2 + (1 - 2*1)(X-1) - (X-1)^2 = [-2, 2], (s = 1 \text{ in } (33)). \tag{37}$$

From the above it follows that $\Box f(X) \subseteq I_i, i = 1, \ldots, 4$ and also that the width of the inclusion for the range is dependent on expression that is evaluated for the function. Since there is an infinite number of possible expressions

$$f(x) = x - x^2 + p(x) - p(x), \text{ where } p(x) \text{ is any expression} \tag{38}$$

for $f(x)$ it is of interest to develop a theory for expressions that deliver narrow inclusions. This was done by a number of researchers whose work up to about 1983 was collected in the monograph [58].

It should be noted that a number of the expressions have second order convergence with respect to the width of the interval when evaluated as interval expressions. That is,

$$w(f(X)) - w(\Box f(x)) = O(w(X))^2 \tag{39}$$

The *centered forms* first suggested by Moore [37] for polynomials and rational functions are such expression. The centered form essentially develops a function at a point c in the domain interval X to form a new expression, which then is evaluated as an interval expression.

A large number of centered forms are possible. We only mention one possibility for rational functions that does not require the explicit computation of derivatives [66]. If

$$f(x) = p(x)/q(x) = \sum_{i=0}^{n} a_i x^i / \sum_{i=0}^{m} b_i x^i \tag{40}$$

is a rational function and $k = \max(n, m)$ then the polynomials $p(x)$ and $q(x)$ are developed using Horner's rule at c such that

$$p(x) = p(c) + t(x)(x - c)$$
$$q(x) = q(c) + u(x)(x - c)$$

and where the values $p(c)$, $q(c)$ and the coefficients t_i and u_i of

$$t(x) = \sum_{i=0}^{n-1} \left(\sum_{j=i+1}^{n} a_j c^{j-i-1} \right) x^i = \sum_{i=0}^{n-1} t_i x^i$$
$$u(x) = \sum_{i=0}^{m-1} \left(\sum_{j=i+1}^{m} b_j c^{j-i-1} \right) x^i = \sum_{i=0}^{m-1} u_i x^i$$

are calculated explicitly as part of the Horner process.

If we now write $f(x)$ as

$$f(x) = f(c) + (x - c)\frac{r(x)}{q(x)} \tag{41}$$

then $r(x)$ has to obey the relation

$$r(x)(x - c) = p(x) - f(c)q(x).$$

Using the above representation for $p(x)$ and $q(x)$ we get

$$r(x)(x - c) = p(c) + t(x)(x - c) - f(c)(q(c) + u(x)(x - c))$$

and finally

$$r(x) = t(x) - u(x)f(c) = \sum_{i=0}^{k-1} \left(\sum_{j=i+1}^{k} (a_j - f(c)b_j)c^{j-i-1} \right) x^i = \sum_{i=0}^{k-1} r_i x^i$$

where the undefined coefficients a_j or b_j are set to zero. Algorithmically we have $r_i = t_i - u_i f(c)$, $i = 1, \dots, k - 1$ where t_i and u_i were calculated by the Horner process. The natural interval extension of (41) therefore provides an outer estimate of $\Box f(X)$ and it is part of the class of centered forms formed by Moore's definition.

If the inclusion Y for $\Box f(X)$ computer by either the natural interval extension of a function $f(x)$ or by one of the centered forms is too wide (which can be estimated by considering $w(X)$) then the inclusion can be improved by subdividing the domain X and then applying either of the forms over the subintervals. This idea was carried further by Skelboe [69] who developed a subdivision strategy where subinterval which can be shown not to contain the upper or the lower bound of the function $f(x)$ over X can be deleted from further consideration.

In [3] an interesting idea on the choice of an optimal developing point is presented.

One interesting aspect of the theory of centered forms is that depending on the width of the domain interval X different forms are recommended.

In a number of cases the exact range is computed for an interval expression. This is exploited in for example [61].

5 An engineering interval example

In the introduction it was pointed out that electrical engineering was the engineering discipline that embraced interval tools first and to the greatest depth. It is therefore natural to take a simple example of the use of interval analysis from this discipline. With this in mind we consider how to determine bounds for the gain and terminal impedances for an audio pad circuit given tolerances for the components of the circuit [19] (see also [70]).

A schematic of the audio pad circuit is given in Figure 1. This kind of circuit is used when connecting audio equipment to match signal levels and meet impedance requirements.

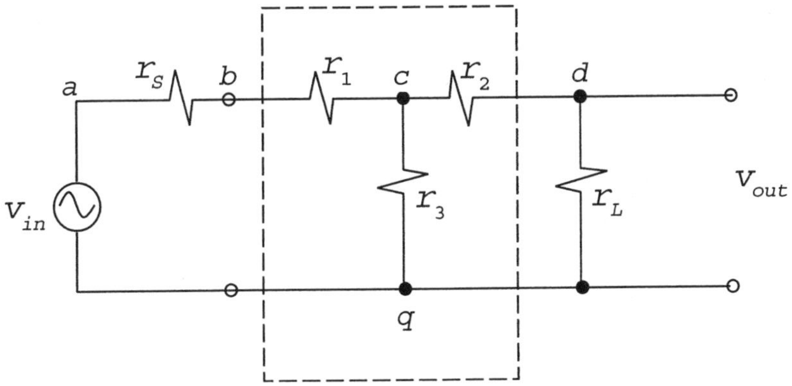

Fig. 1. The audio pad circuit

The circuit was analyzed using the ladder method. Currents, impedances and voltages are denoted by lower case letters and subscripted either by $s, 1, 2, 3, L$ to denote the element through which or over which the quantity is calculated or ab etc. to denote points between which the quantity is calculated. Upper case letters are used to denote interval inclusions of the above quantities. Since the aim here is to demonstrate the use of interval tools we simply list the result of the analysis in a tabular form.

Equation	Remark
$i_L = v_{out}/r_L = r_3$	current through r_L and r_3
$v_{cd} = i_3 r_3 = (v_{out} r_3)/r_L$	voltage across r_3
$v_{cq} = v_{cd} + v_{out} = v_{out}(1 + r_3/r_L)$	voltage across r_2
$i_2 = v_{cq}/r_2 = v_{out}(1 + r_2/r_L)/r_2$	the current through r_2
$i_1 = i_2 + i_3 = v_{out}(1 + r_2/r_L)/r_2 + v_{out}/r_L$ $= v_{out}(r_L + r_3 + r_2)/(r_2 r_L)$	current through r_1
$v_{ac} = i_1(r_1 + r_S)$ $= v_{out}(r_L + r_2 + r_3)(r_1 + r_S)/(r_2 r_L)$	r_1 and r_3 are in series
$v_{in} = v_{out}((r_L + r_2 + r_3)(r_1 + r_S)$ $+r_2(r_L + r_3))/(r_2 r_L)$.	

The gain is defined as

$$g = \frac{v_{out}}{v_{in}} = \frac{r_2 r_L}{((r_L + r_2 + r_3)(r_1 + r_S) + r_2(r_L + r_3)}. \tag{42}$$

The input impedance is give by

$$z_{in} = \frac{v_{in}}{i_1} = (r_1 + r_S) + \frac{r_2(r_L + r_3)}{r_L + r_2 + r_3}. \tag{43}$$

The output impedance is derived by setting v_{in} to zero and calculating the resistance looking into the output terminals.

The effective resistance r_{eff} of r_1, r_2, r_S is $r_{eff} = r_2(r_1 + r_S)/(r_1 + r_2 + r_S)$ which gives the output impedance as

$$z_{out} = r_3 + \frac{r_2(r_1 + r_3)}{r_1 + r_2 + r_S}. \tag{44}$$

The basic problem is now to calculate r_1, r_2 and r_3 given r_S and r_L such that specifications for z_{in}, z_{out} and g are satisfied and such that $\hat{r}_1, \hat{r}_2$ and $\hat{r}_3$ are given as standard resistors with as broad standard tolerances as possible.

A particular example is now presented.

We require that $g = 0.1$ with a 20% tolerance and $z_{in} = 600\Omega$ and $z_{out} = 600\Omega$ with 10% tolerance. That is, if the calculated results are in the tolerance intervals we accept the circuit with the calculated components. We have that $r_S = 50\Omega \pm 5\%$ and $r_L = 150\Omega \pm 5\%$. Using the three equations (42), (43) and (44) we get a system of three equations in the unknowns r_1, r_2 and r_3. Solving this we have $r_1 = 325\Omega$, $r_2 = 375\Omega$ and $r_3 = 412.5\Omega$ using the main values for the given components. The closest standard value components would give $\hat{r}_1 = 330\Omega$, $\hat{r}_2 = 390\Omega$ and $\hat{r}_3 = 430\Omega$.

Several questions now arise

1. How close are we to the specified gain and impedances assuming exact values for the standard value resistors $\hat{r}_1, \hat{r}_2$ and $\hat{r}_3$?
2. If $\hat{r}_1, \hat{r}_2$ and $\hat{r}_3$ are given with tolerances how close do these have to be for reasonable tolerances for gain and impedances to be maintained.

3. Are the specifications met for components $\hat{r}_1, \hat{r}_2$ and $\hat{r}_3$ with $\pm 5\%$ tolerances?

In this example $r_S \in R_S = [47.5, 52.5]$ and $r_L \in R_L = [142.5, 147.5]$. We are first interested in outer estimates for the range of (42), (43) and (44) over these intervals assuming exact values for the selected components which would provide some answers to the first question. The calculations are

$$\begin{aligned}
G(R_S, R_L, \hat{r}_1, \hat{r}_2, \hat{r}_3) &= \frac{390 R_L}{(R_L + 390 + 430)(330 + R_S) + 390(R_L + 430)} \\
&= [0.0921, 0.1.05], \\
Z_{in}(R_S, R_L, \hat{r}_1, \hat{r}_2, \hat{r}_3) &= (330 + R_S) + \frac{390(R_L + 430)}{R_L + 390 + 430} = [605, 621], \\
Z_{out}(R_S, R_L, \hat{r}_1, \hat{r}_2, \hat{r}_3) &= 430 + \frac{390(330 + R_S)}{330 + 390 + R_S} = [620, 625]
\end{aligned}$$

where the results are rounded outwards so that the interval bounds are represented by three digit quantities. Since the calculated values of the gain and the impedances are within the tolerances specified the circuit would be acceptable. Unfortunately, the components selected also come with tolerances. At the 5% tolerance level the components have values in the intervals $\hat{R}_1 = [313.5, 346.5]$, $\hat{R}_2 = [370.5, 409.5]$ and $\hat{R}_3 = [408.5, 451.5]$. Recalculating we get

$$\begin{aligned}
G(R_S, R_L, \hat{R}_1, \hat{R}_2, \hat{R}_3) &= [0.0805, 0.121], \\
Z_{in}((R_S, R_L, \hat{R}_1, \hat{R}_2, \hat{R}_3) &= [561, 670], \\
Z_{out}((R_S, R_L, \hat{R}_1, \hat{R}_2, \hat{R}_3) &= [573, 675]
\end{aligned}$$

and we note that results are not acceptable, but close to acceptable. Since one of the problems with interval arithmetic is overestimation of the results we try to improve the calculations. In this case we can rearrange the formulas as follows:

$$\begin{aligned}
\frac{1}{G} &= \frac{(R_L + \hat{R}_2 + \hat{R}_3)(\hat{R}_1 + R_S)}{\hat{R}_2 R_L} + 1 + \frac{\hat{R}_3}{R_L}, \\
Z_{in} &= \hat{R}_1 + R_S + \frac{1}{1/\hat{R}_2 + 1/(R_L + \hat{R}_3)}, \\
Z_{out} &= \hat{R}_3 + \frac{1}{1/\hat{R}_2 + 1/(R_S + \hat{R}_2)}.
\end{aligned}$$

With these modifications and inverting the result from the calculation of $1/G$ the results are

$$\begin{aligned}
G(R_S, R_L, \hat{R}_1, \hat{R}_2, \hat{R}_3) &= [0.0842, 0.114], \\
Z_{in}(R_S, R_L, \hat{R}_1, \hat{R}_2, \hat{R}_3) &= [582, 644], \\
Z_{out}(R_S, R_L, \hat{R}_1, \hat{R}_2, \hat{R}_3) &= [591, 654]
\end{aligned}$$

and these results fall within the acceptable range specified above.

We note that in the last two expression the interval variables occur only once and to the first power. A result by Skelboe [69] then states that the range is computed by the interval evaluation of the expression (see also Theorem 4.2 in [58]). This illustrates the importance of selecting a good expression for a given formula.

6 New types of algorithms

One interesting feature of interval arithmetic is that it can be used to develop new algorithms that are not simply extensions of algorithms in real arithmetic. An example of this is the interval Newton method first developed in [37]. Let $f(x)$ be given and suppose that we want to find the points ξ where $f(\xi) = 0$ in a given interval X_0. If we choose a point $c \in X_0$ and expand $f(x)$ at c we obtain

$$f(x) = f(c) + (x - c)f'(\theta) \tag{45}$$

where $\theta \in x \vee c$ (the operation $\vee$ was defined by (20)). If $x = \xi$, a root, then

$$0 = f(c)(\xi - c)f'(\theta). \tag{46}$$

Rearranging we have

$$\xi = c - f(c)/f'(\theta). \tag{47}$$

Since we only know that $\theta \in x \vee c \subseteq X_0$ it follows that if we replace θ by X_0 on the right hand side of (46) we get the following interval equation

$$X_1 = c - f(c)/f'(X_0) \tag{48}$$

where $f'(X_0)$ is an interval evaluation of $f'(x)$ (and not the derivative of the interval equation $f(X)$, a concept which has not been defined here). Although the point c can be chosen arbitrarily in X_0 it is most often chosen to be the midpoint. This then leads to the interval Newton iteration

$$X_{n+1} = X_n \cap \{m(X_n) - f(m(X_n))/f'(X_n)\}, \quad n = 0, 1, \ldots \tag{49}$$

with X_0 specified as the initial interval.

The method has some interesting properties.

1. If a zero, ξ, of f exists in X_0 then $\xi \in X_n$ for all n, see [37]. This means that all the zeroes in the initial interval X_0 are retained in subsequent interval iterates.
2. If X_n is empty for some n then f has no zeros in X. [37]

If the zeros of $f : \mathbf{R}^n \to \mathbf{R}^n$ are to be computed then the iteration (49) can be generalized to n dimensions. Again let $X_0 \in \mathbf{I}^n$ be an initial interval containing a zero $\xi \in \mathbf{R}^n$. Consider an expansion of f around $c \in X_0$ in the following form

$$f(x) = f(c) + f'(\theta)(x - c) \tag{50}$$

where $f'(\theta)$ is the matrix of partial derivatives of $f(x)$ evaluated at some point $x \vee c$ (where the operator $\vee$ is applied componentwise). In the same manner as in the one-dimensional case x is set to the zero, obtaining

$$0 = f(c) + f'(\theta)(\xi - c) \tag{51}$$

Rearranging we get a equation

$$f'(\theta)(\xi - c) = -f(c). \tag{52}$$

We do not know θ which means that we have to replace it by X_0 and for simplicity we set $c = m(X_0)$.

This leads to the basic n-dimensional interval Newton iteration

1. Choose an initial interval X_0.
2. Compute an interval inclusion Z to the solution of the linear interval system of equations
$$f'(X_i)(Y - X_i) = -f(m(X_i)), \tag{53}$$
in Y.
3. Set $X_{n+1} = X_n \cap Y$.
4. If termination condition is not satisfied go to 2.

As noted earlier the set of solutions to the linear interval equations (53) is not an interval which means that we have to seek interval supersets to (53) in order to remain within the interval domain. Depending on the method for finding the superset different interval Newton variants results. These are surveyed in [63] where the combination of these methods with subdivision strategies is also emphasized.

A number of termination conditions are possible and they are for example discussed in [63].

Further properties of interval iterations for the solution of equations can be found for example in [43].

We should briefly mention the application of constraint satisfaction to interval Newton methods for polynomial systems as discussed by [74]. Impressive speed-ups of algorithms are obtained by applying constraint based iterations on the equations reducing the widths of the component intervals together with interval Newton steps frequently being able to reject subintervals not containing solutions much faster than with only the interval Newton method and subdivision.

7 Applications of interval analysis to global optimization

Interval methods have a natural application to global optimization.

Global optimization aims at solving problems of the form

$$\min f(x) \text{ subject to } x \in D \subseteq \mathbf{R}^n \tag{54}$$

where f is the objective function and D the feasible set, i.e. the set where the solution to the problem, a set X^* consisting of one or more elements, is to be found. The feasible set D has must be an interval or approximated by one or more intervals. In the terminology of global optimization these intervals are often called *boxes*.

The advantages of interval methods for global optimization problems are that (see also [74])

- they are robust, stable and reliable,
- they do not rely on starting points,
- they can prove existence and uniqueness of solutions,
- they can easily accommodate external constraints.

The interval based global optimization methods are generally based on subdivision of boxes combined with box shrinking and box rejection. As the boxes are subdivided they are entered onto a list for possible further processing.

A number of subdivision strategies have been considered:

- Moore [37] suggested to subdividing all coordinate directions generating 2^n new boxes to be put on the list
- Moore [38] refined this to subdivision of one coordinate direction at a time in a cyclical manner
- Hansen [17] suggested subdivision of the coordinate direction with the longest box edge.

The ordering of the boxes on the list is also important and convergence statements about the methods depend on this ordering.

A number of references relating to interval methods for global optimization are available such as [59,16,27]. Of recent interest can be mentioned the article [63] in the *Handbook of Global Optimization.*

8 Applications of interval analysis to computer graphics and CAGD

Interval methods can be applied in computer graphics in ray-tracing. The intersection calculation between an implicit surface and a ray results in a problem of finding either one (the smallest) root or all the roots of a function $f(x) = 0$, see [15]. Using interval arithmetic techniques an interval result can be computed that is guaranteed to contained the intersections avoiding anomalies in the rendering process (see [25] for a discussion of the problem).

Further discussions on the use of interval arithmetic for implicit surface rendering, in contouring algorithms and in planarity estimation is found in [39,72] where it is combined with subdivision techniques in order to improve the results.

Recently there has been a great deal of interest in applying interval methods to CAGD. These applications are mainly based on the use of interval polynomials in the Bernstein and the Bézier form. One can mention the application to intersections of surfaces [22], to the offsets of Bézier curves [67] and the approximation of curves by interval Bézier curves [68].

9 Applications of interval analysis to exact computations

Todays computers all use a form of fixed precision floating point arithmetic most often satisfying a standard set by IEEE [23,24]. As noted in an earlier section, this can lead to incorrectly computed results due to accumulated roundoff errors. One way to avoid this is to implement exact arithmetic modeling real calculations. If the computer arithmetic is exact then a computer implementation can not introduce errors.

An area where it is easy to see that numerical errors can have serious impact is in geometric computations due to the strong interplay between numerical results and logical tests.

As an example consider the testing for intersection between two line segments. This test is frequently a part of algorithms in geometric modeling, computer graphics, GIS and computational geometry, to name a few areas. The test ultimately depends on the ability to decide if a point is on a line segment or not. If the line segment is defined by two endpoints a, b then the test reduces to the determinant

$$D = \begin{vmatrix} X_1 & a_1 & b_1 \\ X_1 & a_1 & b_1 \\ X_1 & a_1 & b_1 \end{vmatrix} . \tag{55}$$

Depending on whether $D \gtreqless 0$ the point x is to the left, on or to the right of the oriented line from a to b. To be able to guarantee the topology of the geometry the determinant (55) has to be evaluated exactly, normally a computationally expensive process.

In order to achieve better performance when implementing an exact computation of a function, Karasick et al. [26] suggested to first use interval arithmetic as an *interval filter* to evaluate the function and to get conclusive exact results for the cases which can be decided on the basis of the interval result.

Typically a function $f(x)$ has to be evaluated and the for a given value α and the result has to be compared to a given value β. If the evaluation is exact then the test $f(\alpha) \leq \beta$ is exact and the subsequent computation can proceed along the correct path. If the function evaluation is inexact then the path of the subsequent computation could easily proceed along the incorrect branch, particularly if the argument α was chosen so that the function value was close to β. Evaluating $f([\alpha, \alpha])$ as an interval computation an interval B is computed with the property that $f(\alpha) \in B$.

The interval B can therefore be tested and

- if $\beta < B$ then $\beta < f(\alpha)$ (guaranteed),
- if $B < \beta$ then $f(\alpha) < \beta$ (guaranteed),
- if $\beta \in B$ then both $\beta = f(\alpha)$ and $\beta \neq f(\alpha)$ is possible.

The interval computation therefore is able to provide an exact decision in the first two cases. In the last case a more expensive exact computation is required.

An extensive discussion to the use of interval filters is found in the recent thesis by Pion [54].

10 Conclusion

In this article some aspects of interval analysis were discussed without claiming to be complete either with respect to the theory or to the possible areas of application. A complete survey would only have been valid for a short time for it is expected that new theoretical results will continue to be developed and new applications and application areas will be found. When researchers and commercial applications no longer are satisfied with results that might or might not be correct it is also expected that interval methods will become central to computing since they can provide guaranteed results.

References

1. Alefeld, G. (1968). Intervallrechnung über den komplexen Zahlen und einige Anwendungen. Ph. D. thesis, University of Karlsruhe.
2. Alefeld, G. and J. Herzberger (1983). *Introduction to Interval Computations.* New York: Academic Press.
3. Baumann, E. (1987). Optimal centered forms. *Freiburger Intervall-Berichte 87/3*, Institut für Angewandte Mathematik, Universität Freiburg, pp. 5-21.
4. Bohlender, G., H. Böhm, E. Kaucher, R. Kirchner, U. Kulisch, S. Rump, Ch. Ullrich and W. von Gudenberg (1981). PASCAL-SC: A Pascal for Contemporary Scientific Computation. IBM Report RC 9009.
5. Chen, R. and A. C. Ward (1992). Introduction to interval matrices in design. *ASME Design Theory and Methodology 42*, 221-227.
6. Corliss, G. (1989). Survey of interval algorithms for ordinary differential equations. *Journal of Applied Mathematics and Computation 31*, 112-120.

7. Dimargonoas, A. D. (1995) Interval analysis of vibrating systems. *Journal of Sound and Vibration 183*. 739-749.
8. Feng, F., L.-S. Shieh and G. Chen (1997). Model conversions of uncertain linear systems using interval multipoint Padé approximation. *Applied Math. Modeling 21*, 233-244.
9. Flores, J. (1999). Complex fans: a representation for vectors in polar form with interval attributes. *ACM Transactions on Mathematical Software 25*, 129-156.
10. Forsythe, G.E. and C. B. Moler (1967). *Computer Solution of Linear Algebraic Systems*. Englewood Cliffs: Prentice-Hall.
11. Gargantini, I. and P. Henrici (1972). Circular arithmetic and the determination of polynomial zeros. *Numer. Math. 18*, 305-320.
12. Garloff, J. and K.-P. Schwierz (1980). A Bibliography on Interval-Mathematics. *J. Comput. Appl. Math. 6*, 67-79.
13. Garloff, J. (1985). Interval mathematics. A bibliography. *Freiburger Interval-Berichte 85/6*, 1-122.
14. Garloff, J. (1987). Bibliography on interval mathematics. Continuation. *Freiburger Interval-Berichte 87/2*, 1-50.
15. Hanrahan, P. (1989). A survey of ray-surface intersection algorithms. In: *An Introduction to Ray Tracing*, A. Glassner ed. New York: Academic Press.
16. Hansen, E. (1992). *Global Optimization using Interval Analysis*. New York: Marcel Dekker.
17. Hansen, E.R. (1980). Global optimization using interval analysis - the multidimensional case. *Numerische Mathematik 34*, 247-270.
18. Henderson, T. C., T. M. Sobh, F. Zana, B. Brüderlin, and C. Y. Hsu (1994). Sensing strategies based on manufacturing knowledge. In *1994 ARPA Image Understanding Workshop*. University of California Riverside, Monterey, 1109-1113.
19. Herron, A. (1982). Private communication.
20. Horst, R. and M. Panos, M. (eds.) (1995). *Handbook of Global Optimization*. Dordrecht, Kluwer Academic Publishers.
21. http://www.cs.utep.edu/interval-comp/main.html
22. Hu, C.-Y., T. Maekawa, N. M. Patrikalakis and X. Ye (1997). Robust interval algorithm for surface intersections. *Computer-Aided Design 29*, 617-627.
23. IEEE (1985). IEEE standard for binary floating-point arithmetic. IEEE Standard 754-1985. New York, IEEE. IEEE
24. IEEE (1987). IEEE standard for radix-independent floating-point arithmetic. IEEE Standard 854-1987, New York, IEEE.
25. Kalra, D. and A. H. Barr (1989). Guaranteed ray intersections with implicit surfaces. *Computer Graphics 23*, 297-306.
26. Karasick, M., D. Lieber, and L. Nackman (1991). Efficient Delaunay triangulation using rational arithmetic. *ACM Transactions of Graphics 10*, 71-91.
27. Kearfott, R.B. (1996). *Rigorous Global Search: Continuous Problems*. Dordrecht, Kluwer Academic Publishers.
28. Kearfott, R.B. and V. Kreinovich (1996). *Applications of Interval Computations*. Dordrecht, Kluwer Academic Publishers.
29. Klatte, R. and Ch. Ullrich (1980). Complex sector arithmetic. *Computing 24*, 139-148.
30. Kolev, L. V. (1993). *Interval Methods for Circuit Analysis*. Singapore, World Scientific.

31. Kreinovich V. (ed.) (1995). *Reliable Computing, Supplement.* Extended Abstracts of APIC'95 International Workshop on Applications of Interval Computations, El Paso, Texas.
32. Kulpa, K., A. Pownuk and I. Skalna (1998). Analysis of linear mechanical structures with uncertainties by means of interval methods. *Computer Assisted Mechanics and Engineering Sciences 5*, 443-477.
33. Leenaerts, D. M. W. (1990). Application of interval analysis for circuit design. *IEEE Transactions on Circuits and Systems 37*, 803-807.
34. Malan, S., M. Milanese and T. Taragna (1997). Robust analysis and design of control systems using interval arithmetic. *Automatica 33*, 1363-1372.
35. Milanese, M., J. Norton, N. Piet-Lahanier and E. Walter (eds.) (1996). *Bounding Approaches to System Identification.* New York, Plenum Press.
36. Moore. R. E. (1962). Interval arithmetic and automatic error analysis in digital computing. Ph. D. Dissertation, Stanford University.
37. Moore, R. (1966). *Interval Analysis.* Englewood Cliffs: Prentice Hall.
38. Moore, R.E. (1979). *Methods and Applications of Interval Analysis.* Philadelphia: SIAM.
39. Mudur, S.P. and P. A. Koparkar, P.A. (1984). Interval methods for processing geometric objects. *IEEE Computer Graphics and Applications 4, No. 2*, 7-17.
40. Muhanna, R. L. and R. L. Mullen (1999). Formulation of fuzzy finite-element methods for solid mechanics problems. *Computer-Aided Civil and Infrastructure Engineering 14*, 107-117.
41. Mullen, R. L. and R. L. Muhanna (1999). Bounds of structural response for all possible loading combinations. *Journal of Structural Engineering 125*, 98-106.
42. Nesterov, V. M. (ed.) (1995). *Reliable Computing.* Institute of New Technologies in Education, St. Petersburg-Moscov, Kluwer Academic Publishers.
43. Neumaier, A. (1990). *Interval Methods for Systems of Equations.* Cambridge: Cambridge University Press.
44. Nickel, K. (ed.) (1975). *Interval Mathematics.* Lecture Notes in Computer Science 29, Berlin: Springer-Verlag.
45. Nickel, K. (ed.) (1980). *Interval Mathematics 1980.* New York: Academic Press.
46. Nickel, K. (ed.) (1985). *Interval Mathematics 1985.* Lecture Notes in Computer Science 212. Berlin: Springer-Verlag.
47. Oppenheimer, E. P. and A. N. Michel (1988). Application of interval analysis techniques to linear systems: Part I - Fundamental Results. *IEEE Transactions on Circuits and Systems 35*, 1129-1138.
48. Oppenheimer, E. P. and A. N. Michel (1988). Application of interval analysis techniques to linear systems: Part II - The interval matrix exponential function. *IEEE Transactions on Circuits and Systems 35*, 1230-1242.
49. Oppenheimer, E. P. and A. N. Michel (1988). Application of interval analysis techniques to linear systems: Part III - Initial value problems. *IEEE Transactions on Circuits and Systems 35*, 1243-1256.
50. Petković, M. S. and L. D. Petković (1998). Complex Interval Arithmetic and Its Applications. Berlin, Wiley-VCH.
51. Piazzi, A. and G. Marro, G. (1996). Robust stability using interval analysis. *International Journal of Systems Science 27*, 1381-1390.
52. Piazzi, A. and A. Visoli (1997). An interval algorithm for minimum-jerk trajectory planning of robot manipulators. In *Proc. of the 36th Conference on Decision and Control.* San Diego Calif. U. S. A., 1924-1927.

53. Piazzi, A. and A. Visoli (1998). Global minimum-time trajectory planning of mechanical manipulators using interval analysis. *International Journal on Control 71*, 631-652.
54. Pion S.: De la géométrie algorithmique au calcul géométrique. Thèse, Univers-sit'e de Nice Sophia-Antipolis.
55. Qun, L. and J. Rokne (1996). A circular splitting search algorithm for systems of complex equations. *Journal of Computational and Applied Mathematics 75*, 119-129.
56. Rao, S. S. and L. Berke (1997). Analysis of uncertain structural systems using interval analysis. *Journal of American Institute of Aeronautics and Astronautics 35*, 727-735.
57. Ratschek, H. (1970). Die binären Systeme der Intervallmathematik. *Computing 6*, 295-308.
58. Ratschek, H. and J. Rokne (1984). *Computer Methods for the Range of Functions.* Chichester: Ellis Horwood.
59. Ratschek, H. and J. Rokne (1988). *New Computer Methods for Global Optimization.* Chichester: Ellis Horwood.
60. Ratschek, H. and J. Rokne (1993). Experiments using interval analysis for solving a circuit design problem. *Journal of Global Optimization 3*, 501-518.
61. Ratschek, H. and J. Rokne (1994). Box-sphere intersection tests. *Computer-Aided Design 26*, 579-584.
62. Ratschek, H. and J. Rokne (1995). Formulas for the width of interval products. *Reliable Computing 1*, 9-14.
63. Ratschek, H. and J. Rokne (1995). Interval methods. In [20], 751-828.
64. Ratschek, H. and W. Sauer (1982). Linear interval equations. *Computing 28*, 105-115 (1982).
65. Ris, F. (1972). Interval Analysis and Applications to Linear Algebra. Ph. D. Thesis, Oxford university (1972).
66. Rokne, J. (1985). A low complexity rational centered form. *Computing 34*, 261-263.
67. Sederberg, T.W. and D. B. Buehler (1992). Offsets of polynomial Bézier curves: Hermite approximation with error bounds. In *Mathematical Methods in Computer Aided geometric Design II,* T. Lyche and L. Scummaker (eds.), San Diego: Academic Press 549-566.
68. Sederberg, T.W. and R. T. Farouki (1992). Approximation by interval Bézier curves. *IEEE Computer Graphics and Applications 12, No. 5*, 87-95.
69. Skelboe, S. (1974). Computation of rational interval functions. *BIT 14*, 87-95.
70. Skelboe, S. (1979). True worst-case analysis of linear electrical circuits by interval arithmetic. *IEEE Transactions on Circuits and Systems 26*, 874-879.
71. Stahl, V. (1997). Error reduction of the Taylor centered form by half and an inner estimation of the range. *Reliable Computing 3*, 411-420.
72. Suffern, K.G. and E. D. Fackerell (1991). Interval methods in computer graphics. *Computers and Graphics 15*, 331-340.
73. Sun Microsystems (2000). http://access1.sun.com/workshop6ea/docs/mr/READMEs/interval_arithmetic.html, Sum Microsystems, Palo Alto, Ca.
74. van Hentenryck, P., L. Michel and Y. Deville (1997). *Numerica: a modeling language for global optimization.* Cambridge, MA: MIT.
75. van Hentenryck, P., D. McAllester and D. Kapur (1997). Solving polynomial systems using a branch and prune approach. *Siam J. Numer. Anal. 34*, 797-827.

76. Warmus, M. (1956). Calculus of approximations. *Bull. Acad. Polon, Sci. Cl. III 4*, 253-259.

Interval and Ellipsoidal Uncertainty Models

Andrzej Bargiela

Department of Computing, The Nottingham Trent University,

Nottingham NG1 4BU, UK

Abstract. In this Chapter, we present results derived in the context of state estimation of a class of real-life systems that are driven by some poorly known factors. For these systems, the representation of uncertainty as confidence intervals or the ellipsoids offers significant advantages over the more traditional approaches with probabilistic representation of noise. While the filtered-white-Gaussian noise model can be defined on grounds of mathematical convenience, its use is necessarily coupled with a hope that an estimator with good properties in idealised noise will still perform well in real noise. With good knowledge of the plant and its environment, a sufficiently accurate approximation to the probability density function can be obtained, but shortage of prior information or excessive computing demands normally rule out this option. A more realistic approach is to match the noise representation to the extent of prior knowledge. The relative merits of interval and ellipsoidal representations of noise are discussed in a set theoretic setting and are illustrated using both a simple synthetic example and a real-life scenario of state estimation of a water distribution system.

Keywords. System modelling, uncertainty, confidence limit analysis, interval analysis, ellipsoid methods

1 The system model

We start by introducing the deterministic model of the system

$$g(\mathbf{x}) = \mathbf{z} \tag{1}$$

where $\mathbf{x}$ is an n-dimensional state vector and $\mathbf{z}$ is an m-dimensional measurement vector. In the deterministic equation it is assumed that the true measurement vector $\mathbf{z}^t$ is approximated well by the observed measurement vector $\mathbf{z}^o$. Conversly, in the non-deterministic or uncertain model, all that is assumed is that the true measurement vector is contained in the region bounbed by $\mathbf{z}^l$ and $\mathbf{z}^u$. The coordinates of vectors $\mathbf{z}^l$ and $\mathbf{z}^u$ are defined as $z^l_i = z^o_i - e_i$ and $z^u_i = z^o_i + e_i$ for all i=1,2,...n. A measurement set M is defined as a collection of variables in the system for which real metered values or measurement estimates (pseudomeasurements) are available. A distinction is made between the measurement set *M* and the collection of values that this set would produce for a particular operating state. This measurement set at a particular instant of time, or for a particular assumed operating state, will produce a measurement vector $\mathbf{z}^o \in \mathbf{R}^m$, where *m* is cardinality of *M*. With the measurement set *M* producing measurement vector $\mathbf{z}^o$, the set of feasible measurement vectors is given by:

$$Z(M, \mathbf{z}^l, \mathbf{z}^u) := \{ \mathbf{z}^o \in \mathbf{R}^m : z_i^l \leq z_i \leq z_i^u, i=1,...,m \} \quad (2)$$

where *m* is the cardinality of *M* and $\mathbf{z}^l$ and $\mathbf{z}^u$ are defined as above. $Z(M, \mathbf{z}^l, \mathbf{z}^u)$ defines a region of $\mathbf{R}^m$ in which the true measurement vector is contained. This region is the smallest that can be obtained within the limits of accuracy of the measurement set. In this format, the system equation (1) is replaced with the following set inclusion:

$$g(\mathbf{x}) \in Z(M, \mathbf{z}^l, \mathbf{z}^u) \quad (3)$$

following from the assumption that the true measurement vector is unknown but constrained in $Z(M, \mathbf{z}^l, \mathbf{z}^u)$. This gives the set of feasible state vectors, $X(M, \mathbf{z}^l, \mathbf{z}^u)$, for measurement set *M* and measurement vector $\mathbf{z}^o$, as:

$$X(M, \mathbf{z}^l, \mathbf{z}^u) := \{ \mathbf{x} \in \mathbf{R}^n : g(\mathbf{x}) \in Z(M, \mathbf{z}^l, \mathbf{z}^u) \} \quad (4)$$

Equation (3) will be referred to as the **uncertain system equation** with $X(M,\mathbf{z}^l,\mathbf{z}^u)$, of (4), representing the state uncertainty set. For the uncertain system equation there is no unique operating state that can be calculated. All that can be defined is a set of possible operating states resulting from the set of possible measurement vectors. No preference is placed on these, all are assumed to be equally likely. This reflects the lack of preference for a particular measurement vector in $Z(M,\mathbf{z}^l,\mathbf{z}^u)$ Although lack of a unique estimate of $\mathbf{x}$ is at first worrying, we can reassure ourselves by noticing that engineering design is largely a matter of tolerancing for adequate performance in the worst case. Also, this method does not make any unrealistic assumptions about the probabilistic properties of the measurement data, its expected values or its probabilistic variation. The unknown-but-bounded treatment of measurement uncertainty leads to this simple and flexible presentation of state estimate uncertainty. Feasible state estimates are specified by a sharply defined set, $X(M, \mathbf{z}^l, \mathbf{z}^u)$. This fits neatly with many intended engineering uses of state estimates. Questions such as: Is the system operating in an acceptable range? Has the system failed? If so, where is the fault located?; can be answered more easily and categorically when the range of possible operating states can be clearly defined.

When faced with measurement uncertainty in state estimation the most common response of engineers and researchers has been to try and produce estimates that best fit the measurement data in some way. The attraction of providing a single 'optimal' point-value in n-dimensional state space must however be ballanced by the lack of indication of how accurate this estimate is. It has long been recognised that the notion of optimality of such estimates is inherently linked to the implicit assumptions about the measurement noise, which are rarely satisfied. In which case, no particular 'optimal' state estimate can be identified.

To make the uncertainty in state estimates more accessible to engineering practice, uncertainty intervals or confidence limits, similar to those for measurement values, can be derived in the following way. Let

$$x_i^l := \min_{\mathbf{x} \in X(M, \mathbf{z}^l, \mathbf{z}^u)} x_i, \quad i=1,\ldots,n$$

$$x_i^u := \max_{\mathbf{x} \in X(M, \mathbf{z}^l, \mathbf{z}^u)} x_i, \quad i=1,\ldots,n \tag{5}$$

The vectors $\mathbf{x}^l$ and $\mathbf{x}^u$ will provide lower and upper bounds on the state vector $\mathbf{x}$ in the same way that $\mathbf{z}^l$ and $\mathbf{z}^u$ did for the measurement vector. For each individual variable, the interval (x_i^l , x_i^u) is referred to as the **uncertainty interval** for the ith variable and x_i^l and x_i^u are referred to as **confidence limits**. These uncertainty intervals or confidence limits are as tight as can be achieved with the given measurement uncertainty. Calculating these bounds - the process referred to as confidence limit analysis – is discussed in the subsequent sections. If $X^*(M, \mathbf{z}^l, \mathbf{z}^u)$ is the set defined by these bounds, ie

$$X^*(M, \mathbf{z}^l, \mathbf{z}^u) := \{ \mathbf{x} \in \mathbf{R}^n : x_i^l \leq x_i \leq x_i^u , i=1,\ldots,n \} \tag{6}$$

then it must be noted that $X^*(M, \mathbf{z}^l, \mathbf{z}^u)$ may not be the same as $X(M, \mathbf{z}^l, \mathbf{z}^u)$. Clearly, $X(M, \mathbf{z}^l, \mathbf{z}^u) \subseteq X^*(M, \mathbf{z}^l, \mathbf{z}^u)$, but not every combination of values that are each feasible for individual state variables are feasible for a feasible state vector. Let $Z^*(M, \mathbf{z}^l, \mathbf{z}^u)$ be the subset of $\mathbf{R}^m$ onto which $X(M, \mathbf{z}^l, \mathbf{z}^u)$ is mapped by g(.) the system function. Then

$$Z^*(M, \mathbf{z}^l, \mathbf{z}^u) := \{ \mathbf{z} \in \mathbf{R}^m : \mathbf{z}=g(\mathbf{x}), \mathbf{x} \in X(M, \mathbf{z}^l, \mathbf{z}^u) \} \tag{7}$$

$Z^*(M, \mathbf{z}^l, \mathbf{z}^u) \subseteq Z(M, \mathbf{z}^l, \mathbf{z}^u)$, but these two sets are unlikely to be equal. There may be a $\mathbf{z} \in Z(M, \mathbf{z}^l, \mathbf{z}^u)$ for which there is no $\mathbf{x}$ (neither in $X(M, \mathbf{z}^l, \mathbf{z}^u)$ nor $\mathbf{R}^n$) for which $g(\mathbf{x})=\mathbf{z}$. In other words, there may be vectors in $Z(M, \mathbf{z}^l, \mathbf{z}^u)$ that are inconsistent for g(.). These two remarks are illustrated in Figure 1 for the 2-dimensional case.

2 Confidence limit analysis

The process of calculating uncertainty bounds for the state estimates, which result from the measurement and pseudomeasurement uncertainty, is referred to as confidence limit analysis. Based on the model of uncertainty, described in the previous section, mathematical methods for calculating these confidence limits are now presented.

It can now be seen how the confidence limit analysis can be formulated as a series of mathematical optimisation problems. For each of the independent state variables, $i=1,\ldots,n$

$$x_i^l = \min x_i \qquad \text{subject to } \mathbf{x} \in X(M, \mathbf{z}^l, \mathbf{z}^u) \tag{8}$$

$$x_i^u = \max x_i \qquad \text{subject to } \mathbf{x} \in X(M, \mathbf{z}^l, \mathbf{z}^u) \tag{9}$$

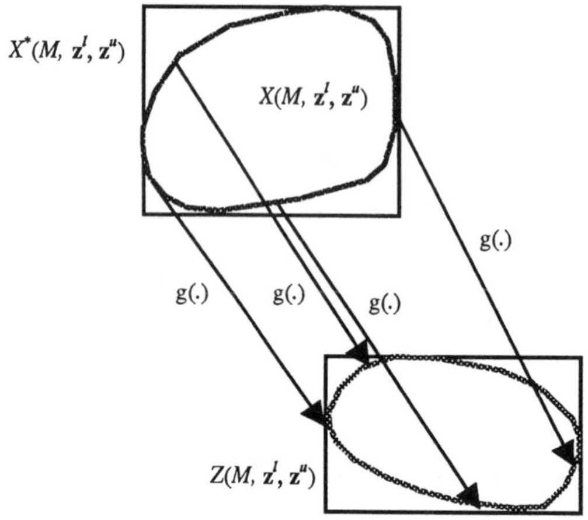

$X(M, \mathbf{z}^l, \mathbf{z}^u)$ – The state uncertainty set

$X^*(M, \mathbf{z}^l, \mathbf{z}^u)$ – The smallest box containing the state uncertainty

$Z(M, \mathbf{z}^l, \mathbf{z}^u)$ – The measurement uncertainty set

$Z^*(M, \mathbf{z}^l, \mathbf{z}^u)$ – The image of the state uncertainty set when mapped by g(.)

Figure 1 Relationship between X^* and X, and between Z^* and Z

The nature of the system equations g(.) means that the optimisation problems of (8) and (9) are non-linear. Therefore the choice of optimisation technique to be used is not at all clear. With n, the number of state variables, and m, the number of measurements, confidence limit analysis requires $2n$ non-linear optimisations each subject to $2m$ constraints (the $2m$ constraints are supplied by the lower and upper bounds on the measurement uncertainty). For real-life systems there may be several hundred state variables and several hundred measurements and pseudomeasurements. Therefore, confidence limit analysis is a highly computationally intensive task which requires efficient optimisation techniques.

In the reminder of this section confidence algorithms are presented. These fall into two categories: non-linear and linearised methods.

2.1 Monte Carlo method

In normal use, deterministic state estimators produce one state estimate for one measurement vector. Used in this way they give no indication of how a state estimate may vary in response to variations in the measurement values. Alternatively, if a deterministic state estimator is used repeatedly for a whole range of measurement vectors then some indication of state estimate variability is provided. It is this idea that forms the basis to the Monte Carlo approach to confidence limit analysis. A large number of feasible state estimates are generated, as randomly as possible, and from these the state estimate confidence limits are estimated. The larger the number of random feasible state estimates the more reliable the confidence limits.

Let $\mathbf{z}^i$ be a measurement vector, selected randomly from the set $Z(M, \mathbf{z}^l, \mathbf{z}^u)$ of all feasible measurement vectors, and let $\mathbf{x}^i$ be a deterministic state estimate calculated from $\mathbf{z}^i$. $\mathbf{x}^i$ is a feasible state estimate if $g(\mathbf{x}^i) \in Z(M, \mathbf{z}^l, \mathbf{z}^u)$. This follows from the definition of feasible vectors given in equation (4). It must be noted that $g(\mathbf{x}^i)$ is not necessarily equal to $\mathbf{z}^i$. If $\mathbf{z}^i$ is not a consistent vector, then there is no state vector $\mathbf{x}$ for which $g(\mathbf{x}^i) = \mathbf{z}^i$. In fact, if $Z^*(M, \mathbf{z}^l, \mathbf{z}^u)$ is defined as in (7), then $Z^*(M, \mathbf{z}^l, \mathbf{z}^u) = Z(M, \mathbf{z}^l, \mathbf{z}^u)$ only when M is a minimal measurement set (ie if M is an observable set and has no observable subset). For sequence, $\mathbf{z}^1, \ldots, \mathbf{z}^k$, of measurement vectors selected randomly from $Z(M, \mathbf{z}^l, \mathbf{z}^u)$, a sequence of sets $X^1, \ldots, X^k$ can be defined, with

$$X^j := \{ \mathbf{x}^j \in \mathbf{R}^n : g(\mathbf{x}^i) \in Z(M, \mathbf{z}^l, \mathbf{z}^u) \text{ for some } i \in \{1,\ldots,j\}\}, j=1,\ldots,k \quad (10)$$

where $\mathbf{x}^j$ is the state estimate calculated from $\mathbf{z}^j$. X^j is the set of feasible state estimates generated by the sequence of measurement vectors $\mathbf{z}^1, \ldots, \mathbf{z}^j$. This sequence of sets is such that $X^j \subseteq X^k \subseteq X(M, \mathbf{z}^l, \mathbf{z}^u)$ for all j=1,...,k-1, as only feasible state estimates are contained in X^j. For a large number, k, of randomly selected measurement vectors it can be assumed that X^k is approximately equal to $X(M, \mathbf{z}^l, \mathbf{z}^u)$. In other words, as $k \to \infty$, $X^k \to X(M, \mathbf{z}^l, \mathbf{z}^u)$.

Before an algorithm description is given, we need to point out that the actual estimator used is of no importance, provided that it is unbiased and that it can guarantee convergence in a high proportion of cases. All state estimates are checked for feasibility before being used to update X^j. A sequence of random measurement vectors can be selected from $Z(M, \mathbf{z}^l, \mathbf{z}^u)$ by using a random number generator. For example, a sequence of random numbers, $r_1^j, \ldots, r_m^j$, scaled to be between 0.0 and 1.0, can be generated and used to construct the measurement vector $\mathbf{z}^j$

$$z_i^j = z_i^l + r_i^j . (z_i^u - z_i^l), i=1,\ldots,m \quad (11)$$

where $\mathbf{z}^l$ and $\mathbf{z}^u$ are the lower and upper bounds for $Z(M, \mathbf{z}^l, \mathbf{z}^u)$. $\mathbf{z}^{j+1}$ can be constructed in a similar way from a new sequence of random numbers. Throughout the computations, for X^j , only two vectors need to be stored, these are $\mathbf{x}^{jl}$ and $\mathbf{x}^{ju}$, the lower and upper bounding vectors for the current set of feasible state estimates, X^j. These vectors are updated whenever a new feasible vector, not contained in any of the X^j 's, is found.

The Monte Carlo confidence limit algorithm

1. Select a large number, k (to limit the number of simulations) and set $i = 0$.
2. Set $i = i + 1$.
3. Select a sequence of m random numbers, $r_1^{\,i},\ldots, r_m^{\,i}$, and use these to construct a random measurement vector $\mathbf{z}^i$ from $Z(M, \mathbf{z}^l, \mathbf{z}^u)$ as described in (11)
4. Calculate a state estimate, $\mathbf{x}^i$ from $\mathbf{z}^i$. If $g(\mathbf{x}^i) \in Z(M, \mathbf{z}^l, \mathbf{z}^u)$, then use $\mathbf{x}^i$ tu update $\mathbf{x}^{il}$, $\mathbf{x}^{iu}$. Otherwise reject $\mathbf{x}^i$ as infeasible.
5. If $i < k$, go back to step 2. Otherwise stop.

Monte Carlo method is obviously slow computationally, but despite this it is useful in some situations. The condition that only feasible state estimates are used to update $\mathbf{x}^{il}$ and $\mathbf{x}^{iu}$ makes the procedure mathematically reliable and ensures that these bounds can be attained. The method can be used as a yardstick, against which the accuracy of all other confidence limit algorithms can be compared. Unfortunately, the method is impractical in many real-time applications.

Other more practical methods are described in subsequent sections. Firstly the problem is linearised and a linear version of the state uncertainty model is presented in section 2.2. In sections 2.3, 2.4 and 2.5, three confidence limit algorithms are presented. These are: the linear programming method; the sensitivity method and the ellipsoid method.

2.2 Linearised confidence limit analysis

Suppose that $\hat{\mathbf{x}}$ is the state estimate calculated from the measurement vector $\mathbf{z}^o$, where $\mathbf{z}^o$ is defined as in section 1. The non-linear system function, g(.), can be linearised around $\hat{\mathbf{x}}$ using a first order Taylor approximation to give:

$$g(\mathbf{x}) \approx g(\hat{\mathbf{x}}) + J.\,(\mathbf{x} - \hat{\mathbf{x}}) \qquad (12)$$

for all state vectors close to $\hat{\mathbf{x}}$. In (12), J is the Jacobian matrix evaluated at $\hat{\mathbf{x}}$. g(.) can be linearised around any state vector, $\hat{\mathbf{x}}$, this need not necessarily be the state estimate for $\mathbf{z}^o$. It is better, however, to use a value for $\hat{\mathbf{x}}$ that is in some way central to the set of feasible state vectors. This is because, the approximation used in (12) is more accurate for values of **x** for which $\| \mathbf{x} - \hat{\mathbf{x}} \|$ is small. The best estimate available for the centre of $X(M, \mathbf{z}^l, \mathbf{z}^u)$ is the state estimate calculated from $\mathbf{z}^o$. In section 1, a feasible state vector was described as one for which $g(\mathbf{x}) \in Z(M, \mathbf{z}^l, \mathbf{z}^u)$. This condition can be linearised using (12) to give a linear approximation, $X^1(M, \mathbf{z}^l, \mathbf{z}^u)$, of the state uncertainty set $X(M, \mathbf{z}^l, \mathbf{z}^u)$.

This is defined as follows:

$$X^1(M, \mathbf{z}^l, \mathbf{z}^u) := \{ \mathbf{x} \in \mathbf{R}^n : g(\hat{\mathbf{x}}) + J.\,(\mathbf{x} - \hat{\mathbf{x}}) \in Z(M, \mathbf{z}^l, \mathbf{z}^u) \} \tag{13}$$

$X^1(M, \mathbf{z}^l, \mathbf{z}^u)$ will be referred to as the linearised state uncertainty set. Substituting **dx** for $\mathbf{x} - \hat{\mathbf{x}}$ and using the definition of $Z(M, \mathbf{z}^l, \mathbf{z}^u)$ given in (2), $X^1(M, \mathbf{z}^l, \mathbf{z}^u)$ can be redefined as:

$$X^1(M, \mathbf{z}^l, \mathbf{z}^u) := \{ \mathbf{x} \in \mathbf{R}^n : \mathbf{x} = \hat{\mathbf{x}} + \mathbf{dx},\ \mathbf{z}^l - g(\hat{\mathbf{x}}) \leq J.\mathbf{dx} \leq \mathbf{z}^u - g(\hat{\mathbf{x}}) \} \tag{14}$$

It is easy to see that these two definitions, (13) and (14), of $X^1(M, \mathbf{z}^l, \mathbf{z}^u)$ are equivalent. For the reasons given in section 1, the set $X^1(M, \mathbf{z}^l, \mathbf{z}^u)$ will not be calculated explicitly. Rather, the smallest 'box' or orthotope containing $X^1(M,\mathbf{z}^l,\mathbf{z}^u)$ is sought. This set will be denoted by $X^{1*}(M, \mathbf{z}^l, \mathbf{z}^u)$ and referred to as the linearised state uncertainty box. Following the definition of $\mathbf{x}^l$ and $\mathbf{x}^u$ in (5), lower and upper limits for $X^1(M, \mathbf{z}^l, \mathbf{z}^u)$ can be defined as follows:

$$x_i^{1\,l} := \min_{\mathbf{x} \in X^1(M, \mathbf{z}^l, \mathbf{z}^u)} x_i, \quad i=1,\ldots,n$$

$$x_i^{1\,u} := \max_{\mathbf{x} \in X^1(M, \mathbf{z}^l, \mathbf{z}^u)} x_i, \quad i=1,\ldots,n \tag{15}$$

The definition of state uncertainty has been linearised in this way to allow confidence limit algorithms based on linear programming methods. Calculating the bounding vectors of $X^1(M, \mathbf{z}^l, \mathbf{z}^u)$ can easily be formulated as a linear programming problem. To allow this, some new notation is introduced:

$$\mathbf{dz}^l := \mathbf{z}^l - g(\hat{\mathbf{x}}) \tag{16}$$

$$\mathbf{dz}^u := \mathbf{z}^u - g(\hat{\mathbf{x}}) \tag{17}$$

$$DZ(M, \mathbf{z}^l, \mathbf{z}^u) := \{ \mathbf{dz} \in \mathbf{R}^m : g(\hat{\mathbf{x}}) + \mathbf{dz} \in Z(M, \mathbf{z}^l, \mathbf{z}^u) \} \tag{18}$$

$$DX^1(M, \mathbf{z}^l, \mathbf{z}^u) := \{ \mathbf{dx} \in \mathbf{R}^n : \hat{\mathbf{x}} + \mathbf{dx} \in X^1(M, \mathbf{z}^l, \mathbf{z}^u) \} \tag{19}$$

$DX^1(M, \mathbf{z}^l, \mathbf{z}^u)$ is just the set $X^1(M, \mathbf{z}^l, \mathbf{z}^u)$ shifted by $\hat{\mathbf{x}}$, $DZ(M, \mathbf{z}^l, \mathbf{z}^u)$ is the measurement uncertainty set shifted by $g(\hat{\mathbf{x}})$ and $\mathbf{dz}^l$ and $\mathbf{dz}^u$ represent 'tightest' lower and upper bounds for the set $DX^1(M, \mathbf{z}^l, \mathbf{z}^u)$. Then, the ith element of $\mathbf{dx}^l$ can be found by solving the linear programming problem

$$minimise\ dx_i \tag{20}$$
$$subject\ to\ \mathbf{dz}^l \leq J.\mathbf{dx} \leq \mathbf{dz}^u$$

Similarly, the ith element of $\mathbf{dx}^u$ can be found by solving the corresponding linear programming problem

$$maximise\ dx_i \tag{21}$$
$$subject\ to\ \mathbf{dz}^l \leq J.\mathbf{dx} \leq \mathbf{dz}^u$$

Hence by performing *2n* linear programs, the vectors $\mathbf{dx}^l$ and $\mathbf{dx}^u$ can be constructed. Once $\mathbf{dx}^l$ and $\mathbf{dx}^u$ have been calculated, it is a simple matter to construct the bounds

$$\mathbf{x}^{1l} = \hat{\mathbf{x}} + \mathbf{dx}^l \tag{22}$$

$$\mathbf{x}^{1u} = \hat{\mathbf{x}} + \mathbf{dx}^u \tag{23}$$

Three special cases can be identified:

(i) $g(\hat{\mathbf{x}}) = \mathbf{z}^o$ where $\mathbf{z}^o$ is the measurement vector from which $\hat{\mathbf{x}}$ was calculated as a state estimate. When this situation occurs, $\mathbf{dz}^u = -\mathbf{dz}^l = \mathbf{e}^z$, in other words, $Z(M, \mathbf{z}^l, \mathbf{z}^u)$ is symmetric about $g(\hat{\mathbf{x}})$. This symmetry is carried over to the linearised state uncertainty set , hence $\mathbf{dx}^u = -\mathbf{dx}^l$. This means that only *n* linear programming problems need to be solved, those in (19) say. It must be noted that a general measurement vector $\mathbf{z}^o$ will suffer from inconsistency, which means that there will be no state vector, **x**, in $\mathbf{R}^n$ for which g(**x**) is equal to $\mathbf{z}^o$.

(ii) *M* is an observable set and no subset of *M* is observable. In other words, *M* is a minimal measurement set. Consequently, all **z** in $Z(M, \mathbf{z}^l, \mathbf{z}^u)$ are consistent and, in particular, $\mathbf{z}^o$ is consistent. Hence, there is an $\hat{\mathbf{x}} \in \mathbf{R}^n$ for which $g(\hat{\mathbf{x}})=\mathbf{z}^o$. If $\hat{\mathbf{x}}$ can be found, only *n* linear programs need be performed (this follows from case *(i)*). Furthermore, when *J* is the Jacobian matrix for *M*, evaluated at $\hat{\mathbf{x}}$, and *M* is a minimal measurement set, *J* is square and non-singular. Hence, there exists an inverse, J^{-1}, for *J*. The lower and upper bounds for $DX^1(M, \mathbf{z}^l, \mathbf{z}^u)$, and hence for $X^1(M, \mathbf{z}^l, \mathbf{z}^u)$, can be calculated from the individual rows of J^{-1}, without help from linear programming methods. Before this can be done, we prove two lemmas:

Lemma 1 Let *M* be a minimal measurement set, $\mathbf{z}^o$ be an observed measurement vector from *M*, $\hat{\mathbf{x}}$ be the state estimate for $\mathbf{z}^o$ and *J* be the non-singular Jacobian matrix defined by *M* and $\hat{\mathbf{x}}$. Then, for $DX^1(M, \mathbf{z}^l, \mathbf{z}^u)$, defined as in (19)

$$DX^1(M, \mathbf{z}^l, \mathbf{z}^u) := \{\ \mathbf{dx} \in \mathbf{R}^n : \mathbf{dx} = J^{-1}.\mathbf{dz},\ \ \mathbf{dz}^l \leq J.\mathbf{dx} \leq \mathbf{dz}^u)\ \} \tag{24}$$

Proof By definitions (16), (17), (18) and (19), $\mathbf{dx} \in DX^1(M, \mathbf{z}^l, \mathbf{z}^u)$ if and only if $\mathbf{dz}^l \leq J.\mathbf{dx} \leq \mathbf{dz}^u$. Putting $\mathbf{dz} = J.\mathbf{dx}$, $\mathbf{dz}^l \leq J.\mathbf{dx} \leq \mathbf{dz}^u$ when and only when $J^{-1}.\mathbf{dz} = \mathbf{dx}$ and $\mathbf{dz}^l \leq \mathbf{dz} \leq \mathbf{dz}^u$. Hence, $DX^1(M, \mathbf{z}^l, \mathbf{z}^u)$ is equal to the set { $\mathbf{dx} \in \mathbf{R}^n$: $\mathbf{dx} = J^{-1}.\mathbf{dz}$, $\mathbf{dz}^l \leq J.\mathbf{dx} \leq \mathbf{dz}^u$) } and the lemma is proved.

Lemma 2 Let M be a minimal measurement set, $\mathbf{z}^o$ be an observed measurement vector from M, $\hat{\mathbf{x}}$ be the state estimate for $\mathbf{z}^o$ and let J be the non-singular Jacobian matrix defined by M and $\hat{\mathbf{x}}$. For i=1,...,n, the ith element of the lower bounding vector for $DX^1(M, \mathbf{z}^l, \mathbf{z}^u)$, dx_i^l, is given by

$$dx_i^l = \mathbf{a}^i.\mathbf{dz}^*, \text{ where}$$

$$dz_j^* = \{ \begin{array}{ll} dz_j^u & \text{if } a_j^i < 0.0, \\ dz_j^l & \text{otherwise}\} \end{array} \quad (25)$$

where $\mathbf{a}^i$ is the ith row of the matrix J^{-1}.

Proof By definition (20), dx_i^l is the minimum value for dx_i subject to $\mathbf{dx}$ being a member of $DX^1(M, \mathbf{z}^l, \mathbf{z}^u)$. Using Lemma 1, this is the same as the minimum value of $(J^{-1}.\mathbf{dz})_i$ subject to $\mathbf{dz}^l \leq \mathbf{dz} \leq \mathbf{dz}^u$. For any $\mathbf{dz}$, the product $(J^{-1}.\mathbf{dz})_i = \mathbf{a}^i.\mathbf{dz}$, where $\mathbf{a}^i$ is the the ith row of the matrix J^{-1}. As to $\mathbf{dz}^l \leq \mathbf{dz} \leq \mathbf{dz}^u$ is the only constraint on the elements of the vector $\mathbf{dz}$, the minimum value of $\mathbf{a}^i.\mathbf{dz}$ is equal to the sum of minimum values of $a_j^i.dz_j$ for j=1,...,n. The minimum value of $a_j^i.dz_j$, subject to of $dz_j^l \leq dz_j \leq dz_j^u$ is just $a_j^i.dz_j^u$ or $a_j^i.dz_j^l$, depending whether a_j^i is less than or greater than 0.0, respectively. When $a_j^i = 0.0$, $a_j^i.dz_j = 0.0$ as well, so such elements do not contribute to the value of dx_i^l. The lemma now follows.

Once the inverse of J has been calculated, Lemma 2 can be applied to each element of $\mathbf{dx}^l$ in turn, providing a straightforward way of calculating this vector that does not rely on optimisation methods. Because of the symmetry in this situation, the upper bound for $DX^1(M, \mathbf{z}^l, \mathbf{z}^u)$, $\mathbf{dx}^u$, is equal to the negative of the lower bound. So, Lemma 2 need only be applied n times. The bounding vectors, $\mathbf{x}^{1l}$ and $\mathbf{x}^{1u}$, are found by adding $\hat{\mathbf{x}}$ to $\mathbf{dx}^l$ and $\mathbf{dx}^u$, by equations (22) and (23).

(iii) $DX^1(M, \mathbf{z}^l, \mathbf{z}^u)$, and hence $X^1(M, \mathbf{z}^l, \mathbf{z}^u)$, may be empty, even when $Z(M, \mathbf{z}^l, \mathbf{z}^u)$ is non-empty. If this situation occurs then $Z(M, \mathbf{z}^l, \mathbf{z}^u)$ is said to be inconsistent. This is reflected in the bounding vectors, by $x_i^{1u} \leq x_i^{1l}$ for at least one i in 1,...,n.

The linearised state uncertainty set, $X^1(M, \mathbf{z}^l, \mathbf{z}^u)$, is only an approximation to the true state uncertainty set, $X(M, \mathbf{z}^l, \mathbf{z}^u)$. The question – how good an approximation is it? - now arises. To answer this question, an upper bound on the difference between x_i^{1u} and x_i^u – the upper limits on the feasible values for the ith state variable in $X^1(M, \mathbf{z}^l, \mathbf{z}^u)$ and $X(M, \mathbf{z}^l, \mathbf{z}^u)$, respectively – is derived. This is furnished by the following lemma.

Lemma 3 For a measurement uncertainty set $Z(M, \mathbf{z}^l, \mathbf{z}^u)$, where M is a minimal measurement set and $\mathbf{z}^o$ is an observed measurement vector derived from M. Let $\hat{\mathbf{x}}$ be the state estimate calculated from $\mathbf{z}^o$ and let J be the Jacobian matrix calculated at $\hat{\mathbf{x}}$. For the ith state variable, $i=1,\ldots,n$,

$$| x_i^u - x_i^{1u} | \leq \| \mathbf{a}^i \| . \| \boldsymbol{\varepsilon} \| \tag{26}$$

where x_i^u and x_i^{1u} are the upper limits for the ith state variable in the true and linearised state uncertainty sets respectively, $\mathbf{a}^i$ is the ith row of J^{-1} and $\boldsymbol{\varepsilon}$ is a vector for which:

$$\| \boldsymbol{\varepsilon} \| = O(\| \mathbf{x} - \hat{\mathbf{x}} \|^2) \tag{27}$$

Here, $\mathbf{x}$ is a feasible state vector in $X(M, \mathbf{z}^l, \mathbf{z}^u)$.

Proof The first point to note is that as M is a minimal measurement set, J^{-1} and $\mathbf{a}^i$ are well defined. The lemma is proved in two cases; for $x_i^u \geq x_i^{1u}$ and $x_i^u < x_i^{1u}$, respectively.

Case (i): $x_i^u \geq x_i^{1u}$. Let $\mathbf{x}^*$ be a vector in $X(M, \mathbf{z}^l, \mathbf{z}^u)$ for which the ith state variable attains its upper bound, ie $\mathbf{x}^*$ is a vector for which $x_i^* = x_i^u$. As $\mathbf{x}^*$ is a feasible member of $X(M, \mathbf{z}^l, \mathbf{z}^u)$ there is a measurement vector, $\mathbf{z}^*$ in $Z(M,\mathbf{z}^l,\mathbf{z}^u)$ for which $g(\mathbf{x}^*)=\mathbf{z}^*$. A vector $\mathbf{x}^{**}$ can be defined equal to $J^{-1}.(\mathbf{z}^*-(\hat{\mathbf{x}}))+\hat{\mathbf{x}}$. This is just the vector in $X^1(M, \mathbf{z}^l, \mathbf{z}^u)$ associated with $\mathbf{z}^*$ under the relationship given in the definition of $X^1(M, \mathbf{z}^l, \mathbf{z}^u)$ by (13). The vector $\mathbf{x}^{**}$ is a member of $X^1(M, \mathbf{z}^l, \mathbf{z}^u)$, so $x_i^{**} \leq x_i^{1u}$, as x_i^{1u} is the maximum value that the ith state variable can take in $X^1(M, \mathbf{z}^l, \mathbf{z}^u)$. Hence

$$| x_i^u - x_i^{1u} | \leq | x_i^u - x_i^{**} | = | x_i^* - x_i^{**} | \tag{28}$$

Attention is now focused on the difference $| x_i^* - x_i^{**} |$. From the definition of $\mathbf{x}^{**}$,

$$| x_i^* - x_i^{**} | = | x_i^* - [J^{-1}.(\mathbf{z}^*-g(\hat{\mathbf{x}})) + \hat{\mathbf{x}}]_i | \tag{29}$$

Since vector $\mathbf{z}^*$ in the *RHS* of this equation is defined as equal to $g(\mathbf{x}^*)$, the Taylor approximation of (12), gives

$$\mathbf{z}^* = g(\mathbf{x}^*) = g(\hat{\mathbf{x}}) + J.(\mathbf{x}^* - \hat{\mathbf{x}}) + \boldsymbol{\varepsilon} \tag{30}$$

for a vector $\boldsymbol{\varepsilon}$ of order $O(\| \mathbf{x} - \hat{\mathbf{x}} \|^2)$. Substituting for $\mathbf{z}^*$ in (29) and cancelling gives

$$| x_i^* - x_i^{**} | = | x_i^* - [\mathbf{x}^* + J^{-1}.\boldsymbol{\varepsilon}]_i | \tag{31}$$

The ith element of $J^{-1}.\boldsymbol{\varepsilon}$ is simply $\mathbf{a}^i.\boldsymbol{\varepsilon}$, where $\mathbf{a}^i$ is the ith row of J^{-1}. Combining (28), (31) with this last remark, and using the *Cauchy-Schwartz* inequality it follows that $| x_i^u - x^1{}_i^u | \leq \| \mathbf{a}^i \| . \| \boldsymbol{\varepsilon} \|$, which proves the lemma in this case.

Case (ii): $x_i^u < x^1{}_i^u$. Let $\mathbf{x}^+$ be a vector in $X^1(M, \mathbf{z}^l, \mathbf{z}^u)$ for which the ith variable attains its upper limit. If $\mathbf{z}^+$ is defined as equal to $g(\hat{\mathbf{x}}) - J.(\mathbf{x}^+ - \hat{\mathbf{x}})$ and $\mathbf{x}^{++}$ is the state vector in $X(M, \mathbf{z}^l, \mathbf{z}^u)$ for which $g(\mathbf{x}^{++}) = \mathbf{z}^+$ (this exists as M is a minimal measurement set and so $\mathbf{z}^+$ must be consistent), then a similar argument to that in case (i) gives:

$$| x^1{}_i^u - x_i^u | \leq | x_i^+ - x_i^{++} | \leq \| \mathbf{a}^i \| . \| \boldsymbol{\varepsilon} \| \tag{32}$$

where $\mathbf{a}^i$ is the ith row of J^{-1} and $\| \boldsymbol{\varepsilon} \|$ is of order $O(\|\mathbf{x}^{++} - \hat{\mathbf{x}}\|^2)$. The lemma is therefore proven.

It can be assumed, without the loss of generality, that the Jacobian martix is scaled so that $\|J\|$ is of order unity. When this is the case, and when J is not ill-conditioned, $\|\mathbf{a}^i\|$ will also be of order unity. This means that the maximum discrepancy between the upper limits for any of the state variables in $X(M, \mathbf{z}^l, \mathbf{z}^u)$ and $X^1(M, \mathbf{z}^l, \mathbf{z}^u)$, respectively, is of order $O(\|\mathbf{x} - \hat{\mathbf{x}}\|^2)$, where $\mathbf{x}$ is a feasible vector in $X(M, \mathbf{z}^l, \mathbf{z}^u)$. In other words, the accuracy of the linearised state uncertainty set is of the same magnitude as that of the Taylor approximation in (12) and so the discrepancy between the $\mathbf{x}^u$ and $\mathbf{x}^{1u}$ will not rise significantly when the confidence limit analysis problem is treated in this linearised form. It should be noted that the results of Lemma 3 provide only an upper bound for the discrepancy between in $X(M, \mathbf{z}^l, \mathbf{z}^u)$ and $X^1(M, \mathbf{z}^l, \mathbf{z}^u)$, the true magnitude of this discrepancy will, in most cases, be much less than this bound.

Lemma 3 was stated for a minimal measurement set. For an over-determined measurement set the state uncertainty sets will be smaller. So, the bounds of Lemma 3 can also be applied in the over-determined case.

2.3 Linear programming method

The general result that the confidence limits on state variables can be found by solving $2n$ linear programs with $2m$ constraints (equations (20), (21), (22) and (23)) is significant but a direct application of the revised simplex, or any similar linear programming algorithm, is likely to be too time consuming. In this section an alternative format for this problem is presented. It will be assumed that the measurement set M is both observable and over-determined. If M is not observable, then the uncertainty set is unbounded. If M is a minimal observable measurement set, then confidence limit analysis can be performed more efficiently using the method given in special case (ii) of the previous section.

Without loss of generality, it can be assumed that the elements of the measurement set M are ordered so that the first n elements correspond to an observable set of measurements. For any measurement vector, $\mathbf{z} \in Z(M, \mathbf{z}^l, \mathbf{z}^u)$, two new vectors can be defined. These are $\mathbf{z}^{\mathbf{n}} \in \mathbf{R}^{\mathbf{n}}$, the vector containing these first n elements of vector $\mathbf{z}$ and $\mathbf{z}^{\mathbf{m-n}} \in \mathbf{R}^{\mathbf{m-n}}$, the vector containing the remaining m-n elements. In the same way we define $\mathbf{dz}^{\mathbf{n}}$, $(\mathbf{dz}^{\mathbf{n}})^l$, $(\mathbf{dz}^{\mathbf{n}})^u \in \mathbf{R}^{\mathbf{n}}$, and $\mathbf{dz}^{\mathbf{m-n}}$, $(\mathbf{dz}^{\mathbf{m-n}})^l$, $(\mathbf{dz}^{\mathbf{m-n}})^u \in \mathbf{R}^{\mathbf{m-n}}$. New matrices J^{n} and $J^{\mathrm{m-n}}$ can also be defined, J^{n} consisting of the first n rows of J and $J^{\mathrm{m-n}}$ the remaining rows.

Lemma 4 The maximisation of (21) is equivalent to

$$\begin{aligned} &\textit{maximise} && \mathbf{a}^{\mathbf{i}}.\mathbf{dz}^{\mathbf{n}} && (33)\\ &\textit{subject to} && (\mathbf{dz}^{\mathbf{n}})^l \leq \mathbf{dz}^{\mathbf{n}} \leq (\mathbf{dz}^{\mathbf{n}})^u \\ & && (\mathbf{dz}^{\mathbf{m-n}})^l \leq J^{\mathrm{m-n}}(J^{\mathrm{n}})^{-1}\mathbf{dz}^{\mathbf{n}} \leq (\mathbf{dz}^{\mathbf{m-n}})^u \end{aligned}$$

where $\mathbf{a}^{\mathbf{i}}$ is the ith row of $(J^{\mathrm{n}})^{-1}$.

Proof The first point to note is that the observability of the first n measurements ensures that J^{n} is non-singular. So $\mathbf{a}^{\mathbf{i}}$ and $J^{\mathrm{m-n}}(J^{\mathrm{n}})^{-1}$ are well defined. Let $\mathbf{dz}^{\mathbf{n}} \in \mathbf{R}^{\mathbf{n}}$ be a feasible vector by the condition in (33). That is $(\mathbf{dz}^{\mathbf{n}})^l \leq \mathbf{dz}^{\mathbf{n}} \leq (\mathbf{dz}^{\mathbf{n}})^u$ and $(\mathbf{dz}^{\mathbf{m-n}})^l \leq J^{\mathrm{m-n}}(J^{\mathrm{n}})^{-1}\mathbf{dz}^{\mathbf{n}} \leq (\mathbf{dz}^{\mathbf{m-n}})^u$. As J^{n} is non-singular, there is a unique $\mathbf{dx}' \in \mathbf{R}^{\mathbf{n}}$ such that $J^{\mathrm{n}}.\mathbf{dx}' = \mathbf{dz}^{\mathbf{n}}$. So, the first constraint of (33) can be written as $(\mathbf{dz}^{\mathbf{n}})^l \leq J^{\mathrm{n}}.\mathbf{dx}' \leq (\mathbf{dz}^{\mathbf{n}})^u$.

Also, $J^{\mathrm{m-n}}.\mathbf{dx}' = J^{\mathrm{m-n}}(J^{\mathrm{n}})^{-1}\mathbf{dz}^{\mathbf{n}}$ which means, by the second constraint of (33), that $(\mathbf{dz}^{\mathbf{m-n}})^l \leq J^{\mathrm{m-n}}.\mathbf{dx}' \leq (\mathbf{dz}^{\mathbf{m-n}})^u$. Consequently, $\mathbf{dx}'$ is acertained to satisfy the constraints of (21).

Conversely, let $\mathbf{dx} \in \mathbf{R}^{\mathbf{n}}$ be a vector satisfying the constraints of (21), and let $(\mathbf{dz}^{\mathbf{n}})' = J^{\mathrm{n}}.\mathbf{dx}$. As the vector $\mathbf{dx}$ is feasible, $(\mathbf{dz}^{\mathbf{n}})^l \leq (\mathbf{dz}^{\mathbf{n}})' \leq (\mathbf{dz}^{\mathbf{n}})^u$. Also $(\mathbf{dz}^{\mathbf{m-n}})^l \leq J^{\mathrm{m-n}}(J^{\mathrm{n}})^{-1}(\mathbf{dz}^{\mathbf{n}})' \leq (\mathbf{dz}^{\mathbf{m-n}})^u$. That is, $(\mathbf{dz}^{\mathbf{n}})'$ is feasible in (33). It can bee seen therefore that there is one-to-one correspondence between the feasible $\mathbf{dx}$ in (21) and the feasible $\mathbf{dz}^{\mathbf{n}}$ in (33). More precisely, $\mathbf{dx}$ is feasible in (21) if and only if $J^{\mathrm{n}}.\mathbf{dx}$ is feasible in (33). To complete the proof, we need to show that for all feasible $\mathbf{dx}$ by (21) the two cost functions are the same. Let $\mathbf{dz}^{\mathbf{n}}$ be feasible by (33) and $\mathbf{dx}' = (J^{\mathrm{n}})^{-1}\mathbf{dz}^{\mathbf{n}}$. It is easy to see that $dx'_i = \mathbf{a}^{\mathbf{i}}.\mathbf{dz}^{\mathbf{n}}$, where $\mathbf{a}^{\mathbf{i}}$ is the ith row of $(J^{\mathrm{n}})^{-1}$, which completes the proof.

In just the same way, it can be shown that the minimisation

$$\begin{aligned} &\textit{minimise} && \mathbf{a}^{\mathbf{i}}.\mathbf{dz}^{\mathbf{n}} && (34)\\ &\textit{subject to} && (\mathbf{dz}^{\mathbf{n}})^l \leq \mathbf{dz}^{\mathbf{n}} \leq (\mathbf{dz}^{\mathbf{n}})^u \\ & && (\mathbf{dz}^{\mathbf{m-n}})^l \leq J^{\mathrm{m-n}}(J^{\mathrm{n}})^{-1}\mathbf{dz}^{\mathbf{n}} \leq (\mathbf{dz}^{\mathbf{m-n}})^u \end{aligned}$$

is equivalent to the minimisation given in (20).

As before, the bounds $\mathbf{dx}^l$ and $\mathbf{dx}^u$ can be constructed for $DX^1(M, \mathbf{z}^l, \mathbf{z}^u)$ by performing $2n$ maximisations and minimisations of this form. Also the bounds $\mathbf{x}^{1l}$ and $\mathbf{x}^{1u}$ for $X^1(M, \mathbf{z}^l, \mathbf{z}^u)$ can be evaluated by adding $\hat{\mathbf{x}}$ to $\mathbf{dx}^l$ and $\mathbf{dx}^u$.

This formulation of the problem has an important advantage over the formulation of (21). While the linear program of (21) has $2m$ constraints the form of (33) has only $2(m-n)$. In many real-life systems measurement redundancy is low so $m-n<<m$. A disadvantage of the second formulation is that it requires the inversion of the martix J^n. However J^n need only to be inverted once while the maximisations and minimisations are carried out $2n$ times. So, with an efficient matrix inversion scheme this disadvantage quickly disappears.

The linear programming confidence limit algorithm

1. Select an observable subset of M containing n measurements. This is the minimal measurement set and is denoted by M'. Order M with the elements of M' appearing first.
2. Re-order $\mathbf{dz}^l$ and $\mathbf{dz}^u$ according to the new ordering of M. Assemble $(\mathbf{dz}^n)^l$, $(\mathbf{dz}^n)^u$, $(\mathbf{dz}^{m-n})^l$, $(\mathbf{dz}^{m-n})^u$, J^n and J^{m-n}.
3. Factorise J^n and calculate $J^{m-n}(J^n)^{-1}$.
4. For each variable, i=1,...,n, calculate $\mathbf{a}^i$, the ith row of $(J^n)^{-1}$ and carry out the maximisation in (33) using a linear programming method. The resultant value of $\mathbf{a}^i.\mathbf{dz}^n$ is the ith element of $\mathbf{dx}^u$. Similarly, carry out the minimisation in (34), to obtain the ith element of $\mathbf{dx}^l$.
5. Add $\mathbf{dx}^l$ and $\mathbf{dx}^u$ to $\hat{\mathbf{x}}$ to obtain $\mathbf{dx}^{1l}$ and $\mathbf{dx}^{1u}$.

Example 1: In order to illustrate the operation of the linear programming algorithm let us consider a simple system described by the following inequalities

$$-1 < -x_1 + x_2 < 1$$

$$1 < x_1 + x_2 < 3$$

$$0.5 < x_2 < 3$$

These are depicted in Figure 2. Clearly, for the linear system the linearisation error is zero (thus there is no need for dx and dz notation). Consequently it is expected that the linear programming method will calculate the bounding box on the state uncertainty set, $X^*(M, \mathbf{z}^l, \mathbf{z}^u)$, that is identical to the one produced by the Monte Carlo simulations (given a sufficiently large number of Monte Carlo iterations).

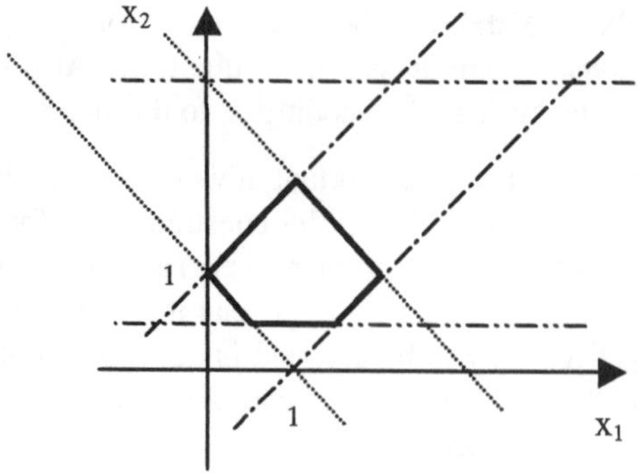

Figure 2. State uncertainty set defined by three double inequalities

We select the first two inequalities as the minimum measurement set, thus defining $(\mathbf{z^n})^l$, $(\mathbf{z^n})^u$, $(\mathbf{z^{m-n}})^l$, $(\mathbf{z^{m-n}})^u$, J^n and J^{m-n}.

$$(\mathbf{z^n})^l=[-1\ 1]^T,\ (\mathbf{z^n})^u=[1\ 3]^T,\ (\mathbf{z^{m-n}})^l=[0.5]^T,\ (\mathbf{z^{m-n}})^u=[3]^T,$$

$$J^n=\begin{bmatrix}-1 & 1\\ 1 & 1\end{bmatrix} \text{ and } J^{m-n}=[0\ 1]$$

So, the corresponding martices $(J^n)^{-1}$ and $J^{m-n}(J^n)^{-1}$ are

$$(J^n)^{-1}=\begin{bmatrix}-0.5 & 0.5\\ 0.5 & 0.5\end{bmatrix} \text{ and } J^{m-n}(J^n)^{-1}=[0.5\ 0.5]$$

The cost function, $\mathbf{a^i}.\mathbf{z^n}$, for i=1 is $\mathbf{a^1}.\mathbf{z^n}= -0.5z_1^n + 0.5z_2^n$, so minimising it with respect of $\mathbf{z^n}$, subject to constraints (33), gives

$$(z_1^n)^l=1 \text{ and } (z_2^n)^l=1$$

and maximising the cost function gives

$$(z_1^n)^u=-1 \text{ and } (z_2^n)^u=3$$

For i=2 we minimise $\mathbf{a^2}.\mathbf{z^n}= 0.5z_1^n + 0.5z_2^n$, to obtain

$$(z_1^n)^l=-1+s \text{ and } (z_2^n)^l=2-s$$

(where the parameter s, $0\leq s\leq 1$, signifies the degenerate LP case)

The maximisation of $\mathbf{a^2}.\mathbf{z^n}$ gives

$$(z_1^n)^u=1 \text{ and } (z_2^n)^u=3$$

Evaluating now the equations $x_i^l=\mathbf{a^i}.(\mathbf{z^n})^l$, and $x_i^u=\mathbf{a^i}.(\mathbf{z^n})^u$, for i=1 and i=2 we have $x_1^l=[-0.5\ 0.5][1\ 1]^T=0$, $x_1^u=[-0.5\ 0.5][-1\ 3]^T=2$, $x_2^l=[0.5\ 0.5][-1+s\ 2-s]^T=0$ and $x_2^u=[0.5\ 0.5][1\ 3]^T=2$, which is the excact solution: $0\leq x_1\leq 2$ and $0.5\leq x_2\leq 2$.

2.4 Sensitivity matrix method

A large class of applications, as exemplified by on-line decision support, requires that the confidence limit analysis procedure calculates uncertainty bounds in real-time. The linear programming method described in the previous section, while being efficient, may fall short of the real-time requirement for large-scale applications. In this section an alternative method, referred to as sensitivity matrix method, is considered.

When the measurement set is minimal (ie it is observable and contains no observable subset), the linearised uncertainty bounds can be calculated without recourse to a linear programming procedure, as discussed in case (ii) in section 2.2. So in these circumstances, confidence limit analysis can be carried out much more rapidly than in the general case when the linear programming algorithm of the previous section has to be used. The shortcut can be taken (when M is minimal) because the Jacobian matrix J is square and invertible. So, any $\mathbf{dx} \in DX^1(M, \mathbf{z}^l, \mathbf{z}^u)$ is given by $J^{-1}.\mathbf{dz}$ for some $\mathbf{dz} \in DZ^1(M, \mathbf{z}^l, \mathbf{z}^u)$. In general, M is over-determined and so J is an m by n matrix of rank n. When J is of this form it has no inverse. The lack of inverse is not due to a shortage of information, rather, there is a surfeit of measurement data. As there is no proper inverse for J, a pseudo-inverse must be used. Let $\mathbf{dx} \in \mathbf{R}^n$ and $\mathbf{dz} \in \mathbf{R}^m$, for which $J.\mathbf{dx}=\mathbf{dz}$, then $J^T J.\mathbf{dx}=J^T.\mathbf{dz}$ and so

$$\mathbf{dx} = (J^T J)^{-1} J^T.\mathbf{dz} \qquad (35)$$

This equation is well defined because when J is of rank n, $J^T J$ is both square and invertible. The matrix $(J^T J)^{-1} J^T$ is the pseudo-inverse that will be used when $(J)^{-1}$ is not well defined. In fact, when $(J)^{-1}$ is well defined $(J^T J)^{-1} J^T$ is equal to J^{-1}. $(J^T J)^{-1} J^T$ is referred to as the **sensitivity matrix** as its $(i,j)^{th}$ element relates the sensitivity of the ith element of the state vector to changes in the jth element of the measurement vector.

A new approximate linearised state uncertainty set, $X^2(M, \mathbf{z}^l, \mathbf{z}^u)$, can be defined as follows:

$$X^2(M, \mathbf{z}^l, \mathbf{z}^u) := \{\mathbf{x} \in \mathbf{R}^n : \mathbf{x} = \hat{\mathbf{x}} + \mathbf{dx},\ \mathbf{dx} = (J^T J)^{-1} J^T.\mathbf{dz},\ \mathbf{dz} \in DZ(M, \mathbf{z}^l, \mathbf{z}^u)\} \qquad (36)$$

This set is an approximation of the linearised state uncertainty set, $X^1(M, \mathbf{z}^l, \mathbf{z}^u)$. The following lemma highlights the relationship between $X^1(M, \mathbf{z}^l, \mathbf{z}^u)$ and $X^2(M, \mathbf{z}^l, \mathbf{z}^u)$.

Lemma 5 For $X^2(M, \mathbf{z}^l, \mathbf{z}^u)$ defined as in (36) and $X^1(M, \mathbf{z}^l, \mathbf{z}^u)$, defined as in (13),

$$X^1(M, \mathbf{z}^l, \mathbf{z}^u) \subseteq X^2(M, \mathbf{z}^l, \mathbf{z}^u) \qquad (37)$$

and when M is a minimal observable measurement set

$$X^1(M, \mathbf{z}^l, \mathbf{z}^u) = X^2(M, \mathbf{z}^l, \mathbf{z}^u) \qquad (38)$$

Proof Let $\mathbf{dx}+\hat{\mathbf{x}} \in X^1(M, \mathbf{z}^l, \mathbf{z}^u)$, then $J.\mathbf{dx} \in DZ(M, \mathbf{z}^l, \mathbf{z}^u)$ by (14) and (18). Put $\mathbf{dz}=J.\mathbf{dx}$. The element of $X^2(M, \mathbf{z}^l, \mathbf{z}^u)$ generated by $\mathbf{dz}$ is just a vector $\mathbf{x}=\hat{\mathbf{x}}+(J^TJ)^{-1}J^T.\mathbf{dz} = \hat{\mathbf{x}}+(J^TJ)^{-1}J^T J.\mathbf{dx} = \hat{\mathbf{x}}+\mathbf{dx}$. Hence $\hat{\mathbf{x}}+\mathbf{dx} \subseteq X^2(M, \mathbf{z}^l, \mathbf{z}^u)$ and (37) follows. Proof of the second part, (38) follows from the observation that when M is minimal, $(J^TJ)^{-1}J^T$ symplifies to J^{-1}.

When M is over-determined there may be vectors $\mathbf{dz} \in Z(M, \mathbf{z}^l, \mathbf{z}^u)$ that are inconsistent. For such a vector there can be no $\mathbf{dx} \in \mathbf{R^n}$ with $J.\mathbf{dx}=\mathbf{dz}$. So, in particular $J(J^TJ)^{-1}J^T.\mathbf{dz}$ is not equal to $\mathbf{dz}$. It may even be that $J(J^TJ)^{-1}J^T.\mathbf{dz}$ is not contained in $DZ(M, \mathbf{z}^l, \mathbf{z}^u)$. It is these vectors that account for the difference between $X^1(M, \mathbf{z}^l, \mathbf{z}^u)$ and $X^2(M, \mathbf{z}^l, \mathbf{z}^u)$. That is, $\mathbf{dx}+\hat{\mathbf{x}} \in X^2(M, \mathbf{z}^l, \mathbf{z}^u)-X^1(M, \mathbf{z}^l, \mathbf{z}^u)$ if and only if the state vector $\mathbf{dx}=(J^TJ)^{-1}J^T.\mathbf{dz}$ for some $\mathbf{dz} \in DZ(M, \mathbf{z}^l, \mathbf{z}^u)$ with $J(J^TJ)^{-1}J^T.\mathbf{dz}$ not a member of $DZ(M, \mathbf{z}^l, \mathbf{z}^u)$. Although $X^2(M, \mathbf{z}^l, \mathbf{z}^u)$ is not identical to $X^1(M, \mathbf{z}^l, \mathbf{z}^u)$ it can be used to form bounds for the linearised state uncertainty set. Although these bounds are less tight than the ones obtained with the linear programming method, at least they do not rule out any feasible state vector from the uncertainty box.

Bounding vectors for $X^2(M, \mathbf{z}^l, \mathbf{z}^u)$, denoted $\mathbf{x}^{2l}$ and $\mathbf{x}^{2u}$, can be defined analogously to $\mathbf{x}^{1l}$ and $\mathbf{x}^{1u}$ for the true linearised state uncertainty set $X^1(M, \mathbf{z}^l, \mathbf{z}^u)$. The following algorithm provides a way of calculating these vectors.

Sensitivity matrix confidence limit algorithm

1. Set $i = 0$.
2. Factorise the matrix J^TJ (This can be done using an augmented matrix formulation so as to preserve the condition number of the matrix J).
3. Set $i = i + 1$.
4. Calculate $\mathbf{b}^\mathbf{i}$, the ith row of the sensitivity matrix $(J^TJ)^{-1}J^T$ (This can be done efficiently taking into account the sparsity of J and using the augmented matrix based factorisation of step 2).
5. Put $x^{2u}_i = \mathbf{b}^\mathbf{i}.\mathbf{dz}^+ + \hat{x}_i$, where

$$dz_j^+ = \begin{cases} dz_j^u & \text{if } b_j^i > 0.0 \\ dz_j^l & \textit{otherwise} \end{cases} \tag{39}$$

Put $x^{2l}_i = \mathbf{b}^\mathbf{i}.\mathbf{dz}^- + \hat{x}_i$, where

$$dz_j^- = \begin{cases} dz_j^l & \text{if } b_j^i > 0.0 \\ dz_j^u & \textit{otherwise} \end{cases} \tag{40}$$

6. If $i < n$, go back to step 3. Otherwise stop.

This algorithm bears striking resemblance to the method of calculating the bounds for $X^1(M, z^l, z^u)$ described in special case (ii) in section 2.2. In fact, justification of equations (39) and (40) follows as a corollary to lemma 2. Although, in this case, there is not necessarily symmetry about $\hat{x}$ as $g(\hat{x})$ may not equal z^o.

Example 2: We apply here the sensitivity matrix confidence limit algorithm to the system of inequalities considered in Example 1.

The pseudoinverse matrix $(J^TJ)^{-1}J^T$ is

$$(J^TJ)^{-1}J^T = \begin{bmatrix} -0.5 & 0.5 & 0 \\ 0.333 & 0.333 & 0.333 \end{bmatrix}$$

so, $\mathbf{b}^1$=[-0.5 0.5 0] and $\mathbf{b}^2$=[0.333 0.333 0.333].

Processing **z**, according to (39) and (40), using the first row, $\mathbf{b}^1$, we obtain

$\mathbf{z}^-=[1.0\ 1.0\ 0.5]^T$ and $\mathbf{z}^+ = [-1.0\ 3.0\ 3.0]^T$ which produces

$$x_1^l = \mathbf{b}^1\mathbf{z}^- = [-0.5\ \ 0.5\ \ 0]\ [1.0\ 1.0\ 0.5]^T = 0$$

$$x_1^u = \mathbf{b}^1\mathbf{z}^+ = [-0.5\ \ 0.5\ \ 0]\ [-1.0\ 3.0\ 3.0]^T = 2$$

and using the second row, $\mathbf{b}^2$, we have $\mathbf{z}^- = [-1.0\ 1.0\ 0.5]^T$ and $\mathbf{z}^+ = [1.0\ 3.0\ 3.0]^T$ which gives

$$x_2^l = \mathbf{b}^2\mathbf{z}^- = [0.333\ 0.333\ 0.333]\ [-1.0\ 1.0\ 0.5]^T = 0.167$$

$$x_2^u = \mathbf{b}^2\mathbf{z}^+ = [0.333\ 0.333\ 0.333]\ [1.0\ 3.0\ 3.0]^T = 2.333$$

The bounding box on the state uncertainty set, $0 \leq x_1 \leq 2$ and $0.167 \leq x_2 \leq 2.33$, as calculated here, is larger than the one obtained with the Monte Carlo and the linear programming algorithm. The widening of bounds along the $\mathbf{x}_2$ direction is caused by the inherent feature of the pseudoinverse, that of attempting to ballance the sum of distances from x_i^l and x_i^u to <u>all</u> upper- and lover- bound constraints respectively. By contrast, the linear programming and the Monte Carlo algorithms are concerned only with the 'active' constraints for any given value of the state vector, thus ignoring the redundant constraints.

Interval arithmetic formalism for the sensitivity matrix method

The computation of the bounds on individual state variables, implemented in Step 5 of the above algorithm (equations (39) and (40)), can be expressed using the formalism of interval arithmetic. The idea of defining algebras over the sets of intervals, rather than over the sets of real numbers, dates back to late 50's, [25] and it was subsequently developed by Moore, [16], as a framework for the quantification of the mathematically accurate results using finite precision computations. The idea being that any real number can be represented by a pair of adjacent machine numbers which define an interval containing the original number. As the computations proceed, the upper- and lower- limits of the interval containing the result are evaluated conservatively so as to ensure that all feasible results are included.

For a scalar variable, an interval [v] is defined by its upper and lower limits as $[v^l,v^u]$ and it represents a subset of **R**. The set of all intervals of **R** is denoted here as **IR**, so that $[v]\in \mathbf{IR}$. For an *n*-dimensional vector variable, an interval **[v]** is defined as a Cartesian product of intervals of the individual coordinates. That is $[\mathbf{v}]=[v_1^l,v_1^u]\mathrm{x}[v_2^l,v_2^u]\mathrm{x}...\mathrm{x}[v_n^l,v_n^u]=[\mathbf{v}^l,\mathbf{v}^u]$ which defines a hyperbox in the $\mathbf{R}^n$ space. The set of all boxes of $\mathbf{R}^n$ is denoted as $\mathbf{IR}^n$, so that $[\mathbf{v}]\subseteq \mathbf{R}^n$ and $[\mathbf{v}]\in \mathbf{IR}^n$.

Here we will confine ourselves to two basic arithmetical operations on intervals. These are: addition of and multiplication of intervals. The constructive definitions of these operations are as follows:

$$[v] + [w] = [v^l,v^u] + [w^l,w^u] = [v^l+w^l,v^u+w^u] \quad (41)$$

$$[v].[w] = [v^l,v^u].[w^l,w^u] =$$

$$=[\min(v^l.w^l,v^l.w^u,v^u.w^l,v^u.w^u), \max(v^l.w^l,v^l.w^u,v^u.w^l,v^u.w^u)] \quad (42)$$

where [v], [w], $[v].[w]\in \mathbf{IR}$ and v^l, v^u, w^l, $w^u\in \mathbf{R}$. The multiplication of an interval, $[v]\in$ **IR,** by a scalar, $a\in \mathbf{R}$, is a special case of (42) with the scalar, a, represented as a point-interval [a,a].

The above formulas can be generalised to vector intervals (hyperboxes) to give

$$[\mathbf{v}] + [\mathbf{w}] = [\mathbf{v}^l,\mathbf{v}^u] + [\mathbf{w}^l,\mathbf{w}^u] = [\mathbf{v}^l+\mathbf{w}^l,\mathbf{v}^u+\mathbf{w}^u]=$$

$$([v_1^l+w_1^l,v_1^u+w_1^u], [v_2^l+w_2^l,v_2^u+w_2^u],...,[v_n^l+w_n^l,v_n^u+w_n^u])^T \quad (43)$$

$$[\mathbf{v}].[\mathbf{w}] = [v_1].[w_1] + [v_2].[w_2] + ... + [v_n].[w_n]=$$

$$= [v_1^l,v_1^u].[w_1^l,w_1^u] + [v_2^l,v_2^u].[w_2^l,w_2^u] + ... + [v_n^l,v_n^u].[w_n^l,w_n^u] =$$

$$= [\min(v_1^l.w_1^l,v_1^l.w_1^u,v_1^u.w_1^l,v_1^u.w_1^u), \max(v_1^l.w_1^l,v_1^l.w_1^u,v_1^u.w_1^l,v_1^u.w_1^u)] +$$

$$+ [\min(v_2^l.w_2^l,v_2^l.w_2^u,v_2^u.w_2^l,v_2^u.w_2^u), \max(v_2^l.w_2^l,v_2^l.w_2^u,v_2^u.w_2^l,v_2^u.w_2^u)] +...$$

$$+ [\min(v_n^l.w_n^l,v_n^l.w_n^u,v_n^u.w_n^l,v_n^u.w_n^u), \max(v_n^l.w_n^l,v_n^l.w_n^u,v_n^u.w_n^l,v_n^u.w_n^u)]=$$

$$[\sum_{i=1}^{n} \min(v_i^l.w_i^l,v_i^l.w_i^u,v_i^u.w_i^l,v_i^u.w_i^u), \sum_{i=1}^{n} \max(v_i^l.w_i^l,v_i^l.w_i^u,v_i^u.w_i^l,v_i^u.w_i^u)] \quad (44)$$

where $[\mathbf{v}]$, $[\mathbf{w}]\in \mathbf{IR}^n$, $\mathbf{v}^l,\mathbf{v}^u,\mathbf{w}^l,\mathbf{w}^u\in \mathbf{R}^n$, $[\mathbf{v}].[\mathbf{w}]\in \mathbf{IR}$ and $v_i^l,v_i^u,w_i^l,w_i^u\in \mathbf{R}$.

As before, the multiplication of an interval $[\mathbf{v}] \in \mathbf{IR}^n$, by a vector, $\mathbf{a} \in \mathbf{R}^n$ can be seen as a special case of (44) with the vector, **a**, is represented as a vector point-interval [**a**, **a**].

From the definitions (41)-(44), it is clear that the sensitivity matrix method, implemented through the formulas (39) and (40), can be written using the interval arithmetic formalism as

$$[x_i] = [\mathbf{b}^i].[\mathbf{dz}] + [\hat{x}_i] \quad , i=1,...,n \tag{45}$$

where $[\mathbf{b}^i]$ is a vector point-interval formed by the ith row of the sensitivity matrix $(J^T J)^{-1} J^T$ and $[\hat{x}_i]$ is a point-interval $[\hat{x}_i, \hat{x}_i]$.

The hyperbox $[\mathbf{x}]=([x_i])$, i=1,...,n, is therefore identical with the set $X^2(M, \mathbf{z}^l, \mathbf{z}^u)$. This result gives an interesting insight into the interval arithmetic based estimation of confidence limits in the context of linear programming and Monte Carlo methods.

Example 3: Continuing with the system of inequalities considered in Example 1 we can formulate the interval equations for the sensitivity matrix method. The point-interval vectors corresponding to the two rows of the pseudoinverse matrix are

$$\mathbf{b}^1=[[-0.5, -0.5]\ [0.5, 0.5]\ [0, 0]]$$

$$\mathbf{b}^2=[[0.333, 0.333]\ [0.333, 0.333]\ [0.333, 0.333]]$$

and the interval vector **z** is

$$[\mathbf{z}]=[[-1, 1]\ [1, 3]\ [0.5, 3]]$$

Substituting the above to (45), the intervals for the state variables, x_1 can be calculated as follows

$$[x_1] = [\mathbf{b}^1].[\mathbf{z}] = [-0.5, -0.5][-1, 1] + [0.5, 0.5][1, 3] + [0, 0][0.5, 3] =$$

$$= [-0.5, 0.5] + [0.5, 1.5] + [0, 0] = [0, 2]$$

and

$$[x_2] = [\mathbf{b}^2].[\mathbf{z}] = [0.333, 0.333][-1, 1] + [0.333, 0.333][1, 3] + [0.333, 0.333][0.5, 3] =$$

$$= [-0.333, 0.333] + [0.333, 1.0] + [0.167, 1.0] = [0.167, 2.333]$$

As expected, the result is identical to that calculated by the sensitivity matrix method (Example 2). However, it must be borne in mind that the mathematical elegance of the interval arithmetic is paid for with the increased computational burden. While the interval arythmetic calculations, listed above, required 48 multiplications, 48 logical operations and 8 additions, those in Example 2 were accomplished with 12 multiplications, 12 logical operations and 8 additions.

2.5 Ellipsoid method

An alternative to confidence limit analysis using sensitivity matrix, which historically preceeded this technique, is a method based on the iterative shrinking ellipsoids. The method has been first reported in [23] and has been considered by other researchers eg [8], [17], [3]. The technique is referred here as **ellipsoid method**.

Mathematically, an ellipsoid, $\mathbf{E}^t$ is a region of space defined as follows:

$$\mathbf{E}^t := \{\mathbf{x} \in \mathbf{R}^n : (\mathbf{x}-\mathbf{x}^t)^T P_t^{-1}(\mathbf{x}-\mathbf{x}^t) \leq 1.0\} \quad (46)$$

for some $\mathbf{x}^t \in \mathbf{R}^n$ and some symmetric and positive-definite matrix P_t , of dimension n by n. $\mathbf{E}^t$ is therefore, a region of $\mathbf{R}^n$ centered on $\mathbf{x}^t$. The aim of the ellipsoid method is to start with a large ellipsoid (usually a n-dimensional sphere) that contains the whole state uncertainty set, and then to generate a sequence of ellipsoids, decreasing in size, leading to one that fits the state uncertainty set as tightly as possible. Using an ellipsoid as an approximation to the state uncertainty set provides a simple and concise description of what can be a very complicated set. The algorithm itself, has the advantages of being sequential, mathematically and conceptually simple and can be very fast computationally.

We consider here the application of the Schweppe's ellipsoid algorithm to confidence limit analysis. The first point to note is that it is a linear method and so the state uncertainty set to be approximated is the set $X^1(M, \mathbf{z}^l, \mathbf{z}^u)$, for a measurement set M and measurement vector $\mathbf{z}^o$. An ellipsoid that is certain to contain $X^1(M, \mathbf{z}^l, \mathbf{z}^u)$ is used as the starting ellipsoid $\mathbf{E}^t$. This ellipsoid may be the n-dimensional sphere centered at the state estimate $\hat{\mathbf{x}}$ (this is the state estimate generated by $\mathbf{z}^o$) with a suitably large diameter, α. In this case $P_0 = \alpha.\mathrm{I}$, where I is the n by n identity matrix. The 'observations' in confidence limit analysis are the linearised measurement constraints provided by (14), which can be re-written as

$$\mathbf{z}^l - g(\hat{\mathbf{x}}) + J.\ \hat{\mathbf{x}} \leq J.\mathbf{dx} \leq \mathbf{z}^u - g(\hat{\mathbf{x}}) + J.\ \hat{\mathbf{x}} \quad (47)$$

for all $\mathbf{x}$ in $X^1(M, \mathbf{z}^l, \mathbf{z}^u)$. In this equation, $\mathbf{z}^l - g(\hat{\mathbf{x}}) + J.\ \hat{\mathbf{x}}$ and $\mathbf{z}^u - g(\hat{\mathbf{x}}) + J.\ \hat{\mathbf{x}}$ are constant vectors and so can be pre-calculated. (47) represents m constraints, bounding $J.\mathbf{dx}$ above and below. Each of these is taken in turn and used to modify the current ellipsoid.

Suppose that the tth constraint is being used to update the t-1st ellipsoid, $t \in \{2,...,m\}$, and that $\mathbf{E}^{t-1}$ contains $X^1(M, \mathbf{z}^l, \mathbf{z}^u)$. The region $\mathbf{F}^t$ bounded by this constraint also contains the uncertainty set $X^1(M, \mathbf{z}^l, \mathbf{z}^u)$. So $X^1(M, \mathbf{z}^l, \mathbf{z}^u)$ is contained in the intersection of these two regions as shown in Figure 3. A new ellipsoid, $\mathbf{E}^t$, can be produced which contains the intersection of $\mathbf{E}^{t-1}$ and the region $\mathbf{F}^t$ bounded by the constraint's hyperplanes. $\mathbf{E}^t$ is the ellipsoid $\{\mathbf{x} \in \mathbf{R}^n : (\mathbf{x}-\mathbf{x}^t)^T P_t^{-1}(\mathbf{x}-\mathbf{x}^t) \leq 1.0\}$, where

$$\mathbf{x}^t = \mathbf{x}^{t-1} + (\rho_t \nu_t / (e_t^z)^2)\, P'_{t-1}\, .\mathbf{a}^t \quad (48)$$

$$P_t = (1 + \rho_t - (\rho_t \nu_t /((e_t^z)^2 + \rho_t\, g_t)))\, P'_{t-1} \tag{49}$$

$$P'_{t-1} = (I + (\rho_t /((e_t^z)^2 + \rho_t\, g_t))\, P_{t-1}\, \mathbf{a}^t\, (\mathbf{a}^t)^T)\, P_{t-1} \tag{50}$$

$$g_t = (\mathbf{a}^t)^T\, P_{t-1}\, \mathbf{a}^t) \tag{51}$$

$$\nu_t = 0.5(\, z_t^u + z_t^l) - (g(\mathbf{x}^{t-1}))_t + (J.\mathbf{dx})_t + (\mathbf{a}^t)^T.\mathbf{x}^{t-1} \tag{52}$$

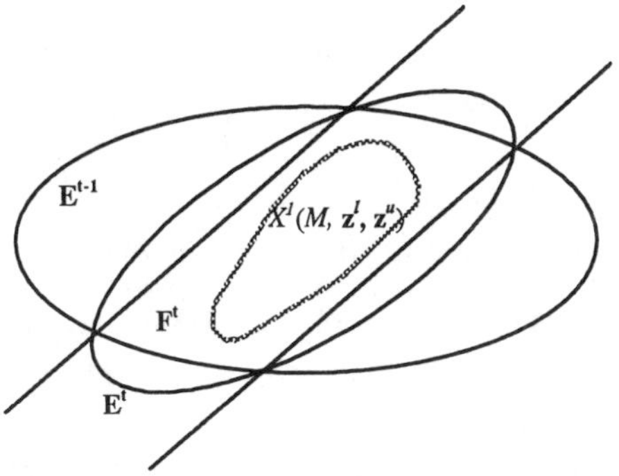

$\mathbf{F}^t$ - the region bounded by the hyperplaines of the constraint

$\mathbf{E}^t$ - the updated ellipsoid

Figure 3 Ellipsoid update (in 2-dimensions)

In these equations, P_{t-1} and $\mathbf{x}^{t-1}$ are the positive definite martix and centre vector, respectively, for the previous ellipsoid, $\mathbf{E}^{t-1}$, and ρ_t can be any non-negative real value. The value, e_t^z used in these equations, is the tth element of the measurement error vector, $\mathbf{e}^z = \mathbf{z}^u - \mathbf{z} = \mathbf{z} - \mathbf{z}^l = 0.5(\mathbf{z}^u - \mathbf{z}^l)$. It should be noted that, despite the fact that (46) refers to P_t^{-1} matrix in its inverted form and (48) to (52) do not, no matrix inversion is involved in the algorithm. In fact, the matrix P_t^{-1} need never be known as all updating is performed using matrices P_{t-1} and P_t.

The parameter ρ_t, which appears in several updating formulae, has not yet been fixed. For any non-negative, real ρ_t, the updated ellipsoid, $\mathbf{E}^t$, will contain the intersection of $\mathbf{E}^{t-1}$ with the region $\mathbf{F}^t$ bounded by the tth measurement constraints. For $\rho_t = 0.0$, it is easily seen that $\mathbf{E}^t$ is just the same as $\mathbf{E}^{t-1}$. So this choice of ρ_t leads to no improvement. A more reasonable choice for ρ_t would be the value that minimises the size of $\mathbf{E}^t$ in some way. Such a selection strategy gives the algorithm better convergence properties. In [8] there are two suggestions. The first involves the solution of a quadratic equation in ρ_t and produces the ellipsoid of minimum volume. The second requires the solution of a cubic equation and minimises the sum of squares of the semi-axis in $\mathbf{E}^t$.

On termination of the algorithm, the confidence limits for each variable are easily calculated from the final positive-definite matrix, P_t, and the final centre vector $\mathbf{x}^t$. These are:

$$x^{1u}_i = x_i^t + \sqrt{P_t(i,i)} \qquad (53)$$

$$x^{1l}_i = x_i^t - \sqrt{P_t(i,i)} \qquad (54)$$

Ellipsoidal confidence limit algorithm

1. Set t=0 and $P_0=\alpha I$.
2. Set t=t+1.
3. Calculate g_t and ν_t from equations (51) and (52).
4. Find ρ_t that minimises the volume of the new ellipsoid by solving the following quadratic equation in ρ_t:

   $$(p-1)g_t^2 \cdot \rho_t^2 + ((2p-1).(e_t^z)^2 - g_t + \nu_t)g_t \cdot \rho_t + (e_t^z)^2(p.((e_t^z)^2-\nu_t^2)-g_t) = 0$$
5. Calculate P'_{t-1} from (50).
6. Update the state variable $\mathbf{x}^t$, as per equation (48).
7. Update P_t, equation (49).
8. If not all constraints have been processed yet than repeat from step 2.
9. If the volume of the ellipsoid has been reduced by less than a pre-specified ratio than stop, otherwise reset the constraints counter, t=0, and repeat from step 2.

In some situations, tight bounds can be found by processing each measurement constraint only once, in which case only m steps are required. However, published research suggests that further reduction in the bounds is often possible by re-processing some or all of the constraints (Belforte, Bona, 1985). Also, the variation of the order in which the constraints are processed has the effect on the rate of convergence of the algorithm.

Although the computational complexity of the ellipsoid algorithm compares favourably with other linear confidence limit routines, its effectiveness in bounding the state uncertainty set can be poor. Other approximate methods, that can bound $X^1(M, \mathbf{z}^l, \mathbf{z}^u)$ more effectively with similar computational cost, can be easily derived [6], [17]. The reason for a poor performance of the ellipsoid method when the Jacobian, J, of the uncertain system equation, g(.), is sparse is illustrated in Figure 4. Since J is sparse, each measurement constraint only bounds a few of the variables. In the ellipsoid algorithm, constraints are considered individually and so can only improve confidence limits on the few variables that they bound explicitly. Using a 2-dimensional example in which the observation hyperplanes only bound one of the variables (ig, variables are bound in the horizontal direction by the hyperplanes but not in the vertical direction). As a result, the new ellipsoid, , produces tighter confidence limits than in the horizontal direction but looser ones in the vertical direction. When this idea is extended to many dimensions, only a few of which are bound by each constraint, it is easy to see that at each iteration the majority of the variables will have their confidence limits increased and only a minority will have them reduced [3].

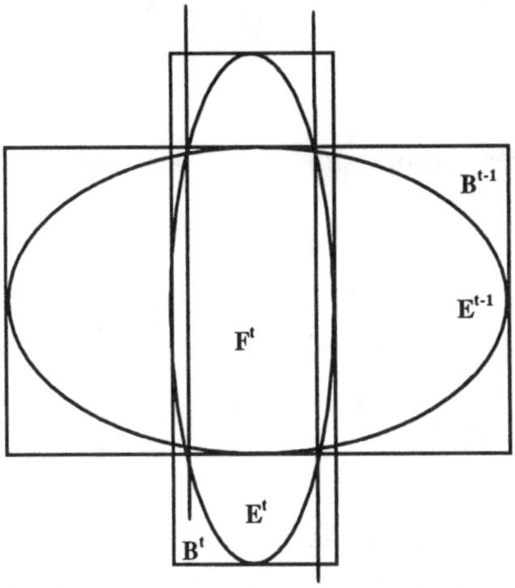

The new bounds, marked by $\mathbf{B}^t$, are tighter then $\mathbf{B}^{t-1}$ in the horizontal direction but are not as tight in the vertical direction

Figure 4. Ellipsoid update does not always lead to improvement in all variables

Example 4: The system of inequalities introduced in Example 1 is now processed using the ellipsoid algorithm. As in the previous three examples we note that for the linear system the **dx** variable is identical to **x** so the equation (52) simplifies to $\nu_t = 0.5(z_t^u + z_t^l) + (\mathbf{a}^t)^T.\mathbf{x}^{t-1}$.

The initial ellipsoid is assumed to be a hypersphere of radius $\sqrt{10}$ centred at $\mathbf{x}^0$=[0, 0]. The measurement upper- and lower- bound vectors are $\mathbf{z}^l$=[-1 1 0.5]^T and $\mathbf{z}^u$=[1 3 3]^T and the matrices J and P_0 are

$$J=\begin{bmatrix}-1 & 1\\ 1 & 1\\ 0 & 1\end{bmatrix} \text{ and } P_0=\begin{bmatrix}10 & 0\\ 0 & 10\end{bmatrix}$$

Processing the three constraints in the order t=1,2,3 we obtain, rounded to two places after decimal point

$$g_1=20.00 \quad v_1=0.00 \qquad P_1=\begin{bmatrix} 7.26 & 5.09 \\ 5.09 & 7.26 \end{bmatrix} \qquad \mathbf{x}^{1l}=\begin{bmatrix} -2.69 \\ -2.69 \end{bmatrix} \quad \mathbf{x}^{1u}=\begin{bmatrix} 2.69 \\ 2.69 \end{bmatrix}$$

$$g_2=24.70 \quad v_2=2.00 \qquad P_2=\begin{bmatrix} 2.18 & -0.21 \\ -0.21 & 2.18 \end{bmatrix} \qquad \mathbf{x}^{2l}=\begin{bmatrix} -0.62 \\ -0.62 \end{bmatrix} \quad \mathbf{x}^{2u}=\begin{bmatrix} 2.33 \\ 2.33 \end{bmatrix}$$

$$g_3=2.18 \quad v_3=0.89 \qquad P_3=\begin{bmatrix} 2.48 & -0.18 \\ -0.18 & 2.48 \end{bmatrix} \qquad \mathbf{x}^{3l}=\begin{bmatrix} -0.74 \\ -0.30 \end{bmatrix} \quad \mathbf{x}^{3u}=\begin{bmatrix} 2.41 \\ 2.44 \end{bmatrix}$$

and reversing the order of processing of the constraints to t=3,2,1 gives

$$g_3=10.00 \quad v_3=1.75 \qquad P_3=\begin{bmatrix} 12.33 & 0.00 \\ 0.00 & 3.12 \end{bmatrix} \qquad \mathbf{x}^{3l}=\begin{bmatrix} -3.51 \\ -0.46 \end{bmatrix} \quad \mathbf{x}^{3u}=\begin{bmatrix} 3.51 \\ 3.07 \end{bmatrix}$$

$$g_2=15.44 \quad v_2=0.69 \qquad P_2=\begin{bmatrix} 5.46 & -2.51 \\ -2.51 & 3.25 \end{bmatrix} \qquad \mathbf{x}^{2l}=\begin{bmatrix} -1.89 \\ -0.38 \end{bmatrix} \quad \mathbf{x}^{2u}=\begin{bmatrix} 2.78 \\ 3.22 \end{bmatrix}$$

$$g_1=13.73 \quad v_1=-0.97 \qquad P_1=\begin{bmatrix} 2.17 & 0.22 \\ 0.22 & 1.63 \end{bmatrix} \qquad \mathbf{x}^{1l}=\begin{bmatrix} -0.57 \\ -0.18 \end{bmatrix} \quad \mathbf{x}^{1u}=\begin{bmatrix} 2.37 \\ 2.37 \end{bmatrix}$$

Although both results contain the linearised state uncertainty set $X^1(M, \mathbf{z}^l, \mathbf{z}^u)$, which is defined by

$$\mathbf{x}^l=\begin{bmatrix} 0.00 \\ 0.17 \end{bmatrix} \text{ and } \mathbf{x}^u=\begin{bmatrix} 2.00 \\ 2.33 \end{bmatrix},$$

the quality of bounding the state variables is clearly inferior to that produced by the linear programming and the sensitivity matrix methods.

This is a particularly worrying characteristic of the ellipsoid technique because it is quite difficult to decide which order of processing of the constraints will result in the most rapid convergence of the estimates. It is also worth pointing out that the criterion of non-increasing the volume of the elipse, when progressing from one iteration to the next, does not necessarily imply the tightening of the bounds on all state variables, as can be seen from the transition from t=2 to t=3 in the example above. The consecutive ellipsoids generated when processing the constraints in the order t=1,2,3 and t=3,2,1 are given in Figure 5.

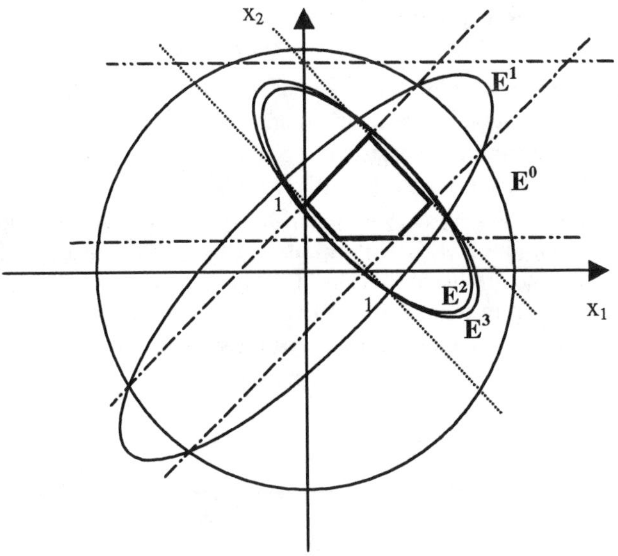

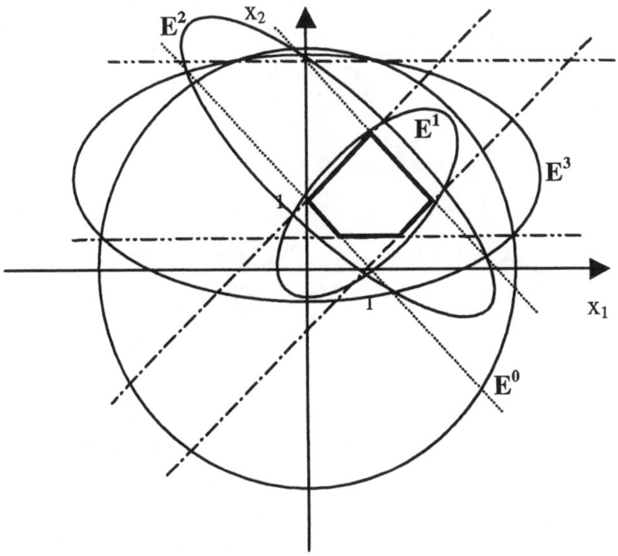

Figure 5. Confidence limits ellipsoids generated by processing the constraints in the order

t=1, 2, 3 and t=3, 2, 1 respectively

3 Real-life application

The uncertainty models from the previous section are illustrated here in the context of state estimation of water distribution networks. A network represented diagrammatically in Figure 6 consists of 65 nodes, 92 pipes and 5 inflow points. The inflow points are the reservoirs at nodes 60 and 160, a pumping station at node 68 and two water supplies from a high pressure zone through pressure-reducing valves at nodes 3 and 26. Pipe data: - length [m]; - diameter [m]; and C-values (conductivity); are listed in Table 1. This data, together with the reference pressure measurement in node 160 and five inflow measurements in nodes 3, 26, 60, 68 and 160 allows calculation of the system state (a 70-dimensional vector of 65 nodal pressures and 5 inflows in fixed-pressure nodes).

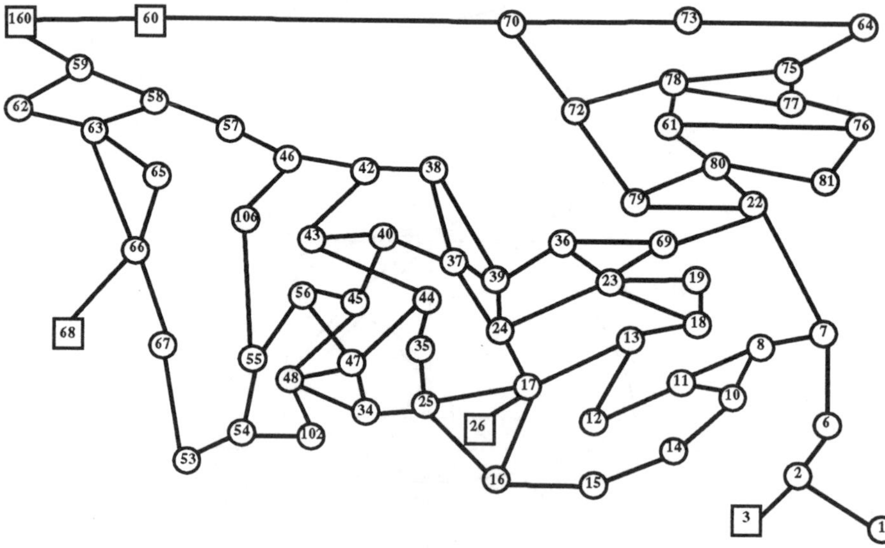

Figure 6. Water distribution network used for testing the confidence limits calculations

Two measurement sets are considered, both for the same operating state. The first measurement set, *M1*, is a minimal measurement set, consisting of nodal consumption values in all but one of the nodes, an inflow measurement for each of the inflow nodes and one reference pressure measurement at node 160. The second measurement set, *M2*, consists of all measurements contained in the set *M1* together with 4 additional pressure and 4 flow measurements.

TABLE 1. Pipe data for the test network

Pipe	*Len.*	*Dia.*	*C*	*Pipe*	*Len.*	*Dia.*	*C*	*Pipe*	*Len.*	*Dia.*	*C*
1-2	800	0.200	140	48-34	900	0.175	80	72-70	1770	0.225	47
2-3	400	0.300	165	48-102	310	0.250	70	70-73	3090	0.356	46
2-6	400	0.300	165	102-54	660	0.225	140	73-64	410	0.225	80
6-7	970	0.300	165	54-53	480	0.225	158	64-75	420	0.225	80
7-8	300	0.225	135	54-55	380	0.225	159	75-78	350	0.094	170
8-10	350	0.225	171	55-56	190	0.225	145	75-77	400	0.150	145
8-11	350	0.125	60	56-45	610	0.125	55	61-76	1470	0.150	81
11-10	360	0.150	105	45-48	1060	0.175	80	70-60	2200	0.356	100
11-12	180	0.125	60	44-43	230	0.250	80	160-59	370	0.381	50
12-13	200	0.175	115	43-40	380	0.250	80	59-58	630	0.300	50
10-14	710	0.225	110	40-45	580	0.168	120	58-57	730	0.300	118
14-15	225	0.225	110	40-37	320	0.168	120	57-46	260	0.300	85
15-16	310	0.225	90	37-24	390	0.200	145	46-42	250	0.300	85
16-17	590	0.094	80	37-39	280	0.168	120	42-38	430	0.300	145
17-13	740	0.175	115	39-24	300	0.150	90	42-43	720	0.250	80
13-18	250	0.225	158	37-38	270	0.200	145	46-106	720	0.200	145
18-19	330	0.300	145	38-39	550	0.300	229	106-55	700	0.142	137
22-7	1510	0.225	96	39-36	210	0.300	127	56-47	550	0.225	145
22-69	120	0.300	145	36-23	180	0.300	112	53-67	220	0.225	160
69-23	420	0.150	90	36-69	147	0.300	145	67-66	270	0.225	160
23-24	300	0.300	145	22-79	160	0.225	80	66-63	800	0.225	110
24-17	650	0.200	158	79-80	340	0.150	90	66-65	210	0.125	60
17-25	330	0.175	127	80-22	390	0.200	145	65-63	590	0.125	60
25-16	630	0.225	104	80-81	1220	0.150	139	63-58	2050	0.150	40
26-17	360	0.300	80	81-76	600	0.150	145	63-62	770	0.225	110
25-34	780	0.175	80	76-77	670	0.150	116	62-59	350	0.225	110
25-35	320	0.225	119	77-78	150	0.150	116	23-19	430	0.300	145
35-44	710	0.225	90	78-61	460	0.094	170	23-18	760	0.117	60
44-47	520	0.250	70	61-80	530	0.150	145	66-68	440	0.300	170
47-34	610	0.225	145	72-78	1100	0.142	105	60-160	270	0.381	32
47-48	540	0.250	70	72-79	600	0.225	60				

As it is pointed out in Section 1.1, the true measurement vector, $\mathbf{z}^t$, rarely coincides with the observed one, $\mathbf{z}^o$. The discrepancy is caused by meter noise affecting real measurements and the inaccuracy of estimates, which are used as pseudomeasurements. In order to reflect this reality the observed measurement values, $\mathbf{z}^o$, are generated in the following way. Firstly, a true operating state, $\mathbf{x}^t$, listed in column 2 of Table 3, is assumed and a true measurement vector, $\mathbf{z}^t$, is calculated as $g(\mathbf{x}^t)$. The true measurement values, $\mathbf{z}^t$, are listed in column 2 of Table 2. The observed measurement values, $\mathbf{z}^o$, listed in column 3, are selected randomly from within the range $[\mathbf{z}^l, \mathbf{z}^u]$, in accordance with (2). The bounds $[\mathbf{z}^l,\mathbf{z}^u]$ were defined in terms of relative variability of $\mathbf{z}^t$ as follows: 50% for $\mathbf{z}^t$ <0.5 *l/s*; 40% for 0.5< $\mathbf{z}^t$ <1.0 *l/s*; 30% for 1.0< $\mathbf{z}^t$ <5.0 *l/s*; 20% for 5.0< $\mathbf{z}^t$ <10.0 *l/s* and 10% for $\mathbf{z}^t$ >10.0 *l/s*. This corresponds to real-life situation where measurement values are not exact but are contained within the range specified by the accuracy of meters. The state vector, $\hat{\mathbf{x}}$, calculated from the observed measurement vector, $\mathbf{z}^o$, is shown in column 3 of Table 3. The difference between this state estimate and the true state should be noted. It is caused solely by the addition of the simulated measurement errors and shows how noise corrupted measurement data affects system state.

TABLE 2. Measurement data

Node	*True*	*Observed*	*Node*	*True*	*Observed*	*Node*	*True*	*Observed*
Reference pressure [m]			25	2.71	2.21	67	1.87	2.45
160	144.77	144.75	61	0.42	0.22	69	4.52	6.21
			34	7.40	5.96	70	2.18	1.74
Inflows [l/s]			35	2.58	2.48	72	11.51	11.70
26	65.00	65.54	36	1.92	2.44	73	2.77	2.47
3	31.00	31.43	37	2.98	2.20	75	1.32	1.39
60	34.00	33.46	38	2.36	1.64	76	5.37	4.83
160	45.00	45.85	39	0.65	1.00	77	1.16	0.93
68	31.00	30.93	40	6.77	7.48	78	1.35	1.27
			102	2.13	2.14	79	1.91	2.52
Consumptions [l/s]			42	8.03	7.68	80	2.64	2.56
1	4.85	5.00	43	3.51	4.56	81	2.79	2.21
2	6.77	5.65	44	1.89	1.97	106	1.74	2.55
6	2.09	1.94	45	1.10	1.62	26	0.26	0.31
7	1.64	0.98	46	2.73	3.19	3	7.95	8.91
8	1.16	0.78	47	10.80	12.82	60	0.58	0.79
10	9.64	8.56	48	2.95	2.47	68	2.46	2.55
11	0.34	0.29	53	0.67	0.84			
12	0.35	0.53	54	4.54	4.79	*M2 – pressure measurements*		
13	0.50	0.79	55	10.83	9.47	7	140.08	140.08
14	6.54	5.38	56	0.78	0.85	44	139.85	139.85
15	3.18	4.40	57	0.16	0.12	66	141.42	141.42
16	2.01	1.46	58	5.68	4.38	80	140.10	140.10
17	8.51	10.27	59	2.88	2.92			
18	8.71	9.69	62	2.94	3.66	*M2 - flow measurements*		
19	0.00	0.00	63	10.46	10.54	22-69	-7.13	-7.13
22	2.35	2.36	64	3.75	2.99	42-38	-0.16	-0.16
23	0.47	0.49	65	3.84	3.99	7-22	1.94	1.94
24	1.73	2.01	66	2.15	2.76	56-45	0.30	0.30

TABLE 3. True and estimated state vector

*State*Node	*TrueState*	*Estimate*	*State*Node	*TrueState*	*Estimate*	*State*Node	*TrueState*	*Estimate*
(1)-1	140.11	140.04	*(25)*-38	140.33	140.15	*(49)*-69	140.31	140.14
(2)-2	140.23	140.17	*(26)*-39	140.33	140.15	*(50)*-70	143.88	143.90
(3)-6	140.20	140.14	*(27)*-40	140.06	139.84	*(51)*-72	140.25	140.10
(4)-7	140.15	140.08	*(28)*-102	140.07	139.86	*(52)*-73	141.78	141.87
(5)-8	140.02	139.96	*(29)*-42	140.33	140.15	*(53)*-75	140.73	140.76
(6)-10	139.94	139.89	*(30)*-43	140.07	139.85	*(54)*-76	139.97	139.95
(7)-11	140.02	139.95	*(31)*-44	140.06	139.85	*(55)*-77	140.36	140.34
(8)-12	140.38	140.21	*(32)*-45	140.02	139.80	*(56)*-78	140.35	140.32
(9)-13	140.41	140.23	*(33)*-46	140.45	140.27	*(57)*-79	140.27	140.11
(10)-14	139.91	139.84	*(34)*-47	139.96	139.75	*(58)*-80	140.24	140.10
(11)-15	139.93	139.85	*(35)*-48	139.99	139.78	*(59)*-81	139.97	139.95
(12)-16	140.05	139.95	*(36)*-53	140.92	140.65	*(60)*-106	140.34	140.13
(13)-17	141.81	141.58	*(37)*-54	140.21	139.99	*(61)*-26	144.37	144.18
(14)-18	140.36	140.18	*(38)*-55	140.05	139.85	*(62)*-3	140.34	140.28
(15)-19	140.36	140.18	*(39)*-56	140.03	139.83	*(63)*-60	144.82	144.81
(16)-22	140.30	140.13	*(40)*-57	140.71	140.56	*(64)*-160	144.77	144.75
(17)-23	140.36	140.18	*(41)*-58	141.11	141.00	*(65)*-68	141.88	141.59
(18)-24	140.40	140.22	*(42)*-59	143.48	143.39			
(19)-25	140.24	140.09	*(43)*-62	142.84	142.66	***Inflows [l/s]***		
(20)-61	140.23	140.10	*(44)*-63	141.79	141.53	*(66)*-26	65.00	65.54
(21)-34	139.94	139.74	*(45)*-64	141.11	141.20	*(67)*-3	31.00	31.43
(22)-35	140.17	140.00	*(46)*-65	141.32	141.01	*(68)*-60	34.00	33.46
(23)-36	140.33	140.15	*(47)*-66	141.71	141.42	*(69)*-160	45.00	45.85
(24)-37	140.32	140.14	*(48)*-67	141.25	140.97	*(70)*-68	31.00	30.95

The first set of results concerns the state uncertainty sets, $X(M1, \mathbf{z}^l, \mathbf{z}^u)$ and $X^1(M1, \mathbf{z}^l, \mathbf{z}^u)$, for the minimal measurement set, $M1$, as calculated by the Monte Carlo confidence limit algorithm and the linear programming confidence limit algorithm. Results for the sensitivity matrix algorithm, $X^2(M1, \mathbf{z}^l, \mathbf{z}^u)$, are not included because for a minimal measurement set these are identical to those using linear programming algorithm, as explained in Lemma 5.

Rather than trying to visualise the state vectors themselves we focus our attention on their variability, $(x_i^u - x_i^l)/2$, around the average value, $(x_i^u + x_i^l)/2$, for each variable $i \in \{1,\dots,n\}$. Figure 7 depicts the following state uncertainty variability sets $\Delta X(M1, \mathbf{z}^l, \mathbf{z}^u)$ and $\Delta X^1(M1, \mathbf{z}^l, \mathbf{z}^u)$:

$$\Delta X(M1, \mathbf{z}^l, \mathbf{z}^u) := \{\, \boldsymbol{\delta}\mathbf{x} \in \mathbf{R}^{+n} : \boldsymbol{\delta}\mathbf{x} = |\mathbf{dx}|,\ \hat{\mathbf{x}} + \mathbf{dx} \in X(M1, \mathbf{z}^l, \mathbf{z}^u) \,\} \tag{55}$$

$$\Delta X^1(M1, \mathbf{z}^l, \mathbf{z}^u) := \{\, \boldsymbol{\delta}\mathbf{x} \in \mathbf{R}^{+n} : \boldsymbol{\delta}\mathbf{x} = |\mathbf{dx}|,\ \hat{\mathbf{x}} + \mathbf{dx} \in X^1(M1, \mathbf{z}^l, \mathbf{z}^u) \,\} \tag{56}$$

The Monte Carlo and linear programming results demonstrate the scale of the potential error in state estimates for a system with no measurement redundancy. Pressure errors are in excess of 2.0 [m] in the region of the network that is most distant from the reference pressure node. In fact, the majority of pressure errors are over 1.0 [m] with only those nodes close to node 160 having relatively tight uncertainty bounds. This indeed confirms the intuitive understanding of the relationship between the uncertainty bounds and the accuracy and location of measurements in the system.

The linear programming results correlate well with the Monte Carlo results; with the linear programming error bounds being no more than 5% off the bounds calculated by the Monte Carlo method for all but the most distant nodes from the reference pressure e.i. 1, 2, 3, 6, 7, 8, 10, 11, 14, 15, 16, 17 and 26, for which the discrepancy is less then 15% (the corresponding state variables are 1, 2, 3, 4, 5, 6, 7, 10, 11, 12, 13, 61 and 62). These observations lead to the conclusion that no significant accuracy is lost in linearising the uncertainty model and justify the use of linearised confidence limit algorithms.

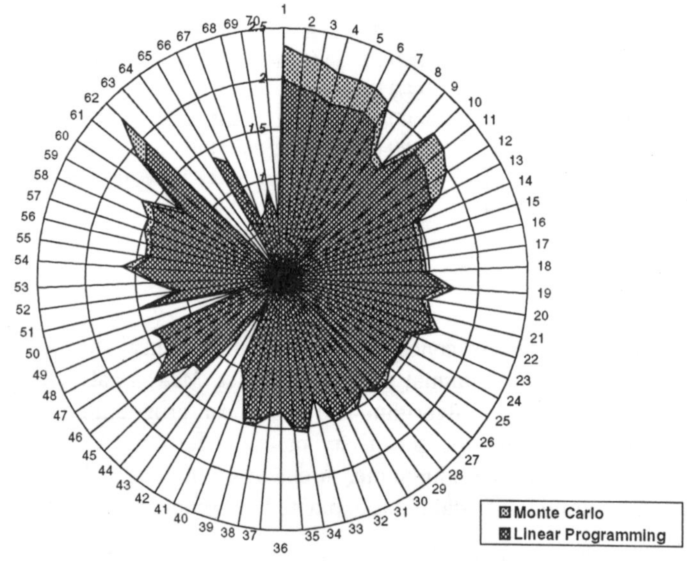

Figure 7. State uncertainty variability sets for Monte Carlo and Linear Programming methods

The results obtained with the Elipsoid method are difficult to assess objectively as they depend on the order of processing of the measurements. Figure 8 illustrates two sets of results obtained by varying the order of processing of the measurements. In either case it is clear that the Elipsoid method produces results that are inferior to those obtained with the linear programming method.

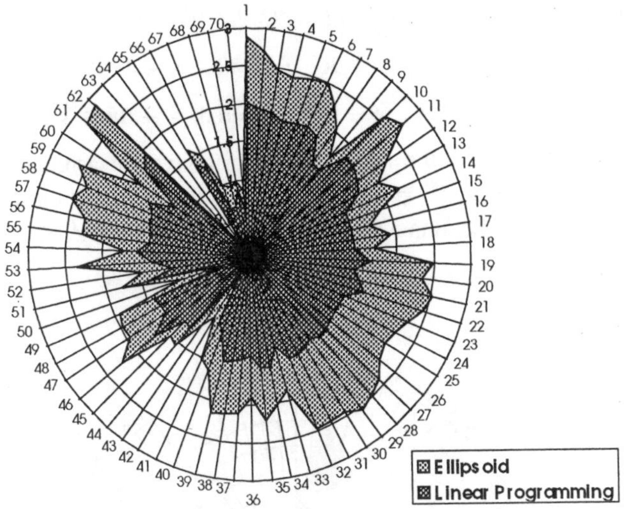

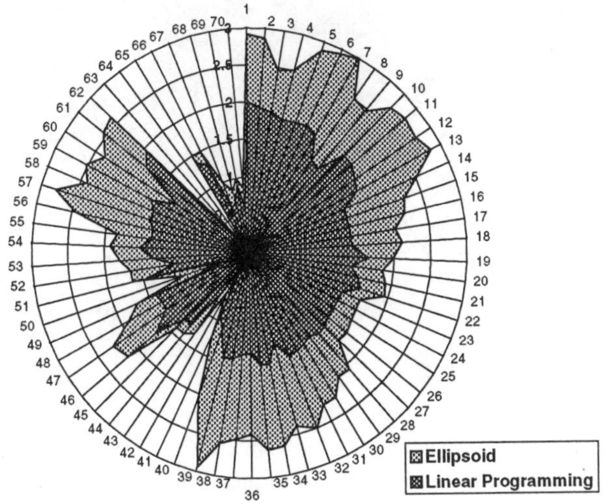

Figure 8. State uncertainty variability sets for two different runs of the Ellipsoid method compared to the result of the Linear Programming method.

The second set of results concerns the state uncertainty sets $X^1(M2, \mathbf{z}^l, \mathbf{z}^u)$ and $X^2(M2, \mathbf{z}^l, \mathbf{z}^u)$ calculated by the linear programming and the sensitivity matrix methods using the augmented measurement set, *M2*. Because of the huge computational requirements of the Monte Carlo method, when the increased number of measurements raises the chances of generating infeasible state vectors, MC method was deemed impractical and was not attempted. As with the previous measurement set, the results are analysed in terms of the state uncertainty variability sets $\Delta X^1(M2, \mathbf{z}^l, \mathbf{z}^u)$ and $\Delta X^2(M2, \mathbf{z}^l, \mathbf{z}^u)$. These are presented in Figure 9.

$$\Delta X^1(M2, \mathbf{z}^l, \mathbf{z}^u) := \{\ \delta\mathbf{x} \in \mathbf{R}^{+n} : \delta\mathbf{x} = |\mathbf{dx}|,\ \hat{\mathbf{x}} + \mathbf{dx} \in X^1(M2, \mathbf{z}^l, \mathbf{z}^u)\ \} \tag{57}$$

$$\Delta X^2(M2, \mathbf{z}^l, \mathbf{z}^u) := \{\ \delta\mathbf{x} \in \mathbf{R}^{+n} : \delta\mathbf{x} = |\mathbf{dx}|,\ \hat{\mathbf{x}} + \mathbf{dx} \in X^2(M2, \mathbf{z}^l, \mathbf{z}^u)\ \} \tag{58}$$

Since for both methods the state estimate for a given measurement vector, $\mathbf{z}^o$, is $\hat{\mathbf{x}}$, and since $\Delta X^1(M2, \mathbf{z}^l, \mathbf{z}^u) \subseteq \Delta X^2(M2, \mathbf{z}^l, \mathbf{z}^u)$, the results demonstrate the important result presented in Lemma 5 that $X^1(M2, \mathbf{z}^l, \mathbf{z}^u) \subseteq X^2(M2, \mathbf{z}^l, \mathbf{z}^u)$.

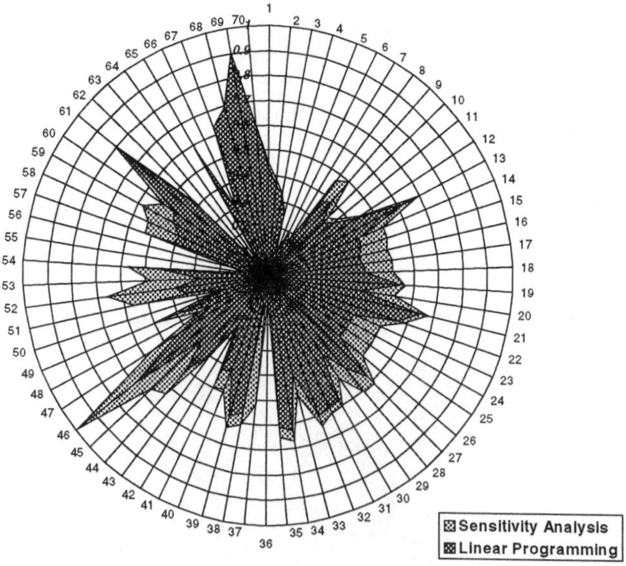

Figure 9. State uncertainty variability sets for Sensitivity Analysis and Linear Programming methods

Considering the individual variables of the state vector it is clear that the linear programming and the sensitivity matrix results are closely correlated and that the overestimation of the state uncertainty variability set with the sensitivity matrix method is confined to approx. 30% of the range of the variables. Again it is interesting to see that the uncertainty bounds have been tightened considerably in the vicinity of the additional meters.

4 Conclusions

State estimation of nonlinear systems is a challenging task that represents a large class of real-life problems. Unfortunately, traditional point-estimate solutions tend to be inadequate because of both, the sensitivity of solutions to the inaccuracy of input data and our inherently approximate knowledge of systems.

It is now widely recognised that, in order to be able to reason about systems' behaviour in presence of uncertainties, it is necessary to abstract from detailed point-estimate solutions to more general, coarser-grain interval-estimates. Such interval-based modelling represent a more credible description of reality, because the predictions or retrodictions based on such models are explicit about the limits of our understanding of systems.

Clearly, while the interval-estimation represents a positive development in terms of building abstract knowledge, the information content of interval-estimates is maximised when the bounds on the individual variables of the state vector (confidence limits) are as tight as possible. We have considered in this chapter one non-linear interval-estimation technique (Monte Carlo) and 3 linear techniques (Linear Programming, Sensitivity Matrix and Ellipsoid) which have been applied to the linearised system model. A rigorous analysis of the effect of the linearisation of the system model on the state uncertainty set has been performed. The analysis shows that the linearisation has only a second-order effect on the shape of the state uncertainty set and, as such, it is an acceptable simplification leading to much more efficient estimation techniques.

Of the three linear interval-estimators the Linear Programming technique produces tightest bounds on the state uncertainty set and is followed closely by the Sensitivity Matrix method. The Ellipsoid method is somewhat disappointing in that it produces a rather conservative bounds and it is sensitive to the order of processing of the constraints. We conclude therefore that the Ellipsoid technique is best suited for rapid, rough approximation of the state uncertaintry set, while the other two techniques are preferable for calculation of tight and consistent bounds on the state uncertainty set. The Sensitivity Matrix method is particularly appealing because of its computational efficiency.

The results demonstrate that the concept of *granular computing* is pertinent to a broad and important class of real life systems. The recognition of information granularity is a recognition of uncertainty in systems modelling and, as such, it can be seen as a progression from *information processing* to *knowledge building*.

ACKNOWLEDGEMENTS. The research reported in this chapter is a result of several research projects funded by the Engineering and Physical Sciences Research Council (EPSRC, UK) and the industry. Recent support from the University of Alberta, Canada, in the form of Visiting Professorship, which enabled compilation of this chapter, is gratefuly acknowledged.

References

1. Bargiela, A. (1985). An algorithm for observability determination in water systems state estimation, *Proc. IEE*, Vol. 132, Pt. D, No. 6, 245-249.
2. Bargiela, A., Hainsworth, G.D. (1989). Pressure and flow uncertainty in water systems, *ASCE J. Water Res. Planning and Management*, 115, 2, 212-229.
3. Bargiela, A. (1994). Ellipsoid method for quantifying the uncertainty in water system state estimation, *Proc. IEE Colloquium on Modelling Uncertain Systems*, Vol. 1994/105, 10/1-10/3.
4. Bargiela, A. (1998). Uncertainty – A key to better understanding of systems, *(Plenary lecture) Proc. European Simulation Symposium ESS'98*, ISBN 1-56555-147-8, 11-19.
5. Bargiela, A. (2000). Operational decision support through confidence limit analysis and pattern classification, *(Plenary lecture) Proc. 5th Int. Conf. Computer Simulation and AI*, Mexico City.
6. Belforte, G., Bona, B. (1985). An improved parameter identification algorithm for signals with unknown-but-bounded errors, *Proc. IFAC/IFORS.*, York.
7. Cichocki, A., Bargiela, A. (1997). Neural networks for solving linear inequality systems, *Parallel Computing*, Vol. 22, No. 11, 1455-1475.
8. Fogel, E., Huang, Y.F. (1982). On the value of information system identification – bounded noise case, *Automatica*, Vol. 18, No. 2, pp.
9. Gabrys, B., Bargiela, A. (1999). Neural networks based decision support in presence of uncertainties, *ASCE J. of Water Res. Planning and Management*, Vol. 125, 5, 272-280.
10. Gabrys, B., Bargiela, A. (1999). Analysis of uncertainties in water systems using neural networks, *Measurement and Control*, Vol. 32, No. 5, 145-147.
11. Hainsworth, G.D. (1988). Measurement uncertainty in water distribution telemetry systems, *PhD thesis, The Nottingham Trent University.*

12. Hartley, J.K., Bargiela, A. (1997). Parallel state estimation with confidence limit analysis, *Parallel Algorithms and Applications*, Vol., 11, No. 1-2, 155-167.

13. Jaulin, L., Walter, E. (1996). *Guaranteed nonlinear set estimation via interval analysis*, in: *Bounding approaches to system identification* (Milanese *et al.*, eds.), Plenum Press.

14. Kandel, A. (1992). *Fuzzy expert systems*, CRC Press.

15. Mo, S.H., Norton, J.P. (1988). Parameter bounding identification algorithms for bounded-noise records, *Proc. IEE,* Vol. 135, Pt D, No. 2.

16. Moore, R.E. (1966). *Interval analysis*, Prentice-Hall, Englewood Cliffs, NJ.

17. Norton, J.P. (1986). *An Introduction to Identification*, Academic Press.

18. Pedrycz, W., Vukovich, G. (1999). Data-based design of fuzzy sets, *Journal of Fuzzy Logic and Intelligent Systems,* Vol. 9, No. 3.

19. Pedrycz, W., Gomide, F. (1998). *An introduction to Fuzzy Sets. Analysis and Design*, MIT Press, Cambridge, MA.

20. Pedrycz, W., Smith M.H., Bargiela, A. (2000), A Granular Signature of Data, *Proc. NAFIPS'2000*.

21. Ratschek, A., Rokne, J. (1988), *New computer methods for global optimization*, Ellis Horwood Ltd., John Wiley & Sons, New York.

22. Rokne, J.G., *Interval arythmetic and interval analysis: An introduction*, in: *Granular Computing* (Pedrycz ed.), Elsevier.

23. Schweppe, F.C. (1973). *Uncertain dynamic systems*, Prentice-Hall, Englewood Cliffs, NJ.

24. Sterling, M.J.H., Bargiela, A. (1984). Minimum-norm state estimation for computer control of water distribution systems, *Proc. IEE*, Part D, Vol. 131, 57-63.

25. Warmus, M. (1956). *Calculus of approximations*, Bull. Acad. Polon. Sci., Cl III 4, 253-259.

26. Zadeh, L.A. (1979). Fuzzy sets and information granularity, in Gupta, M.M., Ragade, R.K., Yager, R.R. (eds.), *Advances in Fuzzy Set Theory and Applications,* Kluwer Academic Publishers.

Nonlinear Bounded-Error Parameter Estimation Using Interval Computation

*L. Jaulin *#E. Walter

*Laboratoire des Signaux et Systemes, CNRS-Supelec-Universite de Paris-Sud, Plateau de Moulon, 91192 Gif-sur-Yvette, France

#on leave from Laboratoire d'Ingeniere des Systemes Automatises
Universite d'Angers, 49045 Angers, France

Abstract. This paper deals with the estimation of the parameters of a model from experimental data. The aim of the method presented is to characterize the set $\mathbb{S}$ of all values of the parameter vector that are acceptable in the sense that all errors between the experimental data and the corresponding model outputs lie between known lower and upper bounds. This corresponds to what is known as bounded-error estimation or membership-set estimation. Most of the methods available to give guaranteed estimates of $\mathbb{S}$ rely on the hypothesis that the model output is linear in its parameters, contrary to the method advocated here which can deal with nonlinear models. This is made possible by the use of the tools of interval analysis, combined with a branch-and-bound algorithm. The purpose of the present paper is to show that the approach can be cast into the more general framework of granular computing.

1 Introduction

In the exact sciences, in engineering, and increasingly in the human sciences too, mathematical models are used to describe, understand, predict or control the behavior of systems. These mathematical models often involve unknown parameters that should be identified (or calibrated, or estimated) from experimental data and prior knowledge or hypotheses (see, e.g., [WP97]).
Let y_k, $k \in \{1, \ldots, k_{max}\}$ be the data collected on the system to be modeled. They form the data vector $\mathbf{y} \in \mathbf{R}^{k_{max}}$. Denote the vector of the unknown parameters of the model by $\mathbf{p} = (p_1, p_2, \ldots, p_n)^T \in \mathbf{R}^n$ Then the corresponding model is expressed by M($\mathbf{p}$). For any given $\mathbf{p}$, M($\mathbf{p}$) generates a vector model output

$\mathbf{y}_m(\mathbf{p})$ homogeneous to the data vector $\mathbf{y}$. Estimating $\mathbf{p}$ from $\mathbf{y}$ is one of the basic tasks of statisticians. This is usually done by minimizing some cost function $j(\mathbf{p})$, for instance a norm of $\mathbf{y} - \mathbf{y}_m(\mathbf{p})$. The Euclidean norm is most commonly used, leading to what is known as least-square estimation. It corresponds to maximum-likelihood estimation of $\mathbf{p}$ under the hypothesis that the data points y_k , $k \in \{1, \ldots, k_{\max}\}$, are independently corrupted by an additive Gaussian measurement noise with zero mean and covariance independent of k. Many other cost functions may be considered, depending on the information available on the noise corrupting the data. The minimization of these cost functions usually lead to a point estimate of $\mathbf{p}$, *i.e.*, a single numerical value for each parameter. Except in a few special cases, the minimization of the cost function is difficult and one can seldom guarantee that a global minimizer has been found. Moreover, the characterization of the uncertainty on the estimate of $\mathbf{p}$ is usually performed, if at all, by using asymptotic properties of maximum-likelihood estimators, which is not appropriate when the number $k_{\max}$ of data points is very small, as is often the case in biology for example. An attractive alternative is to resort to what is known as *bounded-error estimation* or *set-membership estimation*. In this context (see, *e.g.*, [Wal90], [Nor94], [Nor95], [MNPLW96] and the references therein), a vector $\mathbf{p}$ is feasible if and only if all errors $e_k(\mathbf{p})$ between the data points y_k and the corresponding model outputs $y_{m,k}(\mathbf{p})$ lie between known lower bounds $\underline{e}_k$ and upper bounds $\bar{e}_k$, which express the confidence in the corresponding measurements y_k. Let $\mathbb{S}$ be the set of all values of $\mathbf{p}$ that are feasible, *i.e.*,

$$\mathbb{S} = \{\mathbf{p} \in \mathbb{R}^n \mid \text{for all } k \in \{1, \ldots, k_{\max}\},\ y_k - y_{m,k}(\mathbf{p}) \in [\underline{e}_k, \bar{e}_k]\}. \quad (1)$$

Some methods only look for a value of $\mathbf{p}$ in $\mathbb{S}$, but then the size and shape of $\mathbb{S}$, which provide useful information about the uncertainty on $\mathbf{p}$ that results from the uncertainty in the data, remain unknown. This is why one should rather try to characterize $\mathbb{S}$. When $\mathbf{y}_m(\mathbf{p})$ is linear, $\mathbb{S}$ is a convex polytope, which can be characterized exactly [WPL89]. In the nonlinear case, the situation is far more complicated if one is looking for a *guaranteed* characterization of $\mathbb{S}$. The algorithm SIVIA (for Set Inverter Via Interval Analysis) [JW93a], [JW93b], [JW93c] nevertheless makes it possible to compute guaranteed estimates of $\mathbb{S}$ in many situations of practical interest, by combining a branch-and-bound algorithm with techniques of interval computation. The purpose of this chapter is to present the resulting methodology in the framework of granular computing. The chapter is organized as follows. In Section 2, the very few notions of interval analysis required to understand SIVIA are recalled. Section 3 presents this algorithm in the context of granular computing. Its application to nonlinear bounded-error estimation is illustrated on a simple example in Section 4.

2 Interval computation

Interval computation was initially developed (see [Moo79]) to quantify the uncertainty of results calculated with a computer using a floating point number representation, by bracketing any real number to be computed between two numbers that could be represented exactly. It has found many others applications, such as global optimization and guaranteed solution of sets of nonlinear equations and/or inequalities [Han92], [HDM97]. Here, we shall only use interval computation to test whether boxes in parameter space are inside or outside $\mathbb{S}$.

An interval $[x] = [\underline{x}, \overline{x}]$ is a bounded compact subset of $\mathbb{R}$. The set of all intervals of $\mathbb{R}$ will be denoted by $\mathbb{IR}$. A vector interval (or box) $[\mathbf{x}]$ of $\mathbb{R}^n$ is the Cartesian product of n intervals. The set of all boxes of $\mathbb{R}^n$ will be denoted by $\mathbb{IR}^n$. The basic operations on intervals are defined as follows

$$\begin{aligned}
[x]+[y] &= [\underline{x}+\underline{y}, \overline{x}+\overline{y}], \\
[x]-[y] &= [\underline{x}-\overline{y}, \overline{x}-\underline{y}], \\
[x]*[y] &= [\min\{\underline{x}\underline{y}, \underline{x}\overline{y}, \overline{x}\underline{y}, \overline{x}\overline{y}\}, \max\{\underline{x}\underline{y}, \underline{x}\overline{y}, \overline{x}\underline{y}, \overline{x}\overline{y}\}], \\
1/[y] &= [1/\overline{y}, 1/\underline{y}] \quad \text{(provided that } 0 \notin [y]), \\
[x]/[y] &= [x]*(1/[y]) \quad (0 \notin [y]).
\end{aligned}$$

All continuous basic functions such as sin, cos, exp, sqr... extend easily to intervals by defining $f([x])$ as $\{f(x)|x \in [x]\}$. As an example, $\sin([0,\pi/2]) * [-1,3] + [-1,3] = [0,1]*[-1,3]+[-1,3] = [-1,3]+[-1,3] = [-2,6]$. Note that some properties true in $\mathbb{R}$ are no longer true in $\mathbb{IR}$. For instance, the property $x - x = 0$ translates into $[x]-[x] \ni 0$, and the property $x^2 = x * x$ translates into $[x]^2 \subset [x]*[x]$.

Example 1 *If* $[x] = [-1,3]$, $[x]-[x] = [-4,4]$, $[x]^2 = [0,9]$ *and* $[x]*[x] = [-3,9]$.

Moore has shown [Moo66] that if a sequence of basic operations on real numbers is replaced by the same sequence on intervals containing these real numbers, then the interval result contains the corresponding real result. ■

Example 2 *Since* $[0,2]+([2,3]*[4,5])-[2,3] = [0,2]+[8,15]-[2,3] = [8,17]+[-3,2] = [5,19]$, *the result obtained by performing the same computation on real numbers belonging to these intervals is guaranteed to belong to* $[5,19]$. *For instance* $1+(2*5)-2 = 9 \in [5,19]$. ■

For any function f from $\mathbb{R}^n$ to $\mathbb{R}$ which can be evaluated by a succession of elementary operations, it is thus possible to compute an interval that contains the range of f over a box of $\mathbb{IR}^n$. Depending on the formal expression of the function the interval obtained may differ, but it is always guaranteed to contain the actual range. This is illustrated by the following example for $n = 1$.

Example 3 *The function defined by* $f(x) = x^2 - x$ *can equivalently be defined by* $f(x) = (x - 1/2)^2 - 1/4$. *An interval containing the range of* f *over* $[x] = [-1, 3]$ *can thus be obtained in the two following ways.*

$$\begin{aligned} [x]^2 - [x] &= [-1,3]^2 - [-1,3] = [0,9] + [-3,1] = [-3,10], \\ \left([x] - \tfrac{1}{2}\right)^2 - \tfrac{1}{4} &= \left[-\tfrac{3}{2}, \tfrac{5}{2}\right]^2 - \tfrac{1}{4} = \left[0, \tfrac{25}{4}\right] - \tfrac{1}{4} = \left[-\tfrac{1}{4}, 6\right]. \end{aligned} \tag{2}$$

The first result is a pessimistic approximation of the range, whereas the second one gives the actual range. This is due to the fact that in the first expression $[x]$ *appears twice, and that the actual value of* x *in the two occurrences are assumed to vary independently within* $[x]$. *It is thus advisable to write the functions in such a way as to minimize the number of occurrences of each variable.* ■

Let f be a function from $\mathbb{R}^n$ to $\mathbb{R}$ and $[a, b]$ an interval. Interval computation will provide *sufficient* conditions to guarantee either that

$$\forall \mathbf{x} \in [\mathbf{x}] \in \mathbb{IR}^n, \; f(\mathbf{x}) \in [a, b], \tag{3}$$

or that

$$\forall \mathbf{x} \in [\mathbf{x}] \in \mathbb{IR}^n, \; f(\mathbf{x}) \notin [a, b]. \tag{4}$$

Of course, when part of $f([\mathbf{x}])$ belongs to $[a, b]$ and part does not, no conclusion can be reached using these conditions. Although the box $[\mathbf{x}]$, considered as a set of real vectors, is not even countable, these conditions will be tested in a finite number of steps. For this purpose, an enclosure $[f]$ of the range $f([\mathbf{x}])$ will be computed using interval computation.

- if $[f] \subset [a, b]$ then any $\mathbf{x}$ in $[\mathbf{x}]$ satisfies (3),
- if $[f] \cap [a, b] = \emptyset$, then no $\mathbf{x}$ in $[\mathbf{x}]$ satisfies (3), or equivalently all of them satisfy (4).

Note that if these two tests take the value *false*, then no conclusion can be drawn.

These basic principles will be used to test whether a given box $[\mathbf{p}]$ in parameter space is either inside or outside $\mathbb{S}$, where $\mathbb{S}$ is given by (1). It suffices to compute an enclosure $[e_k]$ of $y_k - y_{m,k}([\mathbf{p}])$ for all $k \in \{1, \ldots, k_{\max}\}$, using interval computation.

- if for all k's in $\{1, \ldots, k_{\max}\}$, $[e_k] \subset [\underline{e}_k, \bar{e}_k]$, then $[\mathbf{p}] \subset \mathbb{S}$.
- if there exists k in $\{1, \ldots, k_{\max}\}$, such that $[e_k] \cap [\underline{e}_k, \bar{e}_k] = \emptyset$, then, $[\mathbf{p}] \cap \mathbb{S} = \emptyset$.

Again, the fact that boxes are considered instead of real vectors will allow the exploration of the whole space of interest in a guaranteed way in a finite number of steps, as opposed to Monte-Carlo methods that only sample a finite number of points and can thus not guarantee their results. When no conclusion can be reached for the box $[\mathbf{p}]$, it may be split into subboxes on which the tests will be reiterated, as in the next section.

3 SIVIA

The principle of the *Set Inverter Via Interval Analysis* algorithm is to split the initial problem of characterizing $\mathbb{S}$ into a sequence of more manageable tasks, each of which is solved using interval computation. SIVIA thus pertains to the framework of granular computing.

The parameter vector $\mathbf{p}$ is assumed to belong to some (possibly very large) search domain $[\mathbf{p}_0]$. Let ε be a positive number to be chosen by the user; ε will be called the *level of information granularity.* A *layer* associated with $\mathbb{S}$ at level ε is a list $\mathbb{G}(\varepsilon)$ of nonoverlapping colored boxes $[\mathbf{p}]$ that cover $[\mathbf{p}_0]$, where[1],

1. $[\mathbf{p}]$ is *black* if it is known that $[\mathbf{p}] \subset \mathbb{S}$,
2. $[\mathbf{p}]$ is *grey* if it is known that $[\mathbf{p}] \cap \mathbb{S} = \emptyset$,
3. $[\mathbf{p}]$ is *white* in all other cases.

Note that some white boxes may actually satisfy either $[\mathbf{p}] \subset \mathbb{S}$ or $[\mathbf{p}] \cap \mathbb{S} = \emptyset$, because of the pessimism of interval computation. Each box of $\mathbb{G}(\varepsilon)$ is called a *granule.* The knowledge required to color it may have been obtained by

[1]These colors are, of course, arbitrary. They have been chosen to facilitate the visualization of the boundary of $\mathbb{S}$.

using the techniques explained in Section 2. Moreover, all white boxes in $\mathbb{G}(\varepsilon)$ are such that their widths are smaller than or equal to ε.

The principle of a procedure to build a layer $\mathbb{G}(\varepsilon_0)$ associated with the set $\mathbb{S}$ is described by the following algorithm. As most granular algorithms, it generates a sequence of layers $\mathbb{G}(\varepsilon)$ indexed by the level of information granularity ε for decreasing values of ε. Starting from a very coarse representation associated with a large ε corresponding to the width of the initial search domain, it stops when ε becomes smaller than or equal to ε_0. In what follows, $w([\mathbf{p}])$ is the *width* of $[\mathbf{p}]$, *i.e.*, the length of its largest side, and splitting a box means cutting it into two subboxes, perpendicularly to its largest side.

SIVIA(in: $[\mathbf{p}_0]$, ε_0; out: $\mathbb{G}$));
$\varepsilon := w([\mathbf{p}_0])$; paint $[\mathbf{p}_0]$ white; $\mathbb{G} := \{[\mathbf{p}_0]\}$;
while $\varepsilon \geq \varepsilon_0$, do
    $\varepsilon := \varepsilon/2$;
    while there exists a white $[\mathbf{p}]$ of $\mathbb{G}$ such that $w([\mathbf{p}]) \geqslant \varepsilon$;
        split $[\mathbf{p}]$ into $[\mathbf{p}_1]$ and $[\mathbf{p}_2]$;
        if attempt to prove that $[\mathbf{p}_1] \subset \mathbb{S}$ is a success,
            then paint $[\mathbf{p}_1]$ black;
        if attempt to prove that $[\mathbf{p}_1] \cap \mathbb{S} = \emptyset$ is a success,
            then paint $[\mathbf{p}_1]$ grey;
        otherwise, paint $[\mathbf{p}_1]$ white;
        if attempt to prove that $[\mathbf{p}_2] \subset \mathbb{S}$ is a success,
            then paint $[\mathbf{p}_2]$ black;
        if attempt to prove that $[\mathbf{p}_2] \cap \mathbb{S} = \emptyset$ is a success,
            then paint $[\mathbf{p}_2]$ grey;
        otherwise, paint $[\mathbf{p}_2]$ white;
    endwhile;
endwhile;
return $\mathbb{G}$.

Denote the union of all black boxes by $\mathbb{S}^-$ and the union of all black and white boxes by $\mathbb{S}^+$. At each iteration of the external while loop, one then has

$$\mathbb{S}^- \subset \mathbb{S} \subset \mathbb{S}^+. \tag{5}$$

Example 4 *Consider the set* $\mathbb{S}$ *described by*

$$\mathbb{S} = \{\mathbf{p} \in \mathbb{R}^2 \mid p_1^2 + p_2^2 \in [1,2]\}. \tag{6}$$

For $[\mathbf{p}_0] = [-3,3] \times [-3,3]$ *and* $\varepsilon_0 = 0.04$*,* SIVIA *generates the sequence of layers described by Figure 1. Each of these layers corresponds to the state of*

$\mathbb{G}$ *at Step 13. The hierarchical properties of these layers are schematically portrayed by the information pyramid of Figure 2.* ■

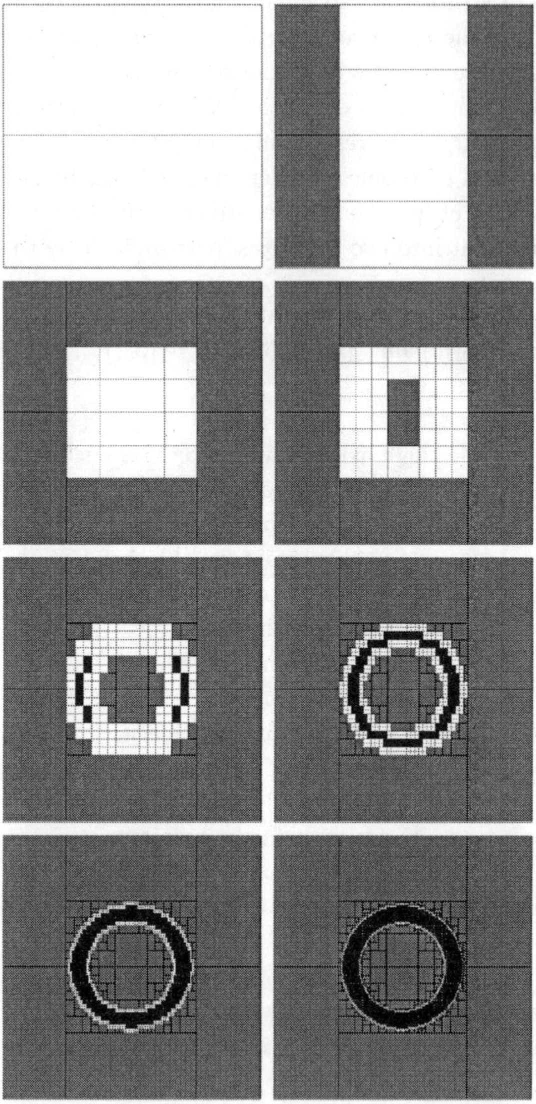

Figure 1: Sequence of layers generated by SIVIA *in Example 4; the black granules are inside* $\mathbb{S}$*, the grey granules are outside* $\mathbb{S}$*, nothing is known about the white granules; the layers correspond to the pavings generated by* SIVIA *for* $\varepsilon \in \{2^2, 2, 2^{-1}, \ldots, 2^{-4}\}$

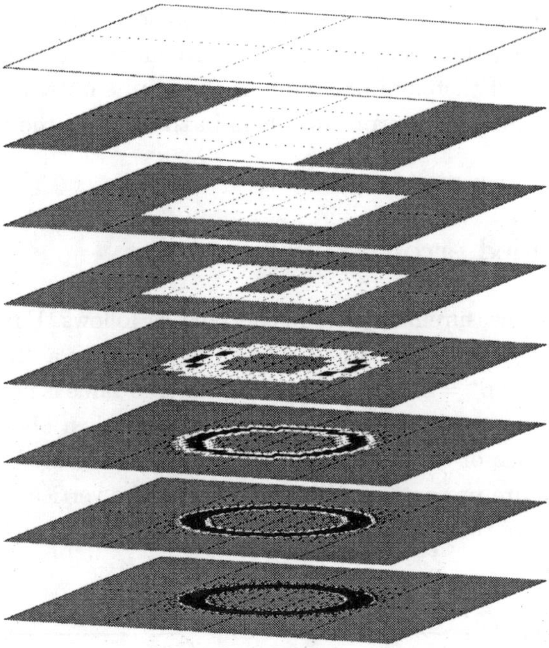

Figure 2: Information pyramid associated with Example 4; the layers are as in Figure 1

Remark 1 *The complexity of* SIVIA *is exponential with respect to the number of parameters dim* **p**. *In practice, dim* **p** *should typically be less than five. However, when the volume of the solution set* $\mathbb{S}$ *is small (as is the case when* $\mathbb{S}$ *is a finite number of points), contraction methods, inserted in* SIVIA, *make it possible to deal with a larger number of parameters (typically, less than fifty). These contraction methods are based on linear interval techniques [Neu90], [Han92] and on interval constraint propagation [Cle87], [Dav87], [BO97], [Jau00].* ■

4 Application to bounded-error estimation

Consider a model where the relation between the parameter vector **p** and the model output is given by

$$y_m(\mathbf{p}, t) = \sin(p_1 * (t + p_2)). \tag{7}$$

We choose a two-parameter model to facilitate illustration, but note that the method applies to higher dimensions without modification. This specific example was chosen to show that the methodology advocated in this paper was able to handle unidentifiable models without requiring an identifiability analysis. We shall therefore first present the results obtained with SIVIA on simulated data, before interpreting them in the light of the notion of identifiability.

4.1 Bounded error estimation

Ten simulated measurements were generated as follows. First ten noise-free measurements y_k^* were computed for $k = 1, 2, \ldots, 10$, at a "true" value of the parameter vector $\mathbf{p}^* = (1, 2)^{\mathrm{T}}$. Of course, this true value is not communicated to the estimation procedure. Noisy data y_k were then obtained by adding to y_k^* realizations of a random noise uniformly distributed in $[-1, 1]$. The resulting data are presented in Figure 3, where the vertical bars indicate all possible values of the noise-free data when the $\underline{e}_k$'s are taken equal to -1 and the $\bar{e}_k$'s to 1.

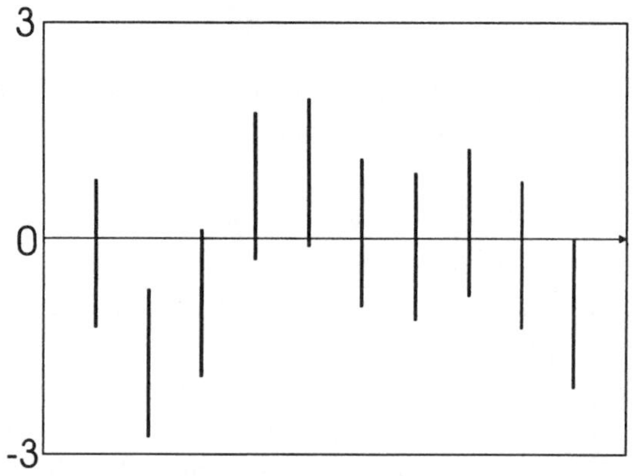

Figure 3: Data bars for the parameter estimation example; each $\mathbf{p}$ *in* $\mathbb{S}$ *is such that the corresponding model output crosses all these bars*

The feasible parameter set $\mathbb{S}$ is thus the set of all $\mathbf{p}$'s that satisfy

$$\left\{ \begin{array}{lll} \sin(p_1 * (1 + p_2)) & \in & [y_1 - 1, y_1 + 1], \\ \sin(p_1 * (2 + p_2)) & \in & [y_2 - 1, y_2 + 1], \\ \vdots & \vdots & \vdots \\ \sin(p_1 * (10 + p_2)) & \in & [y_{10} - 1, y_{10} + 1]. \end{array} \right. \tag{8}$$

SIVIA was used to characterize the part of $\mathbb{S}$ located in three search domains $[\mathbf{p}_0]$ of increasing size. In all cases, the accuracy parameter ε_0 was taken equal to 0.008. The results, obtained in less than 10 seconds on a Pentium 133 MHz computer, are summarized by Figure 4, where the sizes of the frames are those of the search domains, respectively $[0,2] \times [-\pi, \pi]$, $[0,4] \times [-2\pi, 2\pi]$ and $[0,8] \times [-4\pi, 4\pi]$ for subfigures (a), (b) and (c).

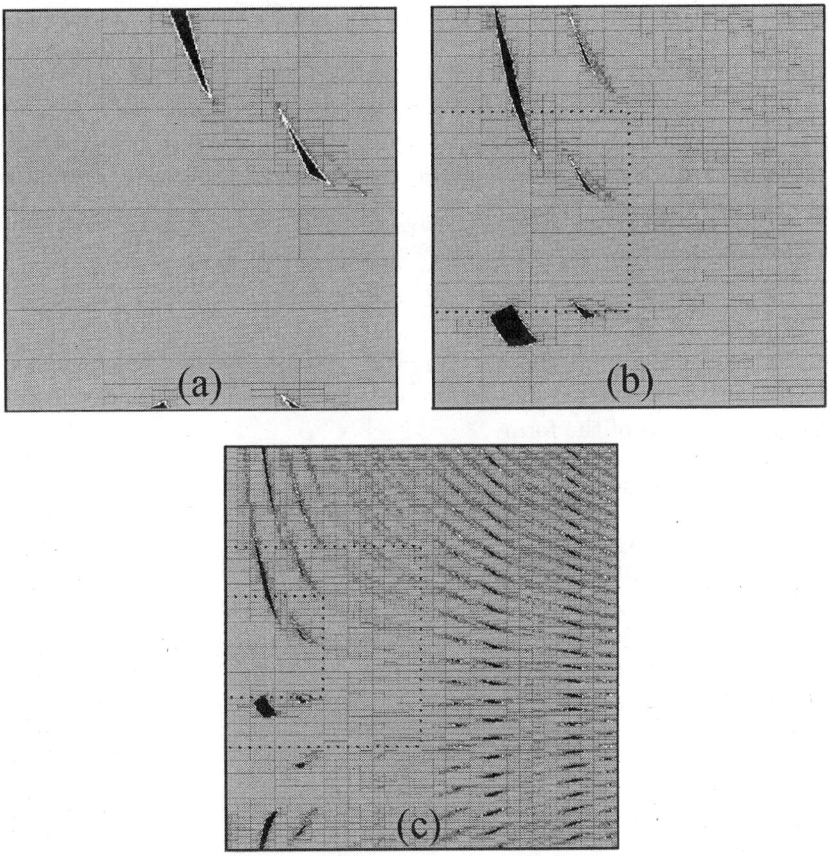

Figure 4: Characterizations of the part of $\mathbb{S}$ in $[\mathbf{p}_0]$ obtained by SIVIA *for increasing sizes of $[\mathbf{p}_0]$; dotted lines in subfigures (b) and (c) indicate the correspondence between the frames of the subfigures*

4.2 Identifiability analysis

As is readily apparent from Figure 4, $\mathbb{S}$ consists of many disconnected components. Part of this fact may be explained by an identifiability analysis [Wal82]. Roughly speaking, a model is said to be uniquely (or globally) identifiable if there is a single value of the parameter vector associated to any given behavior of the model output. From the equation of the model (7), it is clear that any two vectors $\mathbf{p}$ and $\mathbf{q}$ of $\mathbb{R}^2$ such that there exists $\ell \in \mathbb{Z}$ such that

$$\forall t \in \mathbb{R} \mid p_1(t + p_2) = q_1(t + q_2) + 2\ell\pi \tag{9}$$

will lead to exactly the same behavior of the model. Since (9) is equivalent to

$$\begin{cases} p_1 & = & q_1 \\ p_1 p_2 & = & q_1 q_2 + 2\ell\pi \end{cases}, \tag{10}$$

any $\mathbf{q}$ such that $q_1 = p_1$ and $q_2 = p_2 - \frac{2\ell\pi}{p_1}$ will lead to the same behavior as $\mathbf{p}$. The model considered is therefore not uniquely identifiable, and if $\mathbf{p} \in \mathbb{S}$ then all vectors $\mathbf{q}$ of the form

$$\mathbf{q}(\ell) = (p_1, p_2 - \frac{2\ell\pi}{p_1})^{\mathrm{T}}, \ell \in \mathbb{Z}, \tag{11}$$

are also in $\mathbb{S}$. This is why all the components of $\mathbb{S}$ are piled up with a pseudo periodicity given by $2\pi/p_1$. Note that this identifiability analysis was not required for the estimation of the parameters by the method described in this chapter. Moreover, this method allows us to obtain all models with similar behaviors, and not just those that behave *identically*.

5 Conclusions

Bounded-error estimation is an attractive alternative to more conventional approaches to parameter estimation based on a probabilistic description of uncertainty. When bounds are available on the maximum acceptable error between the experimental data and the corresponding model output, it is possible to characterize the set of all acceptable parameter vectors by bracketing it between inner and outer approximations. Most of the results available in the literature require the model output to be linear in the unknown parameters to be estimated and only compute an outer approximation. The method described in this chapter is one of the very few that can be used in the nonlinear case. It provides both inner and outer approximations, which is important because the distance between these approximations is a precious indication about the quality of the description provided. We hope to have shown that the resulting methodology falls naturally into the framework of granular computing. The quality of the approximation obtained depends of the level of granularity, and a compromise must of course be struck between accuracy of description and complexity of representation.

Many important points could not be covered in this introductory material and still form the subject of ongoing research. They include the extension of the methodology to the estimation of the state vector of a dynamical system (or to the tracking of time-varying parameters) [KJW98], and the robustification of the estimator against outliers, *i.e.*, against data points for which the error should be much larger than originally thought, because, for instance, of sensor failure [JWD96], [KJWM99].

References

[BO97] F. Benhamou and W. Older. Applying interval arithmetic to real, integer and Boolean constraints. *Journal of Logic Programming*, pages 1–24, 1997.

[Cle87] J. C. Cleary. Logical arithmetic. *Future Computing Systems*, 2(2):125–149, 1987.

[Dav87] E. Davis. Constraint propagation with interval labels. *Artificial Intelligence*, 32:281–331, 1987.

[Han92] E. R. Hansen. *Global Optimization using Interval Analysis.* Marcel Dekker, New York, 1992.

[HDM97] P. Van Hentenryck, Y. Deville, and L. Michel. *Numerica. A Modeling Language for Global Optimization.* MIT Press, Boston, 1997.

[Jau00] L. Jaulin. Interval constraint propagation with application to bounded-error estimation. *Automatica*, to appear in 2000.

[JW93a] L. Jaulin and E. Walter. Guaranteed nonlinear parameter estimation from bounded-error data via interval analysis. *Math. and Comput. in Simulation*, 35:1923–1937, 1993.

[JW93b] L. Jaulin and E. Walter. Guaranteed nonlinear parameter estimation via interval computations. *Interval Computation*, pages 61–75, 1993.

[JW93c] L. Jaulin and E. Walter. Set inversion via interval analysis for nonlinear bounded-error estimation. *Automatica*, 29(4):1053–1064, 1993.

[JWD96] L. Jaulin, E. Walter, and O. Didrit. Guaranteed robust nonlinear parameter bounding. In *Proc. CESA'96 IMACS Multiconference (Symposium on Modelling, Analysis and Simulation)*, pages 1156–1161, Lille, July 9-12, 1996.

[KJW98] M. Kieffer, L. Jaulin, and E. Walter. Guaranteed recursive nonlinear state estimation using interval analysis. In *Proc. 37th IEEE Conference on Decision and Control*, pages 3966–3971, Tampa, December 16-18, 1998.

[KJWM99] M. Kieffer, L. Jaulin, E. Walter, and D. Meizel. Nonlinear identification based on unreliable priors and data, with application to robot localization. In A. Garulli, A. Tesi, and A. Vicino, editors, *Robustness in Identification and Control*, pages 190–203, LNCIS 245, London, 1999. Springer.

[MNPLW96] M. Milanese, J. Norton, H. Piet-Lahanier, and E. Walter (Eds). *Bounding Approaches to System Identification.* Plenum Press, New York, 1996.

[Moo66] R. E. Moore. *Interval Analysis.* Prentice-Hall, Englewood Cliffs, New Jersey, 1966.

[Moo79] R. E. Moore. *Methods and Applications of Interval Analysis.* SIAM Publ., Philadelphia, 1979.

[Neu90] A. Neumaier. *Interval Methods for Systems of Equations.* Cambridge University Press, Cambridge, 1990.

[Nor94] J. P. Norton (Ed.). Special issue on bounded-error estimation: Issue 1. *Int. J. of Adaptive Control and Signal Processing*, 8(1):1–118, 1994.

[Nor95] J. P. Norton (Ed.). Special issue on bounded-error estimation: Issue 2. *Int. J. of Adaptive Control and Signal Processing*, 9(1):1–132, 1995.

[Wal82] E. Walter. *Identifiability of state space models.* Springer-Verlag, Berlin, 1982.

[Wal90] E. Walter (Ed.). Special issue on parameter identification with error bounds. *Mathematics and Computers in Simulation*, 32(5&6):447–607, 1990.

[WP97] E. Walter and L. Pronzato. *Identification of Parametric Models from Experimental Data.* Springer-Verlag, London, 1997.

[WPL89] E. Walter and H. Piet-Lahanier. Exact recursive polyhedral description of the feasible parameter set for bounded-error models. *IEEE Trans. on Automatic Control*, 34(8):911–915, 1989.

Random-Sets: Theory and Applications

Javier Nuñez-Garcia and Olaf Wolkenhauer

Control Systems Centre, UMIST, PO Box 88, Manchester, UK

Abstract. The relevance, applicability and importance of fuzzy sets is generally linked to successful applications in the domain of engineering, especially when subjective notions are modelled and matched with data. For problems in which uncertainty has been modelled using probability theory in the past, discussions on what approach is right, frequently conclude that both should complement each other. In the present text, we consider such synergy of fuzzy sets, probability and possibility distributions provided by the concept of a random-set. Following a brief review of basic mathematical and semantic aspects of random-set theory, we introduce an application of the theory to time series analysis.

1 Definition of a Random-Set and its Distribution

Usually the outcomes of a random experiment are expressed as qualitative concepts such as whether the dice produces an even or odd number. These random outcomes (i.e. determined in part by random factors) form the space of elementary events Ω. Other events, formed by combinations of elementary events, i.e. subsets of Ω, may also be interesting for the study of the experiment. They may thus be organised in a structure of subsets σ-algebra on Ω and denoted by σ_Ω. This structure ensures (by imposing certain conditions to the subsets) a predictable behaviour in operations with subsets such as complement, union, intersection, etc. and then allows us to measure how likely an event is by introducing a probability measure Pr_Ω in the measurable space (Ω, σ_Ω). The choice of σ_Ω depends on the kind of experiment under consideration. The tuple $(\Omega, \sigma_\Omega, Pr_\Omega)$ is called a probability space and it summarises the experiment or process.

A random variable is a rule that associates for each which elementary event $\omega \in \Omega$ an element $\mathbf{x}(\omega)$ in a space Ξ, in which the elements are organised by a σ-algebra, σ_Ξ . The aim of using a random variable is to generate a probability measure on (Ξ, σ_Ξ) such that the probability space $(\Xi, \sigma_\Xi, Pr_\mathbf{x})$ is the mathematical description of the experiment as well as the original probability space $(\Omega, \sigma_\Omega, Pr_\Omega)$. The benefit arises when (Ξ, σ_Ξ) is a well known measurable space where mathematical tools such as integration are established. One of the most commonly used measurable space is $(\mathbb{R}, \mathcal{B})$, where $\mathcal{B}$ is the σ-algebra of Borel (called Borel algebra), i.e. it it generated from the topological space of the open subsets of $\mathbb{R}$.

For the formal definition of a random variable, let $(\Omega, \sigma_\Omega, Pr_\Omega)$ be a probability space and (Ξ, σ_Ξ) a measurable space. Every $\sigma_\Omega - \sigma_\Xi$ measurable mapping $\mathbf{x} : \Omega \to \Xi$ is called a random variable. A measurable mapping verifies that $\forall A \in \sigma_{\Xi} \Rightarrow \mathbf{x}^{-1}(A) \in \sigma_{\Omega}$. If $\Xi = \mathbb{R}$ and $\sigma_{\Xi} = \mathcal{B}$ then it is called a numerical random variable. The distribution or probability law of $\mathbf{x}$ is defined as $Pr_{\mathbf{x}} = Pr_{\Omega} \circ \mathbf{x}^{-1}$. This means that an event $A \in \sigma_{\Xi}$ has the probability

$$\begin{aligned} Pr_{\mathbf{x}}(A) &= Pr_{\Omega} \circ \mathbf{x}^{-1}(A) \\ &= Pr_{\Omega}\left(\mathbf{x}^{-1}(A)\right) \\ &= Pr_{\Omega}\{\omega :\ \mathbf{x}(\omega) \in A\}. \end{aligned}$$

The definition of a random-set runs in the same way as a random variable, the only difference is that it associates every elementary event $\omega \in \Omega$ to an event of $\mathcal{U}$ which is a set of subsets of Ξ i.e. it associates for $\omega \in \Omega$ a subset of Ξ. This means that a random-set is a multi-valued mapping between Ω and Ξ.

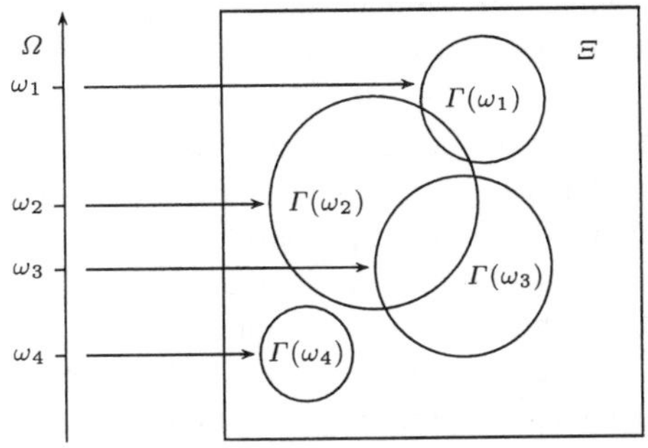

Fig. 1. A random-set seen as a multi-valued mapping onto Ξ.

Let $(\Omega, \sigma_{\Omega}, Pr_{\Omega})$ be a probability space and let $(\mathcal{U}, \sigma_{\mathcal{U}})$ be a measurable space where $\mathcal{U}$ is a set of subsets of Ξ, i.e. $\mathcal{U} \subseteq \mathcal{P}(\Xi)$, power set of Ξ and $\sigma_{\mathcal{U}}$ is a σ-algebra defined on $\mathcal{U}$. Then a $\sigma_{\Omega} - \sigma_{\mathcal{U}}$-measurable mapping $\Gamma : \Omega \to \mathcal{U}$ is defined a random-set.

Γ maps elementary events of Ω to elements of $\mathcal{U}$, so really it is a random variable between the probability space $(\Omega, \sigma_{\Omega}, Pr_{\Omega})$ and the measurable space $(\mathcal{U}, \sigma_{\mathcal{U}})$. Then, as we define the distribution or the probability law of a random variable, we define the distribution of a random-set by $Pr_{\Gamma} = Pr_{\Omega} \circ \Gamma^{-1}$ i.e.

$$\begin{aligned} Pr_{\Gamma}(\mathcal{A}) &= Pr_{\Omega} \circ \Gamma^{-1}(\mathcal{A}) \\ &= Pr_{\Omega}\{\omega : \Gamma(\omega) \in \mathcal{A}\} \quad \forall \mathcal{A} \in \sigma_{\mathcal{U}}. \end{aligned}$$

Note that a random-set can be seen as a random variable from Ω to $\mathcal{U}$ or as a multi-valued mapping from Ω to Ξ since $\Gamma(\omega) \in \mathcal{U} \Rightarrow \Gamma(\omega) \subseteq \Xi$.

As an example, we design an experiment consisting of reading a text on a wall while we walk backwards, stopping when the text becomes unreadable. The random variable $\mathbf{x}$ that we use for this study will associate to each possible result of the experiment a real number which is the distance r between the reader and the wall which; leading to the space $\Xi = \mathbb{R}^+$. As σ-algebra we use the well known Borel algebra $\mathcal{B}^+$ on $\mathbb{R}^+$. Then the measurable space in which the experiment will be carried out is $(\mathbb{R}^+, \mathcal{B}^+)$. The aim is to search for a probability measure $Pr_{\mathbf{x}}$ such as the space $(\mathbb{R}^+, \mathcal{B}^+, Pr_{\mathbf{x}})$ of probability explains the experiment. Of course the distribution of $\mathbf{x}$, $Pr_{\mathbf{x}}$, is unknown and it has to be determined by an estimating process based on a sample of repetitions of the experiment.

Some techniques for estimation hold *a priori* assumptions about the distribution of $\mathbf{x}$. They estimate the parameters of the distribution by applying methods such as maximum likelihood to the sample of data. Other techniques such as non-parametric smoothing methods do not hold any assumption about the distribution of $\mathbf{x}$.

Let us use the same experiment but this time we associate with each possible outcome an interval of $\mathbb{R}^+$, such that $[0, r)$ where $r \in \mathbb{R}^+$. Now the association rule is a random-set (in this case is called random interval) as defined previously. We have the spaces $\Xi = \mathbb{R}^+$ and $\mathcal{U} = \{[0, r) : r \in \mathbb{R}^+\}$. We need to find a σ-algebra $\sigma_{\mathcal{U}}$ on $\mathcal{U}$ with the aim of creating a structure with some necessary features to define a probability measure. In the next section we review the ideas of some authors of how to construct a σ-algebra on these kind of spaces. In the same way as for random variables we can estimate the probability distribution of the random-set Γ, Pr_Γ, from the sample of data obtained by repeating the experiment a number of times. The probability space $(\mathcal{U}, \sigma_{\mathcal{U}}, Pr_\Gamma)$ is the mathematical model that explains the experiment.

Note that the subset associated to each outcome is a single element of $\mathcal{U}$ which means that it is a random variable. On the other hand it can be viewed as a multi-valued mapping since a element of $\mathcal{U}$ is a subset of Ξ i.e. an interval of $\mathbb{R}^+$.

2 Hypermeasurable Spaces

In this section we want to detail the measurable space $(\mathcal{U}, \sigma_{\mathcal{U}})$. Although it plays a fundamental role in the definition of a random-set, it does not seem clearly established and several authors have used different definitions.

$\mathcal{U}$ is a set of subsets included in Ξ and is chosen must be according to the type of process that is being studied. Then the σ-algebra on $\mathcal{U}$, $\sigma_{\mathcal{U}}$, may be a set of sets of subsets of Ξ. This implies a very complicated structure of subsets to identify and work with. Note also that $\sigma_{\mathcal{U}} \subseteq P(\mathcal{U}) \subseteq \mathcal{P}(\mathcal{P}(\Xi))$ since $\mathcal{U} \subseteq \mathcal{P}(\Xi)$. $(\mathcal{U}, \sigma_{\mathcal{U}})$ is termed a hypermeasurable space.

What follows is a review of definitions of the σ-algebra on $\mathcal{U}$ which have been used. Pei-Zhuang Wang and then Qing-De Li [8] defined for $x \in \Xi$,

$\dot{x} = \{A \in \mathcal{P}(\Xi) : x \in A\}$ and $\dot{\Xi} = \{\dot{x} : x \in \Xi\}$. The σ-algebra $\sigma_{\dot{\Xi}}$ generated by $\dot{\Xi}$ is denoted by $\mathcal{B}_0$, and $(\mathcal{P}(\Xi), \mathcal{B}_0)$ is called a hypermeasurable space. Goodman [4] defines $(\mathcal{P}(\Xi), \mathcal{A}, v)$ and $(G(\Xi), \mathcal{B}, u)$ as equivalent probability spaces where $\mathcal{P}(\Xi)$ is the power set of Ξ, and $\mathcal{G}(\Xi)$ is the set of the indicator functions of all the subsets of Ξ. $\mathcal{A}$ and $\mathcal{B}$ are the corresponding appropriate σ-algebras on $\mathcal{P}(\Xi)$ and $\mathcal{G}(\Xi)$. u and v are probability measures on $\mathcal{A}$ and $\mathcal{B}$ respectively. Goodman's paper can be seen as the seminal paper which established the link between fuzzy sets and random-sets. T. Peng et.al. [13] used a definition first introduced by Matheron [9]. Their work provided some advance by creating the hypermeasurable space from a topological space. This is an important property. We must notice that one of the most useful σ-algebra is the Borel algebra which is generated from the open sets of a topological space. It will give some desirable properties to the hypermeasurable space.

Let $\mathcal{G}$, $\mathcal{F}$ and $\mathcal{K}$ be the sets of all open, closed and compact sets respectively in Ξ. They define a hypermeasurable space for all of these structures of subsets. Suppose $(\Xi, \mathcal{G})$ is a locally compact, Hausdorff space. Matheron introduced three topological structures on subsets of $\mathcal{P}(\Xi)$. For any subset $\mathcal{D}$ and any element P of $\mathcal{P}(\Xi)$, they define

$$\begin{aligned}
\mathcal{D}_P &= \{\ Q \ : \ Q \in \mathcal{D},\ P \cap Q = \emptyset\} \\
\mathcal{D}^P &= \{\ Q \ : \ Q \in \mathcal{D},\ P \cap Q \neq \emptyset\} \\
{}_P\mathcal{D} &= \{\ Q \ : \ Q \in \mathcal{D},\ P \subseteq Q\} \\
{}^P\mathcal{D} &= \{\ Q \ : \ Q \in \mathcal{D},\ P \supseteq Q\}\ .
\end{aligned}$$

With these families of subsets of $\mathcal{P}(\Xi)$ we can construct three hypertopologic spaces: $(\mathcal{F}, \mathcal{T}_f)$, $(\mathcal{G}, \mathcal{T}_g)$ and $(\mathcal{K}, \mathcal{T}_k)$ generated respectively by the sub-bases $\{\mathcal{F}_A, \mathcal{F}^B : A \in \mathcal{G}, B \in \mathcal{K}\}$, $\{{}^A\mathcal{G}, {}_B\mathcal{G} : A \in \mathcal{G}, B \in \mathcal{K}\}$ and $\{\mathcal{K}_A, \mathcal{K}^B : A \in \mathcal{G}, B \in \mathcal{F}\}$. Now we can take the corresponding Borel algebras $\mathcal{B}_f$, $\mathcal{B}_g$ and $\mathcal{B}_k$ generated by these three topologies on $\mathcal{F}$, $\mathcal{G}$ and $\mathcal{K}$ respectively. Thus we obtain three hypermeasurable spaces $(\mathcal{F}, \mathcal{B}_f)$, $(\mathcal{G}, \mathcal{B}_g)$ and $(\mathcal{K}, \mathcal{B}_k)$. In [13] they also define open, closed and compact random-sets corresponding to these three hypermeasurable spaces.

Another choice for $\sigma_{\mathcal{U}}$ is $\mathcal{P}(\mathcal{U})$. We must remember that the power set of the set $\mathcal{U}$ is the largest σ-algebra on $\mathcal{U}$ and $\mathcal{P}(\mathcal{U})$ contains all the definable σ-algebras on $\mathcal{U}$. Unfortunately the Borel algebra on $\mathcal{U}$, defined from the open sets of a topological space, is not equal to the power set $\mathcal{P}(\mathcal{U})$. In [2] Bauer proves this for $\mathcal{P}(\mathbb{R})$ and $\mathcal{B}_{\mathbb{R}}$. Thus some desirable properties of the Borel algebras may not be verified in $\mathcal{P}(\mathcal{U})$. The same happens if we take $(\mathcal{P}(\Xi), \mathcal{P}(\mathcal{P}(\Xi)))$ as a hypermeasurable space.

We are going to construct a σ-algebra for the example of the previous section. Let be the following metric in $\mathcal{U} = \{[0, r) : r \in \mathbb{R}^+\}$

$$d([0, r_1), [0, r_2)) = |r_1 - r_2| \quad \forall\ [0, r_1), [0, r_2) \in \mathcal{U}\ . \tag{1}$$

Note that (1) is the Euclidean metric in $\mathbb{R}$ which implies that d verifies the four conditions to be a metric on $\mathcal{U}$. For any $[0,r_1),[0,r_2),[0,r_3) \in \mathcal{U}$ they are

1. $d([0,r_1),[0,r_2)) = |r_1 - r_2| \geq 0$.
2. $d([0,r_1),[0,r_2)) = |r_1 - r_2| = |r_2 - r_1| = d([0,r_2),[0,r_1)$.
3. $d([0,r_1),[0,r_2)) = |r_1 - r_2| = 0 \Leftrightarrow r_1 = r_2 \Leftrightarrow [0,r_1),[0,r_2)$.
4. $d([0,r_1),[0,r_2)) = |r_1 - r_2| \leq |r_1 - r_3| + |r_3 - r_2| = d([0,r_1),[0,r_3)) + d([0,r_3),[0,r_2)$.

$(\mathcal{U},d)$ is said to be a metric space. Let us define the following families of elements of $\mathcal{U}$

$$\mathcal{T}_r = \{[0,t) : t \leq r\} \ .$$

A d-open ball of center $[0,r_o)$ and radius $\epsilon > 0$ is defined by

$$B_\epsilon([0,r_o)) = \{[0,t) : d([0,t),[0,r_o)) < \epsilon\} = \{[0,t) : |t - r_o| < \epsilon\} \ .$$

Let us proof that $\mathcal{T}_r$ are d-open sets. For any $[0,r_o)$ element of $\mathcal{T}_r$ we have to proof that a positive number ϵ exists such as the ball of centre $[0,r_o)$ and radius ϵ is inside of $\mathcal{T}_r$. This is verified by taking $\epsilon_o = \frac{|r-r_o|}{2}$ since

$$B_{\epsilon_o}([0,r_o)) = \{[0,t) : |t - r_o| < \frac{|r - r_o|}{2}\} \subseteq \mathcal{T}_r \ . \tag{2}$$

Fig. 2 illustrates equation (2)

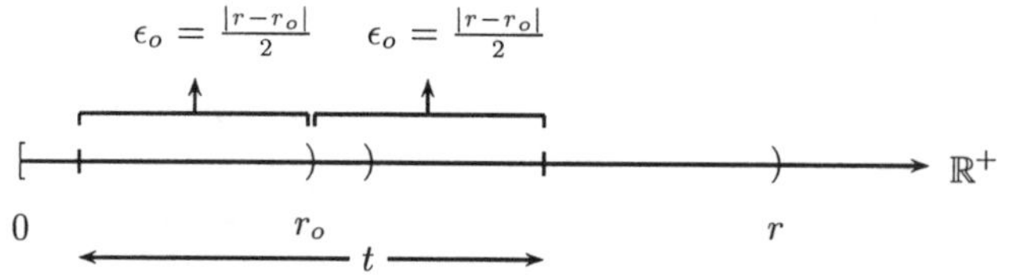

Fig. 2. Representation of the distances of (2).

Then $(\mathcal{U},\mathcal{T})$, where $\mathcal{T}$ is the set of the d-open sets of $\mathcal{U}$, $\{\mathcal{T}_r : r \in \mathbb{R}\}$, is a topological space. The σ-algebra that we are looking for is the Borel algebra generated for this topological space i.e. the σ-algebra generated by the d-open sets, $\sigma(\mathcal{T})$. So the hypermeasurable space in which the random interval is defined is $(\mathcal{U},\sigma(\mathcal{T}))$.

3 Coverage Functions and Set-Valued Statistics

Probability theory develops many others concepts to explore experiments based on a random variable. These are expectation, moments, variance, and so forth; all of them depend on the distribution of the random variable which, in practical problems, is usually unknown. Then statistics are used to estimate these concepts from empirical evidence of the experiment; a random sample of data. This is also the case with random-sets since they are random variables too.

In the following we present a different way for the treatment of these processes by taking advantage of the fact that a random-set is a multi-valued map. Then the analogy between random variables and random-sets followed in our study ends here. The one-point coverage function of a random-set connects possibility theory to random-set theory since it defines a possibility measure on Ξ, which gives the degree of feasibility of the points or subsets of Ξ. The one point-coverage function also depends on the distribution of a random-set (unknown in practical problems) and set-valued statistics are required for its estimation.

For any $A \in \mathcal{U}$, we distinguish three families of subsets

$$\mathcal{C}_A^{\supseteq} \doteq \{C : C \supseteq A,\ C \in \mathcal{U}\} \tag{3}$$

$$\mathcal{C}_A^{\subseteq} \doteq \{C : C \subseteq A,\ C \in \mathcal{U}\}$$

$$\mathcal{C}_A^{\cap} \doteq \{C : C \cap A \neq \emptyset,\ C \in \mathcal{U}\} \tag{4}$$

such that $\mathcal{C}_A^{\supseteq}$ and $\mathcal{C}_A^{\subseteq}$ are subsets of $\mathcal{C}_A^{\cap}$ and $\mathcal{C}_A^{\supseteq} \cap \mathcal{C}_A^{\subseteq} = \{A\}$.

Let us suppose that $(\mathcal{U}, \sigma_{\mathcal{U}})$ is a measurable space such as $\forall A \in \mathcal{U}$, $\mathcal{C}_A^{\supseteq}$, $\mathcal{C}_A^{\subseteq}$, $\mathcal{C}_A^{\cap} \in \sigma_{\mathcal{U}}$. Then (3)-(4) determine the following three measures on Ξ such that $\forall A \subseteq \Xi$ we have

$$\begin{aligned} c^{\supseteq}(A) = Pr_{\Gamma}(\mathcal{C}_A^{\supseteq}) &= Pr_{\Omega}\left(\Gamma(\omega) \in \mathcal{C}_A^{\supseteq}\right) \\ &= Pr_{\Omega}\{\omega : \Gamma(\omega) \supseteq A\} \quad \text{subset coverage function,} \end{aligned}$$

$$\begin{aligned} c^{\subseteq}(A) = Pr_{\Gamma}(\mathcal{C}_A^{\subseteq}) &= Pr_{\Omega}\left(\Gamma(\omega) \in \mathcal{C}_A^{\subseteq}\right) \\ &= Pr_{\Omega}\{\omega : \Gamma(\omega) \subseteq A\} \quad \text{superset coverage func.,} \end{aligned}$$

$$\begin{aligned} c^{\cap}(A) = Pr_{\Gamma}(\mathcal{C}_A^{\cap}) &= Pr_{\Omega}\left(\Gamma(\omega) \in \mathcal{C}_A^{\cap}\right) \\ &= Pr_{\Omega}\{\omega : \Gamma(\omega) \cap A \neq \emptyset\} \quad \text{incidence function.} \end{aligned}$$

The inverse images $\{\omega\colon \Gamma(\omega) \subseteq A\}$ and $\{\omega\colon \Gamma(\omega) \cap A \neq \emptyset\}$ are also called the *lower inverse* and *upper inverse* respectively.

We now focus our attention on the special case of the family of subsets $\mathcal{C}_A^{\supseteq}$ when $A = \{x\}$. Then the family of subsets becomes

$$\mathcal{C}_{\{x\}} = \{C : \{x\} \in C \in \mathcal{U}\} = \{C : x \in C \in \mathcal{U}\} = C_x$$

and the subset coverage function becomes the one-point coverage function of the random-set Γ

$$\begin{aligned} c_\Gamma(x) = Pr_\Gamma(C_x) &= Pr_\Omega \circ \Gamma^{-1}(C_x) \\ &= Pr_\Omega\{\omega : C_x \subset \Gamma(\omega)\} \\ &= Pr_\Omega\{\omega : x \in \Gamma(\omega)\} \quad \forall x \in \Xi . \end{aligned} \tag{5}$$

Or in other words, $\forall x \in \Xi$

$$\begin{aligned} c_\Gamma(x) &\doteq Pr_\Gamma(C_x) \\ &= \int_{C_x} \mathrm{d}Pr_\Gamma \\ &= \int_{\Gamma^{-1}(C_x)} \mathrm{d}Pr_\Omega = \int_\Omega \zeta_{\Gamma^{-1}(C_x)}(\omega)\, \mathrm{d}Pr_\Omega . \end{aligned} \tag{6}$$

Note that $c_\Gamma \colon \Xi \to [0,1]$ defines a fuzzy restriction on Ξ. $\zeta_\Gamma(x)$ is called the characteristic function of coverage of x by Γ. From (5), we can define yet another distribution for subsets of Ξ :

$$C_\Gamma(A) = \sup_{x \in A}\{c_\Gamma(x)\} \qquad \forall A \subset \Xi . \tag{7}$$

Both (5) and (7) are mappings into the unit interval and (7) is called a possibility measure [7] (see Fig. 3).

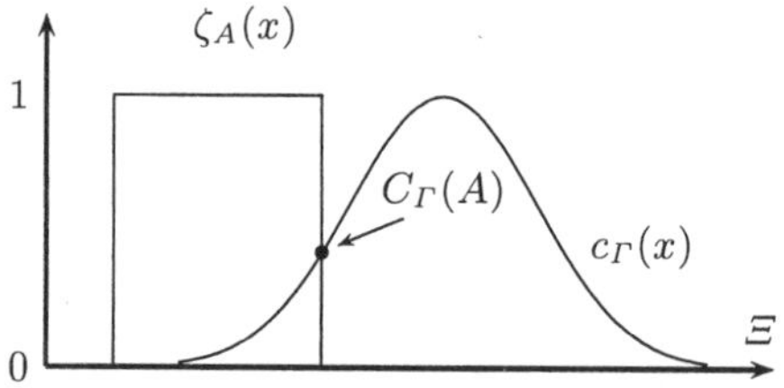

Fig. 3. Possibility of the crisp set A represented as its indicator function $\zeta_A(x)$.

Let the outcomes of a process be in the form of subsets of Ξ and let $C_1, ..., C_n$ be a sequence of random subsets obtained from n realizations of the process. We assume that $C_1, ..., C_n$ is a random-set sample. Then set-valued statistics are required to estimate the one-point coverage function and their properties. An estimator of the one-point coverage function of a random-set obtained from a random-set sample is

$$\hat{c}(x) = \frac{1}{n}\sum_{i=1}^{n} \zeta_{C_i}(x), \quad \forall x \in \Xi \tag{8}$$

where ζ_{C_i} is the indicator function for C_i, $i \in \{1, ..., n\}$.

In [13] Peng et.al. prove that $\hat{c}(\cdot)$ is unbiased and consistent and they give several limit theorems, justifying the use of (8) as an estimator for the one-point coverage function of a random-set (5). They also give some results regarding its properties and they revise the particular case where the random subsets are intervals on the real line.

We now refer to the example introduced in the first section. Suppose that the experiment has been repeated 20 times with different people obtaining the following sample of subsets or intervals of $\mathbb{R}^+$: $[0, r_1), [0, r_2), ..., [0, r_{20})$. The estimator (8) of the one-point coverage function (see Fig. 4) is

$$\hat{c}(x) = \frac{1}{20} \sum_{i=1}^{20} \zeta_{[0,r_i)}(x), \quad \forall x > 0 \; . \tag{9}$$

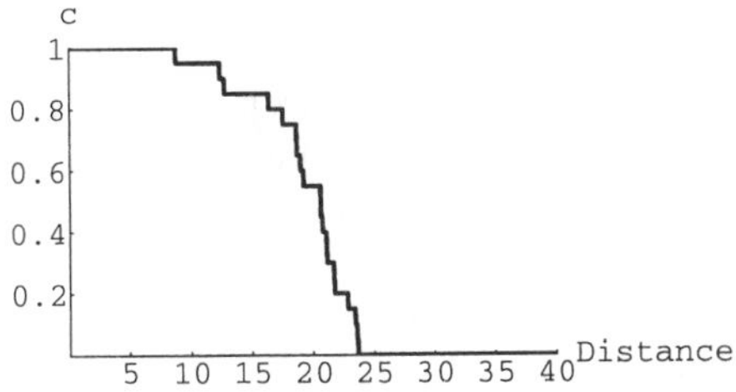

Fig. 4. Sample coverage function of a random interval.

Random intervals such as this example are special cases where the extremes of the interval are random variables and the one-point coverage function can be written in terms of the distributions of these random variables. Membership functions of this type are frequently used in fuzzy systems to represent linguistic knowledge. In the example, the one-point coverage function is the inverse of the distribution function of the random variable $\mathbf{x}$ defined also in section one as the distance between the reader and the wall. This is

$$F(x) = Pr_{\mathbf{x}}([0, x)) = Pr_{\Omega}\{w : \mathbf{x}(\omega) \leq x\} = Pr_{\Omega}(\mathbf{x} \leq x)$$

which is shown in the Fig. 5.

Note the difference between the treatment of an experiment by using classical statistics (where the probability of a subset is calculated in terms of the frequency of sample data that fall into the subset) and set-valued statistics (where the possibility of a point is calculated in terms of the frequency of its incidence in the random-set sample).

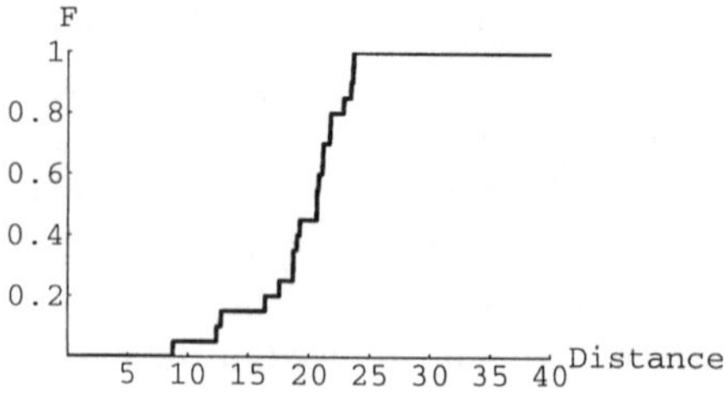

Fig. 5. Sample distribution function of **x**.

4 Conclusions

Problems subject to uncertainty, imposed by subjective and imprecise information, are present in many real world problems and are more and more often considered in engineering applications. Fuzzy set theory has been successfully applied to those problems where a lack of precision exists in the outcome of an experiment, for instance, the definition of a concept by a group of experts, where everyone has a subjective opinion. It is called a fuzzy concept because the borderlines are not clearly defined. Fuzzy mathematics gives good descriptions for those concepts by using fuzzy set membership functions.

Set-valued statistics based on random-set theory introduces a practical method to set up these membership functions from a sample of outcomes such as some different opinions about a concept. This allows us to study the process by applying fuzzy mathematics, which has become a useful extension to probability theory and statistics.

In the previous section we introduced the definition of a random-set and its distribution and the similarity with the definition of a random variable. This similarity allows us to study processes governed by a random-set (i.e. processes with subsets of a space as possible outcomes) by applying probability theory on the $(\mathcal{U}, \sigma_{\mathcal{U}})$ hypermeasurable space. For what follows, our aim is to concentrate on a different problem which is the connection between fuzzy sets or possibility measures with the processes mentioned before. This is achieved through the one-point coverage function of a random-set and set-valued statistics (which provides estimators and properties for practical applications). The one-point coverage function is a single value function with dominion Ξ. For any data point in the space Ξ it provides the level of coverage by the random-set. For example, if a sample of a random-set is based on different opinions of a concept, represented on Ξ, the one-point coverage function gives of a data point a confidence measure verifying the concept. This is clearly the membership function of a fuzzy set that represent the concept. In the example followed through the previous section, (9) gives the membership function of the concept "perceived distance".

To finish this section, we insist on the two different points of view of probability and possibility in the description of a concept. In probability theory, a sample of values subject to the randomness of an experiment is given. The analyst defines a concept or event and the probability of this concept according to the experience is the relative frequency of the values that fall into the definition of the concept. Then we can talk about the probability of a new value verifying the concept. Its probability measures how likely the concept is to occur according to our experience. In the example, we can define "long range vision" by those people that can read the words on the wall for a distance of more than say 22 meters. We can calculate the probability by $Pr_{\mathbf{x}}([22, \infty)) = \frac{\text{number of } r_i \geq 22}{20}$. We then can say that a person has this probability to have good long distance vision. Possibility concentrates overall in the definition of the concept and not in measuring its occurrence. Given some different definition of a concept (sample), the one-point coverage function is the possibility measure or fuzzy set that describes the concept. For any data point in the domain space of the function we can talk about its membership value or its degree in which it verifies the concept. In the example, we represent the concept "view perception of the human eye in the distance" by using the one-point coverage function of a sample of individual opinions or measurements of such a concept (9).

Although fuzzy sets and possibility theories may use probability theory and statistics for their developments, they have different objectives and both can add complementary information in the analysis of a system.

5 An Application to Time Series Analysis

Random set theory had originally been developed as part of stochastic geometry [9,15,10] where it has mainly been used to simulate patterns occurring in natural and technical processes such as paper [3]. The paper by Goodman [4] was the first to investigate the formal relationship with fuzzy logic and was followed by various investigations that linked random-set theory to evidence theory [11,6]. More recently applications to data fusion [5] and time series analysis [14] appeared. In this section we consider an application to time series forecasting.

For our model, we assume a non-linear autoregressive structure of the form $y(k+1) = f(\overline{x}(k))$, where f is an unknown (nonlinear) function describing the process under consideration. $y(k+1)$ is the output of the system at time $k+1$ and $\overline{x}(k)$ is the set of the inputs at k time which consists of past process outputs and process inputs that are relevant in the analysis. For example, if the autoregressive order is p_y and there are two inputs u and e with p_u and p_e relevant steps in the past, $\overline{x}(k)$ can be written as

$$\begin{aligned}\overline{x}(k) = [\ & y(k), y(k-1), ..., y(k-p_y+1), \\ & u(k), u(k-1), ..., u(k-p_u+1), \\ & e(k), e(k-1), ..., e(k-p_e+1)\ ]^T\end{aligned}$$

which is a point belonging to a $p_y + p_u + p_e$ dimensional space, let us say X. $y(k+1) \in Y$ which is a one dimensional space. Then $z(k+1) = [y(k+1), \overline{x}(k)^T]^T \in Y \times X = \Xi$ which is a $1 + p_y + p_u + p_e$ dimensional state space of the process. Subsequently collected training data represents points in that space. For a system subject to uncertainty, a point in the state space could

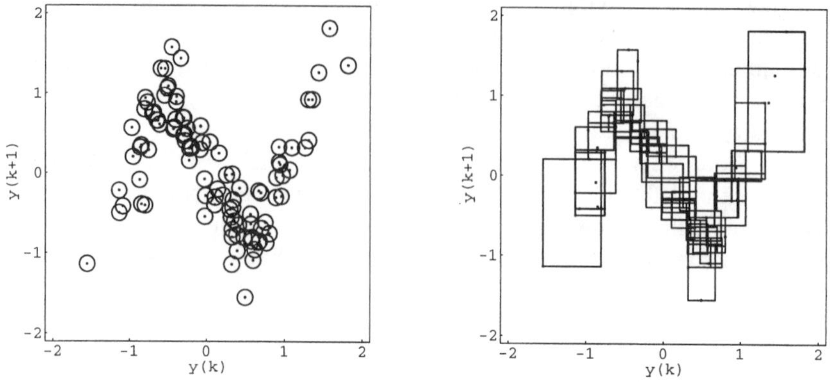

Fig. 6. Two examples of random subsets generated from a sample of data

be located in another position nearby. This suggests the idea of including the neighbourhood of each point in the analysis of the process by generating a random subset for every point of the training data. The random subsets could be balls around every point, with a radius as a subjective parameter depending on the amount of uncertainty in the model. In [14] cuboids around each point were used and in [16] minimal cuboids containing a fixed number of nearest neighbours were used. In this paper we propose to select a number of nearest neighbours and then to generate an ellipsoid for every point by applying principal components analysis to them. Then we can say that our sample of data has become the following sample of random subsets of Ξ

$$C_i = \{z \in \Xi : (z - c_i)^T {S_{u_i}}^{-1}(z - c_i) \leq a_i\}, \quad i = 1, 2, ..., n$$

where c_i and S_{u_i} are the sample mean and the sample covariance matrix respectively of the cluster of nearest neighbours $z_1, z_2, ..., z_{n_i}$ corresponding to the data point i of the state space and a_i is a positive number. The sample covariance matrix is the unbiased estimator of the covariance matrix: $S_{u_i} = \frac{1}{n_i - 1}\sum_{j=1}^{n_i}(z_j - c_i)(z_j - c_i)^T$. Note that $(z - c_i)^T {S_{u_i}}^{-1}(z - c_i)$ is the Mahalonobis distance or probability distance between a data point z and c_i which, under assumption of normality of z, is distributed as $n_i T^2(p, n_i - 1)$ where T^2 is the Hotelling distribution and p is the dimension of the state space $(1 + p_y + p_u + p_e)$. This is equivalent to $\frac{p(n_i-1)}{n_i(n_i-p)} F_{p,n_i-p}$ where F is the F-distribution. Then, given a value for a_i, the C_i's are confidence intervals of the centres c_i's or the compact level set with a distance of less than a_i. Lets see what happens when a_i has different values. Let us take a determined i, for which the number of nearest neighbours is n_i. Then the level of confidence $1-\alpha$ of the C_i according to its sample of data $z_1, z_2, ..., z_{n_i}$ can be calculated by using the following formula

$$F_{p,n_i-p,\alpha} = \frac{a_i n_i(n_i-p)}{p(n_i-1)} \tag{10}$$

then

$$\alpha = P\left(F_{p,n_i-p} > \frac{a_i n_i(n_i-p)}{p(n_i-1)}\right) = \int_{\frac{a_i n_i(n_i-p)}{p(n_i-1)}}^{\infty} F_{p,n_i-p} \,. \tag{11}$$

Equation (11) indicates that the larger is a_i, the smaller the value of the integral is and the higher the level of confidence $1-\alpha$ is, means that the confidence of a new z belonging to C_i is $1-\alpha$. On the other hand this means that the centre c_i becomes a vaguer or a poorer estimator for the data in the neighbourhood C_i. Given a determined level of confidence $1-\alpha_o$ it is also possible to calculate the distance from the centre c_i to the bound of C_i by using (10) and (11). Note that the choice of a_i only influences the size of the ellipsoids but not the shape nor their orientation. In the following practical examples, we have simplified the choice of the a_i, taken all of them as equal which makes sense if we think that the error of the sample is equally distributed in all the regions of the product space. The value for a is 1, which according to the previous comments, provides a medium level of confidence.

The advantage over balls or cuboids used in [14] is that an ellipsoid explains what happens in the neighbourhood of a point more accurately (by taking into account the nearest neighbours) as opposed to a fixed ball or cuboid located around a point. The advantage over the cuboids used in [16] is that an ellipsoid fits the shape of the group of nearest neighbours better. Note that the ellipsoids are local models for their clusters of data.

The larger the number of nearest neighbours generating the ellipsoids is, the more information the local models will provide. The only conditions that we impose on the ellipsoids is that they must fit the general shape of the sample data in addition to the shape of the data which have been used to generate it. The plot on the right in Fig. 7 illustrates the effect of an excessively high number of nearest neighbours leading to ellipsoids that do not fit the data well. Note that the choice of nearest neighbours influences not only the size of the ellipsoids but also the shape and orientation.

Let us suppose that the compact ellipsoids $C_1, ..., C_n$ come from n realization of a random-set as defined before, i.e. they are a random-set sample. We identify elements of the random-set as follows: Ω is the set of all the possible points of the state space $\Xi = \mathbb{R}^{1+p_y+p_u+p_e}$ and σ_Ω the Borel algebra $\mathcal{B}^{1+p_y+p_u+p_e}$ on that space and Pr_Ω is the probability measure of the experiment which is unknown. U is the set of all the open sets in Ξ and $\sigma_\mathcal{U}$ is an appropriate σ-algebra for which we will need to verify that $C_z \in \sigma_\mathcal{U}, \forall z \in \Xi$

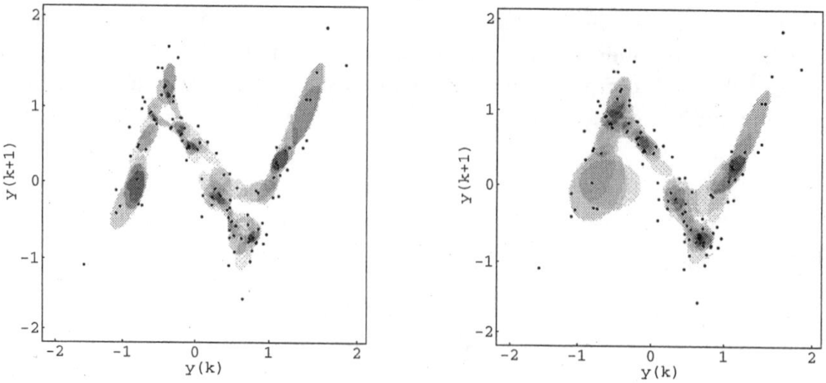

Fig. 7. Random-set sample for 8 (*left*) and 14 (*right*) nearest neighbours.

where $\mathcal{C}_z = \{C : z \in C \in \mathcal{U}\}$. This condition ensures that $\Gamma^{-1}(\mathcal{C}_z) \in \sigma_\Omega$ which is required for the definition of the one-point coverage function of the random-set (5).

Previously we showed that the one-point coverage function of the random-set defines the possibility measure (7) which serves as a measure of the degree of feasibility for any point or subset of Ξ. Given $A \subseteq \Xi$ we can calculate its possibility by

$$C_\Gamma(A) = \sup_{[y,\overline{x}]^T \in A} \left\{ c_\Gamma\left([y, \overline{x}]^T\right) \right\} \tag{12}$$

which is a possibility measure on Ξ. Given an input set $\overline{x}_o \in X$, the marginal one-point coverage function is

$$c_{\Gamma_{\overline{x}_o}}(y) = c_\Gamma([y, \overline{x}_o]^T) \quad \forall y \in Y, \tag{13}$$

where $c_{\Gamma_{\overline{x}_o}} : Y \to [0,1]$ defines the following marginal possibility measure on Y

$$\begin{aligned} C_{\Gamma_{\overline{x}_o}}(B) &= \sup_{y \in B} \left\{ c_{\Gamma_{\overline{x}_o}}(y) \right\} \\ &= \sup_{[y,\overline{x}_o]^T \in B \times \overline{x}_o} \left\{ c_\Gamma\left([y, \overline{x}_o]^T\right) \right\} \quad \forall B \subseteq Y = \mathbb{R}\,. \end{aligned} \tag{14}$$

Note that $c_{\Gamma_{\overline{x}_o}}(y) = C_{\Gamma_{\overline{x}_o}}(\{y\}) \quad \forall y \in Y$, i.e. they are the same function for singles values.

Once we have calculated the estimator of the one-point coverage function $\hat{c}(\cdot)$, we can calculate $\widehat{C}(\cdot)$, $\hat{c}_{\overline{x}_o}(\cdot)$ and $\widehat{C}_{\overline{x}_o}(\cdot)$ as we saw in (12), (13) and (14) respectively. Given a set of input variables $\overline{x}(k_o)$ of the system, we predict the response of the system at $k_o + 1$ time by investigating the distribution $\widehat{C}_{\overline{x}_o}(\cdot)$ which reflects the (un)certainty of the model based on the training data. As this distribution is based on training data, not on an asymptotic probability model, it reflects the confidence of the model more accurately and may in fact be more informative or precise than a single number provided in conventional forecasting methods. However, for numerical comparisons of the accuracy of forecasts, a single value is desirable and we use the standard way to "defuzzify" a distribution, called "centre of gravity method" frequently employed in fuzzy control [7] :

$$\hat{y}(k+1) = \frac{\int_Y y \cdot \hat{c}_{\overline{x}_o}(y)\ dy}{\int_Y \hat{c}_{\overline{x}_o}(y)\ dy} \ . \tag{15}$$

We will use (15) for the first example. In the second example the forecast of $y(k_o+1)$, $\hat{y}(k_o+1)$, is calculated by choosing the most possible point in the output space given a set of inputs, i.e. the point in $\mathbb{R}$ that maximizes the marginal possibility meassure (14) on Y, i.e.

$$\hat{y}(k_o+1) = \left\{ y : y \in Y \wedge \widehat{C}_{\overline{x}(k_o)}(y) = \widehat{C}_{\overline{x}(k_o)}(Y) \right\} \tag{16}$$

being $\widehat{C}_{\overline{x}(k_o)}(Y)$ its possibility. If the maximum is achieved for all the points of an interval of $\mathbb{R}$, the mean of that interval is used as a predictor.

5.1 Non-Linear AR(1) Process

The process that generating the data is from the book [1] and has been used in the paper [16]. The model was first described by Ikoma and Hirota in 1993. It consists of a nonlinear AR(1) dynamic system simulated by the function:

$$y(k+1) = f\big(y(k)\big) + \epsilon(k), \quad f(x) = \begin{cases} 2x-2, & 0.5 \leq \ x, \\ -2x, & -0.5 < \ x < 0.5, \\ 2x+2, & x \leq -0.5 \end{cases} \tag{17}$$

where $\epsilon(k) \sim N(0, \sigma^2)$ with $\sigma = 0.3$, $y(0) = 0.1$ and $k \in \{0, ..., 140\}$.

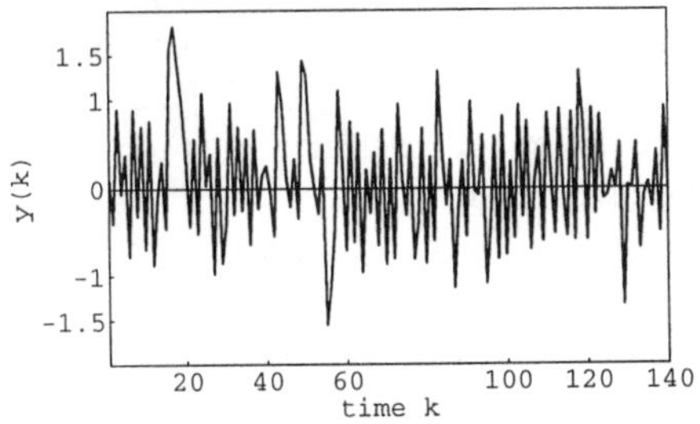

Fig. 8. First-order nonlinear autogregressive process.

We will follow the steps described in the previous section for the analysis of the data. 100 data points are used for the identification the model and 40 for its validation. First we plot the data in the product space which is formed by an output variable $y(k+1)$ and an input autoregressive variable $x = [y(k)]$ i.e. the points are $[y(k+1), y(k)]^T$ or $[y, x]^T$ in the state space Ξ or $Y \times X$ which is $\mathbb{R}^2$ (see Fig. 7). We generate the random-set sample of ellipsoids by using principal components of the eight nearest neighbours of every point as shown in Fig. 7 in the plot on the left. The number of nearest neighbours has been chosen by experience. Similar results would be obtained with a value between 4 and 12. Lower and higher values gave worse results.

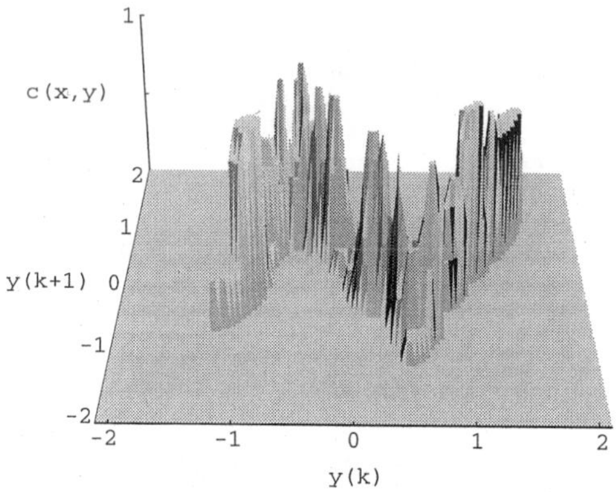

Fig. 9. One-point coverage function estimator $\hat{c}(y, x)$ or $\hat{c}(y(k+1), y(k))$ (*left*) and its contour plot (*right*).

The one-point coverage function estimator $\hat{c}(\cdot)$ from the random-set sample (6) is a function $\hat{c} : \mathbb{R}^2 \to [0, 1]$ (see Fig. 9). This function defines a possibility measure of a two-dimensional process. Suppose that k_o is the actual time. Then the response of the system at this time $y(k_o)$ is known. Consequently the input at $k_o + 1$ will be $\overline{x}_o = y(k_o)$. We use the centre of gravity method (15) to forecast the output at $k_o + 1$, i.e. $\hat{y}(k_o + 1)$.

In the following we consider two examples to illustrate the basic idea of random-set models induced from time-series data. In the first one $k_o = 102$, then the input set is $x(k_o) = y(102) = -0.57$. It is shown as the left vertical line in Fig. 10. The prediction may be calculated for $k = 103$. So the predictor of $y(103)$ is the value of $\mathbb{R}$ given by

$$\hat{y}(103) = \frac{\int_{\mathbb{R}} y \cdot \hat{c}_{y(102)}(y)\ dy}{\int_{\mathbb{R}} \hat{c}_{y(102)}(y)\ dy} = 0.98$$

where the real value for $y(103)$ is 0.92. See Fig. 11 and Fig. 12 for illustration.

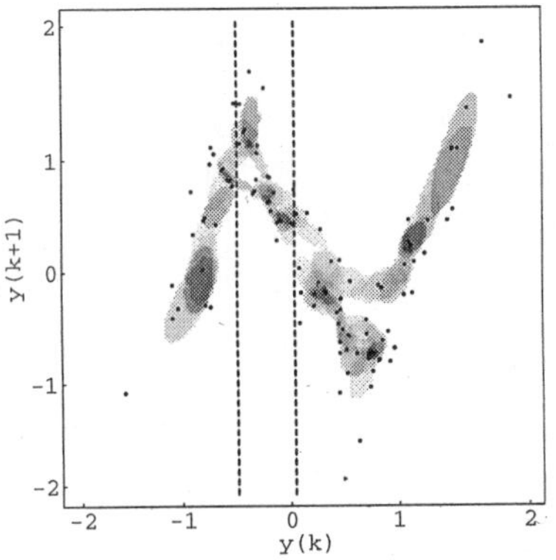

Fig. 10. Input sets for $x_o = y(102) = -0.57$ (*left vertical line*) and $x_o = y(124) = -0.11$ (*right vertical line*). Then used to forecast $y(103)$ and $y(125)$.

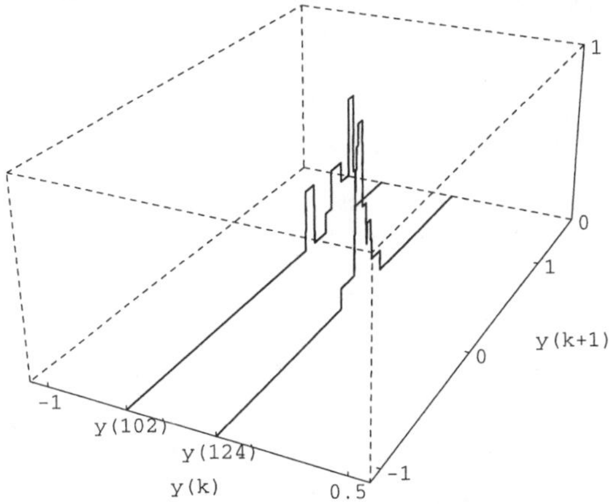

Fig. 11. Marginal coverage function on the state space for the inputs set $x_o = y(102) = -0.57$ and $x_o = y(125) = -0.11$.

In the second example $k_o = 124$, $y(124) = -0.11$ (the right vertical line in Fig. 10). The prediction may be done at $k = 125$. The result obtained is $\hat{y}(125) = 0.34$ while the real value is -0.07. In Fig. 13 on the left, we can see the real process (continued line) and the predictions (dashed line) of the 40 validation data points.

In Fig. 13 right we can see the real process and the predictions by using the three linear models of the function that form the pairwise function (17) without the error variable, i.e. $y(k+1) = f(y(k))$ and $f(x)$ defined as in (17). To compare these two predictions we use the mean square error which is 0.13 for the random-set model and 0.11 for the three linear models.

A problem arises when a given $x_o \in X$ has a zero maginal possibility distribution because there are no ellipsoids which intersect with the line $x =$

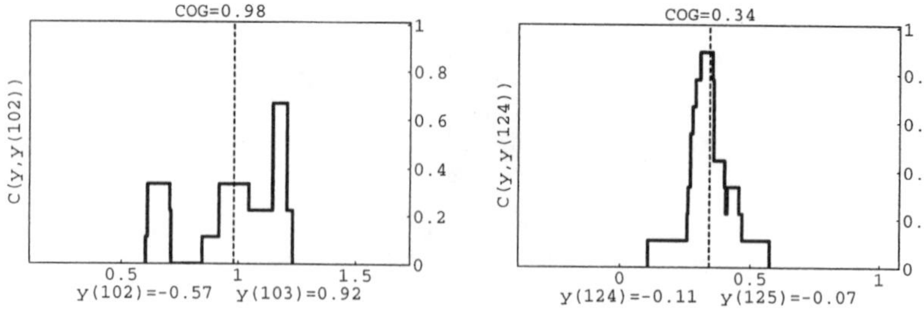

Fig. 12. Marginal coverage function on the output space for the inputs set $x_o = y(102) = -0.57$ and $x_o = y(125) = -0.11$.

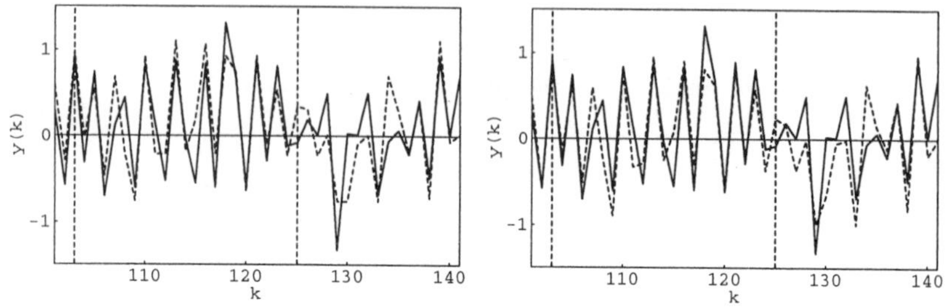

Fig. 13. Predictions (*dash line*) for the 40 data of validation (*left*). Note $k = 102$ and $k = 124$ (*vertical lines*). The predictions by using the 3 pairwise linear models the equation (17) (*right*).

x_o in the space $Y \times X$ (see Fig. 14). There is no information to predict with our model therefore an alternative one must be used. The model works well while the process stays inside an interval around zero ($[-1.2, 1.2]$ depending on how large the error is). In other words, our model is only confident in making predictions for ares in the state space for which training data provided the chance to learn. The model therefore does not extrapolate 'automatically' as conventional models do. When it goes outside the interval, the cumulative error is too large to go back inside the interval and the process becomes unstable. We then have the situation described above. Note in Fig. 14 that for $k = 129$ the value goes out of the region occupied by the model, but for $k = 130$ it goes back inside. This such does not happen for $k = 233$ where the process escapes the interval. For such cases we make predictions by using an ellipsoid merging algorithm developed in [12]. The algorithm creates clusters in the data, following the linear patterns in their shape. These linear structures, which are essentially linear regression models, are then used for predictions.

Fig.15 illustrates the data set, random subsets (ellipsoids) induced by the data and three clusters (large ellipsoids) found by the algorithm. The real values for $y(130)$ and $y(k)$, $k = 234, 235, 236, 237$ are shown by the big black dots and their predictions by circles on the principal axes of the ellipsoids

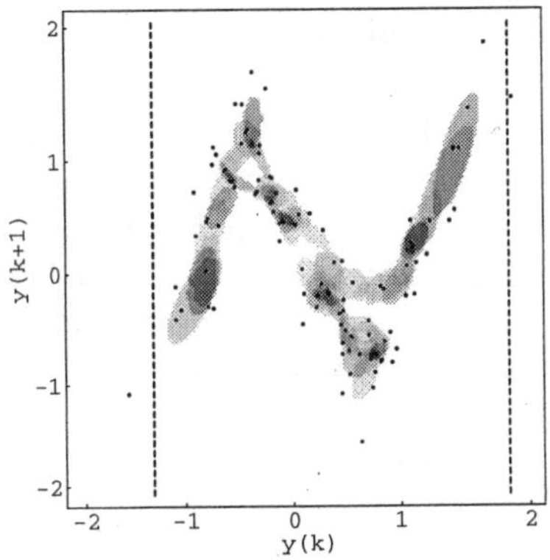

Fig. 14. For $y(129) = -1.32$ (*vertical left line*), $y(233) = -1.38$ and $y(k) \quad \forall \quad k > 233$ the random-set model cannot make predictions.

of the clusters. The dashed lines are the function (17) without error. The vertical lines are for $y(129) = -1.32$ (left) and $y(237) = 1.78$ (right). They correspond to the lines in Fig. 14. Note that the predictions for $y(130)$ and $y(238)$ are on these lines.

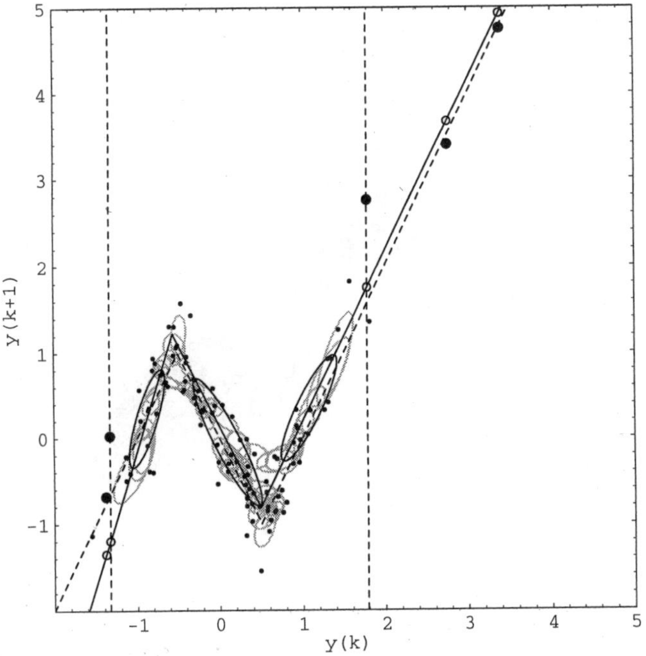

Fig. 15. Clusters and predictions for the data of Fig. 14

5.2 Box-Jenkins Gas Furnace Data

The process consists of introducing a gas flow into a furnace (u input variable) and to measure the response in terms of the CO_2 gas concentration in the outlet gases every nine seconds (y output variable). We have 296 pairs of input-output measurements available. One hundred of them will be used for the modelling process. We choose the model structure $y(k+1) = f(y(k), u(k-2))$. The vector of the input variables $\overline{x}$ has the form $[y(k), u(k-2)]^T$ and then the points in the product space will be as $[y(k+1), y(k), u(k-2)]^T$. This means that the state space $\Xi = Y \times X = \mathbb{R}^3$.

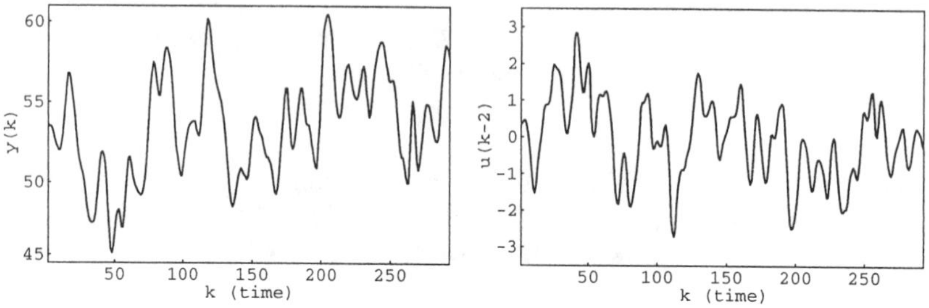

Fig. 16. Box-Jenkins gas furnace data-set.

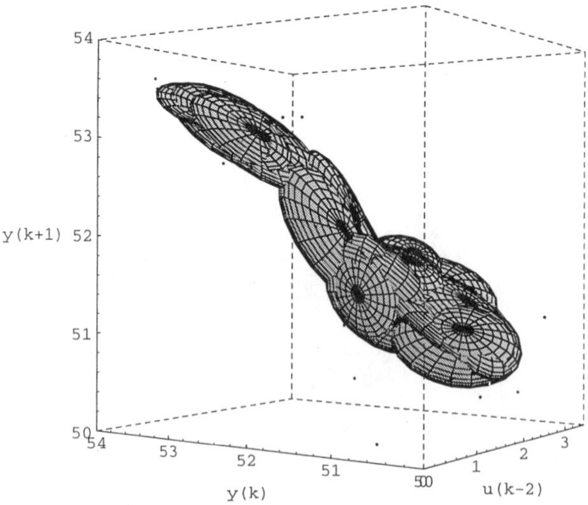

Fig. 17. Gas Furnace Data and the random subsets or ellipsoids on the three dimensional state space

The random-set sample consists of three dimensional ellipsoids or hyper-ellipsoids generated by the principal components of the eight nearest neighbours of every point of the data training (see Fig. 17). Although we have a product space with a higher dimension, the way the modelling process is carried out, is the same as in the previous examples. The one-point coverage function estimated from the sample is a function $\hat{c} : \mathbb{R}^3 \to [0,1]$. This function defines the possibility distribution (7) of a three-dimensional process.

Given a set of inputs $\overline{x}_o = [y(k_o), u(k_o - 2)]^T$, we want to estimate the response of the system at $k_o + 1$. This time we use the mean of the interval of $\mathbb{R}$ given by (16). In Fig. 18 on the left, we can see the vertical line $(y(k), u(k-2)) = (y(159), u(157))$ in the state space, and the ellipsoids which intersect the line. These intersections are intervals in $\mathbb{R}$ and they form the marginal coverage function (13) corresponding to the input set $\overline{x}_o = (y(159), u(157))$ i.e. $\hat{c}_{(y(159),u(157))}(y)$ (see in Fig. 18, right). The predictor $\hat{y}(160)$ of $y(160)$ is also plotted as a dashed line. This predictor has been calculated by using the mean of the interval which maximizes (14) :

$$\hat{y}(160) = \left\{ y : y \in \mathbb{R} \wedge \widehat{C}_{(y(159),u(157))}(y) = \widehat{C}_{(y(159),u(157))}(\mathbb{R}) \right\} .$$

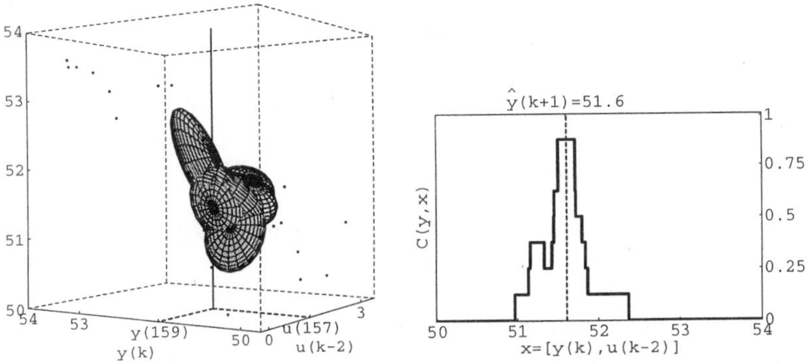

Fig. 18. The random subsets intersecting the input set line $(y(k), u(k-2)) = (y(159), u(157))$ in the product space (*left*). This 3-dimensional figure is equivalent to the 2-dimensional Fig. 10 of the first example. The marginal coverage function for the input set on the output space (*right*).

We used the second order example to demonstrate that the theory and algorithm works equally well for an n-dimensional space and the model (ellipsoids, coverage function, marginal possibilities and defuzzyfication) is obtained as described above. Though there is no numerical optimisation involved, we cannot beat the curse of dimensionality and there is an increment of calculations linked to the increase of dimension.

As in the previous example we also find unpredictable data because the model does not provide information in some areas of the state space where validation data are sparse. In table 1 we show how the number of nearest neighbours and the number of data training affects the number of unpredictable data in the validation process. Remember that we use a total of 293 data partitioned in 2 groups: data training and data validation.

Table 1. Variation of the number of missing forecast depending on the number of nearest neighbours and the number of data training

Nearest neigbours	Data Training	Missing Forecast (Training)	Missing Forecast (Validation)
8	100	25	63
12	100	30	53
8	50	61	112
12	50	63	116

In Fig. 19 the original data set and prediction from the random set model are compared. The grey line and dots are the original data. The black points are the predictions. The circles on the original data indicate the values for which the model is unable to make a forecast (the model that generates these predictions has been created with 100 data training and 8 nearest neighbours). A solution to the problem mentioned above could be to choose another random subset in the product space. In [16], we used cuboids instead of ellipsoids. A cuboid generated by n nearest neighbours covers a larger region than the ellipsoid generated by applying principal components of those n data. At least we can say that the data training will lie inside the cuboids. It does not happen in the same way with ellipsoids (see previous table). On the other hand the use of ellipsoids will give more accuracy since they fit the shape of the data better.

Let us use a random-set model with 8 nearest neighbours and 100 data points for training. This time we update the model at each time k with the present value $y(k)$ before we predict the next value $\hat{y}(k+1)$. The predictions for $k = 101, ..., 293$ are shown in Fig. 20. We have 33 missing predictions. This means a large reduction in comparison to the models which have not been updated. Updating the model takes some computation time, so it may not be able to be used for those processes in which a real time forecast is required.

5.3 Summary and Conclusions

A random-set model may not appear appealing when comparing its accuracy of point-valued forecasts with universal approximators such as neural

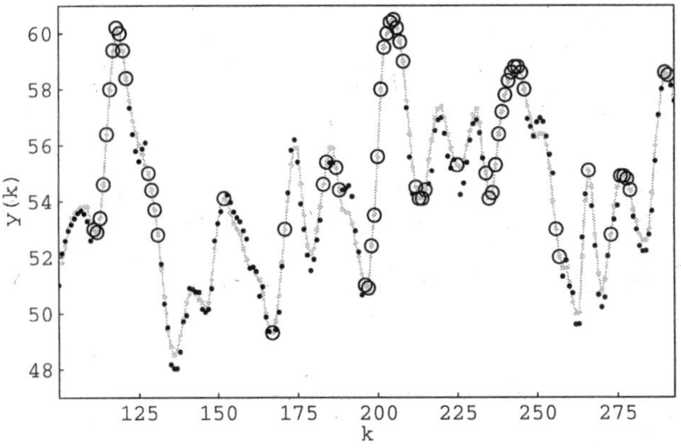

Fig. 19. Gas furnace output data and predictions.

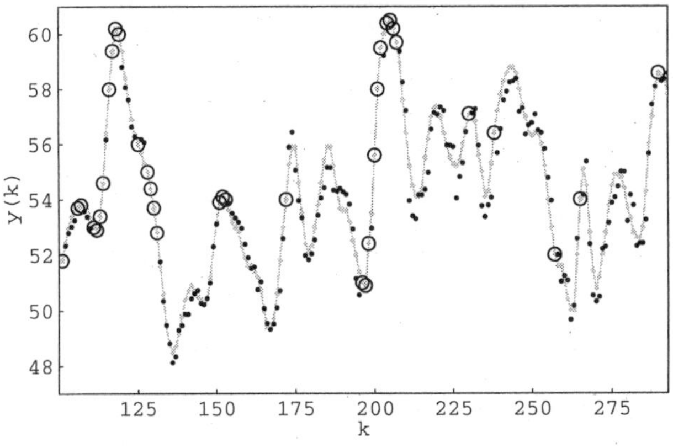

Fig. 20. Gas furnace output data and predictions from an updated model.

networks. The main selling point of the random set model is its ability to quantify the uncertainty of its prediction by means of a possibility distribution. This distribution does not reflect on average the likelihood for a particular value to occur, but instead tells the forecaster the confidence of the model based on the training data available. That is, for regions where no training data were available its confidence level will be 0. As we have seen, we can then resort to conventional regression techniques. Random set models may consequently be particularly useful for problems in which the forecasting is interactive in the sense of a person using, in addition to a quantitative model, his context-dependent expert knowledge. The two examples in the present text were chosen because they are well known and to illustrate the idea that numerical data can induce random subsets from which a model can be build.

References

1. R. Babuska. *Fuzzy Modelling for Control.* Kluwer, 1998.
2. H. Bauer. *Probability Theory and Elements of Measure Theory.* Academic Press, 1981.
3. R.R. Farnood, C.T.J. Dodson, and S.R. Loewen. Modelling flocculation. *Pulp and Paper Science*, 21, No 10:348–356, 1995. Elsevier.
4. I.R. Goodman. Fuzzy sets as equivalence classes of random sets. *Fuzzy Set and Possibility Theory*, pages 327–243, 1982. Pergamon Press.
5. I.R. Goodman et al. *Mathematics of Data Fusion.* Kluwer Academic Publishers, 1997.
6. K. Hestir et al. A random set formalism for evidential reasoning. In *Conditional Logic in Expert Systems*, pages 309–344. Elsevier Science Publishers, 1991.
7. R. Kruse et al. *Foundations of Fuzzy Systems.* John Wiley, 1994.
8. Q.D. Li. The random set and the cutting of random fuzzy sets. *Fuzzy Sets and Systems*, 86:223–234, 1997. Elsevier.
9. G. Matheron. Random sets and integral geometry. 1975.
10. I.S. Molchanov. *Statistics of the Boolean Model for Practitioners and Mathematicians.* Wiley Series in Probability and Statistics. John Wiley & Sons, 1997.
11. H.T. Nguyen. On random sets and belief functions. *Journal of Mathematical Analysis and Applications*, 65:531–542, 1978.
12. J. Nunez and O. Wolkenhauer. Random-sets clustering based on set-similarities. *Control Systems Centre Report*, 878, 1999. http://www.csc.umist.ac.uk/.
13. T. Peng, P. Wang, and A. Kandel. Knowledge acquisition by random sets. *International Journal of Intelligent Systems*, 11:113–147, 1997. John Wiley and Songs.
14. L. Sanchez. A random sets-based method for identifying fuzzy models. *Fuzzy Sets and Systems*, pages 343–354, 1998. Elsevier Science Publishers.
15. D. Stoyan et al. *Stochastic Geometry and Its Applications.* John Wiley & Sons, 1987.
16. O. Wolkenhauer and M. Garcia-Sanz. A random sets statistical approach to system identification. In *AIDA'99*, pages 388–392, 1998.

Rough Sets and Boolean Reasoning

Andrzej Skowron

Institute of Mathematics
Warsaw University
Banacha 2, 02–095, Warsaw,
POLAND
email: skowron@mimuw.edu.pl

Abstract. In recent years we witness a rapid growth of interest in rough set theory and its applications, worldwide. The theory has been followed by the development of several software systems that implement rough set operations, in particular for solving knowledge discovery and data mining tasks. Rough sets are applied in domains, such as, for instance, medicine, finance, telecommunication, vibration analysis, conflict resolution, intelligent agents, pattern recognition, control theory, signal analysis, process industry, marketing, etc.

We introduce basic notions and discuss methodologies for analyzing data and surveys some applications. In particular we present applications of rough set methods for feature selection, feature extraction, discovery of patterns and their applications for decomposition of large data tables as well as the relationship of rough sets with association rules. Boolean reasoning is crucial for all the discussed methods.

We also present an overview of some extensions of the classical rough set approach. Among them is rough mereology developed as a tool for synthesis of objects satisfying a given specification in a satisfactory degree. Applications of rough mereology in such areas like granular computing, spatial reasoning and data mining in distributed environment are outlined.

1 Basic rough set approach

We start by presenting the basic notions of classical rough set approach [41] introduced to deal with imprecise or vague concepts.

Information systems

A data set can be represented by a table where each row represents, for instance, an object, a case, or an event. Every column represents an attribute, or an observation, or a property that can be measured for each object; it can also be supplied by a human expert or user. This table is called an *information system*. More formally, it is a pair $\mathcal{A} = (U, A)$ where U is a non-empty finite set of *objects* called the *universe* and A is a non-empty finite set of *attributes* such that $a : U \to V_a$ for every $a \in A$. The set V_a is called the *value set* of a. By $Inf_B(x) = \{(a, a(x)) : a \in B\}$ we denote the *information signature of x with respect to B*, where $B \subseteq A$ and $x \in U$.

Decision systems

In many cases the target of the classification, that is, the family of concepts to be approximated is represented by an additional attribute called decision. Information systems of this kind are called *decision systems.* A decision system is any system of the form $\mathcal{A} = (U, A, d)$, where $d \notin A$ is the *decision attribute* and A is a set of *conditional attributes* or simply *conditions.*

Let $\mathcal{A} = (U, A, d)$ be given and let $V_d = \{v_1, \ldots, v_{r(d)}\}$. Decision d determines a partition $\{X_1, \ldots, X_{r(d)}\}$ of the universe U, where $X_k = \{x \in U : d(x) = v_k\}$ for $1 \leq k \leq r(d)$. The set X_i is called the *i-th decision class of* $\mathcal{A}$. By $X_d(u)$ we denote the decision class $\{x \in U : d(x) = d(u)\}$, for any $u \in U$.

One can generalize the above definition to a case of decision systems of the form $\mathcal{A} = (U, A, D)$ where the sets $D = \{d_1, ...d_k\}$ of decision attributes and A are assumed to be disjoint. Formally this system can be treated as the decision system $\mathcal{A} = (U, A, d_D)$ where $d_D(x) = (d_1(x), ..., d_k(x))$ for $x \in U$.

The decision tables can be identified with training samples known in Machine Learning and used to induce concept approximations in the process known as supervised learning [28].

Rough set approach allows to precisely define the notion of concept approximation. It is based [41] on the indiscernibility relation between objects defining a partition (or covering) of the universe U of objects. The indiscernibility of objects follows from the fact that they are perceived by means of values of available attributes. Hence some objects having the same (or similar) values of attributes are indiscernible.

Indiscernibility relation

Let $\mathcal{A} = (U, A)$ be an information system, then with any $B \subseteq A$ there is associated an equivalence relation $IND_{\mathcal{A}}(B)$:

$$IND_{\mathcal{A}}(B) = \{(x, x') \in U^2 \; : \; \forall a \in B \; a(x) = a(x')\}$$

$IND_{\mathcal{A}}(B)$ (or, $IND(B)$, for short) is called the *B-indiscernibility relation*, its classes are denoted by $[x]_B$. By X/B we denote the partition of U defined by the indiscernibility relation $IND(B)$.

Now we will discuss what sets of objects can be expressed (defined) by formulas constructed by means of attributes and their values. The simplest formulas, called *descriptors*, are of the form $a = v$ where $a \in A$ and $v \in V_a$. One can consider *generalized descriptors* of the form $a \in S$ where $S \subseteq V_a$. The descriptors can be combined into more complex formulas using propositional connectives. The meaning $\|\varphi\|_{\mathcal{A}}$ in $\mathcal{A}$ of formula φ is defined inductively by

1. if φ is of the form $a = v$ then $\|\varphi\|_{\mathcal{A}} = \{x \in U \; : \; a(x) = v\}$;
2. $\|\varphi \wedge \varphi'\|_{\mathcal{A}} = \|\varphi\|_{\mathcal{A}} \cap \|\varphi'\|_{\mathcal{A}}$; $\|\varphi \vee \varphi'\|_{\mathcal{A}} = \|\varphi\|_{\mathcal{A}} \cup \|\varphi'\|_{\mathcal{A}}$; $\|\neg\varphi\|_{\mathcal{A}} = U - \|\varphi\|_{\mathcal{A}}$.

The above definition can be easily extended to generalized descriptors.

Any set of objects $X \subseteq U$ definable in $\mathcal{A}$ by some formula φ (i.e., X=$\|\varphi\|_{\mathcal{A}}$) is referred to as a *crisp* (exact) set – otherwise the set is *rough* (*inexact, vague*). Vague concepts may be only approximated by crisp concepts; these approximations are defined now [41].

Lower and upper approximation of sets, boundary regions

Let $\mathcal{A} = (U, A)$ be an information system and let $B \subseteq A$ and $X \subseteq U$. We can approximate X using only the information contained in B by constructing the B-*lower* and B-*upper approximations of* X, denoted $\underline{B}X$ and $\overline{B}X$ respectively, where $\underline{B}X = \{x : [x]_B \subseteq X\}$ and $\overline{B}X = \{x : [x]_B \cap X \neq \emptyset\}$.

The lower approximation corresponds to certain rules while the upper approximation to possible rules (rules with confidence greater than 0) for X. The B-lower approximation of X is the set of all objects which can be with certainty classified to X using attributes from B. The set $U - \overline{B}X$ is called the B-*outside region of* X and consists of those objects which can be with certainty classified as not belonging to X using attributes from B. The set $BN_B(X) = \overline{B}X - \underline{B}X$ is called the B-*boundary region of* X thus consisting of those objects that on the basis of the attributes from B cannot be unambiguously classified into X. A set is said to be *rough* (respectively *crisp*) if the boundary region is non-empty (respectively empty). Consequently each rough set has boundary-line cases, i.e., objects which cannot be with certainty classified neither as members of the set nor of its complement. Obviously crisp sets have no boundary-line elements at all. That means that boundary-line cases cannot be properly classified by employing the available knowledge. The size of the boundary region can be used as a measure of the quality of set approximation (in U).

It can be easily seen that the lower and upper approximations of a set are, respectively, the interior and the closure of this set in the topology generated by the indiscernibility relation.

One can consider weaker indiscernibility relations defined by tolerance relations defining coverings of the universe of objects by tolerance (similarity) classes. An extension of rough set approach based on tolerance relations has been used for pattern extraction and concept approximation (see, e.g., [64], [69], [35], [32]).

Quality measures of concept approximation and measures of inclusion and closeness of concepts

We now present some examples of measures of quality approximation as well as of inclusion and closeness (approximate equivalence). These notions are instrumental in evaluating the strength of rules and closeness of concepts as well as being applicable in determining plausible reasoning schemes [45], [53]. Important role is also played by entropy measures (see e.g., [11]).

Let us consider first an example of a quality measure of approximations.

Accuracy of approximation. A rough set X can be characterized numerically by the following coefficient

$$\alpha_B(X) = \frac{|\underline{B}(X)|}{|\overline{B}(X)|},$$

called the *accuracy of approximation*, where $|X|$ denotes the cardinality of $X \neq \emptyset$ and B is a set of attributes. Obviously $0 \leq \alpha_B(X) \leq 1$. If $\alpha_B(X) = 1$, X is *crisp* with respect to B (X is *exact* with respect to B), and otherwise, if $\alpha_B(X) < 1$, X is *rough* with respect to B (X is *vague* with respect to B).

Rough membership function. In classical set theory either an element belongs to a set or it does not. The corresponding membership function is the characteristic function of the set, i.e., the function takes values 1 and 0, respectively. In the case of rough sets the notion of membership is different. The *rough membership function* quantifies the degree of relative overlap between the set X and the equivalence class to which x belongs. It is defined as follows:

$$\mu_X^B(x) : U \to [0,1] \text{ and } \mu_X^B(x) = \frac{|[x]_B \cap X|}{|[x]_B|}.$$

The rough membership function can be interpreted as a frequency–based estimate of $\Pr(y \in X \mid u)$, the conditional probability that object y belongs to set X, given the information signature $u = Inf_B(x)$ of object x with respect to attributes B. The value $\mu_X^B(x)$ measures degree of inclusion of $\{y \in U : Inf_B(x) = Inf_B(y)\}$ in X.

Positive region and its measure. If $X_1, \ldots, X_{r(d)}$ are decision classes of $\mathcal{A}$, then the set $\underline{B}X_1 \cup \ldots \cup \underline{B}X_{r(d)}$ is called the *B–positive region of* $\mathcal{A}$ and is denoted by $POS_B(d)$. The number $|POS_B(d)|/|U|$ measures a degree of inclusion of the partition defined by attributes from B into the partition defined by the decision.

Dependencies in a degree. Another important issue in data analysis is discovering dependencies among attributes. Intuitively, a set of attributes D depends totally on a set of attributes C, denoted $C \Rightarrow D$, if all values of attributes from D are uniquely determined by values of attributes from C. In other words, D depends totally on C, if there exists a functional dependency between values of D and C. Dependency can be formally defined as follows.

Let D and C be subsets of A. We will say that D *depends on* C in a *degree* k $(0 \leq k \leq 1)$, denoted $C \Rightarrow_k D$, if

$$k = \gamma(C, D) = \frac{|POS_C(D)|}{|U|},$$

where $POS_C(D) = POS_C(d_D)$.
Obviously

$$\gamma(C, D) = \sum_{X \in U/D} \frac{|\underline{C}(X)|}{|U|}.$$

If $k = 1$ we say that D *depends totally* on C, and if $k < 1$, we say that D *depends partially* (to a *degree* k) on C. $\gamma(C, D)$ describes the closeness of the partition U/D and its approximation with respect to conditions from C.

The coefficient k expresses the ratio of all elements of the universe which can be properly classified to blocks of the partition U/D by employing attributes C. It will be called the *degree of the dependency.*

Inclusion and closeness in a degree. Instead of classical exact set inclusion inclusion in a degree is often used in the process of deriving knowledge from data. Well known measure of inclusion of two non-empty sets $X, Y \subseteq U$ is described by $|X \cap Y|/|X|$ [2], [45]; their closeness can be defined by

$$min\left(|X \cap Y|/|X|, |X \cap Y|/|Y|\right).$$

2 Searching for knowledge

In this section we discuss problem of concept approximations. We point out that it is also the main goal of strategies searching for knowledge. Next we present selected methods based on rough sets and Boolean reasoning for concept approximation.

2.1 Concept approximation

Searching for concept approximations is a basic task in pattern recognition or machine learning. It is also crucial to knowledge discovery [13], in particular to scientific discovery [73], [23]. For example, scientific discovery [73] is using, as a main source of power, relatively general knowledge, including knowledge to search combinatorial spaces. Hence, it is important to discover efficient searching strategies. This includes the processes of inducing the relevant features and functions over which these strategies are constructed as well as the structure of searching strategy induced from such constructs. The goal of knowledge discovery [60], [23] is to find knowledge that is novel, plausible and understandable. Certainly, these soft concepts should be induced up to a sufficient degree, i.e., their approximations should be induced to specify the main constraints in searching for knowledge. In this sense the concept approximation is the basic step not only for machine learning or for pattern extraction but also for knowledge discovery and scientific discovery. Certainly, in the latter cases the inducing processes of concept approximations are much more complex and searching for such approximations creates a challenge for researchers.

Qualitative process representation, qualitative reasoning, spatial reasoning, perception and measurement instruments, collaboration and communication, embodied agents are only some topics of research directions mentioned in [60] as important for scientific reasoning and discovery. The mentioned above topics are very much in the scope of computing with words [74], [75] and granular computing (see e.g. [53], [65]). Rough set extension called rough mereology (see e.g. [45], [51], [48]) has been proposed as a tool for approximate reasoning to deal with such problems [65], [54]. Schemes of reasoning in rough mereology approximating soft patterns seem to be crucial for making progress in knowledge discovery. In particular this approach has been used to build a calculus on information granules [53], [65] as a foundation for computing with words. Among the discussed issues related to knowledge discovery using this approach there are generalized soft association rules, synthesis of interfaces between sources exchanging concepts and using different languages, problems in spatial reasoning.

Let us now turn back to discussion on inducing concept approximation by using rough set approach. First we recall a generalized approximation space definition introduced in [64]. This definition helps to explain a general approach offered by rough sets for concept approximations.

A *parameterized approximation space* is a system

$$AS_{\#,\$} = (U, I_{\#}, \nu_{\$})$$

where

- U is a non-empty set of objects,
- $I_{\#} : U \rightarrow P(U)$ is an *uncertainty function* and $P(U)$ denotes the powerset of U,
- $\nu_{\$} : P(U) \times P(U) \rightarrow [0,1]$ is a *rough inclusion function*,
- $\#$ and $\$$ are sets of parameters.

The uncertainty function defines for every object x a set of objects, called neighborhood of x, consisting of objects indistinguishable with x or similar to x. The parameters (from $\#$) of the uncertainty function are used to search for relevant neighborhoods with respect to the task to be solved, e.g. concept description.

A constructive definition of uncertainty function can be based on the assumption that some metrics (distances) are given on attribute values. For example, if for some attribute $a \in A$ a metric $\delta_a : V_a \times V_a \longrightarrow [0, \infty)$ is given, where V_a is the set of all values of attribute a then one can define the following uncertainty function:

$$y \in I_a^{f_a}(x) \text{ if and only if } \delta_a(a(x), a(y)) \leq f_a(a(x), a(y)),$$

where $f_a : V_a \times V_a \rightarrow [0, \infty)$ is a given threshold function.

A set $X \subseteq U$ is *definable in* $AS_{\#,\$}$ if it is a union of some values of the uncertainty function.

The rough inclusion function defines a degree of inclusion between two subsets of U [64], [48]. The parameters (from \$) of the rough inclusion function are used to search for relevant inclusion degrees with respect to the task to be solved, e.g. concept description.

For a parameterized approximation space $AS_{\#,\$} = (U, I_{\#}, \nu_{\$})$ and any subset $X \subseteq U$ the lower and the upper approximations are defined by

$LOW\left(AS_{\#,\$}, X\right) = \{x \in U : \nu_{\$}\left(I_{\#}\left(x\right), X\right) = 1\}$,

$UPP\left(AS_{\#,\$}, X\right) = \{x \in U : \nu_{\$}\left(I_{\#}\left(x\right), X\right) > 0\}$, respectively.

Any generalized approximation space consists of a family of approximation spaces creating the search space for data models. Any approximation space in this family is distinguished by some parameters. Searching strategies for optimal (sub-optimal) parameters are basic rough set tools in searching for data models and knowledge.

There are two main types of parameters. The first ones are used to define object sets, called neighborhoods, the second are measuring the inclusion or closeness of neighborhoods.

The basic assumption of the classical rough set approach, shared with other approaches like machine learning, pattern recognition or statistics, is that objects are perceived by means of some features (e.g. formulas being the results of measurement of the form *attribute=value* called descriptors). Hence, some objects can be indiscernible (indistinguishable) or similar to each other. The sets of indiscernible or similar objects expressible by some formulas are called neighborhoods. In the simplest case the family of all neighborhoods create a partition of the universe. In more general case it defines a covering. Formulas defining the neighborhoods are basic building blocks from which the approximate descriptions of other sets (decision classes or concepts) are induced. Usually, like in machine learning, the specification of concepts is incomplete, e.g., given by examples and counterexamples. Having incomplete specification of concepts, one can induce only approximate description of concepts by means of formulas defining the neighborhoods. Hence it follows that it will be useful to have parameterized formulas (e.g. in the simplest case $a > p \wedge b < q$ where a, b are attributes and p, q are parameters) so that by tuning their parameters one can select formulas being relevant for inducing concept approximation. A formula is relevant for concept description if it defines a large neighborhood still included to a sufficient degree in approximated concept. In the simplest case the formulas defining neighborhoods are conjunctions of descriptors. Parameters to be tuned can be of different sort like the number of conjunction connectives in the formula or the interval boundaries in case of discretization of real value attributes. In more general case, these formulas can express the results of measurement or perception of observed objects and represent complex information granules. Among such granules can be decision algorithms labeled by feature value vectors (describing an actual situation in which algorithm should be performed), clusters of such granules defined by their similarity or hierarchical structures of such granules (see e.g. [65]). These complex granules become more and more important for qualitative reasoning, in particular for spatial reasoning [54].

Assuming that a partition (covering) of objects has been fixed, the set approximations are induced by tuning of parameters specifying the degree of set inclusion.

In this way concept approximations are induced from data using rough set approach.

2.2 Discernibility and Boolean reasoning

We have pointed out that rough set approach has been introduced by Z. Pawlak [41] to deal with vague or imprecise concepts. More generally it is an approach for deriving knowledge from data and for reasoning about knowledge derived from data. Searching for knowledge is usually guided by some constraints [23]. A wide class of such constraints can be expressed using rough set setting or its generalizations (like rough mereology [45], or granular computing [53]). Knowledge derived from data by rough set approach consists of different constructs. Among them are basic for rough set approach constructs, called reducts, different kinds of rules (like decision rules or association rules) dependencies, patterns (templates) or classifiers. The reducts are of special importance because all other constructs can be derived from different kinds of reducts using rough set approach. Searching strategies for reducts are based on Boolean (propositional) reasoning [4] because constraints (e.g. related to discernibility of objects) are expressible by propositional formulas. Moreover, using Boolean reasoning data models with the minimum description length [56], [28] can be induced because they correspond to some constructs of Boolean functions called prime implicants (or their approximations). Searching for knowledge can be performed in the language close to data or in a language with more abstract concepts what is closely related to problems of feature selection and feature extraction in Machine Learning or Pattern Recognition [28]. Let us also mention that data models derived from data by using rough set approach are controlled using statistical test procedures (for more details see, e.g., [11], [10]).

In the paper we present illustrative examples showing how the above outlined general scheme is used for deriving knowledge.

Now, it will be important to make some remarks on Boolean reasoning because the most methods discussed later are based on generation of reducts using Boolean reasoning.

Boolean reasoning

The combination of rough set approach with Boolean Reasoning [4] has created a powerful methodology allowing to formulate and efficiently solve searching problems for different kinds of reducts and their approximations.

The idea of Boolean reasoning is based on the construction for a given problem P of a corresponding Boolean function f_P with the following property: the solutions for the problem P can be recovered from prime implicants of f_P. An implicant of a Boolean function f is any conjunction of literals (variables or their negations) such that if the values of these literals are true under an arbitrary valuation v of variables then the value of the function f under v is also true. A prime implicant is a minimal implicant.

Searching strategies for data models under a given partition of objects are based, using rough set approach, on discernibility and Boolean reasoning (see e.g., [35], [32],[62], [69], [70], [49], [50]). This process covers also tuning of parameters like thresholds used to extract relevant partition (or covering), to measure the degree of inclusion (or closeness) of sets, or the parameters measuring the quality of approximation.

It is necessary to deal with Boolean functions of large size to solve real-life problems. However, a successful methodology based on the discernibility of objects and Boolean reasoning has been developed for computing of many important for applications constructs like reducts and their approximations, decision rules, association rules, discretization of real value attributes, symbolic value grouping, searching for new features defined by oblique hyperplanes or higher order surfaces, pattern extraction from data as well as conflict resolution or negotiation. Reducts are also basic tools for extracting from data functional dependencies or functional dependencies in a degree (for references see the papers and bibliography in [62], [37], [49], [50]).

Most of the problems related to generation of the above mentioned constructs are of high computational complexity (i.e., they are NP-complete or NP-hard). This is also showing that most of the problems related to, e.g., feature selection, pattern extraction from data have intrinsic high computational complexity. However, using developed methodology based on discernibility and Boolean reasoning it was possible to discover efficient heuristics returning suboptimal solutions of the problems.

The reported results of experiments on many data sets are very promising. They show very good quality of solutions (expressed by the classification quality of unseen objects and time necessary for solution construction) generated by the heuristics in comparison with other methods reported in literature. Moreover, for large data sets the decomposition methods based on patterns called templates have been developed (see e.g., [35], [32]) as well as a method to deal with large relational databases (see e.g., [31]). The first one is based on decomposition of large data into regular sub-domains which are of size feasible for developed methods. We will discuss this method later. The second, (see e.g., [31]) has shown that Boolean reasoning methodology can be extended to large relational data bases. The main idea is based on observation that relevant Boolean variables for very large formula (corresponding to analyzed relational data base) can be discovered by analyzing some statistical information. This statistical information can be efficiently extracted from large data bases.

Another interesting statistical approach is based on different sampling strategies. Samples are analyzed using the developed strategies and stable constructs for sufficiently large number of samples are considered as relevant for the whole table. This approach has been successfully used for generating different kinds of so called dynamic reducts (see e.g., [3]). It has been used for example for generation of so called dynamic decision rules. Experiments on different data sets have proved that these methods are promising for large data sets.

Our approach is strongly related to propositional reasoning [58] and further progress in propositional reasoning will bring further progress in developing of the discussed methods. It is important to note that the methodology allows to construct heuristics having a very important *approximation property* which can be formulated as follows: expressions (i.e., implicants) generated by heuristics *close* to prime implicants define approximate solutions for the problem [58].

In the sequel we will discuss in more details different kinds of reducts and their applications for deriving different forms of knowledge from data.

2.3 Reducts in information systems and decision systems

We start from reducts of information systems. Given an $\mathcal{A} = (U, A)$, a *reduct* is a minimal set of attributes $B \subseteq A$ such that $IND_{\mathcal{A}}(B) = IND_{\mathcal{A}}(A)$. In other words, a reduct is a minimal set of attributes from A that preserves the original classification defined by the set A of attributes. Finding a minimal reduct is NP-hard [63]; one can also show that for any m there exists an information system with m attributes having an exponential number of reducts. There exist fortunately good heuristics that compute sufficiently many reducts in an acceptable time.

Let $\mathcal{A}$ be an information system with n objects. The *discernibility matrix* of $\mathcal{A}$ is a symmetric $n \times n$ matrix with entries c_{ij} as given below. Each entry consists of the set of attributes upon which objects x_i and x_j differ.

$$c_{ij} = \{a \in A \mid a(x_i) \neq a(x_j)\} \text{ for } i, j = 1, ..., n.$$

A *discernibility function* $f_{\mathcal{A}}$ for an information system $\mathcal{A}$ is a Boolean function of m Boolean variables $a_1^*, ..., a_m^*$ (corresponding to the attributes $a_1, ..., a_m$) defined by

$$f_{\mathcal{A}}(a_1^*, ..., a_m^*) = \bigwedge \left\{ \bigvee c_{ij}^* \mid 1 \leq j \leq i \leq n,\ c_{ij} \neq \emptyset \right\}$$

where $c_{ij}^* = \{a^* \mid a \in c_{ij}\}$. In the sequel we will write a_i instead of a_i^*.

The discernibility function $f_{\mathcal{A}}$ describes constraints which should be preserved if one would like to preserve discernibility between all pairs of discernible objects from $\mathcal{A}$. It requires to keep at least one attribute from each non-empty entry of the discernibility matrix, i.e., corresponding to any pair of discernible objects. One can show [63] that the sets of all minimal sets of attributes preserving discernibility between objects, i.e., reducts correspond to prime implicants of the discernibility function $f_{\mathcal{A}}$.

The intersection of all reducts is the so-called *core*.

In general, the decision is not constant on the indiscernibility classes. Let $\mathcal{A} = (U, A, d)$ be a decision system. The *generalized decision in* $\mathcal{A}$ is the function $\partial_A : U \longrightarrow \mathcal{P}(V_d)$ defined by

$$\partial_A(x) = \{i \mid \exists x' \in U \; x' \, IND(A) \, x \text{ and } d(x') = i\}.$$

A decision system $\mathcal{A}$ is called *consistent (deterministic)*, if $|\partial_A(x)| = 1$ for any $x \in U$, otherwise $\mathcal{A}$ is *inconsistent (non-deterministic)*. Any set consisting of all objects with the same generalized decision value is called the *generalized decision class.*

It is easy to see that a decision system $\mathcal{A}$ is consistent if, and only if, $POS_A(d) = U$. Moreover, if $\partial_B = \partial_{B'}$, then $POS_B(d) = POS_{B'}(d)$ for any pair of non-empty sets $B, B' \subseteq A$. Hence the definition of a decision-relative reduct: a subset $B \subseteq A$ is a *relative reduct* if it is a minimal set such that $POS_A(d) = POS_B(d)$. Decision-relative reducts may be found from a discernibility matrix: $M^d(\mathcal{A}) = (c^d_{ij})$ assuming $c^d_{ij} = c_{ij} - \{d\}$ if $(|\partial_A(x_i)| = 1$ or $|\partial_A(x_j)| = 1)$ and $\partial_A(x_i) \neq \partial_A(x_j)$, $c^d_{ij} = \emptyset$, otherwise. Matrix $M^d(\mathcal{A})$ is called *the decision-relative discernibility matrix of* $\mathcal{A}$. Construction of *the decision-relative discernibility function* from this matrix follows the construction of the discernibility function from the discernibility matrix. one can show [63] that the set of *prime implicants* of $f^d_M(\mathcal{A})$ defines the set of all *decision-relative reducts* of $\mathcal{A}$.

In some applications, instead of reducts we prefer to use their approximations called α-reducts, where $\alpha \in [0, 1]$ is a real parameter. For a given information system $\mathcal{A} = (U, A)$ the set of attributes $B \subseteq A$ is called α-reduct if B has nonempty intersection with at least $\alpha \cdot 100\%$ of nonempty sets $c_{i,j}$ of the discernibility matrix of $\mathcal{A}$.

2.4 Reducts and Boolean reasoning: Examples of applications

We will present examples showing how the rough set methods in combination with Boolean reasoning can be used for solving several KDD problems. A crucial for our approach are rough set constructs called reducts. They are (prime) implicants of suitably chosen Boolean functions expressing discernibility conditions which should be preserved during reduction.

Feature selection

Selection of relevant features is an important problem and has been extensively studied in Machine Learning and Pattern Recognition (see e.g., [28]). It is also a very active research area in the rough set community.

One of the first ideas [41] was to consider the *core* of the reduct set of the information system $\mathcal{A}$ as the source of relevant features. One can observe that relevant feature sets (in a sense used by the machine learning community) can be interpreted in most cases as the decision-relative reducts of decision systems obtained by adding appropriately constructed decisions to a given information system.

Another approach is related to dynamic reducts (for references see e.g., [49]). The attributes are considered relevant if they belong to dynamic reducts with a sufficiently high stability coefficient, i.e., they appear with sufficiently high frequency in random samples of a given information system. Several experiments (see [49]) show that the set of decision rules based on such attributes is much smaller than the set of all decision rules. At the same time the quality of classification of new objects increases or does not change if one only considers rules constructed over such relevant features.

The idea of attribute reduction can be generalized by introducing a concept of *significance of attributes* which enables to evaluate attributes not only in the two-valued scale *dispensable* – *indispensable* but also in the multi-value case by assigning to an attribute a real number from the interval [0,1] that expresses the importance of an attribute in the information table.

Significance of an attribute can be evaluated by measuring the effect of removing the attribute from an information table.

Let C and D be sets of condition and decision attributes, respectively, and let $a \in C$ be a condition attribute. It was shown previously that the number $\gamma(C, D)$ expresses the degree of dependency between attributes C and D, or the accuracy of the approximation of U/D by C. It may be now checked how the coefficient $\gamma(C, D)$ changes when attribute a is removed. In other words, what is the difference between $\gamma(C, D)$ and $\gamma((C - \{a\}, D)$. The difference is normalized and the significance of attribute a is defined by

$$\sigma_{(C,D)}(a) = \frac{(\gamma(C,D) - \gamma(C - \{a\}, D))}{\gamma(C,D)} =$$

$$= 1 - \frac{\gamma(C - \{a\}, D)}{\gamma(C,D)}.$$

Coefficient $\sigma_{C,D}(a)$ can be understood as a classification error which occurs when attribute a is dropped. The significance coefficient can be extended to sets of attributes as follows:

$$\sigma_{(C,D)}(B) = \frac{(\gamma(C,D) - \gamma(C - B, D))}{\gamma(C,D)} =$$

$$= 1 - \frac{\gamma(C - B, D)}{\gamma(C,D)}.$$

Another possibility is to consider as relevant the features that come from approximate reducts of sufficiently high quality.

Any subset B of C can be treated as an *approximate reduct* of C and the number

$$\varepsilon_{(C,D)}(B) = \frac{(\gamma(C,D) - \gamma(B,D))}{\gamma(C,D)} = 1 - \frac{\gamma(B,D)}{\gamma(C,D)},$$

is called an *error of reduct approximation.* It expresses how exactly the set of attributes B approximates the set of condition attributes C with respect to determining D.

Several other methods of reduct approximation based on measures different from positive region have been developed. All experiments confirm the hypothesis that by tuning the level of approximation the classification quality of new objects may be increased in most cases. It is important to note that it is once again possible to use Boolean reasoning to compute the different types of reducts and to extract from them relevant approximations.

Feature extraction

Non-categorical attributes must be discretized in a pre-processing step. The discretization step determines how coarsely we want to view the world. Discretization is a step that is not specific to the rough set approach. A majority of rule or tree induction algorithms require it in order to perform well. The search for appropriate cut-off points can be reduced to finding some minimal Boolean expressions called prime implicants.

Discretization can be treated as a searching for more coarser partitions of the universe still relevant for inducing concept description of high quality. We will also show that this basic problem can be reduced to computing of basic constructs of rough sets, namely reducts of some systems. Hence it follows that we can estimate the computational complexity of the discretization problems. Moreover, heuristics for computing reducts and prime implicants can be used here. The general heuristics can be modified to more optimal ones using konwledge about the problem e.g. natural order of the set of reals, etc. The discretization is only an illustrative example of many other problems with the same property.

The rough set community have been committed to constructing efficient algorithms for (new) feature extraction. Rough set methods combined with Boolean reasoning [4] lead to several successful approaches to feature extraction. The most successful methods are:

- discretization techniques,
- methods of partitioning of nominal attribute value sets and
- combinations of the above methods.

Searching for new features expressed by multi-modal formulae can be mentioned here. Structural objects can be interpreted as models (so called Kripke models) of such formulas and the problem of searching for relevant features reduces to construction of multi-modal formulas expressing properties of the structural objects discerning objects or sets of objects [36].

For more details the reader is referred to the bibliography in [50].

The reported results show that discretization problems and symbolic value partition problems are of high computational complexity (i.e. NP-complete or NP-hard) which clearly justifies the importance of designing efficient heuristics. The idea of discretization is illustrated with a simple example.

Example 21 Let us consider a (consistent) decision system (see Tab. 1(a)) with two conditional attributes a and b and seven objects $u_1, ..., u_7$. The values of the attributes of these objects and the values of the decision d are presented in Tab. 1.

$\mathbf{A}$	a	b	d
u_1	0.8	2	1
u_2	1	0.5	0
u_3	1.3	3	0
u_4	1.4	1	1
u_5	1.4	2	0
u_6	1.6	3	1
u_7	1.3	1	1

(a)

$\Rightarrow$

$\mathbf{A^P}$	a^P	b^P	d
u_1	0	2	1
u_2	1	0	0
u_3	1	2	0
u_4	1	1	1
u_5	1	2	0
u_6	2	2	1
u_7	1	1	1

(b)

Table 1. The discretization process: (a) The original decision system $\mathcal{A}$. (b) The $\mathbf{P}$-discretization of $\mathcal{A}$, where $\mathbf{P} = \{(a, 0.9), (a, 1.5), (b, 0.75), (b, 1.5)\}$

The sets of possible values of a and b are defined by:

$$V_a = [0, 2)\, ; V_b = [0, 4)\, .$$

The sets of values of a and b for objects from U are respectively given by:

$$a(U) = \{0.8, 1, 1.3, 1.4, 1.6\} \text{ and}$$
$$b(U) = \{0.5, 1, 2, 3\}$$

□

A discretization process produces a partition of the value sets of the conditional attributes into intervals. The partition is done so that a consistent decision system is obtained from a given consistent decision system by a substitution of any object's original value in $\mathcal{A}$ by the (unique) name of the interval(s) in which it is contained. In this way the size of the value sets of the attributes may be reduced. If a given decision system is not consistent one can transform it to the consistent decision system by taking the generalized decision instead of the original one. Next it is possible to apply the above method. It will return cuts with the following property: regions bounded by them consist of objects with the same generalized decision. Certainly, one can consider also *soft (impure)* cuts and induce the relevant cuts on their basis (see the bibliography in [49]).

Example 22 The following intervals are obtained in our example system:
$[0.8, 1)$; $[1, 1.3)$; $[1.3, 1.4)$; $[1.4, 1.6)$ for a);
$[0.5, 1)$; $[1, 2)$; $[2, 3)$ for b).

The idea of cuts can be introduced now. Cuts are pairs (a, c) where $c \in V_a$. Our considerations are restricted to cuts defined by the middle points of the above intervals. In our example the following cuts are obtained:
$(a, 0.9)$; $(a, 1.15)$; $(a, 1.35)$; $(a, 1.5)$;
$(b, 0.75)$; $(b, 1.5)$; $(b, 2.5)$.

Any cut defines a new conditional attribute with binary values. For example, the attribute corresponding to the cut $(a, 1.2)$ is equal to 0 if $a(x) < 1.2$; otherwise it is equal to 1. □

Any set P of cuts defines a new conditional attribute a_P for any a. Given a partition of the value set of a by cuts from P put the unique names for the elements of these partition.

Example 23 Let

$$P = \{(a, 0.9), (a, 1.5), (b, 0.75), (b, 1.5)\}$$

be the set of cuts. These cuts glue together the values of a smaller then 0.9, all the values in interval $[0.9, 1.5)$ and all the values in interval $[1.5, 4)$. A similar construction can be repeated for b. The values of the new attributes a_P and b_P are shown in Tab. 1 (b). □

The next natural step is to construct a set of cuts with a minimal number of elements. This may be done using Boolean reasoning.

Let $\mathcal{A} = (U, A, d)$ be a decision system where $U = \{x_1, x_2, \ldots, x_n\}, A = \{a_1, \ldots, a_k\}$ and $d : U \longrightarrow \{1, \ldots, r\}$. We assume $V_a = [l_a, r_a) \subset \Re$ to be a real interval for any $a \in A$ and $\mathcal{A}$ to be a consistent decision system. Any pair (a, c) where $a \in A$ and $c \in \Re$ will be called a *cut on* V_a. Let $\mathbf{P}_a = \{[c_0^a, c_1^a), [c_1^a, c_2^a), \ldots, [c_{k_a}^a, c_{k_a+1}^a)\}$ be a partition of V_a (for $a \in A$) into subintervals for some integer k_a, where $l_a = c_0^a < c_1^a < c_2^a < \ldots < c_{k_a}^a < c_{k_a+1}^a = r_a$ and $V_a = [c_0^a, c_1^a) \cup [c_1^a, c_2^a) \cup \ldots \cup [c_{k_a}^a, c_{k_a+1}^a)$. It follows that any partition $\mathbf{P}_a$ is uniquely defined and is often identified with the set of cuts

$$\{(a, c_1^a), (a, c_2^a), \ldots, (a, c_{k_a}^a)\} \subset A \times \Re.$$

Given $\mathcal{A} = (U, A, d)$ any set of cuts $\mathbf{P} = \bigcup_{a \in A} \mathbf{P}_a$ defines a new decision system $\mathcal{A}^{\mathbf{P}} = (U, A^{\mathbf{P}}, d)$ called $\mathbf{P}$-*discretization of* $\mathcal{A}$, where $A^{\mathbf{P}} = \{a^{\mathbf{P}} : a \in A\}$ and $a^{\mathbf{P}}(x) = i \Leftrightarrow a(x) \in [c_i^a, c_{i+1}^a)$ for $x \in U$ and $i \in \{0, .., k_a\}$.

Two sets of cuts $\mathbf{P}'$ and $\mathbf{P}$ are equivalent, written $\mathbf{P}' \equiv_{\mathbf{A}} \mathbf{P}$, iff $\mathcal{A}^{\mathbf{P}} = \mathcal{A}^{\mathbf{P}'}$. The equivalence relation $\equiv_{\mathcal{A}}$ has a finite number of equivalence classes. Equivalent families of partitions will be not discerned in the sequel.

The set of cuts $\mathbf{P}$ is called $\mathcal{A}$-*consistent* if $\partial_A = \partial_{A^{\mathbf{P}}}$, where ∂_A and $\partial_{A^{\mathbf{P}}}$ are generalized decisions of $\mathcal{A}$ and $\mathcal{A}^{\mathbf{P}}$, respectively. An $\mathcal{A}$-consistent set of cuts $\mathbf{P}^{irr}$ is $\mathcal{A}$-*irreducible* if $\mathbf{P}$ is not $\mathcal{A}$-consistent for any $\mathbf{P} \subset \mathbf{P}^{irr}$. The $\mathcal{A}$-consistent set of cuts $\mathbf{P}^{opt}$ is $\mathcal{A}$-*optimal* if $card(\mathbf{P}^{opt}) \leq card(\mathbf{P})$ for any $\mathcal{A}$-consistent set of cuts $\mathbf{P}$.

It can be shown that the decision problem of checking if for a given decision system $\mathcal{A}$ and an integer k there exists an irreducible set of cuts $\mathbf{P}$ in $\mathcal{A}$ such that $card(\mathbf{P}) < k$ is NP-complete. The problem of searching for an optimal set of cuts $\mathbf{P}$ in a given decision system $\mathcal{A}$ is NP-hard.

Despite these complexity bounds it is possible to devise efficient heuristics that return semi-minimal sets of cuts. The simplest huristics is based on Johnson's strategy. The strategy is first to look for a cut discerning a maximal number of object pairs and then to eliminate all already discerned object pairs. This procedure is repeated until all object pairs to be discerned are discerned. It is interesting to note that this heuristics can be realized by computing the minimal relative reduct of the corresponding decision system. The *"MD heuristic"* is analogous to Johnson's approximation algorithm. It may be formulated as follows:

ALGORITHM: MD-heuristics (A semi-optimal family of partitions)

1. *Construct table* $\mathcal{A}^* = (U^*, A^*, d)$ *from* $\mathcal{A} = (U, A)$ *where* U^* *is the set of pairs* (x, y) *of objects to be discerned by d and* A^* *consists of attribute* c^* *for any cut c and* c^* *is defined by* $c^*(x, y) = 1$ *if and only if c discerns* x *and* y *(i.e.,* x, y *are in different half-spaces defined by c)*; *set* $\mathcal{B} = \mathcal{A}^*$;
2. *Choose a column from* $\mathcal{B}$ *with the maximal number of occurrences of 1's;*
3. *Delete from* $\mathcal{B}$ *the column chosen in Step 2 and all rows marked with 1 in this column;*
4. *If* $\mathcal{B}$ *is non-empty then go to Step 2 else Stop.*

This algorithm searches for a cut which discerns the largest number of pairs of objects (MD-heuristic). Then the cut c is moved from A^* to the resulting set of cuts $\mathbf{P}$; and all pairs of objects discerned by c are removed from U^*. The algorithm continues until U^* becomes empty.

Let n be the number of objects and let k be the number of attributes of decision system $\mathcal{A}$. The following inequalities hold: $card(A^*) \leq (n-1)k$ and $card(U^*) \leq \frac{n(n-1)}{2}$. It is easy to observe that for any cut $c \in A^*$ $O(n^2)$ steps are required in order to find the number of all pairs of objects discerned by c. A straightforward realization of this algorithm therefore requires $O(kn^2)$ of memory space and $O(kn^3)$ steps in order to determine one *cut*. This approach is clearly impractical. However, it is possible to observe that in the process of searching for the set of pairs of objects discerned by currently analyzed cut from an increasing sequence of cuts one can use information about such set of pairs of objects computed for the previously considered cut. The MD-heuristic using this observation [30] determines the best cut (for a given attribute) in $O(kn)$ steps using $O(kn)$ space only. This heuristic is reported to be very efficient with respect to the time necessary for decision rules generation as well as with respect to the quality of unseen object classification.

Let us observe that in the considered case of discretization the new features are of the form $a \in V$, where $V \subseteq V_a$ and V_a is the set of the values of attribute a.

We report some results of experiments on data sets using this heuristic. We would like to comment for example on the result of classification received by application of this heuristic to Shuttle data (Table 3). The result concerning classification quality is the same as the best result reported in [27] but the time is of order better than for the best result from [27]. In the table we present also the results of experiments with heuristic searching for features defined by oblique hyperplanes. This heuristic has been developed using genetic algorithm allowing to tune the position of hyperplane to the optimal one [30]. In this way one can implement propositional reasoning using some background knowledge about the problem.

In experiments we have chosen several data tables with real value attributes from the U.C. Irvine repository. For some tables, taking into account the small number of their objects, we have adopted the approach based on five-fold cross-validation ($CV-5$). The obtained results (Table 3) can be compared with those reported in [9,27] (Table 2). For predicting decisions on new cases we apply only decision rules generated either by the decision tree (using hyperplanes) or by rules generated in parallel with discretization.

Names	**Nr of class.**	**Train. table**	**Test. table**	**Best results**
Australian	2	690×14	CV5	85.65%
Glass	7	214×9	CV5	69.62%
Heart	2	270×13	CV5	82.59%
Iris	3	150×4	CV5	96.00%
Vehicle	4	846×19	CV5	69.86%
Diabetes	2	768×8	CV5	76.04%
SatImage	6	4436×36	2000	90.06%
Shuttle	6	43500×7	14500	99.99%

Table 2. Data tables stored in the UC Irvine Repository

For some tables the classification quality of our algorithm is better than that of the C4.5 or Naive -Bayes induction algorithms [55] even when used with different discretization methods [9,27,15].

Comparing this method with the other methods reported in [27], we can conclude that our algorithms have the shortest runtime and a good overall classification quality (in many cases our results were the best in comparison to many other methods reported in literature).

We would like to stress that inducing of the minimal number of the relevant cuts is equivalent to computing of the minimal reduct of decision system constructed from the discussed above system $\mathcal{A}^*$ [30]. This in turn, as we have

Data tables	Diagonal cuts		Hyperplanes	
	#cuts	quality	#cuts	quality
Australian	18	79.71%	16	82.46%
Glass	14±1	67.89%	12	70.06%
Heart	11±1	79.25%	11±1	80.37%
Iris	7±2	92.70%	6±2	96.7%
Vehicle	25	59.70%	20±2	64.42%
Diabetes	20	74.24%	19	76.08%
SatImage	47	81.73%	43	82.90%
Shuttle	15	99.99%	15	99.99%

Table 3. Results of experiments on Machine Learning data.

shown, is equivalent to the problem of computing of minimal prime implicants of Boolean functions. This is only illustration of a wide class of basic problems of Machine Learning, Pattern Recognition and KDD which can be reduced to problems of relevant reduct computation.

Our next illustrative example concerns symbolic (nominal, qualitative) attribute value grouping. We also present some experimental results of heuristics based on the developed methods in case of mixed nominal and numeric attributes.

In case of symbolic value attribute (i.e., without pre-assumed order on values of given attributes) the problem of searching for new features of the form $a \in V$ is, in a sense, from practical point of view more complicated than the for real value attributes. However, it is possible to develop efficient heuristics for this case using Boolean reasoning.

Let $\mathcal{A} = (U, A, d)$ be a decision table. Any function $P_a : V_a \to \{1, \ldots, m_a\}$ (where $m_a \leq |V_a|$) is called *a partition of* V_a. The *rank of* P_{a_i} is the value $rank(P_{a_i}) = |P_{a_i}(V_{a_i})|$. The family of partitions $\{P_a\}_{a \in B}$ is *consistent with* B ($B-$ *consistent*) iff the condition $[(u, u') \notin IND(B)$ and $d(u) \neq d(u')$ implies $\exists_{a \in B}[P_a(a(u)) \neq P_a(a(u'))]]$ holds for any $(u, u') \in U$. It means that if two objects u, u' are discerned by B and d, then they must be discerned by partition attributes defined by $\{P_a\}_{a \in B}$. We consider the following optimization problem

PARTITION PROBLEM: SYMBOLIC VALUE PARTITION PROBLEM:
Given a decision table $\mathcal{A} = (U, A, d)$ and a set of attributes $B \subseteq A$, search for the minimal $B-consistent$ family of partitions (i.e., such $B-consistent$ family $\{P_a\}_{a \in B}$ that $\sum_{a \in B} rank(P_a)$ is minimal).

To discern between pairs of objects we will use new binary features $a_v^{v'}$ (for $v \neq v'$) defined by $a_v^{v'}(x, y) = 1$ iff $a(x) = v \neq v' = a(y)$. One can apply the Johnson heuristic for the new matrix with these attributes to search for minimal set of new attributes that discerns all pairs of objects from different decision classes. After extracting of these sets, for each attribute a_i we construct a graph $\Gamma_a = \langle V_a, E_a \rangle$ where E_a is defined as the set of all new attributes (propositional variables) found for the attribute a. Any vertex coloring of Γ_a defines a partition of V_a. The colorability problem is solvable in polynomial time for $k = 2$, but remains NP-complete for all $k \geq 3$. But, similarly to discretization, one can apply some efficient heuristic searching for optimal partition.

Let us consider an example of decision table presented in Figure 1 and (a reduced form) of its discernibility matrix (Figure 1).

$\mathcal{A}$	a	b	d
u_1	a_1	b_1	0
u_2	a_1	b_2	0
u_3	a_2	b_3	0
u_4	a_3	b_1	0
u_5	a_1	b_4	1
u_6	a_2	b_2	1
u_7	a_2	b_1	1
u_8	a_4	b_2	1
u_9	a_3	b_4	1
u_{10}	a_2	b_5	1

$\Rightarrow$

$\mathcal{M}(\mathcal{A})$	u_1	u_2	u_3	u_4
u_5	$\mathbf{b}_{b_4}^{b_1}$	$\mathbf{b}_{b_4}^{b_2}$	$\mathbf{a}_{a_2}^{a_1}, \mathbf{b}_{b_4}^{b_3}$	$\mathbf{a}_{a_3}^{a_1}, \mathbf{b}_{b_4}^{b_1}$
u_6	$\mathbf{a}_{a_2}^{a_1}, \mathbf{b}_{b_2}^{b_1}$	$\mathbf{a}_{a_2}^{a_1}$	$\mathbf{b}_{b_3}^{b_2}$	$\mathbf{a}_{a_3}^{a_2}, \mathbf{b}_{b_2}^{b_1}$
u_7	$\mathbf{a}_{a_2}^{a_1}$	$\mathbf{a}_{a_2}^{a_1}, \mathbf{b}_{b_2}^{b_1}$	$\mathbf{b}_{b_3}^{b_1}$	$\mathbf{a}_{a_3}^{a_2}$
u_8	$\mathbf{a}_{a_4}^{a_1}, \mathbf{b}_{b_2}^{b_1}$	$\mathbf{a}_{a_4}^{a_1}$	$\mathbf{a}_{a_4}^{a_2}, \mathbf{b}_{b_3}^{b_2}$	$\mathbf{a}_{a_4}^{a_3}, \mathbf{b}_{b_2}^{b_1}$
u_9	$\mathbf{a}_{a_3}^{a_1}, \mathbf{b}_{b_4}^{b_1}$	$\mathbf{a}_{a_3}^{a_1}, \mathbf{b}_{b_4}^{b_2}$	$\mathbf{a}_{a_3}^{a_2}, \mathbf{b}_{b_4}^{b_3}$	$\mathbf{b}_{b_4}^{b_1}$
u_{10}	$\mathbf{a}_{a_2}^{a_1}, \mathbf{b}_{b_5}^{b_1}$	$\mathbf{a}_{a_2}^{a_1}, \mathbf{b}_{b_5}^{b_2}$	$\mathbf{b}_{b_5}^{b_3}$	$\mathbf{a}_{a_3}^{a_2}, \mathbf{b}_{b_5}^{b_1}$

Fig. 1. The decision table and the discernibility matrix

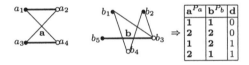

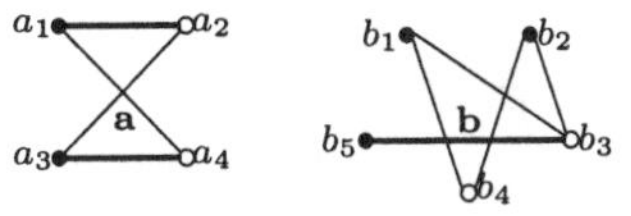

$\Rightarrow$

$\mathbf{a}^{P_a}$	$\mathbf{b}^{P_b}$	d
1	**1**	0
2	**2**	0
1	**2**	1
2	**1**	1

Fig. 2. Coloring of attribute value graphs and the reduced table.

From the Boolean function $f_{\mathcal{A}}$ with Boolean variables of the form $\mathbf{a}_{v_1}^{v_2}$ one can find the shortest prime implicant:

$$\mathbf{a}_{a_2}^{a_1} \wedge \mathbf{a}_{a_3}^{a_2} \wedge \mathbf{a}_{a_4}^{a_1} \wedge \mathbf{a}_{a_4}^{a_3} \wedge \mathbf{b}_{b_4}^{b_1} \wedge \mathbf{b}_{b_4}^{b_2} \wedge \mathbf{b}_{b_3}^{b_2} \wedge \mathbf{b}_{b_3}^{b_1} \wedge \mathbf{b}_{b_5}^{b_3}$$

which can be represented by graphs (see Figure 2).

We can color vertices of those graphs as it is shown in Figure 2. The colors are corresponding to the partitions:

$$P_{\mathbf{a}}(a_1) = P_{\mathbf{a}}(a_3) = 1;$$
$$P_{\mathbf{a}}(a_2) = P_{\mathbf{a}}(a_4) = 2$$
$$P_{\mathbf{b}}(b_1) = P_{\mathbf{b}}(b_2) = P_{\mathbf{b}}(b_5) = 1;$$
$$P_{\mathbf{b}}(b_3) = P_{\mathbf{b}}(b_4) = 2.$$

At the same time one can construct the new decision table (Figure 2).

One can extend the presented approach (see e.g., [33]) to the case when in a given decision system nominal as well as numeric attributes appear. The received heuristics are of very good quality. Experiments for classification methods (see [33]) have been carried over decision systems using two techniques called *"train-and-test"* and *"n-fold-cross-validation"*. In Table 4 some results of experiments obtained by testing the proposed methods: MD (using only discretization based on MD-heurisctic with Johnson approximation strategy [30], [62]) and MD-G (using discretization and symbolic value grouping [32], [62]) for classification quality on some data tables from the "UC Irvine repository" are shown. The results reported in [12] are summarized in columns labeled by S-ID3 and C4.5 in Table 4). Let us note that the heuristics MD and MD-G are also very efficient with respect to the time complexity.

Names of Tables	Classification accuracies			
	S-ID3	C4.5	MD	MD-G
Australian	78.26	85.36	83.69	84.49
Breast (L)	62.07	71.00	69.95	69.95
Diabetes	66.23	70.84	71.09	76.17
Glass	62.79	65.89	66.41	69.79
Heart	77.78	77.04	77.04	81.11
Iris	96.67	94.67	95.33	96.67
Lympho	73.33	77.01	71.93	82.02
Monk-1	81.25	75.70	100	93.05
Monk-2	69.91	65.00	99.07	99.07
Monk-3	90.28	97.20	93.51	94.00
Soybean	100	95.56	100	100
TicTacToe	84.38	84.02	97.7	97.70
Average	78.58	79.94	85.48	87.00

Table 4. Quality comparison of various decision tree methods. Abbreviations: MD: MD-heuristic; MD-G: MD-heuristic with symbolic value partition

In the case of real value attributes one can search for features in the feature set that contains the characteristic functions of half-spaces determined by hyperplanes or parts of spaces defined by more complex surfaces in the multi-dimensional spaces. Genetic algorithms have been applied in searching for semi-optimal hyperplanes [30]. The reported results are showing substantial increase in the quality of classification of unseen objects but at the price of increased time for searching for the semi-optimal hyperplane.

Decision rules

Reducts serve the purpose of inducing *minimal* decision rules. Any such rule contains the minimal number of descriptors in the conditional part so that their conjunction defines the largest subset of a generalized decision class (decision class, if the decision table is deterministic). Hence, information included in conditional part of any minimal rule is sufficient for prediction of the generalized decision value for all objects satisfying this part. The conditional parts of minimal rules define largest object sets relevant for generalized decision classes approximation. It turns out that the conditional parts of minimal rules can be computed (by using Boolean reasoning) as so called reducts relative to objects or local reducts (see e.g., [61], [3]). Once the reducts have been computed, the conditional parts of rules are easily constructed by laying the reducts over the original decision system and reading off the values. In the discussed case the generalized decision value is preserved during the reduction. One can consider stronger constraints which should be preserved. For example, in [67] the constraints are described by probability distributions corresponding to information signatures of objects. Again the same methodology can be used to compute the reducts corresponding to these constraints.

The main challenge in inducing rules from decision systems lies in determining which attributes should be included in the conditional part of the rule. Using the outlined above strategy first the minimal rules are computed. Their conditional parts describe largest object sets (definable by conjunctions of descriptors) with the same generalized decision value in a given decision system. Hence, they create the largest sets still relevant for defining the decision classes (or sets of decision classes when the decision system is inconsistent). Although such minimal decision rules can be computed, this approach can result in set of rules of not satisfactory classification quality. Such detailed rules will be overfit and they will poorly classify unseen cases. Shorter rules should rather be synthesized. Although they will not be perfect on the known cases there is a good chance that they will be of high quality when classifying new cases. They can be constructed by computing approximations of the above mentioned reducts. Approximations of reducts received by drooping some descriptors from the conditional parts of minimal rules define larger sets, not purely included in decision classes but included in a satisfactory degree. It means that these shorter descriptions can be more relevant for decision class (concept) approximation than the exact reducts. Hence, e.g., one can expect that when by dropping the descriptor from the conditional part we receive the description of the object set almost included in the approximated decision class than this descriptor is a good candidate for dropping.

Several other strategies have been implemented. Methods of boundary region thinning [79] are based, e.g., on the idea that sets of objects included in decision classes in satisfactory degree can be treated as parts of the lower approximations of decision classes. Hence the lower approximations of decision classes are enlarged and decision rules generated for them are usually stronger (e.g., they are supported by more examples). The degree of inclusion is tuned experimentally to achieve, e.g., high classification quality of new cases. One can also adopt an idea of dynamic reducts for decision rule generation.

For estimation of the quality of decision classes approximation global measures based on the positive region [61] or entropy [11] are used.

When a set of rules has been induced from a decision system containing a set of training examples, they can be used to classify new objects. However, to resolve conflict between different decision rules recognizing new objects one should develop strategies for resolving conflicts between them when they are voting for different decisions (see the bibliography in [49] and [50]). Recently [71], it has been shown that rough set methods can be used to learn from data the strategy for conflict resolving between decision rules when they are classifying new objects contrary to existing methods using some fixed strategies.

α-reducts and association rules

In this section we discuss a relationship between association rules [2] and approximations of reducts being basic constructs of rough sets [61], [62], [34].

We consider formulas called *templates* being conjunction of descriptors. The templates will be denoted by $\mathbf{T}$, $\mathbf{P}$, $\mathbf{Q}$ and descriptors by D with or without subscripts. By $support_{\mathcal{A}}(\mathbf{T})$ is denoted the cardinality of $\|\mathbf{T}\|_{\mathcal{A}}$ and by $confidence_{\mathcal{A}}(\mathbf{P} \rightarrow \mathbf{Q})$ is denoted the number $support_{\mathcal{A}}(\mathbf{P} \wedge \mathbf{Q})/support_{\mathcal{A}}(\mathbf{P})$.

The, mentioned above, reduct approximations are descriptions of the object sets matched by templates. They describe these sets in an approximate sense expressed by coefficients called support and confidence.

There are two main steps of many developed association rule generation methods for given information system $\mathcal{A}$ and parameters of support s and confidence c:

1. Extraction from data as many as possible templates $\mathbf{T} = D_1 \wedge D_2 ... \wedge D_k$ such that $support_{\mathcal{A}}(\mathbf{T}) \geq s$ and $support_{\mathcal{A}}(\mathbf{T} \wedge D) < s$ for any descriptor D different from descriptors of $\mathbf{T}$ (i.e., generation of maximal templates among those supported by more than s objects);
2. Searching for a partition $\mathbf{T} = \mathbf{P} \wedge \mathbf{Q}$ for any of generated template $\mathbf{T}$ satisfying the following conditions:
   (a) $support_{\mathcal{A}}(\mathbf{P}) < \frac{support_{\mathcal{A}}(\mathbf{T})}{c}$
   (b) $\mathbf{P}$ has the shortest length among templates satisfying the previous condition.

The second step can be solved using rough set methods and Boolean reasoning approach.

Let $\mathbf{T} = D_1 \wedge D_2 \wedge \ldots \wedge D_m$ be a template with $support_{\mathcal{A}}(\mathbf{T}) \geq s$. For a given confidence threshold $c \in (0;1)$ the decomposition $\mathbf{T} = \mathbf{P} \wedge \mathbf{Q}$ is called c-irreducible if $confidence_{\mathcal{A}}(\mathbf{P} \rightarrow \mathbf{Q}) \geq c$ and for any decomposition $\mathbf{T} = \mathbf{P}' \wedge \mathbf{Q}'$ such that $\mathbf{P}'$ is a sub-template of $\mathbf{P}$, we have

$$confidence_{\mathcal{A}}(\mathbf{P}' \rightarrow \mathbf{Q}') < c.$$

Now we are going to explain that problem of searching for c-irreducible association rules from the given template is equivalent to the problem of searching for local α-reducts (for some α) from a decision table. The last problem is a well known problem in rough set theory.

Let us define a new decision table $\mathcal{A}|_{\mathbf{T}} = (U, A|_{\mathbf{T}}, d)$ from the original information system $\mathcal{A}$ and the template $\mathbf{T}$ by

1. $A|_{\mathbf{T}} = \{a_{D_1}, a_{D_2}, ..., a_{D_m}\}$ is a set of attributes corresponding to the descriptors of $\mathbf{T}$ such that $a_{D_i}(u) = \begin{cases} 1 \text{ if the object } u \text{ satisfies } D_i, \\ 0 \text{ otherwise.} \end{cases}$
2. the decision attribute d determines if the object satisfies template $\mathbf{T}$, i.e., $d(u) = \begin{cases} 1 \text{ if the object } u \text{ satisfies } \mathbf{T}, \\ 0 \text{ otherwise.} \end{cases}$

The following facts [62], [34] describe the relationship between association rules and approximations of reducts.

For the given information table $\mathcal{A} = (U, A)$, the template $\mathbf{T}$, the set of descriptors $\mathbf{P}$. The implication $\left(\bigwedge_{D_i \in \mathbf{P}} D_i \longrightarrow \bigwedge_{D_j \notin \mathbf{P}} D_j\right)$ is

1. 100%-irreducible association rule from $\mathbf{T}$ if and only if $\mathbf{P}$ is a reduct in $\mathcal{A}|_{\mathbf{T}}$.
2. c-irreducible association rule from $\mathbf{T}$ if and only if $\mathbf{P}$ is an α-reduct of $\mathcal{A}|_{\mathbf{T}}$, where $\alpha = 1 - (\frac{1}{c} - 1)/(\frac{n}{s} - 1)$, n is the total number of objects from U and $s = support_{\mathcal{A}}(\mathbf{T})$.

One can show, that the problem of searching for the shortest α-reducts is NP-hard [34]. From the above facts it follows that extracting association rules from data is strongly related to extraction from the data reduct approximations [34] being basic constructs of rough sets.

The following example illustrates the main idea of our method. Let us consider the following information table $\mathcal{A}$ with 18 objects and 9 attributes.

Assume the template

$$\mathbf{T} = (a_1 = 0) \wedge (a_3 = 2) \wedge (a_4 = 1) \wedge (a_6 = 0) \wedge (a_8 = 1)$$

has been extracted from the information table $\mathcal{A}$. We have

$$support(\mathbf{T}) = 10 \text{ and } length(\mathbf{T}) = 5.$$

The new constructed decision table $\mathcal{A}|_{\mathbf{T}}$ is presented in Table 5. The discernibility function for $\mathcal{A}|_{\mathbf{T}}$ is of the following form:

$$\begin{aligned} f(D_1, D_2, D_3, D_4, D_5) = {} & (D_2 \vee D_4 \vee D_5) \wedge (D_1 \vee D_3 \vee D_4) \wedge (D_2 \vee D_3 \vee D_4) \\ & \wedge (D_1 \vee D_2 \vee D_3 \vee D_4) \wedge (D_1 \vee D_3 \vee D_5) \\ & \wedge (D_2 \vee D_3 \vee D_5) \wedge (D_3 \vee D_4 \vee D_5) \wedge (D_1 \vee D_5) \end{aligned}$$

$\mathcal{A}$	a_1	a_2	a_3	a_4	a_5	a_6	a_7	a_8	a_9
u_1	0	1	1	1	80	2	2	2	3
u_2	0	1	2	1	81	0	aa	1	aa
u_3	0	2	2	1	82	0	aa	1	aa
u_4	0	1	2	1	80	0	aa	1	aa
u_5	1	1	2	2	81	1	aa	1	aa
u_6	0	2	1	2	81	1	aa	1	aa
u_7	1	2	1	2	83	1	aa	1	aa
u_8	0	2	2	1	81	0	aa	1	aa
u_9	0	1	2	1	82	0	aa	1	aa
u_{10}	0	3	2	1	84	0	aa	1	aa
u_{11}	0	1	3	1	80	0	aa	2	aa
u_{12}	0	2	2	2	82	0	aa	2	aa
u_{13}	0	2	2	1	81	0	aa	1	aa
u_{14}	0	3	2	2	81	2	aa	2	aa
u_{15}	0	4	2	1	82	0	aa	1	aa
u_{16}	0	3	2	1	83	0	aa	1	aa
u_{17}	0	1	2	1	84	0	aa	1	aa
u_{18}	1	2	2	1	82	0	aa	2	aa

$\mathcal{A}\|_{\mathbf{T}}$	D_1	D_2	D_3	D_4	D_5	d
	$a_1 = 0$	$a_3 = 2$	$a_4 = 1$	$a_6 = 0$	$a_8 = 1$	
u_1	1	0	1	0	0	
u_2	1	1	1	1	1	1
u_3	1	1	1	1	1	1
u_4	1	1	1	1	1	1
u_5	0	1	0	0	1	
u_6	1	0	0	0	1	
u_7	0	0	0	0	1	
u_8	1	1	1	1	1	1
u_9	1	1	1	1	1	1
u_{10}	1	1	1	1	1	1
u_{11}	1	0	1	1	0	
u_{12}	1	0	0	1	0	
u_{13}	1	1	1	1	1	1
u_{14}	1	1	0	0	0	
u_{15}	1	1	1	1	1	1
u_{16}	1	1	1	1	1	1
u_{17}	1	1	1	1	1	1
u_{18}	0	1	1	1	0	

Table 5. The example of information table $\mathcal{A}$ and template $\mathbf{T}$ support by 10 objects and the new decision table $\mathcal{A}|_{\mathbf{T}}$ constructed from $\mathcal{A}$ and template $\mathbf{T}$

After simplification we obtain six reducts corresponding to the prime implicants:
$f(D_1, D_2, D_3, D_4, D_5) = (D_3 \wedge D_5) \vee (D_4 \wedge D_5) \vee (D_1 \wedge D_2 \wedge D_3) \vee (D_1 \wedge D_2 \wedge D_4) \vee (D_1 \wedge D_2 \wedge D_5) \vee (D_1 \wedge D_3 \wedge D_4)$ for the decision table $\mathcal{A}|_{\mathbf{T}}$. Thus, we have found from $\mathbf{T}$ six association rules with (100%)-confidence.

If $c = 90\%$ it means that we would like to find α-reducts for the decision table $\mathcal{A}|_{\mathbf{T}}$, where $\alpha = 1 - \frac{\frac{1}{c}-1}{\frac{n}{s}-1} = 0.86$. Hence we would like to search for a set of descriptors that covers at least $\lceil (n-s)(\alpha) \rceil = \lceil 8 \cdot 0.86 \rceil = 7$ elements of the discernibility matrix $\mathcal{M}(\mathcal{A}|_{\mathbf{T}})$. One can see that the following sets of descriptors: $\{D_1, D_2\}$, $\{D_1, D_3\}$, $\{D_1, D_4\}$, $\{D_1, D_5\}$, $\{D_2, D_3\}$, $\{D_2, D_5\}$, $\{D_3, D_4\}$ have nonempty intersection with exactly 7 members of the discernibility matrix $\mathcal{M}(\mathcal{A}|_{\mathbf{T}})$. In Table 6 we present all association rules

$\mathcal{M}(\mathcal{A}\vert_{\mathbf{T}})$	u_2, u_3, u_4, u_8, u_9 $u_{10}, u_{13}, u_{15}, u_{16}, u_{17}$
u_1	$D_2 \vee D_4 \vee D_5$
u_5	$D_1 \vee D_3 \vee D_4$
u_6	$D_2 \vee D_3 \vee D_4$
u_7	$D_1 \vee D_2 \vee D_3 \vee D_4$
u_{11}	$D_1 \vee D_3 \vee D_5$
u_{12}	$D_2 \vee D_3 \vee D_5$
u_{14}	$D_3 \vee D_4 \vee D_5$
u_{18}	$D_1 \vee D_5$

$= 100\% \Longrightarrow$

$D_3 \wedge D_5 \longrightarrow D_1 \wedge D_2 \wedge D_4$
$D_4 \wedge D_5 \longrightarrow D_1 \wedge D_2 \wedge D_3$
$D_1 \wedge D_2 \wedge D_3 \longrightarrow D_4 \wedge D_5$
$D_1 \wedge D_2 \wedge D_4 \longrightarrow D_3 \wedge D_5$
$D_1 \wedge D_2 \wedge D_5 \longrightarrow D_3 \wedge D_4$
$D_1 \wedge D_3 \wedge D_4 \longrightarrow D_2 \wedge D_5$

$= 90\% \Longrightarrow$

$D_1 \wedge D_2 \longrightarrow D_3 \wedge D_4 \wedge D_5$
$D_1 \wedge D_3 \longrightarrow D_3 \wedge D_4 \wedge D_5$
$D_1 \wedge D_4 \longrightarrow D_2 \wedge D_3 \wedge D_5$
$D_1 \wedge D_5 \longrightarrow D_2 \wedge D_3 \wedge D_4$
$D_2 \wedge D_3 \longrightarrow D_1 \wedge D_4 \wedge D_5$
$D_2 \wedge D_5 \longrightarrow D_1 \wedge D_3 \wedge D_4$
$D_3 \wedge D_4 \longrightarrow D_1 \wedge D_2 \wedge D_5$

Table 6. The simplified version of discernibility matrix $\mathcal{M}(\mathcal{A}|_{\mathbf{T}})$ and association rules

corresponding to those sets. Heuristics searching for α-reducts are discussed e.g. in [34].

Decomposition of large data tables

Several methods based on rough sets have been developed to deal with large data tables, e.g., to generate strong decision rules for them. We will discuss one of the methods based on decomposition of tables by using patterns, called templates, describing regular sub-domains of the universe (e.g., they describe large number of customers having large number of common features).

Long templates with large support are preferred in many Data Mining tasks. Several quality functions can be used to compare templates. For example they can be defined by $quality^1_{\mathcal{A}}(\mathbf{T}) = support_{\mathcal{A}}(\mathbf{T}) + length(\mathbf{T})$ and $quality^2_{\mathcal{A}}(\mathbf{T}) = support_{\mathcal{A}}(\mathbf{T}) \times length(\mathbf{T})$. Problems of high quality templates generation (by using different optimization criteria) are of high computational complexity. However, efficient heuristics have been developed for solving them (see e.g., [2,77]), [32]).

Extracted from data templates are used to decompose large data tables. In consequence the decision tree is built with internal nodes labeled by the extracted from data templates, and outgoing from them edges by 0 (false) and 1 (true). Any leaf is labeled by a subtable (subdomain) consisting of all objects from the original table matching all templates or their complements appearing on the path from the root of the tree to the leaf. The process of decomposition is continued until the size of subtables attached to leaves is feasible for existing algorithms (e.g., decision rules for them can be generated efficiently) based on rough set methods. The reported experiments are showing that such decomposition returns interesting patterns of regular subdomains of large data tables (for references see [32], [35], [49] and [50]).

It is also possible to search for patterns that are almost included in the decision classes, i.e., default rules [29]. For a presentation of generating default rules see the bibliography in [49] and [50].

Conclusions

There has been done a substantial progress in developing rough set methods (like methods for extraction from data rules, partial or total dependencies, methods for elimination of redundant data, methods dealing with missing data, dynamic data and others reported e.g., in [6], [7], [8], [16], [18], [24], [29], [30], [37], [49], [50], [52], [81]). New methods for extracting patterns from data (see e.g., [21], [35], [29]), [20], [43]), decomposition of decision systems (see e.g., [35]) as well as a new methodology for data mining in distributed and multiagent systems (see e.g., [48]) have been reported. Recently, rough set based methods have been proposed for data mining in very large relational data bases.

There are numerous areas of successful applications of rough set software systems (see [50] and `http://www.idi.ntnu.no/~aleks/rosetta/` for the ROSETTA system). Many interesting case studies are reported (for references see e.g., [49], [50], [37] and the bibliography in these books, in particular [7], [16], [20], [72], [81]).

We have mentioned some generalizations of rough set approach like rough mereological approach (see e.g., [53], [45]). Several other generalizations of rough sets have been investigated and some of them have been used for real life data analysis (see e.g., [79], [5], [39], [14], [22], [38], [25], [59], [48]).

Finally, we would like to point out that the algebraic and logical aspects of rough sets have been intensively studied since the beginning of rough set theory. The reader interested in that topic is referred to the bibliography in [49].

Acknowledgment. This work has been supported by the Wallenberg Foundation, by the ESPRIT-CRIT 2 project #20288, and by grant from the State Committee for Scientific Research (KBN) of the Republic of Poland.

References

1. Ågotnes, T., Komorowski, J., Loken, T. (1999) Taming large rule models in rough set approaches. Proceedings of the 3rd European Conference of Principles and Practice of Knowledge Discovery in Databases, September 15-18, 1999, Prague, Czech Republic, Lecture Notes in Artificial Intelligence **1704**, Springer-Verlag, Berlin, 193–203
2. Agrawal, R., Mannila, H., Srikant, R., Toivonen, H., Verkano, A. (1996) Fast discovery of association rules. In: Fayyad, U.M., Piatetsky-Shapiro, G., Smyth P., Uthurusamy R. (Eds.), Advances in Knowledge Discovery and Data Mining, The AAAI Press/The MIT Press, 307–328
3. Bazan, J.G. (1998) A comparison of dynamic and non-dynamic rough set methods for extracting laws from decision system. In: [49], 321–365
4. Brown, F.M. (1990) Boolean Reasoning. Kluwer Academic Publishers, Dordrecht
5. Cattaneo, G. (1998) Abstract approximation spaces for rough theories. In: [49], 59–98

6. Cios, J., Pedrycz, W., Swiniarski, R.W. (1998) Data Mining in Knowledge Discovery. Kluwer Academic Publishers, Dordrecht
7. Czyżewski, A. (1998) Soft processing of audio signals. In: [50], 147–165
8. Deogun, J., Raghavan, V., Sarkar, A., Sever, H. (1997) Data mining: Trends in research and development. In: [24], 9–45
9. Dougherty, J., Kohavi, R., Sahami, M. (1995) Supervised and unsupervised discretization of continuous features. In: Proceedings of the Twelfth International Conference on Machine Learning, Morgan Kaufmann, San Francisco, CA
10. Duentsch, I., Gediga, G. (1997) Statistical evaluation of rough set dependency analysis. International Journal of Human-Computer Studies **46**, 589–604
11. Duentsch, I., Gediga, G. (2000) Rough set data analysis. In: Encyclopedia of Computer Science and Technology, Marcel Dekker (to appear)
12. Friedman, J., Kohavi, R., Yun, Y. (1996) Lazy Decision Trees. Proceedings of AAAI-96, 717–724
13. Fayyad, U., Piatetsky-Shapiro, G. (Eds.) (1996) Advances in knowledge discovery and data mining. MIT/AAAI Press, Menlo Park
14. Greco, S., Matarazzo, B., Słowinski, R. (1998) Rough Approximation of a Preference Relation in a Pairwise Comparison Table. In: [50], 13–36
15. Chmielewski, M.R., Grzymala -Busse, J.W. (1994) Global discretization of attributes as preprocessing for machine learning. Proceedings of the Third International Workshop on Rough Sets and Soft Computing (RSSC'94), San Jose State University, San Jose, California, USA, November 10–12, 294–301
16. Grzymała–Busse, J.W. (1998) Applications of the rule induction system LERS. In: [49], 366–375
17. Huber, P.J. (1981) Robust statistics. Wiley, New York
18. Komorowski, J., Żytkow, J. (Eds.) (1997) The First European Symposium on Principles of Data Mining and Knowledge Discovery (PKDD'97). June 25–27, Trondheim, Norway, Lecture Notes in Artificial Intelligence **1263**, Springer-Verlag, Berlin, 1–396
19. Komorowski, J., Pawlak, Z., Polkowski, L., Skowron, A. (1999) Rough sets: A tutorial. In: S.K. Pal and A. Skowron (Eds.), Rough–Fuzzy Hybridization: A New Trend in Decision–Making, Springer-Verlag, Singapore, 3–98
20. Kowalczyk, K. (1998) Rough data modelling, A new technique for analyzing data. In: [49], 400–421
21. Krawiec, K., Słowiński, R., Vanderpooten, D. (1998) Learning decision rules from similarity based rough approximations. In: [50], 37–54
22. Kryszkiewicz, M. (1997) Generation of rules from incomplete information systems. In: [18] 156–166
23. Langley, P., Simon, H.A., Bradshaw, G.L., Żytkow, J.M. (1987) Scientific Discovery, Computational Explorations of the Creative Processes. The MIT Press, Cambridge, Massachusetts
24. Lin, T.Y., Cercone, N. (Eds.) (1997) Rough Sets and Data Mining. Analysis of Imprecise Data. Kluwer Academic Publishers, Boston
25. Lin, T.Y. (1989) Granular computing on binary relations I, II. In: [49], 107–140
26. Marek, V.M., Truszczyński, M. (1999) Contributions to the theory of rough sets. Fundamenta Informaticae **39(4)**, 389–409
27. Michie, D., Spiegelhalter, D.J., Taylor, C.C. (Eds.) (1994) Machine learning, Neural and Statistical Classification. Ellis Horwood, New York
28. Mitchell, T.M. (1997) Machine Learning. Mc Graw-Hill, Portland

29. Mollestad, T., Komorowski, J. (1998) A Rough Set Framework for Propositional Default Rules Data Mining. In: S.K. Pal and A. Skowron (Eds.), Rough – fuzzy hybridization: New trend in decision making, Springer–Verlag, Singapore
30. Nguyen, H.S. (1997) Discretization of Real Value Attributes, Boolean Reasoning Approach. Ph.D. Dissertation, Warsaw University
31. Nguyen, H.S. (1999) Efficient SQL-learning Method for Data Mining in Large Data Bases. Proceedings of the Sixteenth International Joint Conference on Artificial Intelligence (IJCAI'99), 806–811
32. Nguyen, S.H. (2000) Data regularity analysis and applications in data mining. Ph.D. Dissertation, Warsaw University
33. Nguyen, H.S., Nguyen, S.H. (1998) Pattern extraction from data. Fundamenta Informaticae **34**, 129–144
34. Nguyen, H.S., Nguyen, S.H. (1999) Rough sets and association rule generation. Fundamenta Informaticae **40/4** 383–405
35. Nguyen, S.H., Skowron, A., Synak, P. (1998) Discovery of data patterns with applications to decomposition and classification problems. In: [50], 55–97
36. Orłowska, E. (Ed.) (1998) Incomplete Information, Rough Set Analysis. Physica–Verlag, Heidelberg
37. Pal, S.K., Skowron, A. (Eds.) (1999) Rough–fuzzy hybridization: New trend in decision making. Springer–Verlag, Singapore
38. Paun, G., Polkowski, L., Skowron, A. (1996) Parallel communicating grammar systems with negotiations. Fundamenta Informaticae **28/3-4**, 315–330
39. Pawlak, Z. (1981) Information systems – theoretical foundations. Information Systems **6**, 205–218
40. Pawlak, Z. (1982) Rough sets. International Journal of Computer and Information Sciences **11**, 341–356
41. Pawlak, Z. (1991) Rough Sets – Theoretical Aspects of Reasoning about Data. Kluwer Academic Publishers, Dordrecht
42. Pawlak, Z., Skowron, A. (1999) Rough set rudiments. Bulletin of the International Rough Set Society **3/4**, 181–185
43. Piasta, Z., Lenarcik, A. (1998) Rule induction with probabilistic rough classifiers. Machine Learning (to appear)
44. Polkowski, L., Tsumoto, S., Lin, T.Y. (Eds.) Rough Sets: New Developments. Physica-Verlag, Heidelberg (2000) (in print)
45. Polkowski, L., Skowron, A. (1996) Rough mereology: A new paradigm for approximate reasoning. International Journal of Approximate Reasoning **15/4**, 333–365
46. Polkowski, L., Skowron, A. (1996) Adaptive decision-making by systems of cooperative intelligent agents organized on rough mereological principles. Intelligent Automation and Soft Computing, An International Journal **2/2**, 121–132
47. Polkowski, L., Skowron, A. (1998) Towards adaptive calculus of granules. In: Proceedings of the FUZZ-IEEE'98 International Conference, Anchorage, Alaska, USA, May 5–9, 111–116
48. Polkowski, L., Skowron, A. (1998) Rough sets: A perspective. In: [49], 31–58
49. Polkowski, L., Skowron, A. (Eds.) (1998) Rough Sets in Knowledge Discovery 1: Methodology and Applications. Physica-Verlag, Heidelberg
50. Polkowski, L., Skowron, A. (Eds.) (1998) Rough Sets in Knowledge Discovery 2: Applications, Case Studies and Software Systems. Physica-Verlag, Heidelberg

51. Polkowski, L., Skowron, A. (1998) Rough mereological foundations for design, analysis, synthesis, and control in distributive systems. Information Sciences International Journal **104/1-2** Elsevier Science, 129–156
52. Polkowski, L., Skowron, A. (Eds.) (1998) Proc. First International Conference on Rough Sets and Soft Computing (RSCTC'98). Warszawa, Poland, June 22–27, Lecture Notes in Artificial Intelligence **1424** Springer-Verlag, Berlin
53. Polkowski, L., Skowron, A. (1999) Towards adaptive calculus of granules. In: [76], **1**, 201–227
54. Polkowski, L., Skowron, A. (2000) Rough mereology in information systems. A case study: Qualitative spatial reasoning. In: [44] (in print)
55. Quinlan, J.R. (1993) C4.5. Programs for machine learning. Morgan Kaufmann, San Mateo, CA
56. Rissanen, J.J. (1978) Modeling by Shortest Data Description. Automatica **14**, 465-471
57. Roddick J.F., Spiliopoulou, M. (1999) A bibliography of temporal, spatial, and temporal data mining research. Newsletter of the Special Interest Group (SIG) on Knowledge Discovery & Data Mining **1/1**, 34–38
58. Selman, B., Kautz, H., McAllester, D. (1997) Ten challenges in propositional reasoning and search. Proceedings of the Fifteenth International Joint Conference on Artificial Intelligence (IJCAI'97), Japan.
59. Ras, Z.W. (1996) Cooperative knowledge–based systems. Journal of the Intelligent Automation Soft Computing **2/2** (special issue edited by T.Y. Lin), 193–202
60. Shrager, J., Langley, P. (1990) Computational Approaches to Scientific Discovery. In: Shrager, J., Langley, P. (Eds.), Computational Models of Scientific Discovery and Theory Formation, Morgan Kaufmann, San Mateo, 1–25
61. Skowron, A. (1995) Synthesis of adaptive decision systems from experimental data. In: A. Aamodt, J. Komorowski (eds), Proc. of the Fifth Scandinavian Conference on Artificial Intelligence (SCAI'95), May 1995, Trondheim, Norway, IOS Press, Amsterdam, 220–238.
62. Skowron, A., Nguyen, H.S. (1999) Boolean reasoning scheme with some applications in data mining. Proceedings of the 3-rd European Conference on Principles and Practice of Knowledge Discovery in Databases, September 1999, Prague Czech Republic, Lecture Notes in Computer Science **1704**, 107–115
63. Skowron, A., Rauszer, C. (1992) The Discernibility matrices and functions in information systems. In: Słowiński [68], 331–362
64. Skowron, A., Stepaniuk, J. (1996) Tolerance Approximation Spaces. Fundamenta Informaticae **27**, 245–253
65. Skowron, A., Stepaniuk, J., Tsumoto, S. (1999) Information Granules for Spatial Reasoning. Bulletin of the International Rough Set Society **3/4**, 147–154
66. Stepaniuk, J. (1998) Approximation spaces, reducts and representatives. In: [50] 109–126
67. Ślęzak, D. (1998) Approximate reducts in decision tables. In: Proceedings of the Sixth International Conference, Information Processing and Management of Uncertainty in Knowledge-Based Systems (IPMU'96) **3**, July 1–5, Granada, Spain, 1159–1164
68. Słowiński, R. (Ed.) (1992) Intelligent Decision Support – Handbook of Applications and Advances of the Rough Sets Theory. Kluwer Academic Publishers, Dordrecht.

69. Słowiński, R., Vanderpooten, D. (1997) Similarity relation as a basis for rough approximations. In: P. Wang (Ed.): Advances in Machine Intelligence & Soft Computing, Bookwrights, Raleigh NC, 17–33.
70. Słowiński, R., Vanderpooten, D. (1999) A generalized definition of rough approximations based on similarity. IEEE Trans. on Data and Knowledge Engineering (to appear)
71. Szczuka, M. (2000) Symbolic and neural network methods for classifiers construction. Ph.D. Dissertation, Warsaw University
72. Tsumoto, S. (1998) Modelling diagnostic rules based on rough sets. In: Polkowski and Skowron [52], 475–482
73. Valdz-Prez, R.E. (1999) Discovery tools for science apps. Comm ACM **42/11**, 37–41
74. Zadeh, L.A. (1996) Fuzzy logic = computing with words. IEEE Trans. on Fuzzy Systems **4**, 103–111
75. Zadeh, L.A. (1997) Toward a theory of fuzzy information granulation and its certainty in human reasoning and fuzzy logic. Fuzzy Sets and Systems **90**, 111–127
76. Zadeh, L.A., Kacprzyk, J. (Eds.) (1999) Computing with Words in Information/Intelligent Systems **1–2**. Physica-Verlag, Heidelberg
77. Zaki, M.J., Parthasarathy, S. , Ogihara, M., Li, W. (1997) New parallel algorithms for fast discovery of association rules. Data Mining and Knowledge Discovery : An International Journal, special issue on Scalable High-Performance Computing for KDD **1/4**, 343–373
78. Zhong, N., Skowron, A., Ohsuga, S. (Eds.) (1999) Proceedings of the 7-th International Workshop on Rough Sets, Fuzzy Sets, Data Mining, and Granular-Soft Computing (RSFDGrC'99). Yamaguchi, November 9-11, 1999, Lecture Notes in Artificial Intelligence **1711**, Springer-Verlag, Berlin
79. Ziarko, Z. (1993) Variable Precision Rough Set Model. J. of Computer and System Sciences **46**, 39–59
80. Ziarko, W. (ed.) (1994) Rough Sets, Fuzzy Sets and Knowledge Discovery (RSKD'93). Workshops in Computing, Springer–Verlag & British Computer Society, London, Berlin
81. Ziarko, W. (1998) Rough sets as a methodology for data mining. In: L. Polkowski A, Skowron (Eds.), Rough Sets in Knowledge Discovery 1: Methods Applications, Physica-Verlag, Heidelberg, 554–576.

Granulation and Nearest Neighborhoods: Rough Set Approach

T. Y. Lin1,2

[1] Department of Mathematics and Computer Science
San Jose State University, San Jose, California 95192-0103
[2] Berkeley Initiative in Soft Computing,
University of California, Berkeley, California 94720
tylin@cs.sjsu.edu, tylin@cs.berkeley.edu

Abstract. "Nearest" neighborhoods are informally used in many areas of AI and database. Mathematically, a "nearest" neighborhood system that maps each object p a unique crisp/fuzzy subset of data, representing the "nearest" neighborhood, is a binary relation between the object and data spaces. "Nearest" neighborhood consists of data that are semantically related to p, and represents an elementary granule(atoms) of the system under consideration. This paper examines "rough set theory" of these elementary granules. Applications to databases, fuzzy sets and pattern recognition are used to illustrate the idea.

Keywords: Fuzzy set, rough set, neighborhood system, binary relation

1 Introduction

The notion of neighborhoods has been used informally in many areas: In database applications, neighborhoods are used to formulate approximate retrieval or query relaxation [30],[10],[1], [12],[11],[4],[18]. The term has also been referred in pattern recognition, genetic algorithm texts and papers in modal logic [2],[6],[14]. Implicitly, the idea is an important ingredient in generalized rough set theories; the literatures are abundant, just to mention a few [13],[3],[33], [35],[37].

Mathematically neighborhood is an abstraction of ϵ-neighborhoods. It is the the building blocks of topological and pre-topological spaces [34]. The notion of neighborhood systems that we proposed is more general; it involves two universes and imposed no axioms; see Section 2.

Recently, Lotfi Zadeh has been pointing out the importancy of information granulation [43],[44]. The notion was formed, in fact, much earlier. Implicitly, the notion is embedded in fuzzy logic control, explicitly it can be traced back to Zadeh's 1979 paper [45].

In response to Zadeh's call, we have proposed to use neighborhood systems as a mathematical model for crisp/fuzzy granulation [26],[27],[28]. This paper focuses on "sofset" aspect of this theory and represents a summary, updates and some new illustrations to previous studies; here "sofset" (note the spelling) is a technical term of fuzzy sets [18].

The organization of this paper is as follows: First, we review the rough set theory. Then, we take it as a "prototype," and theories of crisp and fuzzy binary relations are investigated. Some applications to databases and pattern recognition are illustrated. Finally, the theory of qualitative fuzzy sets is developed base on nearest neighborhoods.

2 Neighborhood Systems and Granulation

Let us start with a quote from Lotfi Zadeh's keynote speech at FUZZ-IEEE'96 [44].

- *"information granulation involves partitioning a class of objects(points) into granules, with a granule being a clump of objects (points) which are drawn together by indistinguishability, similarity or functionality."*

First we will paraphrase his informal words: "Drawn together" is a special notion of "drawn towards center points(objects)." Note that "center points" could be every point. So "drawn towards center points" is, in fact, more general than "drawn together." In other words, we are localizing the granulation (which is a generalization). Now we paraphrase his words as follows:

- information granulation is a collection of granules, with a granule being a clump of data (points) which are drawn *towards a center object(s)*, which is(are) not necessary in the same universe as data. In other words, each objects (in object space) is associated with a family of clumps (in Data space).

If we interpret the granule as neighborhood, this paraphrase is essentially the neighborhood systems:

Definitions. We are given two universes V, called object space and U, called data (or information) space.

1. As in [46], let M be the membership space, in this paper, it will be either the unit interval $[0,1]$ or the "true or false" set $\{0,1\}$. The set of all crisp/fuzzy sets will be denoted by M^U, where if $M = \{0,1\}$, it is the set of all crisp subsets, and if $M = [0,1]$, it is the set of all possible W-sofsets(See Section 6.1.
2. (crisp/fuzzy) Local neighborhood system: For a *fixed* object p, the family

$$NS_V(U,p) = \{B_p^i \mid \textit{ i runs through the index set }\} \subseteq M^U$$

is called the crisp/fuzzy local neighborhood system at p. Each member B_p^i of $NS_V(U,p)$ is called a crisp/fuzzy *fundamental neighborhood at p.* The smallest one (if it exists) is the "nearest" neighborhood.

3. (crisp/fuzzy)Fundamental neighborhood system: We consider local neighborhood system at every object $p \in V$. Mathematically it is a map:

$$NS_V(U) : p \in V \longrightarrow NS_V(U,p) \subseteq M^U$$

The association is called *the neighborhood system for* V *on* U.
4. (crisp/fuzzy) Binary (nearest) neighborhood system: It is a neighborhood system that consists of nearest neighborhood only. Intuitively it should be called the nearest neighborhood system, but mathematically, it is related to binary relation (Section 4.1, 6.2), so we have called it binary neighborhood system, and the (crisp/fuzzy) nearest neighborhood a (crisp/fuzzy) binary neighborhood. We will call it (crisp/fuzzy) *elementary* neighborhood, because in rough set theory, it is called elementary set. The singleton $NS_V(U,p)$, as well as its element will be denoted by B_p. When $NS_V(U,p)$ is an empty family, we will interpret it as $B_p = \emptyset$.
5. Granulation: The collection of crisp/fuzzy sets,

$$GR_V(U) = \bigcup_p \{NS_V(U,p) \mid NS_V(U,p) \in M^U \ \forall\, p \in V\}$$

is called a (crisp/fuzzy)*granulation* of U by V. Note that B_p^i and B_q^j are considered equal if they are equal as crisp sets or W-sofsets. Locally (at each object) there are multiple granules; so it is a multilevel granulation.
6. A covering is a neighborhood system, if each cover is considered as a neighborhood of each of its points. So covering is a granulation by our definition.
7. Binary granulation: if the neighborhood system that defines granulation is a (crisp/fuzzy) binary neighborhood system, the granulation is called a binary granulation. At each point, there is only one granule; so locally it is single level granulation. Note that an elementary neighborhood is a "cover" with a center (may not unique). So a binary granulation is a partial covering, where each cover has a center.

This paper will focus binary (nearest) neighborhood systems.

3 Rough Sets - Theory of Single Equivalence Relations

Let $V = U$ be the only universe. Let E be an equivalence relation. E partitions the universe into equivalence classes. They are called *elementary sets* in rough set theory. We may also refer them as *elementary granules*. A union of a collection of elementary sets is called a *definable* set The collection of all the equivalence classes of a partition is a classical set, called *quotient set*, denoted by U/E. The map $P : U \longrightarrow U/E$ is called *natural projection*. Note that an elementary granule plays two roles: one as a subset of U, another as an element of U/E. More generally any subset of U has two roles: One as a subset of U, another as an element of power set. We will refer to the second role as the *canonical name* of the first role. The canonical name could be represented by bits or lists.

Given an arbitrary set $X \subseteq U$,

$$\underline{E}(X) = \bigcup_{E_x \subseteq X} E_x,$$
$$\overline{E}(X) = \bigcup_{E_x \cap X \neq \emptyset} E_x, \tag{1}$$

where

$$E_x = \{ y \mid x \; E \; y \}, \tag{2}$$

is the elementary granule (equivalence class) containing x. We are using E_x, instead of usual $[x]_E$, because we are trying to integrate the notations into the theory of single binary relations. The lower approximation $\underline{E}(X)$ is the union of all the elementary granules that are subsets of X. It is the largest definable set contained in X. The upper approximation $\overline{E}(X)$ is the union of all the elementary granules that have a non-empty intersection with X. It is the smallest definable set containing X. In the theory of topological space, lower approximation is the interior points, and upper approximation is the closure. The collection of the elementary sets is a *base* for clopen topology [8]. Pawlak has developed many interesting properties of the lower and upper approximations. For comparisons, we list them in Section 4. For now, we gather the strongest properties that characterize rough set theory abstractly.

3.1 Axiomatic Rough Set Theory

In [?], we showed that Pawlak's lower and upper approximations can be characterized by the following six axioms; they are chosen from those properties developed by Pawalak. Let $X \subseteq U$, and $\sim X = U - X$. Two operators,$L, H : 2^U \longrightarrow 2^U$, are called the lower rough and upper (higher) rough operators, if they satisfy the following axioms.

$$
\begin{aligned}
&\text{Axiom1} \;\; (PU02) H(\emptyset) = \emptyset,\\
&\text{Axiom2} \;\; (PL07) L(X) \subseteq X,\\
&\text{Axiom3} \;\; (PL09) L(X) \subseteq L(L(X)),\\
&\text{Axiom4} \;\; (PU03) H(X \cup Y) = H(X) \cup H(Y),\\
&\text{Axiom5} \;\; (PL01) L(X) = \;\; \sim H(\sim X),\\
&\text{Axiom6} \;\; (RU10) H(L(X)) \subseteq L(X),\\
&\text{Axiom6}' \;\; (RL10)(for\ reference\ only) H(X) \subseteq L(H(X)).
\end{aligned}
$$

The label of each axiom comes from next section. The first five axioms are essentially the axioms of Kuratowski's closure operators [8]. H and L are the closure and interior operators respectively. The axiom (RU10) implies that an open set is also a closed set. (PL10) is an equivalent statement using interior operators. The six axioms characterize rough set theory [15] in the following sense:

Theorem Given the lower and upper rough operators, L and H, that satisfy the six axioms. Then there is an equivalence relation R, such that

1. $L(X) = \underline{E}(X) \ \ \forall \ \ X \subseteq U$
2. $H(X) = \overline{E}(X) \ \ \forall \ \ X \subseteq U$

4 Theory of Single Binary Relations

In this section, we will extend the theory of single equivalence relations to single binary relations. Algebraically binary relation is the natural generalization of equivalence relation. However, geometrically, "the common geometric generalization of a partition is *a covering*. Unfortunately, it is not the geometric equivalence of a binary relation [27]. The equivalence one is a more elaborate concept; it is the notion of binary neighborhood systems. Intuitively, it is a collection of "nearest" neighborhoods. Obviously, the collection of the "nearest" neighborhood at each point (with the non-empty assumption) forms a covering of the universe. However, we should observe that it is more than a covering, it is a collection of covers, in which each cover has a center or centers."

4.1 Basic Notions

A *binary relation* B is a subset of Cartesian product of object and data spaces, in notations, $B \subseteq V \times U$. For each object $p \in V$, we can define a subset,

$$B_p = \{u \mid p\ B\ u,\ \forall\ \in\ U\ \}.$$

In words, B_p consists of all elements u that are related to p by the binary relation B. B_p is called *binary set or binary neighborhood*, since it consists of all those points that are binary related to p. If the binary relation is an equivalence relation, the binary neighborhood is an equivalence class; they are called elementary sets by rough setters. To line up its terminology, we also call B_p *elementary neighborhood* or with granular computing *elementary granule*. The association

$$B : p(\in V) \ \longrightarrow\ B_p \subseteq U$$

is called a binary neighborhood system. In the case of equivalence relation this map is the natural projection.

Instead of using binary relation B, we can define it directly

A *binary neighborhood system* is a mapping N that, to each object $p \in V$, we associate an arbitrary chosen subset, denoted by $N_p \subseteq U$. In notation,

$$N : V \longrightarrow 2^U : p \longrightarrow N_p$$

N_p is the *elementary* neighborhood (also known as binary or basic neighborhood) [25],[26]. The map N or the collection $\{N_p\}$ will be referred to as an *binary neighborhood system for* V *on* U It is clear that given a N, there is a binary relation $B \subseteq V \times U$ and vice versa. This assertion follows immediately from the equation:

$$B \ = \bigcup_{p,u}\{N_p(u) \mid \ \forall\, p \in \ V \ and \, \forall \ u \ \in \ U\}.$$

So from now on B is the binary relation as well as the binary neighborhood system.

For specific binary relations, we have the following:

1. B is a serial system, if $\forall \ p \ \in \ V, \ N(p) \ \neq \ \emptyset$.
2. B is a reflexive system, if $\forall \ p \ \in \ V = U, \ p \ \in \ N(p)$.
3. B is a symmetric system, if $\forall \ p \ , \forall \ q, \ q \ \in \ N(p) \ \Rightarrow \ p \ \in \ N(q)$
4. B is a transitive system, if $\forall \ p, \ \forall \ q, \ and \ \forall \ r$,
$$q \ \in \ N(p) \ and \ r \ \in \ N(q) \ \Rightarrow \ r \ \in \ N(p).$$
5. B is a Euclidean system, if $q \ \in \ N(p), andr \ \in \ N(p), \ \Rightarrow \ r \ \in \ N(q)$.
6. B is a rough set system, if B is an equivalence relation.

4.2 Approximations by a Single Binary Relation

Neighborhood systems are generalizations of topological spaces and rough set theory. We adop both.
Given an arbitrary set $X \subseteq U$,

$$\underline{B}(X) = \bigcup_{B_p \subseteq X} B_p = \{p : \ \exists \ B_p \subseteq X\} = \ the \ interior \ of \ X, \tag{3}$$

$$\overline{B}(X) = \bigcup_{B_p \cap X \neq \emptyset} B_p = \{p : \ \forall B_p \ \ X \cap B_p \ \neq \ \emptyset\} = the \ closure \ of \ X, \tag{4}$$

where

$$B_p = \{y \mid x \ B \ y\}, \tag{5}$$

is the elementary neighborhood of p; note that p may or may not be in B_p. Similar to rough set theory, the lower approximation $\underline{B}(X)$ is the union of all the elementary granules that are subsets of X, and the upper approximation $\overline{B}(X)$ is the union of all the elementary granules that have a non-empty intersection with X. $\underline{B}(X)$ and $\overline{B}(X)$ are precisely the lower and upper approximation in rough set theory, if the elementary neighborhood system is a partition; or the binary relation is an equivalence relation

Lin and Yao has developed some interesting properties of neighborhood approximation: For comparison, we will generalize the rough set list in [41] to the following list: All properties that are valid for $\underline{B}$ and $\overline{B}$ are also valid for $\underline{E}$ and $\overline{E}$. For properties that are specific for rough set only will be so indicated. Let $X, Y \subseteq U$, the lower approximation $\underline{B}$ satisfies properties:

$$
\begin{array}{ll}
\text{(PL01)} & \underline{B}(X) = \ \sim\overline{B}(\sim X), \\
\text{(PL02)} & \underline{B}(U) = U, \\
\text{(PL03)} & \underline{B}(X \cap Y) \subseteq \underline{B}(X) \cap \underline{B}(Y), \\
\text{(RL03)} & (\textit{rough set only})\underline{E}(X \cap Y) = \underline{E}(X) \cap \underline{E}(Y), \\
\text{(PL04)} & \underline{B}(X \cup Y) \supseteq \underline{B}(X) \cup \underline{B}(Y), \\
\text{(PL05)} & X \subseteq Y \Longrightarrow \underline{B}(X) \subseteq \underline{B}(Y), \\
\text{(PL06)} & \underline{B}(\emptyset) = \emptyset, \\
\text{(PL07)} & \underline{B}(X) \subseteq X, \\
\text{(PL08)} & X \subseteq \underline{B}(\overline{B}(X)), \\
\text{(PL09)} & \underline{B}(X) \subseteq \underline{B}(\underline{B}(X)), \\
\text{(RL10)} & (\textit{rough set only})\overline{E}(X) \subseteq \underline{E}(\overline{E}(X)),
\end{array}
$$

and the upper approximation $\overline{B}$ satisfies properties:

$$
\begin{array}{ll}
\text{(PU01)} & \overline{B}(X) = \ \sim\underline{B}(\sim X), \\
\text{(PU02)} & \overline{B}(\emptyset) = \emptyset, \\
\text{(PU03)} & \overline{B}(X \cup Y) \supseteq \overline{B}(X) \cup \overline{B}(Y), \\
\text{(RU03)} & (\textit{rough set only})\overline{E}(X \cup Y) = \overline{E}(X) \cup \overline{E}(Y), \\
\text{(PU04)} & \overline{B}(X \cap Y) \subseteq \overline{B}(X) \cap \overline{B}(Y), \\
\text{(PU05)} & X \subseteq Y \Longrightarrow \overline{B}(X) \subseteq \overline{B}(Y), \\
\text{(PU06)} & \overline{B}(U) = U, \\
\text{(PU07)} & X \subseteq \overline{B}(X), \\
\text{(PU08)} & \overline{B}(\underline{B}(X)) \subseteq X, \\
\text{(PU09)} & \overline{B}(\overline{B}(X)) \subseteq \overline{B}(X), \\
\text{(RU10)} & (\textit{rough set only})\overline{E}(\underline{E}(X)) \subseteq \underline{E}(X),
\end{array}
$$

where $\sim X = U - X$ denotes the set complement of X.

4.3 Approximate Retrieval

This is a binary relation version of the example in [10], [11], [12]. In a Restaurant Database, there is a binary neighborhood system NN, called nearest neighborhood. on each set of attribute values.

1. The binary neighborhood system of LOCATION-attribute:

RESTAURANT	TYPE	LOCATION	PRICE
Wendy	American	West wood	inexpensive
Le Chef	French	West LA	moderate
GreatWall	Chinese	St Monica	moderate
Kiku	Japanese	Hollywood	moderate
South Sea	Chinese	Los Angeles	expensive

Table 1. A Restaurant Database

$NN_{Westwood} = \{Westwood, SantaMonica, WestLA\}$
$NN_{SantaMonica} = \{\ldots\}$

...

2. The binary neighborhood systems of TYPE-attribute:
   $NN_{Japanese} = \{$ Chinese, Japanese $\}$
   $NN_{Chinese} = \{\ldots\}$

   ...
3. The binary neighborhood systems of PRICE-attribute:
   $NN_{moderate} = \{$ moderate, expensive $\}$
   $NN_{expensive} = \{\ldots\}$

   ...

Now, let us consider the following query:

Q_1 : Select a Restaurant which is Japanese, moderate in price and located in Westwood.

The traditional database will return a null answer. Very often, users have to vary his conditions slightly and issue new queries until he finds his answers. For this new system, it will supply an approximate answer:

"Great Wall"

because the three neighborhood relations indicated that

1. "Chinese" is in the nearest neighborhood of "Japanese"(in TYPE)
2. "St Monica" is in the nearest neighborhood of "Westwood" (in LOCATION)
3. Moderate (in PRICE)

5 The Induced Equivalence Relation

Given a binary relation, $B \subseteq V \times U$, or equivalently, a binary neighborhood system:

$$B : V \longrightarrow 2^U : p \longrightarrow B_p$$

This map induced a partition by the inverse map B^{-1}, namely

$$v_1 \equiv v_2, \text{ if } B_{v_1} = B_{v_2} \text{ as two subsets of } U.$$

In words, two objects in V are equivalent, iff they have the same elementary neighborhoods in U (set theoretical equality). We will denote this induced equivalence as E_B. The association from B to E_B induce the following map:

$$2^{V \times U} \longrightarrow SE(U)$$

where SE(U) denote the set of all equivalence relations on U; recall that every subset of the Cartesian product is a binary relation.

This is essentially a "forgetful" map which "forgets" some data semantics captured in the binary relation B. To recapture these semantics, we need to recover the binary relation in the quotient set $Q = V/E_B$. We will consider the special case, $V = U$.

Proposition Let c and d be two elements in Q. Then B induces a binary relation B_Q on Q:

$$c \ B_Q \ d \text{ iff } \exists p \in V \text{ and } \exists q \in U = V \text{ such that } p \ B \ q.$$

This is relatively a new and unexplored area, we have touched lightly on this in several papers [29], [26], [27], [28]; We expect it will play a major role in data mining on OODB. For now, we give a simple application next.

5.1 Classifications of Research Articles by Keywords

In [39], Viveros build a system that can automatically classify research articles in computer science. We can re-express the idea from the prospect of this paper : Let V be a list of research articles, and U be a set of keywords. The goal is to classify V by keywords.

First, we take training data to be V. Namely, each article is labeled with its classification (subfield name). Then each class of articles in V is associated with a subset B_p of key words. Such an association is learned from the training data. In the terminology of this paper, this association defines a binary neighborhood system for V on U. Note that these elementary neighborhoods may have non-empty intersections, so the binary neighborhood system is not a partition on U. Then, we assign the label (of a class of articles) to each elementary neighborhood. Classify an unknown article is then, a process of matching the keyword list with these named elementary neighborhoods.

6 Granulation by Single Fuzzy Binary Relations

Last few section the discussions are in the crisp world. Now, we turn to the fuzzy notion.

6.1 Weighted Soft Sets - Classical Fuzzy Sets

The term "fuzzy set" may take several meanings; to avoid confusing, some technical terms have been introduced [18]. In term of that formulation, a traditional fuzzy set is a W-sofset. Formally,

- A W-sofset (a soft set with weighted memberships) is a mathematical object defined uniquely by a membership function.

This is precisely the definition of traditional fuzzy set; a W-sofset is solely represented by stating its membership function ([46], pp.12). The membership of each point is weighted by a value in the unit interval. The function value at a point p is called the weight (grade) of the membership of p.

6.2 W-Soft Binary Relations

By fuzzify previous discussions (Section 4.1), we have

A *W-soft binary relation* is: Let I be the unit interval [0, 1]. Let FR be a W-soft binary relation whose membership function, denoted by the same notation again, is

$$FR : V \times U \longrightarrow I : (p, u) \longrightarrow r.$$

To each $p \in V$, we associate a W-sofset FN_p whose membership function is,

$$FN_p : U \longrightarrow I \text{ is defined by } FN_p(u) = FR((p, u)).$$

A *W-soft binary neighborhood* is: To each object $p \in V$, we associate a W-sofset, denoted by FB_p. In other words, we have a map

$$NB : V \longrightarrow I^U : p \longrightarrow FB_p,$$

where I^U means all W-sofsets on U. FB_p is the W-soft nearest neighborhood at p (or the W-soft elementary neighborhood or W-soft binary neighborhood). The association FB is called a W-soft binary neighborhood system. To each $p \in V$, we set $FB_p(u) = FR((p, u))$, then FR define a W-soft binary relation. So, as in crisp case, W-soft binary relations and W-soft binary neighborhood systems are equivalent.

Instead of using a formal membership function to represents a W-sofset, we will use a *list of pairs*, the element and its weight, to represent a W-sofset.

6.3 Pattern Recognition Applications

We will use the following convention: The term "character A" means the abstract concept of English alphabet, and the term "sample letter A" means the "physical" printed letter A. In this section, we will use grey scales of pixels to express sample letters. For that purpose, each sample letter is printed in a square area that is partitioned into very fine pixels:

$$pixel\#1,\ pixel\#2,\ ...,\ pixel\#n.$$

Then, we use some instruments to measure the grey scales of these pixels. So each sample letter is represented by an n-dimensional vector of grey scales, n-vector for short. Because of the measuring errors, these grey scales are random variables. Mathematically, it means there is a probability associated with each grey scale. These probabilities come from the distributions. If distributions are not available, we may use grade to estimate its likelihood. We will use the term "weight" to mean either grade or probability depending on which number we are given. The pair of the grey scale and its weight will be referred to as the *weighted grey scale*. So n-vector becomes n-vector of weighted grey scales.

Suppose we have measured 100 samples of a given character, then the n-vector of s-th measurement looks like,

$$(\ (\xi(s_{\#1}), W_\xi(s_{\#1})),\ \ (\xi(s_{\#2}), W_\xi(s_{\#2})), \ldots,\ \ (\xi(s_{\#n}), W_\xi(s_{\#n}))\ )$$

Let s varies from 1 to 100, then each component of a n-vector becomes an array of weighted grey scales(length 100). Array often means an ordered set. In this case the order of the sequence is not important, so instead of n-vector of weighted arrays, we have n-vector of W-sofsets. Each W-sofset contains at most 100 weighted grey scales, since some of these grey scales in the array may coincide.

Crisp Grey Scales Let us make an unrealistic assumption: All sample letters (of a character) are exactly the same and the instruments to measure the gray scale are also the same. Then, in stead of 26 vectors of W-sofsets, we get 26 vectors of grey scales with weights=1. In other words, we have associated each class (representing a character) of the sample letters a n-vector of grey scale. Using our terminology, each character is assigned an elementary neighborhood that consists of one vector. Then a pattern recognition task is to identify an unknown sample with one of these 26 elementary neighborhoods. This consideration is very similar to that of Section 5.1; each n-vector play a similar roles as a list of keywords.

Fuzzy Grey Scales Let $W_{\mathbf{A}}(s_k)$ be the weight of the grey scale $\mathbf{A}(s_k)$ of s-th measurement of a sample letter A at $pixel\#k$. When s varies through 100 samples, we get an arrays of pairs (length 100). Such an array is actually a W-sofset. The n-vector of these W-sofsets will be named by the intended character. So we have associated a character to a n-vector of W-sofsets. The association is a W-soft binary neighborhood system (or equivalently, a W-soft binary relation)

$$A \to (\ \{\mathbf{A}(s_{\#1}), W_{\mathbf{A}}(s_{\#1})\},\ \ \{\mathbf{A}(s_{\#2}), W_{\mathbf{A}}(s_{\#2})\}, \ldots, \{\mathbf{A}(s_{\#n}), W_{\mathbf{A}}(s_{\#n})\}\ )$$
$$B \to (\ \{\mathbf{B}(s_{\#1}), W_{\mathbf{B}}(s_{\#1})\},\ \ \{\mathbf{B}(s_{\#2}), W_{\mathbf{B}}(s_{\#2})\}, \ldots, \{\mathbf{B}(s_{\#n}), W_{\mathbf{B}}(s_{\#n})\}\ )$$

...

$$Z \to (\ \{\mathbf{Z}(s_{\#1}), W_{\mathbf{Z}}(s_{\#1})\}, \ \{\mathbf{Z}(s_{\#2}), W_{\mathbf{Z}}(s_{\#2})\}, \ldots, \{\mathbf{Z}(s_{\#n}), W_{\mathbf{Z}}(s_{\#n})\} \ )$$

For a unknown character, say ξ, we have measurement

$$(\ \{\xi(s_{\#1}), W_{\xi}(s_{\#1})\}, \ \{\xi(s_{\#2}), W_{\xi}(s_{\#2})\}, \ldots, \{\xi(s_{\#n}), W_{\xi}(s_{\#n})\} \ )$$

Then the pattern recognition task is to identify this unknown sample with one of these 26 named vector of W-sofsets. Various techniques have been developed for such identification; one common way is to find the "mean point" of elementary neighborhood in data space. In other words, pattern recognition is to set up a map from object p at V to the "mean point" of the elementary neighborhood of p. The well known "mean points" are the k-mean or c-mean.

7 Qualitative Fuzzy Sets

In this section, we recall some of our efforts in exploring a mathematical notion of fuzzy set that "fully" reflect the intuition of fuzzy sets [18], [16]. We have call such fuzzy sets qualitative fuzzy sets. There are other approach to this notion [38].

7.1 "Elastic" Membership Functions

In a real world fuzzy set, intuitively, each element has an imprecise and vague membership. Traditionally, one uses a membership function to capture such properties. So a fuzzy set is uniquely characterized by its membership function. We feel, however, that such a definition has not fully reflected the intuition of fuzzy set. For example, if we add ϵ, a very small number, to one particular value, then by definition, the fuzzy set is changed, sine the membership function is no longer the same. "... we believe that ideal membership functions for *real world fuzzy sets* are made from elastic material, because fuzzy sets should be able to tolerate certain amount of perturbation or stretching. Then, the important question is

Could elasticity be expressed mathematically?

The answer is "yes." Each perturbation is a new function, so mathematically such an elastic membership function must consist of a set of functions, each of which can be continuously "stretched" into another function with limited broken points. Such a set of functions is crisp/fuzzy neighborhood system on the space of membership functions. [16]. An elementary neighborhood (nearest neighborhood) in the membership function space, as a whole, defines an elastic membership function ; see Section 7.3.

In order to talk about elasticity, we need two mathematical structures that have to be imposed on the universe of discourse U. Namely, it has to be a measurable space at the same it is also a topological space [7], [?]. We also need "derived" universe, namely, the space of membership functions, denoted by $MF(U)$, which consists of all the membership functions that define W-sofsets on U. Since topological structure is very close to neighborhood system, we will skip the review.

Almost and Nearly Everywhere The purpose of using measure is to estimate the "size of broken points." Roughly, a measure is (but not the same) a content (geometric measurements): it is the length if the space X is one dimensional, is area if X is 2-dimensional, and is volume if X is 3-dimensional, and etc. Technically content and measure are different; measure is countably additive, while content is finitely additive.

We say "X has property P almost everywhere" means X has property P at almost every point except small set of "criminal" points, and the collection of such "criminal" points has measure zero.

X has property P almost everywhere, or in notation, "X has P a.e".

is a standard mathematical notion [7]. In this paper, we need to extend a little bit more: The sentence

X has property P nearly everywhere, or in notation, "X has P n.e."

means X has property P at almost every point except some points, and the collection of such points has measure less than given ϵ.

7.2 Granulation by Deformation

Let $FX \in W = MF(U)$ be a membership function. We say a membership function $FY \in W$ is ϵ-deformable from FX iff there exists a map

$$F : W \times I \longrightarrow I$$

such that

1. $F(u, 0) = FX(u)$ n.e. or a.e.
2. $F(u, 1) = FY(u)$ n.e. or a.e.
3. Let the core be $C = FX^{-1}(1)$ and exterior be $E = FX^{-1}(0)$,

$$F(u, t) = u \; \forall \; u \; \in \; C \; or \; E$$

In other words, those points in C and E are pointwise fixed during the deformation. Note that both FX and FY have the same core and exterior n.e. or a.e. ϵ-deformable is a binary relation.

Remark: If ϵ =1 and a.e. is the actual equality, then the deformation, called homotopy, is an equivalence relation, moreover, there is algebraic structure on the equivalence classes (induced by e.g., bounded sum and multiplication of the unit interval) ([36], pp. 24 and 34). The algebraic structure by itself is a worthy topics, we will report it in a different paper.

Further, to each FY, we associate a number, called weight, defined by

$$w(FY) = 1 - ||FY - FX||$$

In particular, we have

$$w(FX) = 1.$$

The collection of all such FY with its weight forms a W-sofset in W; So we have associated every membership function a W-sofset $\subseteq W$, a W-soft nearest neighborhood. Mathematically, such an association defines a *W-soft binary neighborhood system* on W (or equivalently a W-soft binary relation of deformation). We will use it to define qualitative fuzzy sets.

7.3 ϵ-Qualitative Fuzzy Sets

Suppose a W-soft binary neighborhood system is given on $W = MF(U)$. Then,

- An ϵ - qualitative fuzzy set is the mathematical object defined solely by a W-soft elementary neighborhood.
- A qualitative fuzzy set is an epsilon-qualitative fuzzy set with the $\epsilon = 1$.

8 Multilevel Granulation- Neighborhood Systems

A neighborhood system associate each object a family of granules (neighborhoods),so it is a multilevel granulation. Let us generalize some fundamental properties about neighborhood systems in [25] to fuzzy (W-soft) versions.

Propositions & Definitions

1. A crisp/fuzzy subset X is a definable neighborhood if it is a union of fundamental neighborhoods See Section 2; the subset is a definable neighborhood of p, if the union contains a fundamental neighborhood of p.
2. A crisp/fuzzy neighborhood system of an object p, denoted by $NS_V(U, p)$, is the *maximal* family of definable neighborhoods of p. If $NS_V(U, p)$ is an empty family, we simply say that p has no neighborhood.
3. A crisp/fuzzy neighborhood system of U, denoted by $NS_V(U)$, is the collection of $NS_V(U, p)$ for all p in U. For simplicity a crisp/fuzzy set U together with $NS_V(U)$ is called a crisp/fuzzy neighborhood system space (NS-space) or simply neighborhood system.
4. A crisp/fuzzy subset X of U is open if for every object p in X, there is a crisp/fuzzy neighborhood $N_p \subseteq X$. A subset X is closed if its complement is open.

5. $NS_V(U, p)$ and $NS_V(U)$ are open if every neighborhood is open. $NS_U(U)$ is crisp/fuzzy topological, if $NS_U(U)$ are open and U is the usual topological space (for crisp case see [34]. In such a case both $NS_U(U)$ and the collection of open sets are called crisp/fuzzy topology -see next proposition.
6. An object p is a limit point of a set X, if every neighborhood of p contains a point of X other than p In sofset, "belong" means the weight of the membership is positive. The crisp/fuzzy set of all limit points of E is call derived set. E together with its derived set is a closed set.
7. For crisp neighborhood system: $NS_U(U)$ is discrete, if $NS_U(U)$ is the power set. $NS_U(U)$ is indiscrete, if $NS_U(U)$ is a singleton $\{U\}$.
8. A crisp/fuzzy topological space is a crisp/fuzzy neighborhood system space (NS-space is used for both), but not the converse.
9. Intersections and finite unions of closed sets are closed in NS-spaces.
10. In topological spaces, unions and finite intersections of open sets are open. In NS-spaces, unions is open, but intersections may not be open.
11. In a crisp/fuzzy topological space $NS_U(U)$ determines and is determined by the collection of all open sets. This property may or may not be true for crisp/fuzzy neighborhood systems.
12. A family S of crisp/fuzzy subsets is a base for a topology, if every open set is a union of members of S.
13. A family X of crisp/fuzzy subsets is a subbase, if the finite intersections of members of X is a base. Note that the union and intersection of empty family is the whole space and empty set respectively.
14. We can take a crisp/fuzzy neighborhood system as a subbase of some topology. Note that two distinct crisp/fuzzy neighborhood systems may give the same topology.
15. A crisp/fuzzy neighborhood system is a subbase (see below) of a topology. An open set in NS-space is also open in this topology, the converse may not be true.

9 Conclusions

There are two main themes in this paper:

1. Granulation: to understand the granules.
2. Approximation: to use them to approximate a general set.

We use crisp/fuzzy neighborhood systems to granulate the universe; it is a multilevel granulation. However, we have focused on single level granulation; its granules are nearest neighborhoods and can be defined by a binary relation. We found granules in function spaces can be used to define qualitative fuzzy sets. Neighborhood systems is a generalization of rough sets and topological spaces so granules are used to approximate general subset(concept).

Here we found applications in approximate retrieval, classifications of literatures and recognition of printed letters. The approximation aspect of granular Computing (approximations by granules) have profound implications in soft computing. For example,

1. In neural networks, the universal approximator can be interpreted as the approximation of a general W-sofset by fundamental granules. Former is a W-sofset defined by that function, and latter are the W-sofsets whose membership functions are the activation functions [20],[17], [31].
2. In fuzzy logic, the unknown control function is approximated by the W-sofsets of linguistic terms.
3. In genetic algorithm, the environmental pressure is approximated by mutation and cross-over operators.

The investigations of such approximations will be reported in the near future.

References

1. S. Bairamian, *Goal Search in Relational Databases*, Thesis, California State University at Northridge, 1989.
2. T. Back, *Evolutionary Algorithm in Theory and Practice*, Oxford University Press, 1996.
3. Cattaneo, G.: Mathematical foundations of roughness and fuzziness. In: S. Tsumoto, S. Kobayashi, T. Yokomori, H. Tanaka and A.Nakamura (eds.), The fourth International Workshop on Rough Sets Fuzzy Sets, and Machine Discovery, PROCEEDINGS (RS96FD), November 6-8, The University of Tokyo (1996) 241–247
4. W. Chu, Neighborhood and associative query answering, *Journal of Intelligent Information Systems*, **1**, 355-382, 1992.
5. K. Fukunaga, "Statistical Pattern Recognition," *Academic Press,* 1990.
6. K. Engesser, Some connections between topological and Modal Logic, *Mathematical Logic Quarterly*, **41**, 49-64, 1995.
7. Halmos, P., Measure Theory, Van Nostrand, 1950.
8. John Kelly. General topology, 1955.
9. Alexander Hinneburg and Daniel A. Keim "An Efficeint Approach to Clustering in Large Multimedia Databases with Noise." In: Proceedings of the 4th International Conference on Knowledge Discovery Data Mining," Agrawahl, Stolorz, and Pistetsky-Shapiro (eds), Aug 27-31, 1998,58-65.
10. T. Y. Lin, Neighborhood Systems and Relational Database. In: *Proceedings of 1988 ACM Sixteen Annual Computer Science Conference*, February 23-25, 1988, 725
11. Topological Data Models and Approximate Retrieval and Reasoning, in: *Proceedings of 1989 ACM Seventeenth Annual Computer Science Conference*, February 21-23, Louisville, Kentucky, 1989, 453.
12. T. Y. Lin, Neighborhood Systems and Approximation in Database and Knowledge Base Systems, Proceedings of the Fourth International Symposium on Methodologies of Intelligent Systems , Poster Session, October 12- 15, pp. 75-86, 1989.

13. T. Y. Lin, Topological and Fuzzy Rough Sets. In: Decision Support by Experience - Application of the Rough Sets Theory, R. Slowinski (ed.), Kluwer Academic Publishers, 287-304, 1992
14. "A Logic System for Approximate Reasoning via Rough Sets and Topology," Methodologies of Intelligent Systems, Lecture notes in Artificial Intelligence 869, ed. By Z. Ras, and M. Zemankova, 1994, 65-74. Co-author: Q. Liu and Y. Y. Yao)
15. T. Y. Lin, and Q. Liu, Rough Approximate Operators-Axiomatic Rough Set Theory. In: Rough Sets, Fuzzy Sets and Knowledge Discovery, W. Ziarko (ed), Springer-Verlag, 256-260, 1994. Also in: The Proceedings of Second International Workshop on Rough Sets and Knowledge Discovery, Banff, Oct. 12-15, 255-257, 1993.
16. T. Y. Lin and S. Tsumoto, Qualitative Fuzzy Sets Revisited: Granulation on the Space of Membership Functions., Atlanta, July 13-15, 2000
17. "Universal Approximator,-Turing computability," The First Annual International Conference Computational Intelligence & Neuroscience, Proceedings of Second Annual Joint Conference on Information Science, Wrightsville Beach, North Carolina, Sept. 28-Oct. 1, 1995, 157-160.
18. T. Y Lin, A Set Theory for Soft Computing. In: *Proceedings of 1996 IEEE International Conference on Fuzzy Systems*, New Orleans, Louisiana, September 8-11, 1140-1146, 1996.
19. T. Y. Lin, Rough Set Theory in Very Large Databases, Symposium on Modeling, Analysis and Simulation, IMACS Multi Conference (Computational Engineering in Systems Applications), Lille, France, July 9-12, 1996, Vol. 2 of 2, 936-941.
20. "The Power and Limit of Neural Networks," Proceedings of the 1996 Engineering Systems Design and Analysis Conference, Montpellier, France, July 1-4, 1996, Vol. 7, 49-53.
21. T. Y. Lin, and Y. Y. Yao, Mining Soft Rules Using Rough Sets and Neighborhoods. In: Symposium on Modeling, Analysis and Simulation, CESA'96 IMACS Multiconference (Computational Engineering in Systems Applications), Lille, France, 1996, Vol. 2 of 2, 1095-1100, 1996.
22. T. Y. Lin, and M. Hadjimichaelm M., Non-classificatory Generalization in Data Mining. In: Proceedings of The Fourth Workshop on Rough Sets, Fuzzy Sets and Machine Discovery, Tokyo, Japan, November 8-10, 404-411,1996.
23. T. Y. Lin and Rayne Chen, Supporting Rough Set Theory in Very Large Database Using ORACLE RDBMS, Soft Computing in Intelligent Systems and Information Processing, Proceedings of 1996, Asian Fuzzy Systems Symposium, Kenting, Taiwan, December 11-14, 1996, 332-337 (Co-author: R. Chen)
24. T. Y. Lin and Rayne Chen, Finding Reducts in Very Large Databases, Proceedings of Joint Conference of Information Science,Research Triangle Park, North Carolina, March 1-5, 1997, 350-352.
25. T. Y. Lin, Neighborhood Systems -A Qualitative Theory for Fuzzy and Rough Sets. In: *Advances in Machine Intelligence and Soft Computing*, Volume IV. Ed. Paul Wang, 132-155, 1997. Also in *Proceedings of Second Annual Joint Conference on Information Science*, Wrightsville Beach, North Carolina, Sept. 28-Oct. 1, 1995, 257-260, 1995.
26. T. Y. Lin, "Granular Computing on Binary Relations I: Data Mining and Neighborhood Systems." In: Rough Sets In Knowledge Discovery, A. Skoworn and L. Polkowski (eds), Springer-Verlag, 1998, 107-121
27. T. Y. Lin, "Granular Computing on Binary Relations II: Rough Set Representations and Belief Functions." In: Rough Sets In Knowledge Discovery, A. Skoworn and L. Polkowski (eds), Springer-Verlag, 1998, 121-140.

28. T. Y. Lin, "Granular Computing: Fuzzy Logic and Rough Sets. " In: Computing with words in information/intelligent systems, L.A. Zadeh and J. Kacprzyk (eds), Springer-Verlag, 183-200, 1999
29. T. Y. Lin, Data Mining and Machine Oriented Modeling: A Granular Computing Approach Journal of Applied Intelligence, 2000
30. A. Motro, Supporting Goal Queries in Relational Databases. In: Kerschberg (ed) Proceedings of the First International Conference on Expert Database Systems, Charleston, South Carolina, April 1-4, 1986
31. J. Park and I. W. Sandberg, Universal Approximation Using radial-Basis-Function Networks, Neural Computation 3, 1991, pp. 246-257.
32. Z. Pawlak, Rough sets. Theoretical Aspects of Reasoning about Data, Kluwer Academic Publishers, 1991
33. Polkowski, L., Skowron, A., and Zytkow, J., (1995), Tolerance based rough sets. In: T.Y. Lin and A. Wildberger (eds.), *Soft Computing: Rough Sets, Fuzzy Logic Neural Networks, Uncertainty Management, Knowledge Discovery*, Simulation Councils, Inc. San Diego CA, 55–58.
34. W. Sierpenski and C. Krieger, General Topology, University of Torranto Press 1956.
35. Słowínski, R., Vanderpooten, D.: Similarity relation as a basis for rough approximations. ICS Research Report 53/95, Warsaw Unviersity of Technology (1995)
36. E. Spanier, Algebraic Topology,McGraw-Hill.
37. Stefanowski, J.: Using valued closeness relation in classification support of new objects. In: T. Y. Lin, A. Wildberger (eds.), Soft Computing: Rough Sets, Fuzzy Logic, Neural Networks, Uncertainty Management, Knowledge Discovery, Simulation Councils, Inc., San Diego CA=20 (1995) 324–327
38. Helmut Thiele: On the Concept of Qualitative Fuzzy Set, University of Dortmund, Pre-print, 1998
39. M. Viveros, *Extraction of Knowledge from Databases*, Thesis, California State University at Northridge , 1989.
40. Y. Y. Yao and T. Y. Lin, Yao, Y.Y., and Lin, T.Y. Generalization of rough sets using modal logic. Intelligent Automation and Soft Computing, An International Journal, **2**, No. 2, pp. 103-120, 1996.
41. Y. Y. Yao, M. S. K. Wong, and T. Y. Lin, "A Review of Rough Set Models," Rough Sets and Data Mining: Analysis of Imprecise Data, Kluwer Academic Publisher, 1997, 47-75
42. Y. Y. Yao, Binary Relation Based Neighborhood Operators. In: *Proceedings of the Third Annual Joint Conference in Information Sciences*, Research Triangle Park, March 1-5, 169-172, 1997.
43. Lotfi Zadeh, Fuzzy Graph, Rough sets and Information Ganularity. In: *Proceedings of the Third International Workshop on Rough Sets and Soft Computings*, San Jose, Nov. 10-12, 1, 1994
44. Lotfi Zadeh, The Key Roles of Information Granulation and Fuzzy logic in Human Reasoning. In: *1996 IEEE International Conference on Fuzzy Systems*, September 8-11, 1, 1996.
45. L.A. Zadeh, Fuzzy Sets and Information Granularity, in: M. Gupta, R. Ragade, and R. Yager, (Eds), *Advances in Fuzzy Set Theory and Applications*, North-Holland, Amsterdam, 1979, 3-18.
46. H. Zimmerman, Fuzzy Set Theory –and its Applications, Second Ed., Kluwer Acdamic Publisher, 1991.

An Inquiry into the Theory of Defuzzification

Shounak Roychowdhury

Oracle Corporation, 500 Oracle Parkway, Redwood Shores, CA, 94065, USA
Email: sroychow@us.oracle.com

Abstract. Defuzzification is an important operation in the theory of fuzzy sets. It transforms a fuzzy set information into a numeric data information. This operation along with the operation of fuzzification is critical to the design of fuzzy systems as both of these operations provide nexus between the fuzzy set domain and the real valued scalar domain. We need the synergy of both of these domains to solve many of our ill-posed problems effectively. In this paper, we will address the problem of defuzzification, present merits and demerits of various defuzzification strategies that are used in the theory and practice, and in design and implementation of applications involving fuzzy theory, fuzzy control, and fuzzy rule base, and fuzzy inference-based systems. We also present in this paper a simple and yet novel defuzzification mechanism.

Keywords: Defuzzification, fuzzy sets, fuzzy encoding, fuzzy decoding, fuzzy control, and fuzzy applications

1 Introduction

Fuzzy control theory has been a very successful pragmatic application of fuzzy set theory. No doubt that many researchers are actively involved in this area. Fuzzy control theory effectively handles hard control problems (the ones that are difficult to solved by analytical methods) by using expert's knowledge base. The knowledge base is expressed in terms of fuzzy rules and these rules are executed through a fuzzy inference engine as shown in the Figure 1. Also, in the same figure, we observe that there are other two key components: a fuzzifier and a defuzzifier.

The fuzzifier performs the act of fuzzification -- is it transforms a numeric value to a fuzzy set. The fuzzification is denoted by $Fuz(\Re) \rightarrow \Im$, where $\Re$ is the domain of real numbers, and $\Im$ is a domain of fuzzy sets. The converse of fuzzification is process of defuzzification. It transforms a fuzzy set to a numeric value. The defuzzification is denoted by $Def(\Im) \rightarrow \Re$. The inference engine takes a fuzzy set and transforms it by applying a collection of fuzzy rules. In this article we will denote the inference engine as $IE(\Im) \rightarrow \Im$. Mamdani's method and the

Compositional Rule of Inference (CRI) are two major fuzzy inference techniques that are often used in practice.

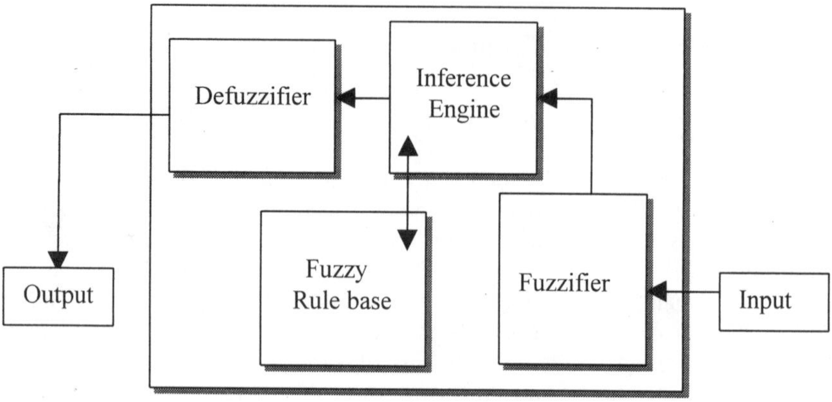

Figure. 1. Fuzzy control system usually has 4 components: Inference Engine, Fuzzy rule base, and Fuzzifier and Defuzzifier.

Thus a generic fuzzy system can be represented by the following transformation of a real number to another real number:

$$\Re \rightarrow Fuz(\Re) \rightarrow \Im \rightarrow IE(\Im) \rightarrow \Im' \rightarrow Def(\Im') \rightarrow \Re .$$

Naturally, the fuzzification and defuzzification processes are like preprocessors and are treated as interfaces in formal fuzzy modeling [16] where these interfaces provide numeric to linguistic conversion and vice versa. Even though fuzzification and defuzzification sound complementary operations, but they do not behave like inverse functions. As noted by Bortolan and Pedrycz in [1] that ideally any form of transformation from linguistic level to numeric level should be recoverable, but is usually not the case due conversion mechanism and granularization effects. They also treat fuzzification and defuzzification as fuzzy decoding and fuzzy encoding process respectively.

$$x' = Def(Fuz(x))$$

Their use of defuzzification operation is geared towards their study of linguistic reconstruction mechanism, and fuzzy encoding and decoding process.

From the perspective of fuzzy control a transformation of defuzzification is acceptable whenever it solves the practical problems. However, conceptually defuzzification operation is still a mysterious transformation that bridges the uncertainty oriented domains and certainty oriented domains. We will discuss a few such issues in this study.

In the section 3, we will also summarize main defuzzification strategies and results that have been proposed by researchers during the last decade. In the section 4 we will provide a new and simple defuzzification process that can be used in practice effectively. This method uses the background information of the universe of discourse. At the end of the paper we provide a brief discussion and summarize with a few concluding notes.

2 Defuzzification problem: we visit again!

In this section we will revisit the defuzzification problem, and critically analyze and interpret the meaning of defuzzification. We would also put forth our viewpoint, and study some of the properties of the defuzzification mechanism.

2.1 Epistemological perspective

Conventionally, defuzzification operation transforms a fuzzy set to a numeric real value. And this view has been used in last so many years quite successfully. However, the epistemological reference behind defuzzification is not so simple.

Certainty deals with that aspect of epistemology where a piece of knowledge cannot be "knowledge" if it is wrong or unclear. There are many epistemological theories [5,8,17] where certainty is measured by the identifying a set of beliefs that can be justified completely, and leaves no room for doubt in the mind of an observer. In this respect, we can say that a real number by its set of properties leaves no choice to a scientific mind, but to believe that it is a concept imbrued with certainty. In this paper, the concepts belonging to this domain are termed as certainty-based concepts.

On the other hand, uncertainty theory deals with those epistemological sources where there can be multifold of beliefs in the observer's mind. Vagueness is due to lack of complete knowledge [5,8], and this induces sets of partial beliefs, for which rational justification may be difficult to present. And as human beings we live by them. In our world of limited knowledge, and understanding we often infer information by using our beliefs and notions, and create vague and amorphous concepts. Fuzzy theory nicely captures these concepts by effectively modeling vagueness. The concepts belonging to this domain are termed in this paper as uncertainty-based concepts.

To understand the role of epistemology in defuzzification we further need to understand the connection between uncertainty and certainty. Defining a transformation from uncertainty-based concept to certainty-based concept is not possible without proving the correctness, and the validity of the uncertainty-based concept. They are two different concepts having different set of properties. Anything that is definite and certain cannot be uncertain and shall not possess any measure of uncertainty. Absolutely justified piece of information that has evidential reasoning form knowledge sources with certainty. Whereas, knowledge source that lacks evidential reasoning cannot be marked by certainty.

Condensing the information that is captured by a fuzzy set into a numeric value is like transforming an uncertainty-based concept to certainty-based concept. Fuzzy set captures a certain an amount of vague knowledge and a numeric value captures absolutely precise piece of knowledge. In this respect, transformation of a vague piece of knowledge into a precise piece of knowledge means to remove all forms of ambiguity by extracting only the true core knowledge source from the former piece of vague imbrued knowledge source. Or through another view, it would mean to compress the vague data source to a pure knowledge source.

In fuzzy literature, the defuzzification process has never been addressed from these viewpoints. Here we follow a path of rational explanation of what defuzzification should really mean in the context of fuzzy theory in general and not just from the perspective of fuzzy logic controllers. From an engineering standpoint we are usually happy with a functional that takes a fuzzy set or a fuzzy number and converts it to a real value and solves our control problems effectively. But we should definitely attempt to secure a scientific answer to this question of defuzzification from the epistemological perspective.

It is possible to map an uncertainty-based concept to another uncertainty-based concept. But it is not feasible to map uncertainty-based concepts to certainty-based concepts. Once the evidential reasoning is presented, truth of knowledge is justified for uncertainty-based concept, then it is no longer an uncertainty based concept, rather becomes certainty-based concept. And that should be treated structurally different.

Defuzzification should be viewed as a transformation from an uncertainty-based concept to another uncertainty-based concept, rather than uncertainty-based concept. Thus, defuzzification should map from a fuzzy set to concept that has the form of a tuple in a Cartesian space of information in terms of defuzzified value (real number) and an associated belief of that value (between 0 and 1). Therefore, the modified defuzzification function should be defined as follows:

$$Def_M(\Im) \rightarrow \Re \times [0,1].$$

Examples of the modified defuzzification values are the tuples such as (20.3, 0.6), and (-75, 0.12) etc., where abscissa is the defuzzified value and the ordinate is its belief value. Therefore, (20.3, 0.6) and (20.3, 0.3) convey different information. What about values with unit belief value? For example, how about (10, 1) tuple? Does it qualify as certainty-based concept? Not really. Here the tuple itself is part of the uncertainty-based concept. We will use this definition in our method that is described in the Section 4 and show how background information can aid in the defuzzification process effectively.

However, we would like to browse through other solutions that have been proposed by various researchers in the field of fuzzy theory. In the following section we will see how a simplified optimization solution has worked for all these years.

2.2 Center of Gravity: a simple solution

Is defuzzification a minimization problem? For most of the practical purposes we think that the minimization of a simple cost function provides us with a solution that we admire. It is simple, and elegant -- the Center of Gravity (ÇOG) defuzzification [13]. The COG method generates a value that is the center of gravity of a fuzzy set. It actually minimizes the membership graded weighted mean of the square of the distance. Consider the discrete fuzzy set:

$$A_f = \left\{ \frac{\mu_1}{x_1}, \cdots \frac{\mu_i}{x_x}, \cdots, \frac{\mu_n}{x_n} \right\},$$

where $\mu_i = \mu(x_i)$ and $\mu(\cdot)$ is a membership function. Let us consider a cost function that is a weighted by its membership function,

$$K(\bar{x}) = \sum_{i=1}^{n} (x_i - \bar{x})^2 \mu(x_i),$$

where $\mu_i = \mu(x_i)$. Minimizing $\mu_i = \mu(x_i)$ by differentiation, we have

$$\frac{dK(\bar{x})}{d\bar{x}} = 2\sum_{i=1}^{n} (x_i - \bar{x})\mu(x_i) = 0,$$

provides the COG, and it is given by:

$$\bar{x} = \frac{\sum_{i=1}^{n} x_i \mu(x_i)}{\sum_{i=1}^{n} \mu(x_i)}.$$

It is clear that although the COG involves an optimization process, it is not responsible for the optimal design of the overall fuzzy systems. Definitely, it has been noticed that a choice of defuzzification always influences the performance of fuzzy systems appreciably, therefore it is necessary to deal the problem of defuzzification as a part of the design process of the fuzzy systems.

2.3 Properties of defuzzification

Runkler and Glesner [23] have provided a set of 13 features that are observed by most of the defuzzification methodologies. Here again, they have assumed that Defuzzification operation takes a fuzzy set and transforms it into a numeric value. We will summarize 13 of their features into 4 core properties.

Property 1: Defuzzification operator always computes to one numeric value.

This property is by the definition of defuzzification. That implies that the defuzzification operator is always injective. Clearly, two fuzzy sets then can have the same defuzzified value. Also, it is assumed that the defuzzified value is always within the support set of the original fuzzy set.

Property 2: The membership function determines the defuzzified value.

The membership function is critical in determining the defuzzified value. Concentration of a fuzzy set monotonically leads to the normal of a normal fuzzy set. Similarly, the dilation operator monotonically leads the defuzzified value away from the normal of the fuzzy set. Note that neither scaling nor translation of fuzzy sets affects the membership function, therefore defuzzified values do not get scaled or translated.

Property 3: The defuzzified value of two triangular-operated fuzzy sets is always contained within the bounds of individual defuzzified values.

If fuzzy set $C_f = T(A_f, B_f)$ where A_f and B_f are fuzzy sets and T is the T-norm, $Def(A_f) \leq Def(C_f) \leq Def(B_f)$, and so is for true for T-conorm (T*) $C_f = T*(A_f, B_f)$

Property 4: In case of prohibitive information, defuzzified value should fall in the permitted zone.

In many application specific situations, strange fuzzy sets may be inferred from the inference engine. As described in [18,26] an application related to robotics that used fuzzy inference had generated two peaked fuzzy sets for which none of the standard defuzzification mechanisms would work. However, in such a case, center of largest area strategies was found to be effective.

3 Defuzzification schemes

In this section we will briefly go though most of the important works that were presented in the last decade. Starting from the standard, ad hoc methods, we then browse through the main results of Yager and Filev [24, 25, 26], Roychowdhury and Wang [20, 21], and Runkler *et. al*. [7, 23].

3.1 The standard methods

There are four most often used defuzzification mechanisms in the fuzzy control theory. They are Mean of Maxima, Computation by the Center of Gravity method, and Center of Means, and the Midpoint of an area procedure [12,13]. Fundamentally these methods focus on geometric area based computation.

- *Mean of Maxima, (MOM)*. This method the computes the center of gravity of the area under the maxima of the fuzzy set.

$$\bar{x} = \frac{\sum_{i=1}^{n} x_i}{\left|\{i \mid \mu_i = \max\{\mu_1, \ldots, \mu_n\}\right|}.$$

It has been noted in [6, 26] that MOM method generates poor steady-state performance, and yields less smooth response curve compared with the COG method.

- The *Center of Gravity (COG)* method computes the defuzzified value based on the equation given below:

$$\bar{x} = \frac{\sum_{i=1}^{n} x_i \mu(x_i)}{\sum_{i=1}^{n} \mu(x_i)}.$$

This method is highly popular and is often used as a standard defuzzification method in experimental as well as industrial controllers. This method is also referred to as the Center of Area (COA) method in the fuzzy literature.

- The *Center of Mean (COM)* procedure computes a defuzzified value of a fuzzy set

$$\bar{x} = x_h$$

where,

$$\sum_{i=\min_j\{j \in M\}}^{h} \mu(x_i) = \sum_{i=h}^{n} \mu(x_i)$$

and

$$M = \{i \mid \mu_i = \max\{\mu_1, \ldots, \mu_n\}\}$$

- The *Midpoint of Area (MOA)* is given by

$$\bar{x} = x_h$$

where,

$$\sum_{i=h}^{n} \mu(x_i) = \sum_{i=1}^{h} \mu(x_i).$$

Most of these methods are ad hoc methods and do not have a very proper scientific reasoning has not been established. However, these methods are quite popular as they are computationally inexpensive and are easy to implement within fuzzy hardware chips. Many researchers have attempted to understand the logic and workings of the defuzzification process. They have provided a set of reasoning and structural basis for defuzzification. Among the well-known methods, Yager and Filev [26] have contributed to the understanding of the process of defuzzification from the perspective of invariant transformations between to different uncertainty paradigms. They have provided many possible solutions, and we will discuss few of them in the following sections. Similarly, Roychowdhury and Wang [21] have attempted to understand the problem of defuzzification from the scope of optimal selection of an element from a fuzzy set. They have used the concepts of interaction, variability, and voting techniques to compute an optimal solution. Learning and clustering techniques [7,9,10] have been used to derive better solutions to cope up be prohibitive information in some of the fuzzy control theory problems. In the next sections, we will explain the foundations in detail.

3.2 Possibility-Probability based defuzzification model

With the advent of fuzzy sets there has been an ongoing debate between the Bayesian scientists and the fuzzy theorists regarding the possibility and probability transformation and other various types of uncertainty [12,13].

To mention in brevity, Klir [14] firmly believes neither of them is stronger nor weaker than the other. He has attempted to prove that the concept of entropy in probability theory and the likewise the concept of non-specificity in the possibility theory have similar roles of measuring of uncertainty-based information. He appreciates the idea that it is possible to transform from possibility distributions to probability distribution and vice-versa.

On the other hand there are supporters of a viewpoint that ignorance and randomness contradict structural homomorphism because they have different characteristics [3]. Recently Dubois and Prade [4] have presented a well-written survey paper on this debatable topic.

3.2.1 BADD and SLIDE

In 1991, Yager and Filev proposed methods based on Basic Defuzzification Distribution (BADD) approach [6]. Their main idea has been to transform a possibility distribution to a probabilistic distribution based on Klir's principle of uncertainty invariance [14]. BADD transforms as concentration or dilation operator to a desired degree depending on δ. The BADD transformation is $F_2(x_i) = (F_1(x_i))^{\delta}$ where $\delta \in (0, \infty)$ and F_1 and F_2 are the original and transformed fuzzy sets, respectively.

The BADD method converts possibility distribution of a normalized fuzzy set to a probability distribution, by using a transformation $\omega(x_i) = K(\mu(x_i))^{\delta}$. Earlier this mapping was proposed and used by Klir to provide a conversion framework between probability and possibility distributions [14]. The probability distribution is given by the following:

$$p_i = \frac{\mu(x_i)^{\delta}}{\sum_{i=1}^{n} \mu(x_i)^{\delta}},$$

after imposing the convexity conditions on the above transformation. They were able to show that the expected value, $E(p_i)$, is given by:

$$\bar{x} = \frac{\sum_{i=1}^{n} x_i \mu(x_i)^{\delta}}{\sum_{i=1}^{n} \mu(x_i)^{\delta}}.$$

BADD method generates COG and MOM for $\delta \to 1$ and $\delta \to \infty$ respectively.

They [25] also proposed SLIDE (Semi linear defuzzification) and an adaptive method that used a linear transformation to transform an original fuzzy set to another fuzzy set that is given by

$$T_{\alpha,\beta} : F_1 \to F_2,$$

where

$$\mu_i = \begin{cases} \eta_i & \text{if } \eta_i \geq \alpha \\ (1-\beta)\eta_i & \text{if } \eta_i < \alpha \end{cases}$$

and η_i, and μ_i are the membership values of fuzzy sets F_1 and F_2 for a particular alpha-cut α, respectively. See [4, 14] for details. Here are some of the transformations that have been proposed in literature. It is clear that the transformation $T_{\alpha,\beta}: F_1 \to F_2$ preserves the shape and the values of the membership function η_i for values of η_i that are equal or higher that α. Thereafter the authors construct a probability distribution by normalizing the sequence of μ_i. The probability distribution is given by the following:

$$P_i = \frac{\mu(x_i)}{\sum_{i=1}^{n} \mu(x_i)},$$

such that

$$\sum_{i=1}^{n} P_i = 1,$$

$$P_i \geq 0, i = (1, n)$$

The properties of SLIDE are proved in [25]. The expected value defined by SLIDE transformation is given by the following:

$$\bar{x} = \frac{(1-\beta)\sum_{i \in L} x_i \mu(x_i) + \sum_{i \in H} x_i \mu(x_i)}{(1-\beta)\sum_{i \in L} \mu(x_i) + \sum_{i \in H} \mu(x_i)},$$

where $L = \{i \mid \eta_i < \alpha, i = (1, n)\}$ and $H = \{i \mid \eta_i > \alpha, i = (1, n)\}$. Another extension of SLIDE is called Modified-SLIDE (M-SLIDE), and interested reader can see [25,26]. Apparently there is not much a significant difference of concepts in both SLIDE and M-SLIDE. In [26] Yager and Filev have discussed the multiplicative transformation which is given by:

$$T_{\alpha,\beta} : \mu_i = \eta_i^{\alpha} s^{\beta},$$

and where $s_i \geq 0$, and is the scaling factor for η_i and $i \in [1, m]$, α and $\beta \geq 0$. Refer to [26] for details.

3.2.2 Constrained defuzzification: Using RAGE

Yager in [24,26] proposed RAGE (Random Generation) method. Similar to earlier methods of BADD and SLIDE a fuzzy set is transformed into a probability distribution. Divide the support set of the fuzzy set in such a way that each of the element fall in a given probability range. And then perform a random experiment to choose the defuzzified element. The viability of this method has been shown in fuzzy control experiments where inferred fuzzy sets exhibit prohibitive information.

3.3 Optimal selection through interacting neighbors

In [21] Roychowdhury and Wang have attempted to explain the concept of defuzzification using the ideas of evolutionary biology. In evolutionary biology the ecology is functionally dependent on the interaction of the different species, and is guided by the Principle of Natural Selection. They proposed a variety of variations functions, $v(\cdot)$, that were derived by modeling the biological concepts of interaction among the neighboring elements in the set. As shown in the Figure 2, all the neighbors (empty circles) interact with the ith element (filled circle) and reach a consensus. The consensus is passed the element i. The defuzzification strategy then selects how the element i should interact with other neighboring elements. As described in the paper, we have:

$$v(x_i) = \frac{f_i(\mu(x_1),\ldots,\mu(x_n))}{\sum_{i=1}^{n} f_i(\mu(x_1),\ldots,\mu(x_n))}.$$

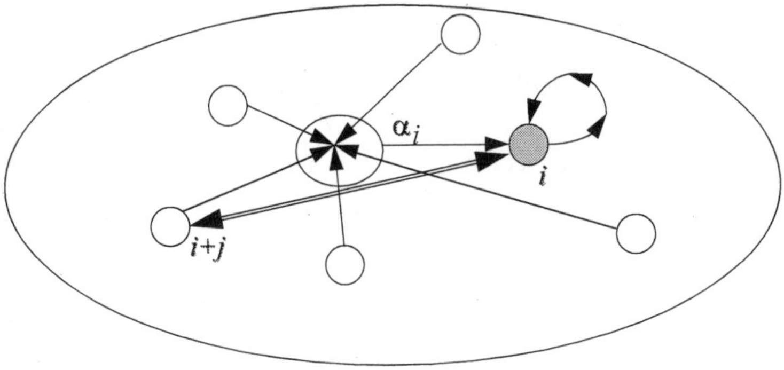

Figure. 2. All neighbors (empty circles) interact with the ith element (filled circle) and reach a verdict α_i. The verdict is then passed to the element i. The defuzzification strategy selects how to the element i should interact with other neighboring elements. The loop on the ith element indicates the self-element contribution $g(\mu_i)$.

and finally by using the quadratic minimization they were able to show the following:

$$\bar{x} = \frac{\sum_{i=1}^{n} x_i v(x_i)}{\sum_{i=1}^{n} v(x_i)}.$$

The above model does generalize COG, and BADD, for $v(x_i) = \mu_i$, and $v(x_i) = \mu_i^{\alpha}$. This is a very generic model that does not need to convert a possibility distribution into a probability distribution and so on, rather uses an alternative approach to understand the problem of defuzzification.

3.3.1 Radial defuzzification

In another paper [22] Roychowdhury *et. al.* extended the above concept of applying cooperative elements of a fuzzy set to a understand the defuzzification process. The method is called the Radial Defuzzification method. In short, the basic idea is to use a Cartesian product of interaction $A_f \times A_f$, called the *i*-space of a fuzzy set A_f such that

$$\mu_{ij} : A_f \times A_f \to [0,1].$$

This means that the ordered pair (i, j) of the *i*-space consists of interaction between the *ith* element and the *jth* element of the same fuzzy set. The interaction value on the ordered pair is denoted by μ_{ij}. It is computed by using membership values of the *ith* element and *jth* element of the fuzzy set, and is given by

$$\mu_{ij} = g(\mu_i, \mu_j).$$

The paper discusses the effect of interacting neighbors with a radial zone of the *i*-space.

Here we will only provide the results of the simplest case when neighbors do not interact, and when only nearest neighbors interact. In the first case, we observe that

$$\mu_{ij} = g(\mu_i, \mu_j) = \mu_i \mu_j.$$

Therefore, the variation function is:

$$v(x) = \frac{\mu_i^2}{\sum_{i=1}^{n} \mu_i^2}.$$

And the defuzzified value is thus given by the following:

$$\bar{x} = \sum_{i=1}^{n} x_i \frac{\mu_i^2}{\sum_{i=1}^{n} \mu_i^2}.$$

We find that the above equation is similar to the BADD method. Notice that the value of $\alpha = 2$.

In another case when nearest neighbors interact we have the following defuzzification result where the variation function $v(x)$ is given by

$$v(x) = \frac{\mu_i(\mu_i + \frac{1}{2}(\mu_{i-1} + \mu_{i+1}))}{\sum_{i=1}^{n} \mu_i(\mu_i + \frac{1}{2}(\mu_{i-1} + \mu_{i+1}))}.$$

And therefore the defuzzified value is:

$$\bar{x} = \frac{\sum_{i=1}^{n} x_i \mu_i(\mu_i + \frac{1}{2}(\mu_{i-1} + \mu_{i+1}))}{\sum_{i=1}^{n} \mu_i(\mu_i + \frac{1}{2}(\mu_{i-1} + \mu_{i+1}))}.$$

In general, in an m-dimensional i-space, the above nearest neighbor defuzzification result can be extended to the following:

$$\bar{x} = \sum_{i=1}^{n} x_i \frac{\mu_i^{m-1}(1+\frac{1}{2}(\mu_{i-1}+\mu_{i+1}))}{\sum_{i=1}^{n}\mu_i^{m-1}(1+\frac{1}{2}(\mu_{i-1}+\mu_{i+1}))}.$$

Refer to [21,22] for details.

3.4 Defuzzification through fuzzy clustering

Genther *et. al* [7]. have proposed another alternative approach to understand the process defuzzification. Their approach is based on fuzzy clustering.

Fuzzy clustering enables us to partition the data into subclasses by generating fuzzy sets. In other words fuzzy membership is assigned to each cluster. Fuzzy c-Means algorithm is a well know fuzzy clustering algorithm [2]. It is an optimization problem dealing with the minimization of the following objective functional:

$$J(U,v) = \sum_{i=1}^{n}\sum_{j=1}^{c} u_{ij}^{m} d(x_i, v_j)^2,$$

where $d(x_i, v_j)$ is a distance function between the cluster center v_j the datum x_i. The membership is calculated by the following:

$$u_{ij} = \frac{1}{\sum_{k=1}^{c}\left(\frac{d(x_i,v_j)^2}{d(x_i,v_j)^2}\right)^{\frac{1}{m-1}}},$$

and $1 \le j \le c$ and $1 \le i \le n$. The centers of the clusters are computed by the following equation.

$$v_j = \frac{\sum_{i=1}^{n} u_{ij}^{m} x_i}{\sum_{i=1}^{n} u_{ij}^{m}} \quad \text{for all } j.$$

This is a standard procedure to compute FCM algorithm. Genther *et.al.* [7] modified the distance function between a datum x_i and the associated membership value $\mu(x_i)$, to the following:

$$d_{ij} = \sqrt{\alpha\frac{(x_i - v_j)}{(x_i - v_j)_{\max}} + (1-\alpha)\frac{(\mu(x_i) - v_j)}{(\mu(x_i) - v_j)_{\max}}}$$

and

$$w_j = \frac{\sum_{i=1}^{n} u_{ij}^m \mu(x_i)}{\sum_{i=1}^{n} u_{ij}^m}.$$

Finally, on obtaining c clusters centers with associated membership values (v_j, w_j), they chose the cluster center k with $w_k = \max_j(w_j)$, and the defuzzified value was its center v_k. Interestingly, by adjusting the parameter α they were able to achieve acceptable defuzzification results for fuzzy sets having prohibitive information.

3.5 Neural network based defuzzification

It has been found in many cases that the choice of defuzzification method can be a critical in designing fuzzy systems. Usually trial and error methods are employed to find out the suitable defuzzifier for a given fuzzy control system. Tuning defuzzification strategy to an application is a desirable option. Halgamuge and Glesner describe in [9] a customizable BADD defuzzifier that uses a transparent neural network to tune defuzzification to different applications. Song and Bortolon [19] have proposed some properties of neural networks that learn the defuzzification process.

Halgamuge[10] describes the CBADD neural network and the learning algorithm that is based on the gradient descent learning and has proved the associated convergence algorithm. By using a learning network, CBADD acts a universal defuzzification function approximator for a fuzzy rule base. He has also shown that CBADD can be approximate most of the standard defuzzification methods as described in the Section 3.1.

4 Defuzzification with background information: *a novel proposal*

The current understanding of the process of defuzzification explicates that the process takes a fuzzy set and converts it to a real number. However, as we have mentioned earlier that a defuzzified value should also have a belief value associated with it. Apart from this additional parameter, we would like to propose that the defuzzification of fuzzy set should be performed only with respect to the universe of discourse in which it really belongs to. This background information will help in our understanding of the fuzzy set that will be defuzzified. We need this universe of discourse because it then provides a relative comparison with respect to the knowledge sources (fuzzy sets) constitute the universe of discourse. Definitely it makes sense to do a relative evaluation of fuzzy set that is to be defuzzified.

In most of the fuzzy systems we use fuzzy inference and a set of fuzzy rules in the fuzzy rule base. Let there be a fuzzy rule base where the ith fuzzy rule is denoted by $F_i : R_i \Rightarrow Q_i \forall i \in (1,n)$ where R_i and Q_i are the antecedent and the consequent fuzzy sets respectively. If R' is an input to the fuzzy rule base, then after using fuzzy inference we obtain Q' as an inferred fuzzy set. Now we would defuzzify Q' without taking into any consideration of the relative knowledge (fuzzy sets) existing in the universe of discourse Q_i s. And that is why many times we face the problem of prohibitive learning.

The proposed mechanism is rather simple in nature. We consider commonality of knowledge Q_i and Q' (denoted by fuzzy sets) by computing the T-norm of both of these fuzzy sets such that $\widetilde{Q}_i = T(Q_i, Q')$. And we compute this for all *i*s. Now let us compute the belief value of defuzzified set by using the following equation:

$$\overline{\mu} = \frac{\sum_{j=1}^{n} x_j \mu(x_j)}{\sum_{j=1}^{n} x_j}.$$

Next step is to compute the defuzzified value of a fuzzy set by the standard COG value, which will be:

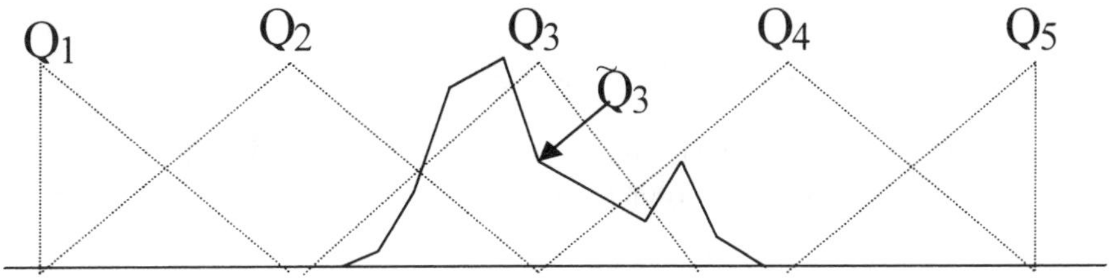

Figure 3.0. In the new defuzzification scheme, the output fuzzy set is compared with the fuzzy sets in the universe of discourse.

$$\overline{x} = \frac{\sum_{i=1}^{n} x_i \mu(x_i)}{\sum_{i=1}^{n} \mu(x_i)}.$$

Therefore following the above two equations we have, $Def_M(\tilde{Q}_i)$, (modified-defuzzification of $\tilde{Q}_i$) is given by $(\bar{x}_i, \bar{\mu}_i)$. Next calculate the maximum of $\bar{\mu}_i$ s, and that is $\bar{\mu}_o = \max(\bar{\mu}_1, ... \bar{\mu}_n)$. And $\bar{x}_o$ is the corresponding value on the fuzzy support set. Thus, the output-defuzzified value would be $(\bar{x}_o, \bar{\mu}_o)$.

By careful observation, it can be noted that this method can take care of prohibitive information defuzzification. Figures 3.0-3.3 illustrate an example. Figure 3.0 shows the original fuzzy set relative to the universe of discourse. Initially we calculate the modified defuzzified value for Q_2 and $\tilde{Q}_3$ (Figure 3.1). Similarly, in Figure 3.2 we compute the modified defuzzified value for Q_3 and $\tilde{Q}_3$. Thereafter, we also compute the same for fuzzy sets Q_4 and $\tilde{Q}_3$ in the following Figure 3.3. Once we calculate these defuzzified values we take the max of the belief values to compute the defuzzified value of the fuzzy set.

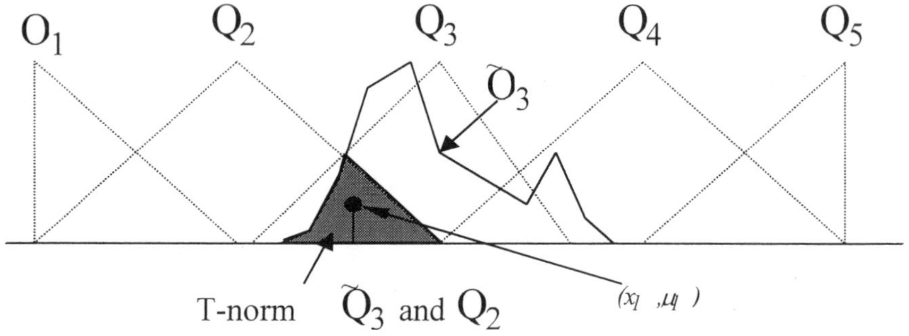

Figure 3.1. Calculate the Modified defuzzification after computing the intersection of the two fuzzy sets.

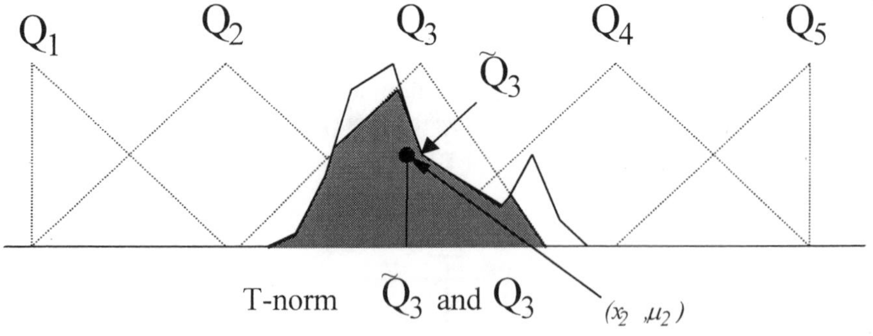

Figure 3.2. Similar to Figure 3.1 compute the modified defuzzified value.

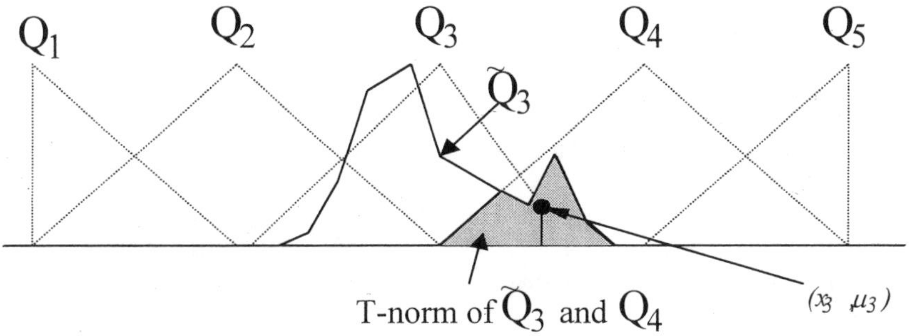

Figure 3.3 Compute the modified defuzzified value shown above.

Example 1 Let the universe of discourse have five fuzzy sets as shown in the above figures, and let its support set range from 0 to 100. The Q_1 be a fuzzy set with the support set from [0, 25], similarly the support sets of other fuzzy sets be Q_2 = [0,50], Q_3 = [25,65], Q_4 = [50,100], and Q_5 = [75, 100]. And the fuzzy set $\tilde{Q}_3$ = [29,72] will be defuzzified. As mentioned in this section we compute the modified-defuzzified values between $\tilde{Q}_3$ and all other fuzzy sets of the universe of discourse. After computation we observe the modified-defuzzified value is 47.34.

5 Discussions

Defuzzification is an interesting topic of research. From a practical standpoint it appears that we have solved defuzzification theory. In the previous sections we briefly scanned though some of the important defuzzification proposals that have been placed in front of the fuzzy research community. Although we have raised some of our concerns regarding the foundational point of view in the section 1, however, it still appears that we need to work on the principles, as the topic has not been fully addressed. Here in this section we would like to briefly browse through the same questions again.

A cluster's prototype represents the generic and common properties of a data set in the cluster. Can we then put this inquiry that defuzzification process is a process of locating a prototype of cluster? Can we abstract the key points from fundamental concepts of statistical pattern recognition, Voronoi diagrams of computational geometry, or vector quantization in study of error codes and signal processing etc. to seek an a convincing answer.

To extract the core knowledge from nebulous information we either strip off all superfluous information and retain what is absolute necessary, or we use some compression mechanism that would take the nebulous information and encode to another piece of information that would be although different, yet would have some usefulness. That knowledge can be treated as defuzzified knowledge. And of course on later decoding of that defuzzified information we should be able to retrieve whatever was encoded in that.

6 Conclusions

In this brief article, we have revisited the defuzzification problem, proposed a few inquiries to the nature of defuzzification. We also did a brief survey of main defuzzification schemes that have been proposed during the last decade. We still find that COG, and MOM remain standard defuzzification operators for most of the fuzzy sets that have been designed either industry or in academia.

We have proposed a novel defuzzification method that is based on the uncertainty-based concept. It is a simple mechanism that uses the background information that is present in the universe of discourse, and we hope it will be find its usage in practical design of fuzzy systems.

References

1. Bortolan, G. and W. Pedrycz (1997), “Reconstruction problem and information granularity,” in the *IEEE Transactions of Fuzzy Systems*, 5, 234-248.

2. Bezdek, J.C., (1981), *Pattern Recognition with Fuzzy Objective Algorithms*, Plenum Press, New York.

3. Driankov, D., H. Hellandroon, and M. Reinfrank (1993), *An Introduction to fuzzy control*, New York, Springer-Verlag.

4. Dubois, D., and H. Prade (1993), “Fuzzy sets and Probability: Misunderstandings, bridges, and gaps,” in the *Proceedings of the 2nd IEEE International Conference on Fuzzy Systems*, San Francisco, CA, 2, 1059-1068.

5. Edwards, P., (1967), (Ed.), *The encyclopedia of philosophy*, Macmillan, New York.

6. Filev, D. P., and R. R. Yager (1991), “A generalized defuzzification method via Bad distributions,” *International Journal of Intelligent Systems*, 6, 687-697.

7. Genther, H., T. A. Runkler, and M. Glesner (1994), “Defuzzification based on fuzzy clustering,” in the *Proceedings of IEE Einternational Conference of Fuzzy Systems*, 1645a-1648.

8. Griffiths, A. P. (1967), (Ed.), *Knowledge and belief*, Oxford university press, Oxford, England.

9. Halgamuge, S. K., and M. Glesner (1994), “Neural networks in designing fuzzy systems for real world application,” *Fuzzy Sets and Systems*, 65, 1-12.

10. Halgamuge, S. K., T. A. Runkler, and M. Glesner (1996), “On neural defuzzification networks,” in the *Proceedings of the IEEE International Conference of Fuzzy Systems*, New Orleans, LA, 463-469.

11. Jaynes, E. T. (1979), "Where do we stand on the maximum entropy?," Levine and Tribus (Eds.) MIT Press, Cambridge, MA.

12. Kosko, B. (1992), "Neural networks and fuzzy systems," Prentice-Hall, Englewood Cliffs, New Jersey.

13. Klir, G. J., and T. A. Folger (1988), "Fuzzy Sets, information, and uncertainty," Prentice-Hall, Englewood cliffs, New Jersey.

14. Klir, G. J. (1990), "A principle of uncertainty and information invariance," *International Journal General Systems*, 17, 249-275, 1990.

15. Lee, C. C. (1990), "Fuzzy Logic in control systems: fuzzy logic controllers-Part II," in the *IEEE Transactions Systems, Man and Cybernetics*, 20.

16. Lygeros, J. (1997), "A formal approach to fuzzy modeling," in the *IEEE Transactions. of Fuzzy Systems*, 5, 317-327.

17. Martins, J. P., and S.C. Shapiro (1984), "Reasoning in multiple belief spaces," in the *Encyclopedia of Artificial Intelligence*, 370-373.

18. Pluger, N, J. Yen, and R. Langari, (1992), "A defuzzification strategy for fuzzy logic controllers employing prohibitive information in command formulation," *in Proc. 1st IEEE Conference on Fuzzy Systems*, San Diego, CA, 717-723.

19. Song, Q, and G. Bortolan (1994), "Some properties of defuzzification neural networks", *Fuzzy Sets and Systems*, vol. 61, 83-89.

21. Roychowdhury, S, and B. H. Wang (1994), "Cooperative neighbors in defuzzification," *Fuzzy Sets and Systems*, 78, 37-49.

22. S. Roychowdhury, B. H. Wang, and S. K. Ahn (1999), "Radial defuzzification," *International Journal of General Systems*, 28, 201-225.

23. Runkler, T. A., and M. Glesner (1993), "A set of axioms for defuzzification strategies –towards a theory of rational defuzzification operators," in the *Proceedings of the 2nd IEEE International Conference of Fuzzy Systems*, San Francisco, CA, 1161-1166.

24. Yager, R. R., and D. P. Filev (1993), "Constrained defuzzification," in the *Proceedings of the Fifth IFSA World Congress*, Seoul, Korea, 1167-1170.

25. Yager, R.R., and D. P. Filev (1993), "SLIDE: a simple adaptive defuzzification method," in the *IEEE Transactions of Fuzzy Systems*, 1, 69-78.

26. Yager, R.R., and D. P. Filev (1994), *Essentials of Fuzzy Modeling*, John Wiley and Sons, Inc..

Fuzzy Partitioning Methods

Christophe Marsala

Université Pierre et Marie Curie, LIP6 - mailbox 169, 4, place Jussieu,
75252 Paris cedex 05, France - email: Christophe.Marsala@lip6.fr

Abstract. In this chapter, we propose two new algorithms to infer automatically a fuzzy partition for the universe of a set of values, when each of these values is associated with a class. These algorithms are based on the use of mathematical morphology operators that are used to filter the given set of values and highlight kernel for fuzzy subsets. Their purpose is to be used in an inductive learning algorithm to construct fuzzy partitions on numerical universes of values.

1 Introduction

In this chapter, we focus on discretization methods of a set of numerical values in the context of inductive learning.

Inductive learning raises a *particular* to the *general*. A set of classes $\mathcal{C}$ is considered, representing a physical or a conceptual phenomenon. This phenomenon is described by means of a set of attributes $\mathcal{A} = \{A_1, ..., A_N\}$. Each attribute A_j can take a value v_{jl} in a given universe X_j. A *description* is a N-tuple of attribute-value pairs (A_j, v_{jl}). Each description is associated with a particular class c_k from the set $\mathcal{C} = \{c_1, ..., c_K\}$ to make up an *instance* (or *example*, or *case*) e_i of the phenomenon. Inductive learning is a process to generalize from a *training set* $\mathcal{E} = \{e_1, ..., e_n\}$ of examples to a general law to bring out relations between descriptions and classes in $\mathcal{C}$.

Given an attribute A_j, a training set $\mathcal{E}$ defines the ordered set $\{v_{j1}, ..., v_{jm_j}\}$ of possible values for A_j. In this paper, to simplify our notation, we denote A_j both the attribute itself, and the set $\{v_{j1}, ..., v_{jm_j}\}$ of all values from the training set of this attribute. Hereafter, for each e_i from $\mathcal{E}$, we denote by $e_i(A_j)$ its value for attribute A_j (*ie.* there exist v_{jl} from A_j such that $e_i(A_j) = v_{jl}$).

An attribute A_j is a numerical attribute if its set of values is a subset of a continuous or infinite universe X_j (for instance, $X_j = \mathbb{R}$ or $X_j = \mathbb{N}$). In this context, given a numerical attribute A_j, the universe X_j of its values has to be partitioned in order to reduce the number of values of A_j.

The problem of the use of numerical attributes in an inductive learning algorithm has been studied in several algorithms (for instance, in [31], or [33]). Several solutions are based on the use a fuzzy representation of these values ([5], [7], [37], [43]). However, this solution gives rise to the complex problem of the generation of a fuzzy representation. A natural idea is to obtain it from experts of the studied domain. But it could be difficult to find experts or expertises for particular kinds of data.

One possible means to obtain a fuzzy representation of the values of an attribute is to infer a fuzzy partition from the data of the training set. There exist various methods to infer a fuzzy partition from a set of data [1]. In an inductive learning scheme, we prefer to use an automatic method which infers a fuzzy partition at each step of the learning algorithm that can be, for instance, an algorithm to build decision trees. Thus, the induced fuzzy partition can be related to the training set of the current step of the decision tree construction. We do not want a highly elaborated method which covers the whole space of training examples as the Krishnapuram's method [21], the neural network method, or genetic algorithm method. These methods cluster the space covered by all the attributes of the data, in one step. We prefer to use an algorithm which builds a fuzzy decision tree to cluster this space.

Thus, we propose to infer a fuzzy partition in an automatic way, in an intermediate step of the construction of the decision tree. This method is easy to implement and gives good results. At the end, the final decision tree will take into account all the attributes involved in the recognition of a class and dependencies of these attributes.

In this chapter, we present a solution based on the use of mathematical morphology operators, and formalized by means of tools from theory of formal language. The implementation of this solution is described in [23].

In Section 2, we present the state of the art of fuzzy partitioning methods. In Section 3, we present a new algorithm to infer a fuzzy partition over a set of numerical values. In Section 4, we propose an extension of this algorithm to extract a reduced set of fuzzy values from a set of fuzzy values. And, finally, we conclude on this method.

2 Discretization of numerical attributes

Classical method to discretize a universe of numerical values search for thresholds in this set. These thresholds will be used to test values under or above them.

In this section, we present several existing methods to find such thresholds: classical methods and fuzzy methods. More details on such methods can be found in several general articles [13], [20], or [36].

2.1 Searching for crisp thresholds

Methods that do not use the distribution of classes on the universe X_j to discretize are not very interesting to study.

Such kinds of discretization are, for instance, based on the splitting of the universe of values into a set of intervals with the same length, or with a length related to a given proportion of examples. In this case, nothing is done to handle really the numerical attribute: this attribute is only considered as a symbolic one. We focus here on methods that use the class associated with each value v_{jl} of attribute A_j.

The discretization method implemented in the CART system [8] is based on the optimization of the value of a criterion: the Shannon entropy measure. The chosen splitting points in X_j are those who enables us to obtain the best value (the lowest one) for that measure. A splitting point generates 2 subsets of values from the whole set: the subset of lower values and the subset of greater values. So, the best splitting point minimizes the value of the entropy measure relatively to the distribution of classes in the 2 created subsets.

This kind of method can be optimized when considering that it is unnecessary to generate all the possible splitting points for a given universe in order to find the best one. It can be proven that the best splitting point is always between two values associated with 2 different classes [14,15].

Moreover, the entropy measure can be substituted with another criterion, for instance the minimum description length principle [15], or with a contrast measure [40]. However, a contrast measure alone could be inappropriate in a learning scheme because it does not handle classes associated with values. Thus, [40] proposed a new measure composed of a measure of entropy and a contrast measure to find the best thresholds.

In literature, there exist a lot of statistical-based methods to find such thresholds. Some of these methods present the advantage of taking into account the distribution of classes related to the numerical universe to find best splitting points.

For instance, the CHIMERGE algorithm introduced by [19] is based on the use of the statistical χ^2 measure to merge intervals. This merging is done until a given threshold for which the χ^2 measure is reached. This algorithm is a bottom-up algorithm that starts from values and ends on intervals.

The FUSINTER algorithm by [35,45] is based on the use of the statistical test of Mood. This test highlights the separability of classes by studying their distribution curve. It enables us to build subsets and to find a set of points that can be compared by means of a quadratic entropy measure.

In the bayesian classifier domain, the discretization algorithm by [30] proceeds by splitting or merging intervals. This process is supervised by a cross validation method.

This algorithm is similar to the one proposed by [17]. Here, neighboring intervals are merged when they are labeled by the same class. Moreover, the authors introduced a new value for the class in addition of existing classes. This new value of the class is called "indecision class" and is used to label intervals when no majority class exists in this interval.

2.2 Searching for fuzzy threshold

The problem with the discretization by means of the construction of crisp thresholds, consists in the importance brought out by such thresholds when they are used. The crispness of such a threshold leads to a decisive answer when tested that produces a lack of flexibility in a learning scheme. In fact, we have to keep in mind that such a threshold is artificial and reflects only the result of the discretization method applied to a particular set of values. For instance, the threshold of 1m72 is found on the numerical universe of human size as the average size of French men. This value splits the French men as small ($< 1m72$) or tall ($\geq 1m72$), but it is rather difficult to separate people with size equal to 1m71 or equal to 1m73.

To limit the influence of a threshold in the determination of a decision, [9] proposed a flexibility in the result of comparison related to this threshold. Like a fuzzy set theory based method, the authors weighted the result of the comparison (below or above) with the estimated probability of this result. This kind of method is also used by [32] to construct probabilistic decision trees.

A fuzzy set theory based method was introduced in [6,22]. In this method, the numerical universe X_j is discretized by means of the Shannon entropy, like in the CART system, into two subsets of values by searching for a threshold $t \in X_j$. This discretization highlights two particular values v_l and v_h from A_j that flank the threshold t: $v_l \leq t < v_h$. These two values are used to define two trapezoidal membership functions of two fuzzy subsets: v_l is the highest value of the kernel of the fuzzy subset *lower than* t and the lowest value of the support of the subset *greater than* t, v_h is the lowest value of the kernel of the fuzzy subset *greater than* t and the highest value of the support of the fuzzy subset *lower than* t. Thus, a fuzzy partition is constructed for X_j.

A similar method is proposed by [18]. However, here, the membership functions are considered as gaussian and not only piecewise linear.

A rather similar method is proposed by [44]. They consider an interval around a threshold as a fuzzy region.

The limitation of this kind of methods lies on the fuzziness of the region created around a threshold. This region is very limited and, when using it, few values will benefit from it.

That is the reason why the search for fuzzy partitioning methods seems to be more promising.

2.3 Searching for fuzzy partitions

In general, fuzzy values on a numerical universe can be obtained by several means.

First of all, they can be given by expert of the domain. But, often either no expert exists to give them, or existing experts cannot express such kind of knowledge or express it poorly, even if they use it. Another means is to use a methodology to construct such a fuzzy partition.

Various methods exist to infer a fuzzy partition. A lot of them are based on statistical theory to find a fuzzy partition.

Questionnaire-based methods construct a membership function by means of answers to basic questions asked to experts or users of the domain. The questionnaire can consist either in simple questions like: "Is this value in the subset or not?", or more complex questions: "Is this value greater than this one or not?" or "Can you associate this value with the more representative subset?". Answers allow the construction of kernels of membership functions by means of a given aggregated method, more or less elaborated (see [1] or [2] for more details).

We are interested here by automatic methods to construct a fuzzy partition that are also used. We do not develop here neural network based methods or genetic methods used to optimize a fuzzy partition because this kind of methods is based on the use of a learning method only to optimize parameters of fuzzy membership functions. Neural networks provide a fuzzy partition of a numerical universe as a function valued by means of the network. It is by itself a learning method [28]. In the same way, genetic methods are based on the optimization of the parameters of membership functions given a previous fuzzy partition (either obtained from experts, or from any other method)[38]. It is an optimization-based method. They can be useful in a pre-treatment of the data in order to highlight fuzzy subsets of values in the numerical universe.

Similarly, in the SAFI software, [37] introduces an automatic method to enhance fuzzy partitions given by an expert. The expert can interactively modify its fuzzy partition in relation with the data from the training set. Constructed fuzzy partitions are thus optimized by means of an entropy measure of fuzzy events. The method is based on a fixed limit, the minimal spread degree, that will enable the kernel of the membership functions to be distant by at least this degree.

In this chapter, we focus only on methods to construct membership functions to help a fuzzy inductive learning process, for instance in the data mining step of knowledge discovery from data [10], [25]. In this context, it is obvious that an automatic and autonomous method is more satisfactory than a parameterized or an expert-based method.

Such a method is introduced by [43] in a fuzzy decision tree construction process. Given intervals of X_j with the same length, a basic triangular function is associated with each interval. An iterative algorithm is used to optimize the kernel by valuing the distance between each training data and the middle of the kernel and by modifying the middles by means of their nearest data. This automatic method is interesting but does not use the distribution of the values of the class on the numerical universe.

Another method is proposed by [42] in a fuzzy decision tree construction process too. This method constructs a fuzzy partition thanks to the Kolmogorov-Smirnov measure[1] normalized by means of a contrast measure.

---

[1] Given two class values c_1 and c_2, the Kolmogorov-Smirnov measure values the maximal distance between the two distributions of probability of each class: $\mathcal{K}(c_1, c_2) = \max_x(|P_{c_1}(x) - P_{c_1}(x)|)$ with P_{c_1} (resp. P_{c_2}) distribution of probability of c_1 (resp. c_2).

The authors justify the use of this measure because it is a non-convex measure and, in consequence, it does not favor non fuzzy partitioning in spite of fuzzy partitioning when it is optimized [41].

In a learning scheme, another interesting method is introduced by [29]. Given a set of numerical values, called *elements*, the mean is valued for each element associated with the same class. This mean value is used as a prototype value: the distance of each element to this mean is valued. This distance is used to associate a membership degree to the class for the element and, thus, to build a fuzzy partition.

3 Fuzzy partitioning of a set of crisp values

In this section, we propose a new method to construct a fuzzy partition of a set of crisp values associated with crisp classes. This method is based on the use of mathematical morphology operators to filter the distribution of classes on the set A_j of values.

First of all, we present the adaptations of these operators in our context. Thus, an algorithm is given to fuzzify a numerical universe. Basic mathematical morphology operators used in this method are recalled in Annex.

3.1 Smoothing a set of values

Usually, the morphological operators are applied on a 2D-picture. They have to be adapted to be used in our context of fuzzy partitioning.

Let A_j be the numerical attribute whose universe X_j has to be partitioned, according to a training set $\mathcal{E}$. We introduce the representation of the distribution of classes on the ordered set $\{v_{j1}, ..., v_{jm_j}\}$ as a word. And we propose the use of rewriting systems, from formal language theory [16], to smooth this word in order to obtain fuzzy modalities of the attributes.

Let $\mathcal{L}$ be an alphabet, each letter of $\mathcal{L}$ representing one of the classes in $\mathcal{E}$. We construct the alphabet $\mathcal{L}_u = \mathcal{L} \cup \{u\}$ with $u \notin \mathcal{L}$. The letter u is a particular letter in the system, we will use it to determine *uncertain* sequences. For any alphabet $\mathcal{A}$, we denote $\mathcal{A}^*$ the set of all possible words composed by letters from $\mathcal{A}$.

For example, let the training set be $\mathcal{E}$ ={(17, cheap), (24, expensive), (29, expensive), (33, expensive), (42, expensive)} with {cheap, expensive} as set of classes and $\{17, 24, 29, 33, 42\}$ as set of values of an attribute (e.g. the attribute "size"). Let $\mathcal{L} = \{c, e\}$, c representing *cheap* and e representing *expensive*. The word defined in $\mathcal{L}^*$ by the training set is *ceeee*.

After this transformation of the training set into a word, we define various ways of using this word.

To construct a fuzzy partition of X_j, we want to obtain sequences of letters from $\mathcal{L}$, as homogeneous as possible, in order to associate them with fuzzy subsets of X_j (constituting thus a fuzzy partitioning of X_j). Each subset will represent a linguistic modality of the attribute (for instance *small* and *big* with the previous example). To obtain such sequences, we present several techniques to alter a word. Our goal is to erase non-representative values within a word in order to smooth it. We use operators inspired by mathematical morphology theory presented previously. To alter a word, these operators are implemented as rewriting systems. Each rewriting system is given as a *transduction*, an *automaton* where each transition is associated with both an input and an output [16]. Basic notions on transductions are recalled in Annex.

Transduction for the erosion and for the dilatation. Let us define a transduction $Er_x = \langle \mathcal{L}_u, \mathcal{L}_u, S_{Er}, I_{Er}, E_{Er}, \delta_{Er} \rangle$ (see Fig. 1). This rewriting

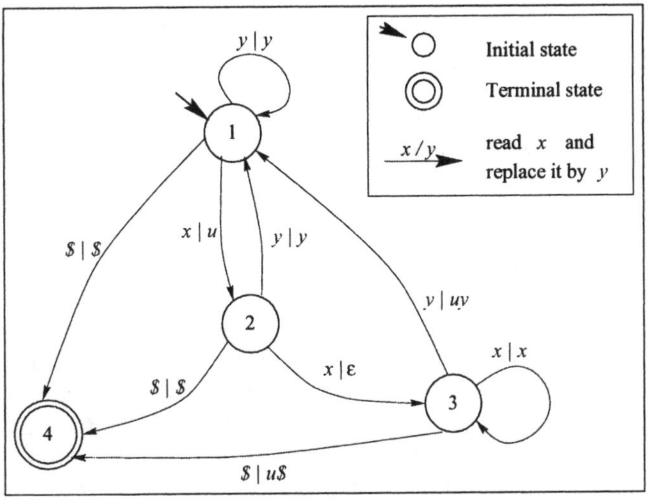

Fig. 1. Transduction for the erosion related to $x \in \mathcal{L}$ with $y \in \mathcal{L}_u$ $y \neq x$

system is used for the erosion of a word with a particular letter $x \in \mathcal{L}$ as structuring element. It corresponds to the reduction of sequences of x in the word. From the word w, $w \in \mathcal{L}_u^*$, we obtain the word $Er_x(w) \in \mathcal{L}_u^*$.

For example, with the word $w = xyyyy$, to find $Er_x(w)$ we use the transduction given in Fig. 1, and we obtain $Er_x(xyyyy) = uyyyy$.

Now, let us define $Di_x = \langle \mathcal{L}_u, \mathcal{L}_u, S_{Di}, I_{Di}, E_{Di}, \delta_{Di} \rangle$ (see Fig. 2), another rewriting system. This system will *dilate* a sequence in a word when this sequence is surrounded with letters u.

It can be proven that for any given word, the computed terminal word is unique [23]. Thus, we are sure that $Er_x(w)$ and $Di_x(w)$ exist for all word $w \in \mathcal{L}_u^*$ and for all letter $x \in \mathcal{L}$, and therefore, for all training sets.

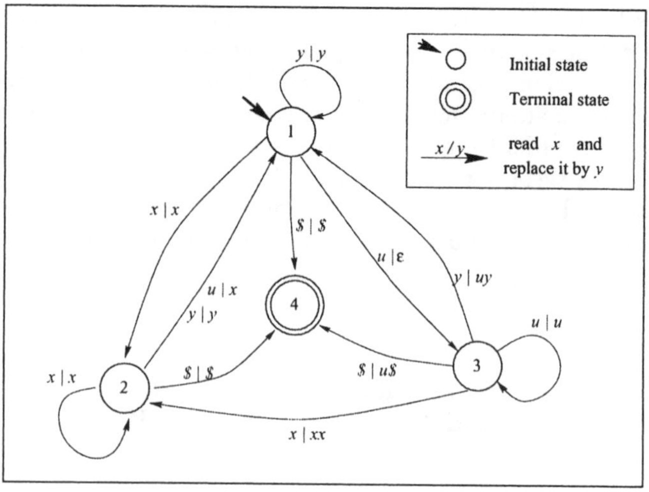

Fig. 2. Transduction for the dilatation related to $x \in \mathcal{L}$ with $y \in \mathcal{L}_u$ $y \neq x$

Moreover, for any word $w \in \mathcal{L}_u^*$, we call $Er_x^k(w)$ (resp. $Di_x^k(w)$), with $k > 0$, the word obtained from w after k consecutive erosions (resp. dilatations).

Now, with these two rewriting systems we will define the two usual operations from the mathematical morphology: the *opening* and the *closure*.

The opening and closure operators. The opening is the composition $Di_x \circ Er_x$ of the two previous operators (*e.g.* rewriting systems). The k-opening ($k \in \mathbb{N}$) of a word $w \in \mathcal{L}_u^*$ with respect to $x \in \mathcal{L}$ is defined as $Op_x^k(w) = Di_x^k(Er_x^k(w))$. The k-opening of a word allows to erase small sequences with length smaller than $2k$. The advantage of this operation is that we can erase all sequences in w with length smaller than a fixed value.

For example, with $w = yyxxxyxyx$, we have $Op_x^1(w) = yyxxxyuyu$ and $Op_x^2(w) = Di_x^2(Er_x^2(w)) = Di_x^2(yyuuuyuyu) = yyuuuyuyu$.

The closure is the composition $Er_x \circ Di_x$ of the two operators of erosion and dilatation. This composition allows to join sequences of letters in a word if these sequences are separated by less than two letters u. The k-closure ($k \in \mathbb{N}$) of the word $w \in \mathcal{L}_u^*$ with respect to $x \in \mathcal{L}$ is defined as $Cl_x^k(w) = Er_x^k(Di_x^k(w))$. With this operator, two sequences separated by less than $2k$ letters u are unified.

For example, with the word $w = uuxxuuuxxuxuuu$, we have $Cl_x^1(w) = uuxxuuuxxxxuuu$ and $Cl_x^2(w) = uuxxxxxxxxxuuu$.

Finally, we introduce the *filter* operator which transforms a word into a sequence of homogeneous series of letters of $\mathcal{L}_u$. In the framework of the utilization of a training set, a filter allows to smooth the training set to deduce a fuzzy partition.

c A filter operator. A filter is a composition of the previously described word-transforming operators. Let $w \in \mathcal{L}_u^*$, $x \in \mathcal{L}$ and $k \in \mathbb{N}$. The k-filter of the word w with respect to x is defined as:

$$\text{if } k = 1 : Fil_x^1(w) = Cl_x^1(Op_x^1(w))$$
$$\text{if } k > 1 : Fil_x^k(w) = Cl_x^k(Op_x^k(Fil_x^{k-1}(w)))$$

The particular combination of these operators has some interesting properties. A filter will allow to *smooth* a fuzzy subset. First, the sequences with a length of $2k$ letters are erased (*ie.* , replaced by the letter u), then we unify sequences separated by $2k$ letters.

3.2 Fuzzy partitioning of a set of values

Now, we present an algorithm to infer a fuzzy partition on a set of numerical values, after the use of the previous rewriting system.

When we apply a filter to the word induced by a training set, we are able to translate small sequences of classes into uncertain sequences. To smooth a training set, we apply a k-filter to it. Then, we obtain a word with large sequences (with a length larger than $2k$). k is a value, empirically fixed, given to the system. In our system, to filter the universe of values of attribute A_j, we choose k equal to 10% of the number of cases in $\mathcal{E}$: $k = \lfloor 0.1 * n \rfloor$.

The sequences of u represent *uncertain* sequences where the classes are highly mixed. Some sequences consist of a single letter x, $x \in \mathcal{L}$. These sequences describe roughly a single class; these sequences do not contain any u character. We call them *certain* sequences, whatever x may be. We will use these uncertain and certain sequences to build a fuzzy partition of the training set T that is related to an attribute. *Certain* sequences of letter x correspond to the kernels of the fuzzy sets of the partition.

Let r be the number of fuzzy modalities we want for the attribute. We select the r largest certain sequences containing one class (Fig. 3).

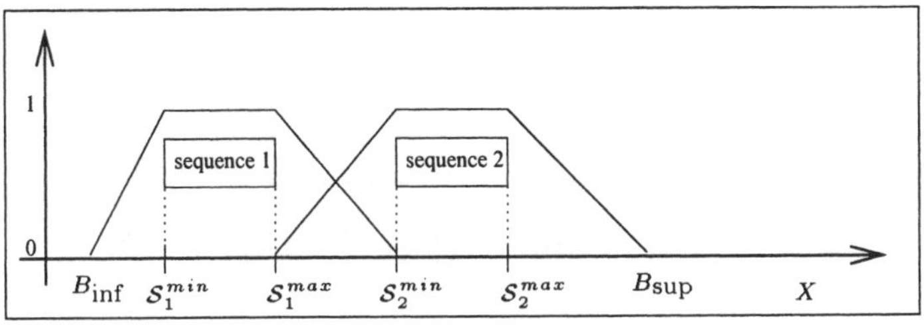

Fig. 3. Fuzzy partition from 2 kernels

To each sequence, we assign intervals from X, for instance $[\mathcal{S}_1^{min}, \mathcal{S}_1^{max}]$ and $[\mathcal{S}_2^{min}, \mathcal{S}_2^{max}]$ when $r = 2$. In the case where we cannot find r such sequences, we can either reduce the number of applied filters, or select fewer sequences. We summarize this in the following algorithm FPMM, Fuzzy Partitioning using Mathematical Morphology, with $r = 2$:

Algorithm 31 (FPMM) *To find a fuzzy partition on X, given a training set $\mathcal{E}$, in 2 fuzzy subsets*

1. *Transform $\mathcal{E}$ into a word w.*
2. *For k fixed, smooth w.*
3. *Find the two largest certain sequences $\mathcal{S}_1$ and $\mathcal{S}_2$.*
4. *We denote by $\mathcal{S}_i^{min}$ (resp. $\mathcal{S}_i^{max}$) the value associated with the first (resp. last) letter of $\mathcal{S}_i$ in X, $i = 1, 2$. $\mathcal{S}_1 \equiv [\mathcal{S}_1^{min}, \mathcal{S}_1^{max}]$ (resp. $\mathcal{S}_2 \equiv [\mathcal{S}_2^{min}, \mathcal{S}_2^{max}]$) with $\mathcal{S}_1^{max} < \mathcal{S}_2^{min}$.*
5. *The fuzzy partition is defined as a family of two fuzzy subsets. The kernel of the first one is $]-\infty, \mathcal{S}_1^{max}]$ and its support is $]-\infty, \mathcal{S}_2^{min}]$. The kernel of the second one is $[\mathcal{S}_1^{max}, +\infty[$ and its support is $[\mathcal{S}_2^{min}, +\infty[$*

3.3 An illustration of the algorithm

Let $\mathcal{E}$ be a training set with a numerical attribute A (e.g. the age), defined on the universe X, and two classes + and –. $\mathcal{E}$ = {(5, –), (7, +), (8, –), (13, –), (14, –), (17, +), (20, –), (21, +), (22, –), (23, –), (25, +), (29, +), (30, +), (35, +), (36, –), (38, +), (40, +) }. We represent $\mathcal{E}$ in a graphical form (Fig. 4).

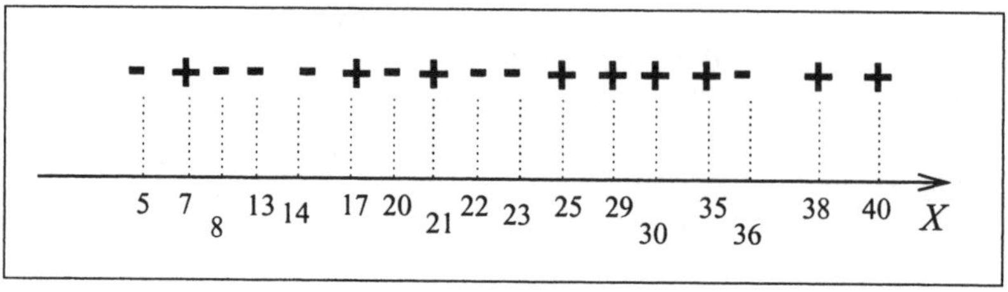

Fig. 4. A training set

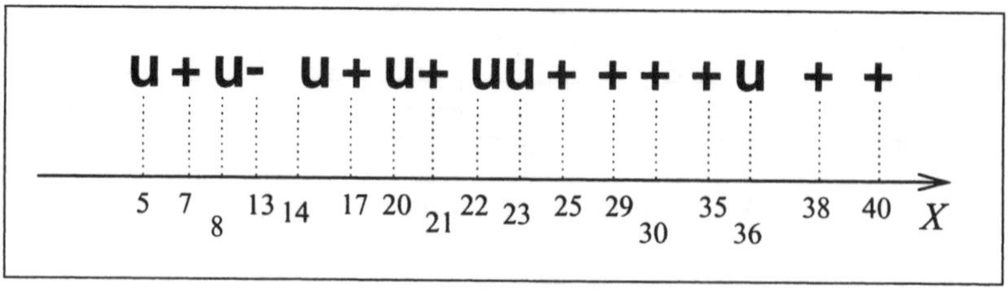

Fig. 5. The training set after an erosion related to –

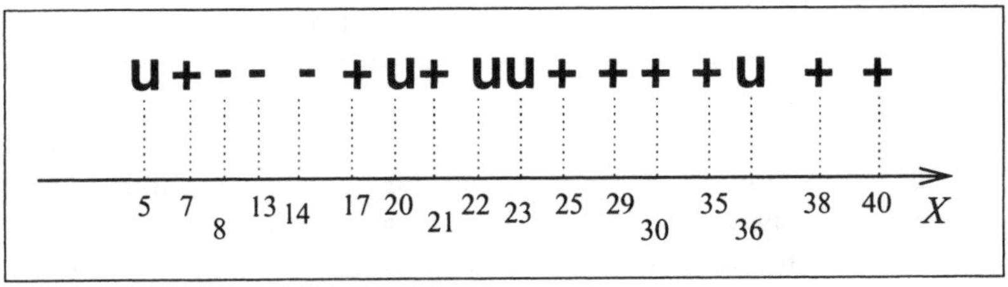

Fig. 6. The training set after an opening related to –

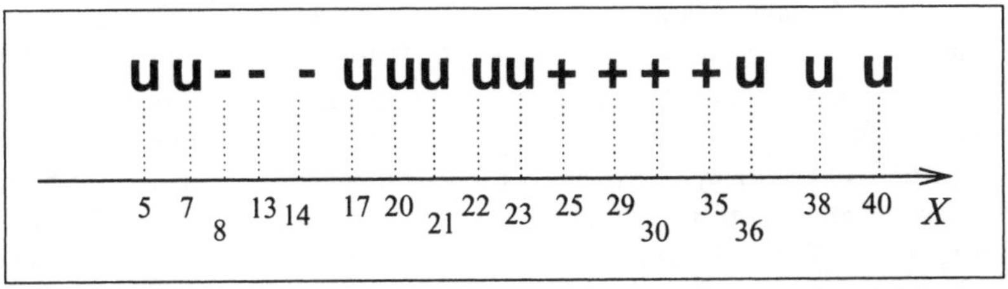

Fig. 7. The training set after an opening related to – followed by an opening related to +

To infer a fuzzy partition on the X, we apply the algorithm FPMM. First, $\mathcal{E}$is transformed into a word. Let $\mathcal{L} = \{+, -\}$, the word associated with $\mathcal{E}$ is w =-+---+-+--++++-++ (see Fig. 4).

We filter w with $k = 1$, and we obtain the filtered word on $\mathcal{L}_u$ where two non uncertain sequences appear: a sequence of - and a sequence of +. We use them as the basis of kernels of two fuzzy subsets (Fig. 8). In Fig. 5, Fig. 6, and Fig. 7 an illustration of the use of morphological operators is presented.

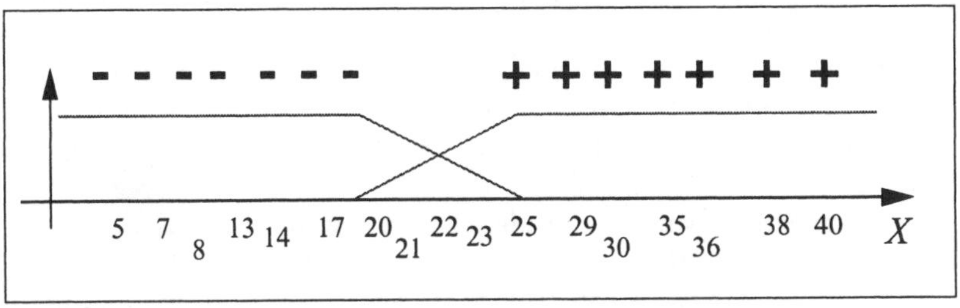

Fig. 8. A fuzzy partition of the training set

3.4 Applications of this algorithm

We implement this algorithm to infer fuzzy partitions for numerical attributes during the construction of a fuzzy decision tree [23].

The software *Salammbô* is based on an extension of the ID3 Algorithm with a fuzzy measure of entropy, the entropy-star [7]. This measure is usable when a set of numerical values is associated with a fuzzy partition over it. Usually, this fuzzy partition is given by an expert of the considered domain. In the software *Salammbô*, fuzzy partitions for numerical attributes from the training set are constructed by means of the algorithm *FPMM*.

Applications in several domain of knowledge discovery ([24–27]) show that the results are more interesting with the fuzzy decision tree based method than with a traditional ID3-based method: fuzzy trees are shorter and generalize better for new cases.

For instance (for more details, see [25]), for the Breiman's waveforms problem [8] where examples are described by means of 21 numeric attributes and where there are 4 classes to recognize. Fuzzy decision trees constructed with the help of the fuzzy partitioning algorithm have an average classification rate (the number of test examples that are well classified by means of the built tree) of 78.2%. In comparison, for the same data, classical decision trees constructed by means of the algorithm C4.5 (see [34]) have an average classification rate of 72.7%.

4 Fuzzy partitioning of a set of fuzzy values

In an inductive learning scheme, the training set could be composed of training cases described by means of fuzzy values for some attributes. Moreover, the class can be a fuzzy value itself and each cases can thus be associated with a degree of membership to each class. The previous algorithm has to be enhanced to take into account such kind of training set.

Thus, we propose here a new algorithm, based on the same kind of considerations as previously given, to find a fuzzy partition for an universe from a fuzzy set of values.

When considering a set of values, a new method is introduced to reduce and to generalize this set. Given an attribute A defined on a numerical universe X, and for each (fuzzy) value $c_k \in \mathcal{C}$, a fuzzy subset μ_{c_k} on X has to be constructed.

This fuzzy subset will be considered as a new fuzzy value v_{c_k} of A that associates values from the training set and knowledge related to the class c_k. This new fuzzy value v_{c_k} is constructed by aggregating all fuzzy values v_l of A from the training set $\mathcal{E}$. Moreover, in this aggregation, each value v_l must be weighted by its membership to the class c_k. Thus, a set of new fuzzy values v_{c_k} is defined on X, for each class c_k.

In a learning scheme, it appears that such a new value, that reflects perfectly the primary set of values, should be too specialized to this set and does not handle impreciseness of the training set values. So, it should be filtered, as in the previous method, to fuzzify it. A morphological method is introduced after the aggregation of values to obtain a set of more general fuzzy values.

In this section, we present a proposition of such a method of fuzzy partitioning from a set of fuzzy values.

Let $\mathcal{E}$ be a training set as previously defined. Let $c \in \mathcal{C}$ be a (fuzzy) class, and let $A \subseteq \mathcal{A}$ be a numerical attribute[2], with X as universe of values. Our goal is to construct a fuzzy partition on X according to $\mathcal{E}$.

4.1 Aggregation of fuzzy values

First of all, we construct a single fuzzy subset of values on X associated with $\mathcal{E}$ for c.

For each case e from $\mathcal{E}$, the membership degree of e to c is denoted by $\delta_c(e)$. Given a value $v \in A$, several cases $e_i \in \mathcal{E}$, $i \in \{1, ..., n\}$ can possess this value, each one associated with a membership degree $\delta_c(e_i)$ to c. We define $\gamma_c(v)$ the membership degree of v to c as:

$$\gamma_c(v) = \max_{\{e_i \mid e_i(A)=v\}} (\delta_c(e_i)).$$

The membership of v to c is defined as the union of the membership degrees to c of all cases that possess this value.

The degree $\gamma_c(v)$ enables us to weight the membership function μ_v. This will produce a new membership function $\mu_{v,c}$ to characterize the simultaneous occurrence of v and c in $\mathcal{E}$. We define:

$$\forall x \in X,\ \mu_{v,c}(x) = \min(\mu_v(x), \gamma_c(v)).$$

The intersection operator is used to reflect the fact that x *belongs to* v and *to* e_i *which itself belongs to* c *with the degree* $\delta_c(e_i)$

All computed degrees $\mu_{v,c}$ for modalities v are aggregated to obtain a new fuzzy value v_c for attribute A. The membership function μ_{v_c} of this new modality reflects the membership of the whole set of training cases to c with regard to A. We define μ_{v_c} as the aggregation of all the membership functions computed to measure the membership of a single value:

$$\forall x \in X,\ \mu_{v_c}(x) = \max_{v \in A}(\mu_{v,c}(x)).$$

At the end, for each class $c \in \mathcal{C}$, a single fuzzy value v_c is obtained for A. Such a value can be used directly, but it can also be filtered to generalize it. The membership function of this fuzzy value is too complex because too close to the training values present in the training set.

In the following, we propose an algorithm to filter and to smooth again such fuzzy value in order to enhance its generalization power.

[2] We recall that we denote by A both the attribute and its ordered set of values belonging to the training set.

4.2 Fuzzy morphological transformation

The FPMM algorithm cannot be applied directly to filter the obtained fuzzy value. In this case, we cannot use a transduction that modifies the letters because the training set is represented for each class and not for the whole set of classes. Thus, no word can be constructed. Here, given a class, the erosion and the dilation should be applied on the membership degrees.

Definition of a granularity level. In order to use morphological operators, we introduce a granularity level that will enable us to define transformation functions on continuous fuzzy sets.

Let v_c be the fuzzy value constructed previously and let μ_{v_c} be its membership function on X. Let a step $s \in \mathbb{R}$ be given. s defines the granularity level we want for the transformation. We define the following ordered set $X_S = \{x_1, x_2, ..., x_S\}$ as:

- $\forall i \in \{1, ..., S\}$, $x_i \in X$ and $\forall i \in \{2, ..., S\}$, $x_i = x_{i-1} + s$.
- x_1 is defined such that: $\forall x \in X$, $x < x_1$ implies $\mu_{v_c}(x) = 0$.
- x_S is defined such that: $\forall x \in X$, $x > x_S$ implies $\mu_{v_c}(x) = 0$.

x_1 is the lowest value of the support of μ_{v_c}, and x_S is the greatest value of the support of μ_{v_c}.

We define the set of corresponding membership degrees $\{\alpha_1, \alpha_2, ..., \alpha_S\}$ as: $\forall i \in \{1, ..., S\}$, $\alpha_i = \mu_{v_c}(x_i)$.

Transformation of membership degrees. Given a value $x \in X_S$, let α be the membership degree of x before the morphological transformation, and let α' be the membership degree of x after the morphological transformation.

The study done in the previous part (part 3.1) highlights the fact that a morphological transformation applied on $x_i \in X_S, i = 2, ..., S-1$ depends only on membership degrees of x_{i-1} and x_{i+1}. Thus, α'_i depends only on α_i, α_{i-1} and α_{i+1}, and the morphological transformation is a function: $f : [0,1] \times [0,1] \times [0,1] \longrightarrow [0,1]$ such that $\alpha'_i = f(\alpha_{i-1}, \alpha_i, \alpha_{i+1})$. The function f has to be defined for each kind of transformation we want to implement.

The properties wanted for such a function are connected to the mathematical morphology theory, as in the previous part. Fuzzy mathematical morphology theories exist, for instance [3,4] or [12]. In [3,4], fuzzy subsets are eroded or dilated according to a structuring element which can be another fuzzy subset. Erosion operator and dilatation operator are implemented by means of a t-norm or a t-conorm.

In our scheme, the structuring element is implicit and can be associated directly to the definition of f. Thus, we propose two functions to erode and to dilate a fuzzy subset. These functions are based on t-norms and can be viewed as particular cases of functions from fuzzy mathematical morphology theory presented in [3,4].

Erosion of a fuzzy set. The function $f_{\mathrm{Er}} : [0,1] \times [0,1] \times [0,1] \longrightarrow [0,1]$, such that $\alpha'_i = f_{\mathrm{Er}}(\alpha_{i-1}, \alpha_i, \alpha_{i+1})$, that enables us to erode a fuzzy set should satisfy the following properties:

i) $\alpha'_i \leq \alpha_i$,
ii) if $\alpha_{i-1} = \alpha_{i+1} = 1$ then $\alpha'_i = \alpha_i$,
iii) $|\alpha'_i - \alpha_i|$ increases when α_{i-1} and α_{i+1} tends to 0.

Property *i)* is required for an erosion-like function. Property *ii)* ensures that an element really inside the kernel of a fuzzy set (*ie.* surrounded by elements in the kernel) will not be eroded. Property *iii)* reflects the fact that the power of the erosion increases when α_{i-1} and α_{i+1} belongs to another class.

A example of function f_{Er} that satisfies these properties is:

$$f_{\mathrm{Er}}(\alpha_{i-1}, \alpha_i, \alpha_{i+1}) = \max(0;\ \alpha_i - \max(\bar{\alpha}_{i-1}, \bar{\alpha}_{i+1})),$$

with $\bar{\alpha} = 1 - \alpha$.

Dilatation of a fuzzy set. In the same way, a function $f_{\mathrm{Di}} : [0,1] \times [0,1] \times [0,1] \longrightarrow [0,1]$, such that $\alpha'_i = f_{\mathrm{Di}}(\alpha_{i-1}, \alpha_i, \alpha_{i+1})$, is defined to implement a dilatation of fuzzy sets. For such a function, the following properties are required:

i) $\alpha'_i \geq \alpha_i$,
ii) if $\alpha_{i-1} = \alpha_{i+1} = 0$ then $\alpha'_i = \alpha_i$,
iii) $|\alpha'_i - \alpha_i|$ increases when α_{i-1} and α_{i+1} tends to 1.

Property *i)* is required for a dilatation-like function. Property *ii)* ensures that an element really outside fuzzy set (*ie.* surrounded by elements outside the support) will not be dilated. Property *iii)* reflects the fact that the power of the dilatation increases when α_{i-1} and α_{i+1} belong to the same class.

A example of function f_{Di} is:

$$f_{\mathrm{Di}}(\alpha_{i-1}, \alpha_i, \alpha_{i+1}) = \min(1;\ \alpha_i + \max(\alpha_{i-1}, \alpha_{i+1})).$$

An illustration of the use of the erosion function and the dilatation function is given in Fig. 11.

Opening and closure of a fuzzy set. As usually, these morphological operators are defined by combination of erosion and dilatation operators.

Here, the opening and the closure of a fuzzy set are the composition of the erosion function and the dilatation function:

$$f_{\mathrm{Op}} = f_{\mathrm{Di}} \circ f_{\mathrm{Er}} \text{ and } f_{\mathrm{Cl}} = f_{\mathrm{Er}} \circ f_{\mathrm{Di}}.$$

An illustration of the use of the opening and the closure is given in Fig. 12.

Filtering a fuzzy set. The FPMM algorithm is extended to be applied to fuzzy sets by means of the introduced functions. The new algorithm is:

Algorithm 41 (FPMM') *To construct a fuzzy partition for a numerical attribute A, defined on X, from a training set $\mathcal{E}$,:*

1. *For each class $c \in \mathcal{C}$, aggregate the values of A to compute v_c.*
2. *Filter each v_c into v'_c by means of the defined morphological functions.*
3. *Each obtained value v'_c is a fuzzy subset of the fuzzy partition of X.*

4.3 Illustration of the algorithm

Let the training set $\mathcal{E}$ be given in Table 1. This training set is a toy set composed of motorbikes described by means of a single attribute: their estimated *average speed* (a fuzzy value), and associated with a class: their level of *price* (cheap or expensive, a fuzzy value).

Motorbike	*500 GSE*	*CBR 1100*	*125 Rebel*	*900 Bandit*
Average speed	normal	around 150	slow	around 180
Cheap	0.7	0.25	1.0	0.5
Expensive	0.3	0.75	0.0	0.5

Table 1. A fuzzy data base

The attribute *average speed* is associated with estimated fuzzy values as *around 150 km/h* or *slow*. Membership functions of these values are given in Fig. 9.

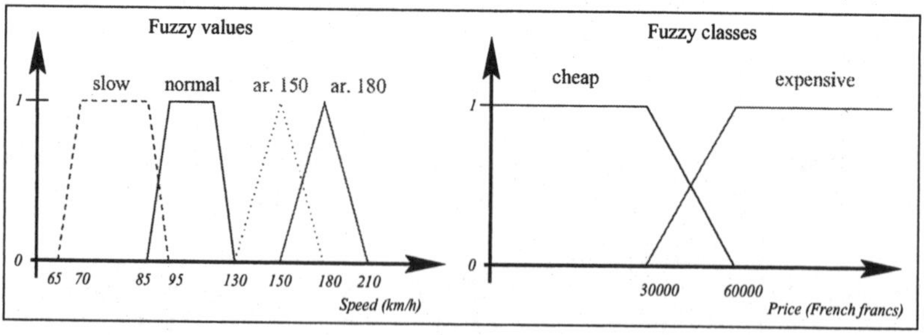

Attribute A is the average speed and is associated with the numerical universe $X = [0, 250]$. It possesses 4 fuzzy values *slow, normal, around 150* et *around 180*. Only one case in the training set is associated with the fuzzy value *normal*, with the membership degree $\mu_{\text{cheap}}(500\ GSE) = 0.7$ and $\mu_{\text{expensive}}(500\ GSE) = 0.3$. Thus, we have $\gamma_{\text{cheap}}(\text{normal}) = 0.7$ and $\gamma_{\text{expensive}}(\text{normal}) = 0.3$. The membership function $\mu_{\text{normal, cheap}}$ is shown in Fig. 10. The result of the aggregation for the class *cheap* is given in Fig. 10.

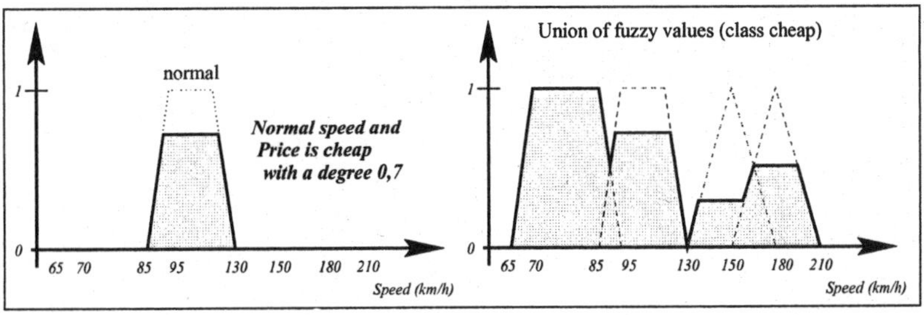

The value v_{cheap} is a continuous membership function here. We decomposed with a level of granularity to define points. For instance, we can use a step s of 10 to define 16 points from Fig. 10. Results of the application of an erosion or a dilatation is given in Fig. 11. Results of application of an opening and

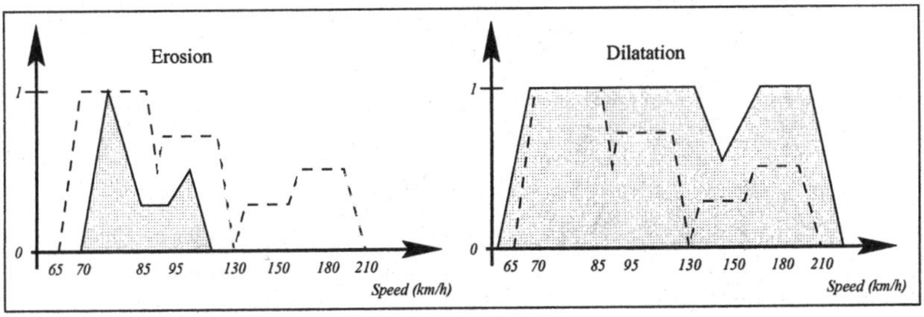

Fig. 11. Erosion and dilatation of fuzzy subsets

closure is given in Fig. 12. In Fig. 13, we present results of filtering for the two classes *cheap* and *expensive*. Thus, 2 new modalities for the attribute *speed* are obtained for each class.

5 Conclusion

In this chapter, we propose two new algorithms to infer automatically a fuzzy partition for the universe of a set of values, when each of these values is associated with a class.

These algorithms are based on the use of mathematical morphology operators that are used to filter the given set of values and highlight kernel for

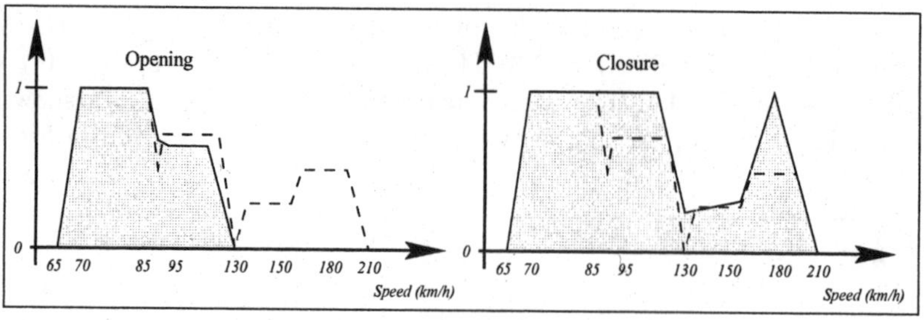

Fig. 12. Opening and closure of fuzzy sets

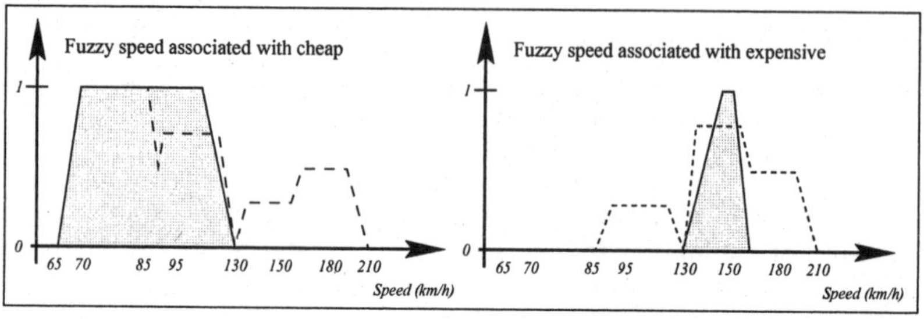

Fig. 13. Filtering a fuzzy set

fuzzy subsets. Their purpose is to be used in an inductive learning algorithm to construct fuzzy partitions on numerical universes of values.

The first algorithm enables us to construct a fuzzy partition from a set of numerical values associated with a class. This algorithm is based on several rewriting systems which are represented as transductions. Each of these rewriting systems implements a mathematical morphology operator. An operator is defined to reduce an arbitrary sequence of letters, and another operator is defined to enlarge a sequence of letters. When these two operators are combined, depending on the order of the composition, we obtain two operators to filter the set of values. Finally, we obtain an algorithm to smooth a word induced by a training set, and, from this word we propose a way to define a fuzzy partition. This algorithm has been implemented and is currently used in several applications of fuzzy decision tree construction.

The second algorithm is a proposition of extension of the first one. It enables us to construct a fuzzy partition from a set of fuzzy values associated with a fuzzy class. This algorithm is also based on mathematical morphology theory but operators are implemented in a different way. Functions are defined to filter the membership function associated with a given class. Finally, we obtain an algorithm to smooth a fuzzy subset induced by a training set, and a way to define a fuzzy partition.

References

1. N. Aladenise and B. Bouchon-Meunier. Acquisition de connaissances imparfaites : mise en évidence d'une fonction d'appartenance. *Revue Internationale de Systémique*, 11(1):109–127, 1997.
2. T. Bilgiç and I. B. Türkşen. Measurement of memberchip functions: Theoretical and empirical work. In D. Dubois and H. Prade, editors, *Fundamentals of fuzzy Sets*, volume 7 of *Handbook of Fuzzy Sets*. Kluwer, 2000.
3. I. Bloch and H. Maître. Constructing a fuzzy mathematical morphology: Alternative ways. In *Proceedings of the Second IEEE International Conference on Fuzzy Systems*, San Francisco, USA, April 1993.
4. I. Bloch and H. Maître. Fuzzy mathematical morphology. *Annals of Mathematics and Artificial Intelligence*, 9(3,4), 1993.
5. B. Bouchon-Meunier and C. Marsala. Learning fuzzy decision rules. In D. D. J. Bezdek and H. Prade, editors, *Fuzzy Sets in Approximate Reasoning and Information Systems*, volume 3 of *Handbook of Fuzzy Sets*, chapter 4. Kluwer Academic Publisher, 1999.
6. B. Bouchon-Meunier, C. Marsala, and M. Ramdani. Arbres de décision et théorie des sous-ensembles flous. In *Actes des 5èmes journées du PRC-GDR d'Intelligence Artificielle*, pages 50–53, 1995.
7. B. Bouchon-Meunier, C. Marsala, and M. Ramdani. Learning from imperfect data. In D. Dubois, H. Prade, and R. R. Yager, editors, *Fuzzy Information Engineering: a Guided Tour of Applications*, pages 139–148. John Wileys and Sons, 1997.
8. L. Breiman, J. H. Friedman, R. A. Olshen, and C. J. Stone. *Classification And Regression Trees*. Chapman and Hall, New York, 1984.
9. C. Carter and J. Catlett. Assessing credit card applications using machine learning. *IEEE Expert*, Fall Issues:71–79, 1987.
10. K. J. Cios, W. Pedrycz, and R. W. Swiniarski. *Data Mining - Methods for Knowledge discovery*. Engineering and Computer Science. Kluwer Academic Publishers, 1998.
11. M. Coster and J.-L. Chermant. *Précis d'analyse d'images*. Presses du CNRS, 1989.
12. B. De Baets, N. Kwasnikowska, and E. Kerre. Fuzzy morphology based on conjunctive uninorms. In M. Mareš, R. Mesiar, V. Novák, J. Ramik, and A. Stupňanová, editors, *Proceedings of the Seventh International Fuzzy Systems Association World Congress*, volume 1, pages 215–220, Prague, Czech Republic, June 1997.
13. J. Dougherty, R. Kohavi, and M. Sahami. Supervised and unsupervised discretization of continuous features. In A. Prieditis and S. Russell, editors, *Machine Learning: Proceedings of the Twelfth International Conference*, San Francisco, CA, 1995. Morgan Kaufmann.
14. U. M. Fayyad and K. B. Irani. On the handling of continuous-valued attributes un decision tree generation. *Machine Learning*, 8(1):87–102, January 1992. *Technical note.*
15. U. M. Fayyad and K. B. Irani. Multi-interval discretization of continuous-valued attributes for classification learning. In *Proceedings of the 13th International Joint Conference on Artificial Intelligence*, volume 2, pages 1022–1027, 1993.

16. S. Ginsburg. *The Mathematical Theory of Context Free Languages.* McGraw-Hill, New-York, 1966.
17. E. G. Henrichon and K.-S. Fu. A nonparametric partitioning procedure for pattern classification. *IEEE Transactions on Computers*, C-18(7):614–624, July 1969.
18. J.-S. R. Jang. Structure determination in fuzzy modeling: a fuzzy CART approach. In *Proceedings of the 3rd IEEE Int. Conf. on Fuzzy Systems*, volume 1, pages 480–485, Orlando, 6 1994. IEEE.
19. R. Kerber. ChiMerge: Discretization of numeric attributes. In *Proceedings of the 10th National Conference on Artificial Intelligence*, pages 123–128. AAAI, 1992.
20. G. J. Klir and B. Yuan. *Fuzzy Sets and Fuzzy Logic. Theory and Applcations.* Prentice Hall, 1995.
21. R. Krishnapuram. Generation of membership functions via possibilistic clustering. In *Proceedings of the 3rd IEEE Int. Conf. on Fuzzy Systems*, volume 2, pages 902–908, Orlando, Florida, June 1994.
22. C. Marsala. Arbres de décision et sous-ensembles flous. Rapport 94/21, LAFORIA-IBP, Université Pierre et Marie Curie, Paris, France, Novembre 1994.
23. C. Marsala. *Apprentissage inductif en présence de données imprécises : construction et utilisation d'arbres de décision flous.* Thèse de doctorat, Université Pierre et Marie Curie, Paris, France, Janvier 1998. Rapport LIP6 n° 1998/014.
24. C. Marsala and B. Bouchon-Meunier. Fuzzy partioning using mathematical morphology in a learning scheme. In *Proceedings of the 5th IEEE Int. Conf. on Fuzzy Systems*, volume 2, pages 1512–1517, New Orleans, USA, September 1996.
25. C. Marsala and B. Bouchon-Meunier. An adaptable system to construct fuzzy decision trees. In *Proc. of the NAFIPS'99 (North American Fuzzy Information Processing Society)*, pages 223–227, New York (USA), June 1999.
26. C. Marsala and N. Martini Bigolin. Spatial data mining with fuzzy decision trees. In N. F. F. Ebecken, editor, *Data Mining*, pages 235–248. WIT Press, 1998. Proceedings of the International Conference on Data Mining, Rio de Janeiro, Sept. 1998.
27. C. Marsala, M. Ramdani, M. Toullabi, and D. Zakaria. Fuzzy decision trees applied to the recognition of odors. In *Proceedings of the IPMU'98 Conference*, volume 1, pages 532–539, Paris, July 1998. Editions EDK.
28. X. Ménage. *Apprentissage pour le contrôle de qualité, approche basée sur la théorie des possibilités et la théorie des réseaux de neurones.* PhD thesis, Université P. et M. Curie, Paris, France, Septembre 1996.
29. H. Narazaki and A. L. Ralescu. An alternative method for inducing a membership function of a category. *International Journal of Approximate Reasoning*, 11(1):1–28, july 1994.
30. M. J. Pazzani. An iterative improvement approach for the discretization of numeric attributes in bayesian classifiers. In U. M. Fayyad and R. Uthurusamy, editors, *Proceedings of the First International conference on Knowledge Discovery and Data Mining*, pages 228–233, Montréal, Québec, Canada, August 1995.
31. J. R. Quinlan. Induction of decision trees. *Machine Learning*, 1(1):86–106, 1986.

32. J. R. Quinlan. Probabilistic decision trees. In R. S. Michalski, J. G. Carbonell, and T. M. Mitchell, editors, *Machine Learning*, volume 3, chapter 5, pages 140–152. Morgan Kaufmann Publishers, 1990.
33. J. R. Quinlan. *C4.5: Programs for Machine Learning*. Morgan Kaufmann, San Mateo, Ca, 1993.
34. J. R. Quinlan. Improved use of continuous attributes in C4.5. *Journal of Artificial Intelligence Research*, 4:77–90, 3 1996.
35. S. Rabaséda-Loudcher. *Contributions à l'extraction automatique de connaissances. Une application à l'analyse clinique de la marche.* PhD thesis, Université Lumière Lyon 2, France, Décembre 1996.
36. S. Rabaséda-Loudcher, M. Sebban, and R. Rakotomalala. Discretisation of continuous attribute: a survey of methods. In *Proceedings of the 2nd annual Joint Conference on Information Sciences*, pages 164–166, Wrightsville Beach, North Carolina, USA, September 1995.
37. M. Ramdani. *Système d'Induction Formelle à Base de Connaissances Imprécises.* PhD thesis, Université P. et M. Curie, Paris, France, Février 1994. Aussi publiée en rapport du LAFORIA-IBP n° TH94/1.
38. E. Sanchez, T. Shibata, and L. A. Zadeh, editors. *Genetic algorithm and fuzzy logic systems. Soft Computing Perspectives.*, volume 7 of *Advances in Fuzzy Systems – Applications and Theory*. World Scientific Publishing Co., 1997.
39. J.-P. Serra. *Image Analysis and Mathematical Morphology.* Academic Press, New York, 1982.
40. T. Van de Merckt. Decision trees in numerical attribute spaces. In *IJCAI-93 Proceedings of the 13th International Joint Conference on Artificial Intelligence*, volume 2, pages 1016–1021, 1993.
41. L. Wehenkel. On uncertainty measures used for decision tree induction. In *Proceedings of the 6th International Conference IPMU*, volume 1, pages 413–418, Granada, Spain, july 1996.
42. L. Wehenkel. Discretization of continuous attributes for supervised learning. variance evaluation and variance reduction. In M. Mareš, R. Mesiar, V. Novák, J. Ramik, and A. Stupňanová, editors, *Proceedings of the Seventh International Fuzzy Systems Association World Congress*, volume 1, pages 381–388, Prague, Czech Republic, June 1997.
43. Y. Yuan and M. J. Shaw. Induction of fuzzy decision trees. *Fuzzy Sets and systems*, 69:125–139, 1995.
44. J. Zeidler and M. Schlosser. Continuous-valued attributes in fuzzy decision trees. In *Proceedings of the 6th International Conference IPMU*, volume 1, pages 395–400, Granada, Spain, july 1996.
45. D. A. Zighed, R. Rakotomalala, and S. Rabaséda. A discretization method of continuous attributes in induction graphs. In R. Trappl, editor, *Cybernetics and systems'96*, volume 2, pages 997–1002. Austrian Society for Cybernetics Studies, 1996.

Annex

Mathematical morphology theory

The fundamental operators from mathematical morphology theory are the erosion operator and the dilatation operator. They are combined to produce the opening operator and the closure operator that can be used to filter a set of bodies. There operators come from the pattern recognition domain and are often used to filter 2D-pictures. More details on these operators and on the mathematical morphology theory can be found in [39] or in [11].

The basic operators: erosion and dilatation. We consider a space of morphological bodies. These two basic operators enable us to modify a morphological body $\mathcal{C}$ (Fig. 14). This modification is related to a *structuring element* s_e. The *erosion* is a particular subtraction of e in $\mathcal{C}$, and the *dilatation* is a particular addition of s_e in $\mathcal{C}$.

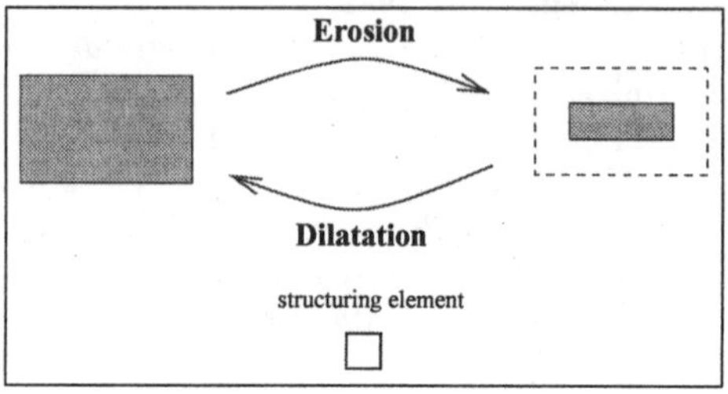

Fig. 14. Mathematical morphology operators

The operators opening and closure. Each of these operators is a combination of the two basic operators.

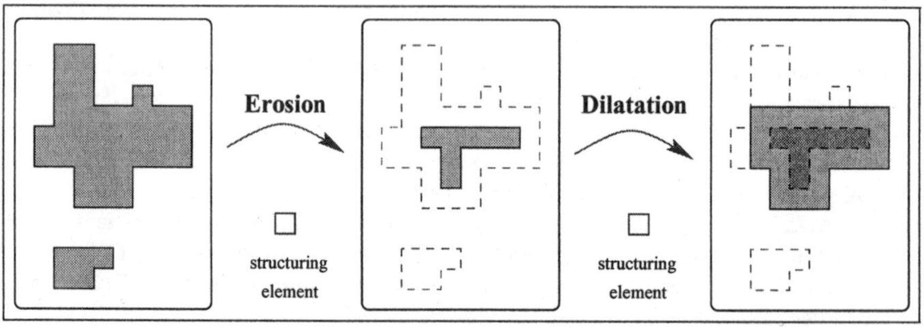

Fig. 15. Opening operator

The opening is the combination of an erosion followed by a dilatation applied to a morphological body, with the same structuring element (Fig. 15). It enables destruction of small bodies in the space, with respect to the size of the chosen structuring element.

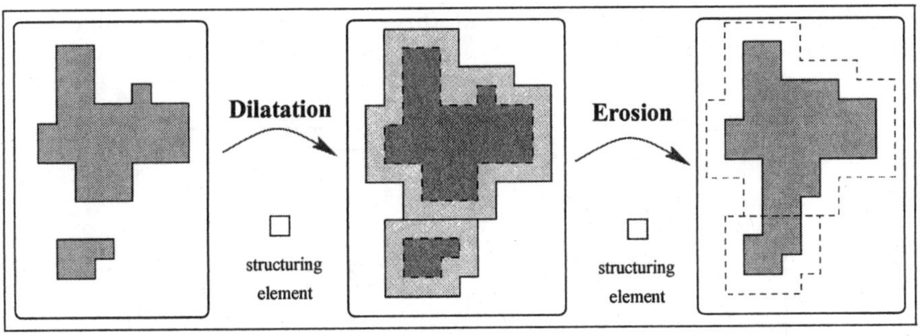

Fig. 16. Closure operator

The closure is the combination of a dilatation followed by an erosion applied to a morphological body, with the same structuring element (Fig. 16). It enables the destruction of small vacuum places occurring in a body, with respect to the size of the chosen structuring element.

The open-close filter. A filter is a combination of openings and closures. It is composed by k successive openings followed by k successive closures ($k = 1, 2, \ldots$) applied to all bodies of the space, with the same structuring element. Thus it enables the destruction of small bodies present in the space with respect to the chosen structuring element. Simultaneously, it enables the filling of small vacuum places occurring in bodies. The value of k enables us to control the power of the modification.

Transductions

Let us recall that a transduction is a 6-tuple $\langle \mathcal{A}, \mathcal{B}, S, I, E, \delta \rangle$ where $\mathcal{A}$ is the input alphabet, $\mathcal{B}$ is the output alphabet, S is a (finite) set of states, $I \subseteq S$ is the set of initial states of the transduction, $E \subseteq S$ is the set of terminal states of the transduction and $\delta \subset S \times \mathcal{A}^* \times \mathcal{B}^* \times S$ is the transition function.

A transduction reads a word $w \in \mathcal{A}^*$ and rewrites it in a corresponding word $w_o \in \mathcal{B}^*$. It proceeds sequentially from the first letter of w to the last one, as a reading head moves letter by letter. The rewriting rules to generate w_o are based on δ. Let $(s_i, z, t, s_j) \in \delta$, with $s_i, s_j \in S, z \subset \mathcal{A}^*$ and $t \subset \mathcal{B}^*$, (s_i, z, t, s_j) is called a *transition* of the transduction. If s_i is the current state and we can read z in w (i.e. z is composed by the successive letters coming just after the reading head), we replace it by t and the current state becomes s_j. A convention is to use \$ to match the end of the input word, and ε (the *null word*) is introduced when nothing has to be written.

For example, let $\mathcal{A} = \{a, b\}, \mathcal{B} = \{0, 1\}$, $S = \{\mathrm{S1}, \mathrm{S2}, \mathrm{S3}\}$, $I = \{\mathrm{S1}\}$, $E = \{\mathrm{S3}\}$ and $\delta = \{(\mathrm{S1}, a, 0, \mathrm{S1})$, $(\mathrm{S1}, b, \varepsilon, \mathrm{S2})$, $(\mathrm{S2}, a, 0, \mathrm{S1})$, $(\mathrm{S2}, b, 1, \mathrm{S2})$, $(\mathrm{S1}, \$, \$, \mathrm{S3})$, $(\mathrm{S2}, \$, \$, \mathrm{S3})\}$

A simple visual representation of this transduction is a graphic form (Fig. 17).

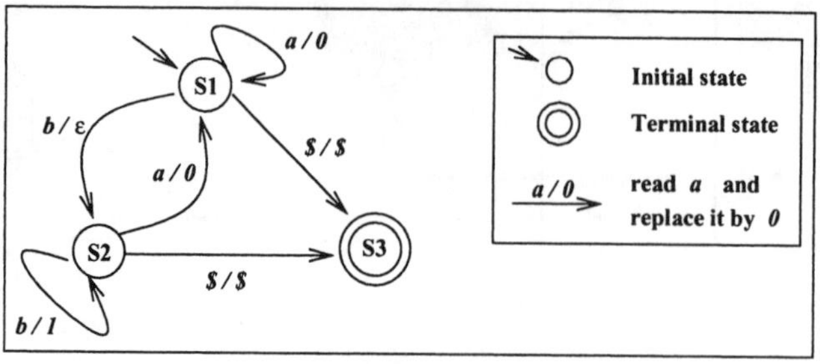

Fig. 17. Example of transduction

Let us rewrite $w = abbaabaab$ with this transduction. After the sequence of states (S1, S1, S2, S2, S1, S1, S2, S1, S1, S2, S3), we have $w_o = 010000$ and, as the current state is S3, a terminal state, and we have nothing more to read in w, the rewriting is done.

A Coding Method to Handle Linguistic Variables

A. Bailón, A. Blanco, M. Delgado, and W. Fajardo

Departamento Ciencias de la Computación e Inteligencia Artificial, Universidad de Granada, E.T.S. Ingeniería Informática, 18071 Granada, Spain

Abstract. We present a coding method for linguistic variables which we have named Incremental Discretization. It allows us to express any fuzzy subset of the universe of discourse of the linguistic variable in binary or bipolar terms. This will permit us to process fuzzy information expressed in linguistic terms using discrete models of Artificial Intelligence. In order to test the effectiveness, we apply this method to the memorization of a set of fuzzy rules using models of discrete associative memories.

Keywords. Linguistic Variable, Associative Memory, Linguistic Rule, Coding of Fuzzy Information, Discrete Coding, Pattern Recognition, Discrete to Linguistic Extension, Fuzzy Sets, Coding Method

1 Introduction

Automated systems for dealing with data normally require a high degree of precision whereas human beings perceive their environment in an imprecise way. We innately use imprecise terms when we reason, therefore using highly precise terms to communicate with an automatic data processing system does not come naturally to us.

It would be more appropriate to design systems which are capable of dealing with information in ways which are more in tune with the way in which the information is supplied by the user through the use of vague and imprecise concepts. We have two options for doing this:

- Design new systems which are capable of dealing with imprecisely expressed data and abandon the traditional systems.
- Design a method which uses a filter in order to accept imprecisely expressed information which has the capacity to convert it into a precise representation and be compatible with the requirements of a traditional system. In this way we can use all the existing systems without the need to design and investigate totally new ones.

From a practical point of view, it is preferable to design a filter which can process the imprecise information entered by the user so that it can be used by a traditional system. The precise output would be presented in a way which is more understandable to the user. (Fig.1). In all cases it would be more desirable if the filter acted independently of the data processing system so that it could be used with any existing traditional system.

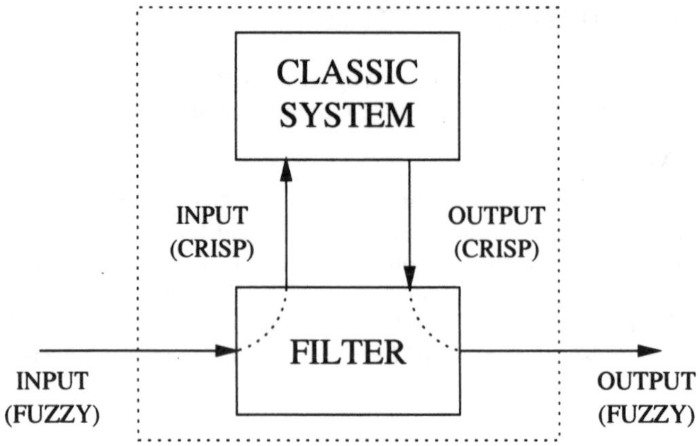

Fig. 1. Input/output filter as a functional element of the system

2 Linguistic Variables

After the initial appearance of the computers, it was soon realized that they had great potential as calculating machines. This gave rise to the expectation that they could be used to solve complex problems, even to the point of emulating the behavior of the human brain.

However, the computer is a machine designed to work with precise numerical values whereas humans use vague and imprecise information in their reasoning processes. In spite of the imprecision of the information, humans are capable of carrying out complex tasks and the results obtained are satisfactory. On the other hand, in the majority of cases, the use of increased precision in computers results in inferior results when compared to the same tasks done by humans. This can be viewed as a manifestation of the incompatibility principle [11] which states that the complexity of a system and the precision with which it operates are inversely proportional, in other words, one cannot tackle highly complex tasks using very precise concepts.

If we wish to use the computer in order to emulate the behavior of the human brain we need to use a system of representation of information similar to that which is employed by the human being, in other words, we must express information using vague and imprecise terms. In this way, communication with the user is simplified because it is carried out using concepts which are familiar to the user. The study of systems which emulate the capacity of man to process information belong to a branch of Artificial Intelligence known as Approximate Reasoning.

Language is used as the medium for human reasoning, the information is expressed in linguistic terms which can be vague and imprecise (high, low, near, a lot...) although theses terms refer to numerical values which are precise. Put another way, although the age of a person can be stated in precise numerical values we can also refer to age with imprecise terms such as:- young, adult, old,.... This leads us to the definition of what we call linguistic variables.

The values of a linguistic variable [9] are not numerical but rather words from a formal language and they are defined as a quintuple.

$$(H, T(H), V, G, M)$$

where:

- H is the name of the variable.
- $T(H)$ is the set of linguistic terms that the variable H can take.
- V is the numeric domain over which the variable H takes as its value.
- G is a syntactic rule which generates the elements t of $T(H)$.
- M is a semantic rule which associates every $t \in T(H)$ with its meaning $M(t)$ which is a fuzzy set of V.

The meaning of value t is a linguistic variable characterized by a compatibility function which associates each element u of V its compatibility in the range $[0,1]$ with the linguistic value t.

Example:

$H \equiv Age$

$T(H) \equiv$young, not young, very young, not very young,..., old, not old, very old, not very old, very very old,...

$V \equiv [0, 100]$

$G \equiv [not]very * (young|old)$

$M(old) = \int_{50}^{100} \left[1 + \left(\frac{u-50}{5}\right)^{-2}\right]^{-1} /u$

$M(young) = \int_{0}^{50} \left[1 + \left(\frac{50-u}{5}\right)^{-2}\right]^{-1} /u$

Fuzzy sets used as models for the linguistic variables allow us to satisfy the need for flexibility in the representation of knowledge, in other words they help to give meaning to the imprecise values of a linguistic variable. Therefore, the process of reasoning based on linguistic terms will be developed by the employment of fuzzy sets.

3 Codifying Linguistic Variables Using the Incremental Discretization Method

Linguistic variables constitute an information representation system similar to the one used by humans. This is the key that will permit us to automatically process vague and imprecise terms using diverse formulas in a computer.

In fact, the aim of this paper is to propose a method of coding that will facilitate this process.

At this point, we must solve the problem of coding the linguistic terms in accordance with a protocol capable of being used by a traditional model of Artificial Intelligence. In order to represent the linguistic terms in a discrete way, we present a coding method for linguistic variables which we have named Incremental Discretization.

Let H be a linguistic variable described by a quintuple:

$$(H, T(H), V, G, M) .$$

This is based on a way of representing fuzzy information which is highly accepted and widespread in reasoning in the presence of fuzzy data [6], [7], [8], any fuzzy subset $\mathcal{F}$ over V is represented in terms of $T(H)$ as:

$$\mathcal{F} \equiv \alpha_1 t_1, \ldots, \alpha_n t_n; \qquad \alpha_i \in [0,1],\ t_i \in T(H)\ i = 1, \ldots, n ,$$

being α_i the compatibility of $\mathcal{F}$ with t_i according to any measure of uncertainty.

Based on this idea, we propose to code the values of the variable as shown in the following:

We suppose that $T(H)$ is composed of n elements. We give $T(H)$ an arbitrary order and each element t of $T(H)$ is associated with a vector dimension m (supposing that we desire a global precision to the order of $1/m$) and $T(H)$ is associated with a dimension vector $m \times n$.

Therefore $\mathcal{F}$ can be coded as a binary vector (valued in $\{0, 1\}$) or bipolar (valued in $\{-1, 1\}$) [1] of dimension $m \times n$:

$$C(\mathcal{F}) = (C_{11}, \ldots, C_{1m}, C_{21}, \ldots, C_{2m}, \ldots, C_{n1}, \ldots, C_{nm}) .$$

In the case of being binary, the components of the vector are:

$$\begin{cases} \alpha_i = 0 \rightarrow C_{ij} = 0;\ i = 1, \ldots, n;\ j = 1, \ldots, m \\ \alpha_i \neq 0 \rightarrow \begin{cases} \exists j \text{ t.q. } \frac{j}{m} \leq \alpha_i < \frac{j+1}{m} \\ C_{il} = \begin{cases} 1 \text{ if } l \leq j \\ 0 \text{ if } l > j . \end{cases} \end{cases} \end{cases}$$

[1] The choice of the type of codification (binary o bipolar) carried out, depends on the type of data (binary or bipolar) which processes the discrete model in relation to the one which will establish the method of codification

Or the same, stated differently is; $C(\mathcal{F})$ is a vector whose components are calculated following an algorithm expressed in the following outline:

1. – From i = 1 until n
   1.1. – From j = 1 until m
      1.1.1. – Si j $\leq$ Integer part of $(\alpha_i \cdot m)$
         1.1.1.1. – Then $C_{ij} = 1$
      1.1.2. – Si j $>$ Integer part of $(\alpha_i \cdot m)$
         1.1.2.1. – Then $C_{ij} = 0$
   1.2. – End
2. – End

For example, let *height* be a linguistic variable with the terms short, medium and tall. With precision 1/10, the value 3/10 medium, 7/10 tall, is coded by:

0000000000 1110000000 1111111000

Likewise, in the case of being bipolar, the components of the vector are:

$$\begin{cases} \alpha_i = 0 \rightarrow C_{ij} = -1;\ i = 1, \ldots, n;\ j = 1, \ldots, m \\ \alpha_i \neq 0 \rightarrow \begin{cases} \exists j \text{ t.q. } \frac{j}{m} \leq \alpha_i < \frac{j+1}{m} \\ C_{il} = \begin{cases} 1 \text{ if } l \leq j \\ -1 \text{ if } l > j\ . \end{cases} \end{cases} \end{cases}$$

Or the same, stated differently is; the components are calculated following an algorithm expressed in the following outline:

1. – From i = 1 until n
   1.1. – From j = 1 until m
      1.1.1. – Si j $\leq$ Integer part of $(\alpha_i \cdot m)$
         1.1.1.1. – Then $C_{ij} = 1$
      1.1.2. – Si j $>$ Integer part of $(\alpha_i \cdot m)$
         1.1.2.1. – Then $C_{ij} = -1$
   1.2. – End
2. – End

For example, let *weight* be a linguistic variable with the terms light, normal and heavy. With precision 1/5, the value 0.2 light, 0.8 normal, is coded by:

+1-1-1-1-1 +1+1+1+1-1 -1-1-1-1-1

Therefore we have a coding system capable of expressing any fuzzy subset over the discourse universe V in terms of the set $T(H)$.

In particular if $A = t_k$ then it is coded in binary [2] with:

$$C_{ij} = \begin{cases} 1 \text{ if } i = k; \ \forall j \\ 0 \text{ in another case} . \end{cases}$$

In the case where we know beforehand that we are going to limit and employ terms of the set $T(H)$, it is redundant to use a high number of bits per label, for this reason it is better to use only one bit ($m = 1$) for each term $t_i \in T(H)$.

When we fix the number of bits used in the coding we are calibrating the scale of sensitivity of the measurements which we wish to work with. The maximum precision must be decided at the start.

On the other hand, in order to codify the value of a linguistic variable using the Incremental Discretization method we must take into account the different modes in which the information expressed can come ie: in compatibility terms, in crisp terms or in purely linguistic terms.

In the following subsections we will show the coding process for each of the three cases:

3.1 Case 1: Information Expressed in Compatibility Terms

The values of the discourse universe V are expressed by experts as compatibility degrees α_i with the elements of the set of terms $T(H)$. Example: Height= 0.4 medium, 0.6 tall.

Then the coding of the linguistic variable, expressed with an accuracy of $1/m$ would be:

$$C(\mathcal{F}) = (C_{11}, \ldots, C_{1m}, C_{21}, \ldots, C_{2m}, \ldots, C_{n1}, \ldots, C_{nm}) ,$$

where $C_{i1}, \ldots, C_{im}$ is the coding of the compatibility of α_i with the term t_i.

3.2 Case 2: Information Expressed in Crisp Terms

Let us suppose that the variable is expressed using a value for the universe of discourse V, obtained using measuring methods which provide crisp information. Example: Weight=85 kg.

The compatibility degrees of the crisp information with the set of terms are calculated using the semantic representation of the linguistic variable. This representation is expressed using membership degrees of the associated fuzzy sets.

[2] In all the examples from this point onwards (due to the similarity that exists between binary and bipolar coding) the linguistic labels will be coded in binary

The coding of the crisp value u of the linguistic variable, with precision $1/m$ will be:

$$C(u) = (C_{11}, \ldots, C_{1m}, C_{21}, \ldots, C_{2m}, \ldots, C_{n1}, \ldots, C_{nm}) ,$$

where $C_{i1}, \ldots, C_{im}$ is the codification of the compatibility $\alpha_i = \mu_i(u)$ with the term t_i.

3.3 Case 3: Information Expressed in Linguistic Terms

The values of the universe of discourse V are expressed as fuzzy subsets of V. Example: Age=old

In this case we must consider α_i to be the compatibility degree of the fuzzy subset $\mathcal{F}$ of V with the term t_i.

The coding of the fuzzy subset u with precision $1/m$ will be:

$$C(\mathcal{F}) = (C_{11}, \ldots, C_{1m}, C_{21}, \ldots, C_{2m}, \ldots, C_{n1}, \ldots, C_{nm}) ,$$

where $C_{i1}, \ldots, C_{im}$ is the codification of the compatibility α_i of $\mathcal{F}$ with the term t_i.

4 Example

Let's suppose that we wish to code the linguistic variable, height. The terms that the variable can have are {short, medium, tall} and they are semantically represented in the universe of discourse $[0.50, 2.50]$ using the fuzzy sets shown in Fig. 2.

The following subsection shows the three cases that can be presented for the coding of the information according to the way in which it is expressed.

4.1 Coding of Information Expressed in Terms of Compatibility

The information which refers to the height of an individual is expressed by experts according to the set of linguistic terms that can take the variable "tall". In other words, the compatibility degree of height is indicated by each one of the linguistic labels "short", "medium" and "tall".

If the information provided is 0.2 short, 0.8 medium and we wish to use a global precision of $1/10$, the coding of this information will be the following:

$$\begin{aligned} C(h) &= (C_{\text{short}\,1}, \ldots, C_{\text{short}\,10}, C_{\text{med.}\,1}, \ldots, C_{\text{med.}\,10}, C_{\text{tall}\,1}, \ldots, C_{\text{tall}\,10}) \\ &= (1100000000\,1111111100\,0000000000) \ . \end{aligned}$$

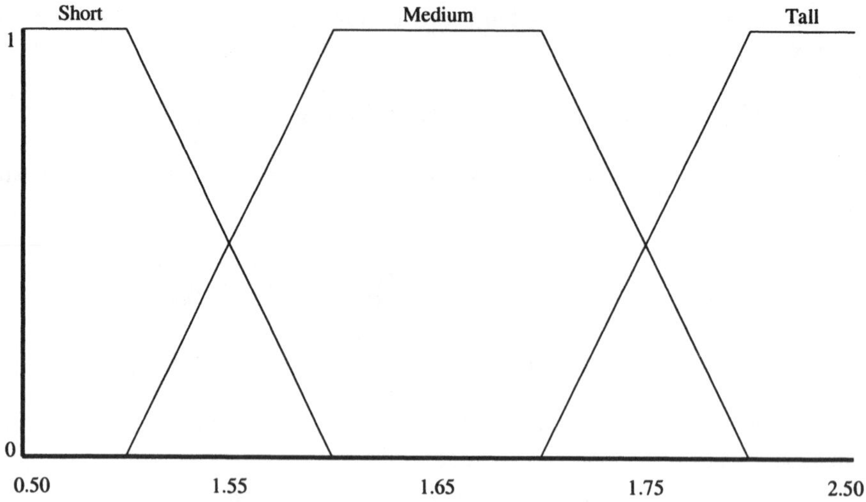

Fig. 2. Height

4.2 Coding of Crisp Information

The height of the individual has been measured using measuring devices that provide crisp information. In this case we have to calculate the compatibility degrees of the measurements obtained with each one of the fuzzy sets which represent the semantic of the linguistic terms of the variable "height".

If the measurement indicates a height value of 1.79, the compatibilities obtained with the associated fuzzy sets are (Fig. 3):

$$\begin{aligned}\alpha_{\text{short}} &= \mu_{\text{short}}\,(1.79) = 0.0\\ \alpha_{\text{med.}} &= \mu_{\text{med.}}\,(1.79) = 0.1\\ \alpha_{\text{tall}} &= \mu_{\text{tall}}\,(1.79) = 0.9\ .\end{aligned}$$

Now, we have the information expressed in degrees of compatibility, 0.9 medium, 0.1 tall. Therefore, in order to code the information in the linguistic variable "height" with a global precision of $1/10$ we proceed with the previous case.

$$\begin{aligned}C\,(h) &= (C_{\text{short}\,1}, \ldots, C_{\text{short}\,10}, C_{\text{med.}\,1}, \ldots, C_{\text{med.}\,10}, C_{\text{tall}\,1}, \ldots, C_{\text{tall}\,10})\\ &= (0000000000\;1000000000\;1111111110)\ .\end{aligned}$$

4.3 Coding of Linguistic Information

It is possible that the information provided about height is expressed as a fuzzy subset over V. In this case it is sufficient to continue considering α_i to be the compatibility degree with the linguistic term medium.

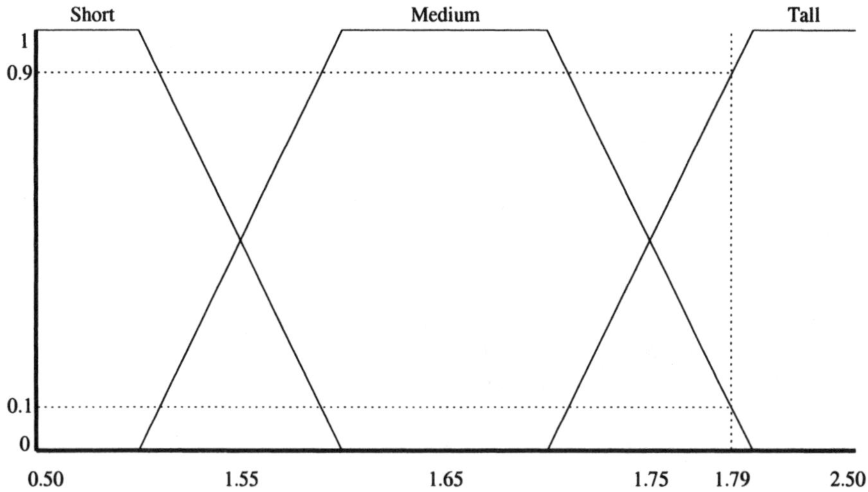

Fig. 3. Crisp Information 1.79

If the information provided is h = medium and we wish to carry out a coding with a precision of 10 bits per term, the resulting vector will be:

$$\begin{aligned} C(h) &= (C_{\text{short}\,1}, \ldots, C_{\text{short}\,10}, C_{\text{med.}\,1}, \ldots, C_{\text{med.}\,10}, C_{\text{tall}\,1}, \ldots, C_{\text{tall}\,10}) \\ &= (0000000000\;1111111111\;0000000000)\ . \end{aligned}$$

In this case it would have been better to use only one bit per term. The resulting coding of this would be:

$$C(h) = (C_{\text{short}}, C_{\text{medium}}, C_{\text{tall}}) = (0\;1\;0)\ .$$

5 Decoding of Results Coded Using Incremental Discretization

The coding method we have presented permits the use of a discrete model for the processing of information expressed in linguistic terms. This allows us to work with information using the same terms that were provided by the human and employs already existing crisp models.

After processing the information with the discrete system, the results obtained are expressed in precise terms. To obtain the desired behavior as a filter (Fig. 1) it is necessary to express the result in linguistic terms which are more understandable to the user.

In order to perform the decoding of the result we must carry out a process which is the inverse of the coding.

The coding process of Incremental Discretization of a value A with respect to the set of terms $t_1, \ldots, t_n$ consists of two steps:

- Step 1: $A \to \alpha_1 t_1, \dots, \alpha_n t_n$
- Step 2: $\alpha_1 t_1, \dots, \alpha_n t_n \to C(A) \equiv (C_{11}, \dots, C_{m1}, \dots, C_{1n}, \dots, C_{mn})$

Therefore, in order to decode the result we have to carry out a process which is the opposite of the two steps used in the coding.

After $C(A) \equiv (C_{11}, \dots, C_{m1}, \dots, C_{1n}, \dots, C_{mn})$ we immediately obtain the compatibilities with respect to the terms $t_1, \dots, t_n$; it is only necessary to follow the inverse process which took place in the coding:

1. – From i = 1 until n
   1.1. – $\alpha_i^{,} = 0$
   1.2. – From j = 1 until m
      1.2.1. – Si $C_{ji} = 1$ then
         1.2.1.1. – $\alpha_i^{,} = \alpha_i^{,} + \frac{1}{m}$
      1.2.2. – End
   1.3. – End
2. – End

Obtaining information based on the compatibilities with the terms is not immediate because the set of terms of a linguistic variable are not calculated within the operations of the set or in the arithmetic of the same fuzzy sets. For example, if R_1 and R_2 are linguistic values of a quantitative variable X whose semantic comes from the real fuzzy numbers r_1 and r_2, it seems logical to consider that the product of R_1 and R_2 is defined by the extended product $r_1 \otimes r_2$.

The problem is that there is no reason for the existence of a label in $T(X)$ whose semantic is $r_1 \otimes r_2$. It is logical to expect that the output of a model has the same characteristics as the input. If we are using linguistic terms as the input we can require that the output is a label that represents a certain linguistic value.

This is a problem of linguistic approximation [10] and it can be solved using any of the accepted methods such as, for example, using a centroid in order to obtain the linguistic label which best represents the fuzzy subset obtained as a result of the model.

6 Coding the Linguistic Rules IF-THEN Using Incremental Discretization

Once the coding method for linguistic variables has been shown, it is worthwhile studying the way in which the expressed rules in linguistic terms are coded. This will allow the representation of a system based on fuzzy rules in such a way that it can be processed by a discrete model.

A fuzzy system can be identified as a system based on rules [5] and a rule is an association of cause-effect patterns which can be coded in a discrete way using Incremental Discretization. The complexity of the rule, which varies according to the number of consequents associated with an antecedent and the connectors which unite them, is reflected in associations of more complex patterns.

Once the method of coding the values of a linguistic variable for the different ways it can be expressed are known, either in compatibility, crisp, or linguistic terms, the coding of the rules is reduced merely to the coding of its antecedent and consequents.

We consider this case to be very simple, because more complex cases are deduced in the way that we shall see in the text below.

Suppose that we are using a fuzzy system represented by rules with fuzzy antecedents and consequents set out as (A, B), where A y B is described using the n and p labels, over which the value of the respective linguistic variables can be taken. This rule is expressed using vectors of characteristics which have as components the n and p linguistic labels over which A and B are evaluated.

The coding will result in a vector of bits which represent the compatibilities of the antecedent A and of the consequent B with the n and p linguistic labels over which they are evaluated A and B respectively.

If the precision in all the linguistic labels of the antecedent A is j and the one for the labels of the consequent B is k, the fuzzy rule (A, B) can be expressed as a pair of vectors of $n \times j$ and $p \times k$ components respectively.

For example, we will take the following set of rules of a system which calculates, according to the position and velocity of a moving object, the braking force necessary to apply to it in order to stop it:

1. If medium positive position and zero velocity then medium negative force.
2. If low positive position and low positive velocity then low negative force.
3. If low positive position and low negative velocity then zero force.
4. If medium negative position and zero velocity then medium positive force.
5. If low negative position and low negative velocity then low positive force.
6. If low negative position and low positive velocity then zero force.
7. If zero position and zero velocity then zero force.

The coding for the above using 4 bits of precision is shown in the table 1.

Table 1. Coding of the rules with 4 bits of precision

	POSITION					VELOCITY			FORCE				
	MN	LN	ZE	LP	MP	LN	ZE	LP	MN	LN	ZE	LP	MP
1	0000	0000	0000	0000	1111	0000	1111	0000	1111	0000	0000	0000	0000
2	0000	0000	0000	1111	0000	0000	0000	1111	0000	1111	0000	0000	0000
3	0000	0000	0000	1111	0000	1111	0000	0000	0000	0000	1111	0000	0000
4	1111	0000	0000	0000	0000	0000	1111	0000	0000	0000	0000	0000	1111
5	0000	1111	0000	0000	0000	1111	0000	0000	0000	0000	0000	1111	0000
6	0000	1111	0000	0000	0000	0000	0000	1111	0000	0000	1111	0000	0000
7	0000	0000	1111	0000	0000	0000	1111	0000	0000	0000	1111	0000	0000

7 Extension of Incremental Discretization Coding for the Representation of Uncertainty

Going back to the coding of linguistic labels, it can be interesting to enrich the information presented to the system with the use of multiple simultaneous measures of uncertainty [8].

In order to show the extension of the coding method for linguistic labels using Incremental Discretization with the simultaneous employment of multiple measures of uncertainty, we are going to focus on the case of combined use of possibility and necessity.

Let $\mathcal{F}$ be a fuzzy subset of V which we represent in terms t_i of $T(H)$ using possibility and necessity as:

$$\mathcal{F} \equiv \alpha_1 t_1, \beta_1 t_1, \ldots, \alpha_n t_n, \beta_n t_n ,$$

being α_i the possibility and β_i the necessity of $\mathcal{F}$ in respect to t_i.

Supposing that $T(H)$ is formed by the elements $t_1, \ldots, t_n$, in an arbitrary order, we associate each term t_i with two m-dimension vectors (supposedly a desired precision $1/m$), one for the possibility and another for the necessity. Therefore, the linguistic variable H is associated with a vector of dimension $2m \cdot n$.

The coding of $\mathcal{F}$ will give the dimension vector $2m \cdot n$:

$$C(\mathcal{F}) = (C^P_{11}, \ldots, C^P_{1m}, C^N_{11}, \ldots, C^N_{1m}, \ldots, C^P_{n1}, \ldots, C^P_{nm}, C^N_{n1}, \ldots, C^N_{nm})$$

C^P_{ij} is calculated the method of Incremental Discretization based on the values of possibility α_i of the terms t_i:

$$\begin{cases} \alpha_i = 0 \rightarrow C^P_{ij} = 0;\ i = 1, \ldots, n;\ j = 1, \ldots, m \\ \alpha_i \neq 0 \rightarrow \begin{cases} \exists j \text{ t.q. } \frac{j}{m} \leq \alpha_i < \frac{j+1}{m} \\ C^P_{il} = \begin{cases} 1 \text{ if } l \leq j \\ 0 \text{ if } l > j \,. \end{cases} \end{cases} \end{cases}$$

Expressed as an algorithm, the value of C^P_{ij} is calculated in accordance with the algorithm:

1. – From i = 1 until n
   1.1. – From j = 1 until m
      1.1.1. – Si j $\leq$ Integer part of $(\alpha_i \cdot m)$
         1.1.1.1. – Then $C^P_{ij} = 1$
      1.1.2. – Si j $>$ Integer part of $(\alpha_i \cdot m)$
         1.1.2.1. – Then $C^P_{ij} = 0$
   1.2. – End
2. – End

Likewise, C_{ij}^N is calculated based on the values of possibility β_i of the terms t_i:

$$\begin{cases} \beta_i = 0 \rightarrow C_{ij}^N = 0;\ i = 1, \ldots, n;\ j = 1, \ldots, m \\ \beta_i \neq 0 \rightarrow \begin{cases} \exists j \text{ t.q. } \frac{j}{m} \leq \beta_i < \frac{j+1}{m} \\ C_{il}^N = \begin{cases} 1 \text{ si } l \leq j \\ 0 \text{ si } l > j\,. \end{cases} \end{cases} \end{cases}$$

The value of C_{ij}^N is calculated using the algorithm:

1. – From i = 1 until n
   1.1. – From j = 1 until m
      1.1.1. – Si j $\leq$ Integer part of $(\beta_i \cdot m)$
         1.1.1.1. – Then $C_{ij}^N = 1$
      1.1.2. – Si j > Integer part of $(\beta_i \cdot m)$
         1.1.2.1. – Then $C_{ij}^N = 0$
   1.2. – End
2. – End

Example: let *Height* be a linguistic variable with terms short, medium and tall. The value of possibility 0.2 short, 0.8 medium and the value of necessity 0.6 medium, 0.4 tall are coded with 5 bits of precision:

$$\overbrace{10000}^{P}\ \overbrace{00000}^{N}\ \overbrace{11110}^{P}\ \overbrace{11100}^{N}\ \overbrace{00000}^{P}\ \overbrace{11000}^{N}$$

Therefore, the system of coding has been extended into a system capable of coding any element of the universe of discourse V according to any two measurements for the treatment of uncertainty. Of course, the system can be easily extended so that it can handle the simultaneous employment of as many representation functions as we wish.

8 Linguistic Associative Memories

In order to demonstrate the use of the coding method for linguistic labels using Incremental Discretization, we are going to see how it is applied in the memorization of information expressed in linguistic terms with discrete associative memories.

An associative memory is a system which allows us to store patterns for posterior recovery based on alterations. (The extent of the alterations may vary)

The characteristics of the associative memories make them particularly suitable for the cases in which the information is altered due to noise or when part of the information is not known.

Most of the models for associative memories are designed to work with discrete information. Coding the linguistic information using Incremental Discretization is especially useful because it will allow us to store fuzzy information in any model of discrete associative memory. Using this method it acts as a filter, the functioning of the associative memories is not altered in any way and all the characteristics are conserved.

To illustrate the ease of use of this method of coding we are going to extend two models of discrete associative memories in order to memorize the rules of a braking system. This system is represented by the following set of rules, the force that is required to brake a moving object when we know its position and velocity (Figs.4 and 5):

1. If medium positive position and zero velocity then medium negative force.
2. If low positive position and low positive velocity then low negative force.
3. If low positive position and low negative velocity then zero force.
4. If medium negative position and zero velocity then medium positive force.
5. If low negative position and low negative velocity then low positive force.
6. If low negative position and low positive velocity then zero force.
7. If zero position and zero velocity then zero force.

FORCE		VELOCITY				
		MN	LN	ZE	LP	MP
POSITION	MN			MP		
	LN		LP		ZE	
	ZE			ZE		
	LP		ZE		LN	
	MP			MN		

Fig. 4. Representation of the rules of the braking system

8.1 Lineal Linguistic Associative Memory

In the following section we are going to see how the implantation of the coding system over a LAM [1] gives rise to a lineal linguistic associative Memory LLAM.

The LAM is a two-layer associative memory model which stores arbitrary pairs of spatial analogous patterns of the type (A_k, B_k), $k = 1, \ldots, m$ using hebbian learning.

The LAM operates in discrete time and its topology is shown in Fig.6.

In order to correctly recover the pairs of patterns stored (A_k, B_k), it is necessary for all the A_k to be normalized and to be mutually orthogonal. In

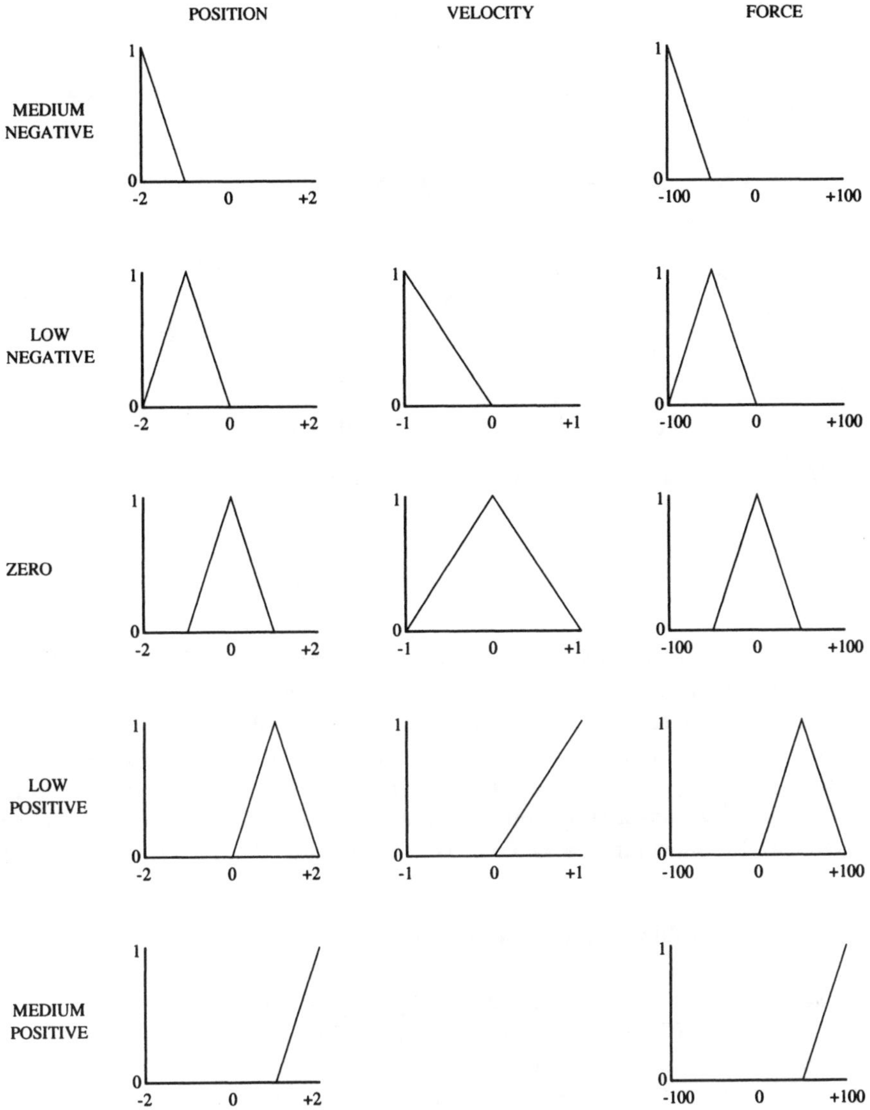

Fig. 5. Fuzzy sets associated with linguistic labels

order to do this, a preprocessing of the pairs that we wish to store is necessary in order to fulfill these conditions.

The normality condition is easily achievable thanks to the fact that the LAM works with analogical patterns. This enables each of the pattern components to be divided by the corresponding normalization factor $\sum_{i=1}^{n}(a_{ki})^2$, with $A_k = (a_{k1}, \ldots, a_{kn})$.

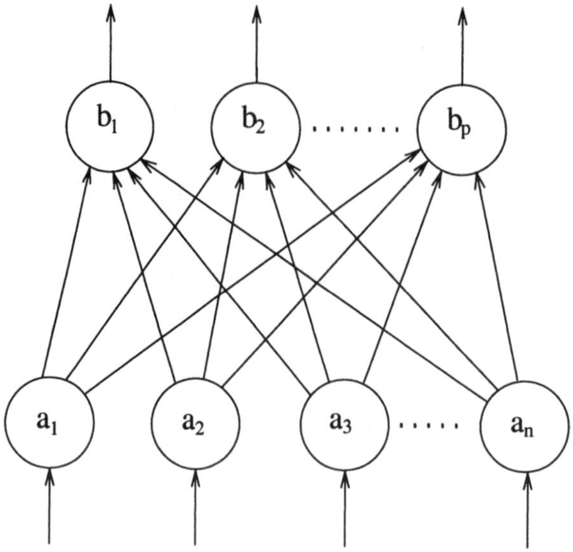

Fig. 6. Lineal Associative Memory

The situation changes regarding orthogonality because a general solution cannot be shown and an individual study is necessary for each case.

To memorize the set of rules, pairs of patterns will have the form (*cause*, *effect*) where the pattern *cause* will be formed by the state relating to the position and velocity and the pattern *effect* will be the force applicable in each case. If we use a single bit per label, the set of pairs of patterns to be memorized is shown in the Table 2. We can see that the patterns *cause* (A_k)

Table 2. Coding of the rules with 1 bit of precision

	POSITION					VELOCITY			FORCE				
	MN	LN	ZE	LP	MP	LN	ZE	LP	MN	LN	ZE	LP	MP
1	0	0	0	0	1	0	1	0	1	0	0	0	0
2	0	0	0	1	0	0	0	1	0	1	0	0	0
3	0	0	0	1	0	1	0	0	0	0	1	0	0
4	1	0	0	0	0	0	1	0	0	0	0	0	1
5	0	1	0	0	0	1	0	0	0	0	0	1	0
6	0	1	0	0	0	0	0	1	0	0	1	0	0
7	0	0	1	0	0	0	1	0	0	0	1	0	0
	CAUSE								EFFECT				

are not orthogonal which is why the system is not capable of storing all the pairs of patterns. One possible solution is to decompose the system into two subsystems in which the patterns *cause* are orthogonal and whose responses are subsequently recombined as shown in the Fig.7.

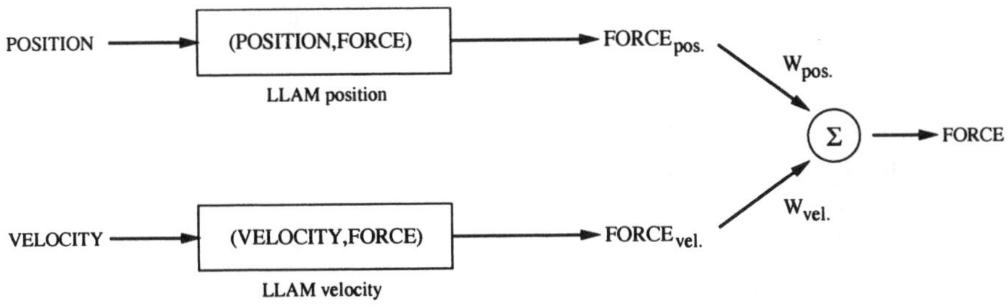

Fig. 7. Braking force according to relative velocity and position

These subsystems would be formed by the following sets of rules whose coding, using 1 bit of precision, is shown in the Tables 3 and 4.

Rules relating to the position of the moving object:

1. If medium positive position then medium negative force.
2. If low positive position then low negative force.
3. If medium negative position then medium positive force.
4. If low negative position then low positive force.
5. If zero position and zero velocity then zero force.

Rules relating to the velocity of the moving object:

1. If low negative velocity then low positive force.
2. If low positive velocity then low negative force.
3. If zero velocity then zero force.

Table 3. Set of rules relating to the position of the moving object coded with 1 bit of precision

	POSITION					FORCE				
	MN	LN	ZE	LP	MP	MN	LN	ZE	LP	MP
1	0	0	0	0	1	1	0	0	0	0
2	0	0	0	1	0	0	1	0	0	0
3	1	0	0	0	0	0	0	0	0	1
4	0	1	0	0	0	0	0	0	1	0
5	0	0	1	0	0	0	0	1	0	0
	CAUSE					EFFECT				

In this way we have achieved a coding which, respecting the conditions imposed by the LAM, allows the system to function with information expressed in linguistic terms. Nevertheless, the fact that a single bit has been used to code the linguistic labels will prevent us from measuring the degree of compatibility with terms which describe the linguistic variables. In order for the model of responses to be in accordance with the different degrees of interpretation we must increase the number of bits used to code the labels.

Table 4. Set of rules relating to the velocity of the moving object coded with 1 bit of precision

	VELOCITY			FORCE				
	LN	ZE	LP	MN	LN	ZE	LP	MP
1	1	0	0	0	0	0	1	0
2	0	0	1	0	1	0	0	0
3	0	1	0	0	0	1	0	0
	CAUSE			EFFECT				

Table 5. Coding of the rules of position with 5 bits of precision

	POSITION					FORCE				
	MN	LN	ZE	LP	MP	MN	LN	ZE	LP	MP
1	00000	00000	00000	00000	11111	11111	00000	00000	00000	00000
2	00000	00000	00000	11111	00000	00000	11111	00000	00000	00000
3	11111	00000	00000	00000	00000	00000	00000	00000	00000	11111
4	00000	11111	00000	00000	00000	00000	00000	00000	11111	00000
5	00000	00000	11111	00000	00000	00000	00000	11111	00000	00000
	CAUSE					EFFECT				

Table 6. Coding of the rules of velocity with 5 bits of precision

	VELOCITY			FORCE				
	LN	ZE	LP	MN	LN	ZE	LP	MP
1	11111	00000	00000	00000	00000	00000	11111	00000
2	00000	00000	11111	00000	11111	00000	00000	00000
3	00000	11111	00000	00000	00000	11111	00000	00000
	CAUSE			EFFECT				

If we use 5 bits to code the labels, the patterns to store are shown in Tables 5 and 6.

In these new sets of patterns, the condition of orthogonality continues to be fulfilled but this does not happen with the normality. In order to normalize them, it is enough to multiply each bit of the vectors *cause* by the normalization factor which, in this case, is $1/\sqrt{5}$.

Once the patterns have been memorized following the learning method of the LAM, the system is ready to provide responses according to the information presented to it.

For example, let us suppose that the position obtained for the moving object is -1.2, which, in linguistic terms, is equivalent to 0.2 medium negative, 0.8 low negative. This information would be coded as:

$$10000\ 11110\ 00000\ 00000\ 00000\ .$$

When the LLAM is provided with the specified pattern (position) as an input, the response of the system (force) is:

$$00000\ 00000\ 00000\ 11111\ 00000\ .$$

This response, which is equivalent to a low positive force, is not exactly the same as that which at first might have been expected of 0.8 low positive, 0.2 medium positive, that is to say:

00000 00000 00000 11110 10000 .

This behavior is due to the fact the LAM on which the LLAM is based is not an interpolator but rather an associator of patterns and, for this reason, only provides previously stored responses. The input patterns which do not coincide with the previously memorized ones are interpreted as erroneous patterns and the most similar stored pattern is sought. In this way, the input pattern corresponding to 0.2 medium negative, 0.8 low negative is interpreted as noisy and the most similar stored pattern is:

00000 11111 00000 00000 00000 .

Therefore, not only is this behavior not erroneous but it is also desirable since it provides the system with greater robustness. If we want the responses of the system to have a precision of 1/5, we need to train it in accordance with this level, i.e. we must carry out a training in which each rule gives rise to 5 patterns.

For example, the rule:

1. If medium positive position then medium negative force

would give rise to the set of patterns shown in the Table 7.

Table 7. If position MP then force MN

POSITION					FORCE				
MN	LN	ZE	LP	MP	MN	LN	ZE	LP	MP
00000	00000	00000	00000	10000	10000	00000	00000	00000	00000
00000	00000	00000	00000	11000	11000	00000	00000	00000	00000
00000	00000	00000	00000	11100	11100	00000	00000	00000	00000
00000	00000	00000	00000	11110	11110	00000	00000	00000	00000
00000	00000	00000	00000	11111	11111	00000	00000	00000	00000
CAUSE					EFFECT				

With the new set of patterns, it is necessary to achieve the normality and orthogonality of all the *cause* patterns again. Once the training has been carried out with all the patterns, given the position 0.2 medium negative, 0.8 low negative, the system response is of a force 0.8 low positive, 0.2 medium positive. This is due to the fact that the system returns an interpolated response for each pattern capable of being recognized from the input presented.

8.2 Classifying Linguistic Associative Memory

The classifying linguistic associative memory is the model which we obtain by applying the coding method on a CLAM.

Classifying Associative Memory. The Classifying Associative Memory CLAM [2] is an associative memory formed by two layers of processing elements which classify bipolar patterns by the criteria of the closest neighbor according to the Hamming distance.

The topology of the memory is dynamic since the number of output processing elements and the number of connections vary according to the number of patterns memorized.

The input/output layer is made up of the same number of bipolar processing elements as components of the patterns to memorize. In this layer the input pattern is presented and, after the recovery process, the closest stored pattern is obtained.

The output layer is competitive and made up of binary processing elements which represent the classes by which the patterns will be classified. For each pattern memorized, an associated processing element will be created in the output layer. A CLAM which still has not memorized any pattern will not have processing elements in this layer.

The two layers are linked by connections whose weights are represented in a matrix which has the same number of rows as the number of stored patterns and the same number of columns as the number of components of the patterns.

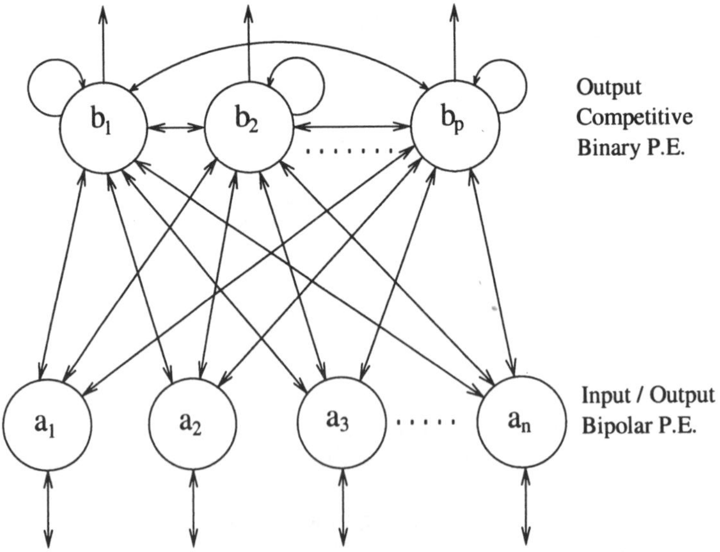

Fig. 8. Classifying Associative Memory

It functions in the following way:

- A possibly incomplete and incorrect pattern arrives at the input/output layer.

- The activation signal is propagated from the input/output layer to the output layer by means of the weight matrix.
- The processing elements of the output layer compete among themselves so that only the one which received the maximum activation signal is active. This PE represents the stored pattern which is closest to the input one. If after the competition, there is more than one active processing element it means that these received the maximum signal. In this case, one of them must be selected by applying some criteria and deactivating the others.
- The signal is propagated to the input/output layer by means of the weight matrix.
- The stored pattern closest to the input pattern is recovered in the input/output layer.

The main features of the model are:

- The dynamic topology makes it adapt to the number of patterns to store.
- All the stored patterns are perfectly recovered.
- It classifies the incomplete or incorrect patterns by the closest stored pattern according to the Hamming distance.
- It does not present spurious states, i.e. a pattern which was not stored beforehand will never be recovered.
- No type of condition is imposed on the patterns to store in order to guarantee their correct memorization.

Classifying Linguistic Associative Memory. The coding system of linguistic labels by Incremental Discretization applied on a Classifying Associative Memory CLAM gives rise to a Linguistic Classifying Associative Memory LCLAM.

As the CLAM memorizes bipolar patterns it is necessary for the linguistic labels to be coded in bipolar.

Going back to the example used with the LAM, the coding of the rules that we wish to memorize will give rise to a set of patterns in which some components will correspond to the cause and others to the effect. For example, using 3 bits of precision we can carry out the following coding:

If position medium positive and zero velocity then medium negative force

$$\underbrace{\overbrace{-1-1-1\ -1-1-1\ -1-1-1\ -1-1-1\ +1+1+1}^{\text{Position}}\ \overbrace{-1-1-1\ +1+1+1\ -1-1-1}^{\text{Velocity}}}_{\text{CAUSE}}$$

$$\underbrace{\overbrace{+1+1+1\ -1-1-1\ -1-1-1\ -1-1-1\ -1-1-1}^{\text{Force}}}_{\text{EFFECT}}\ .$$

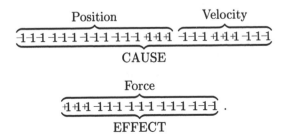

In order to recover a stored pattern, the unknown components in the input pattern will be set to zero while the others will keep their value. For example, if we want to find out the necessary force which must be applied when the moving object is at a low negative position with a low negative velocity, the input pattern will be:

$$\overbrace{\underbrace{\overbrace{\text{-1-1-1 +1+1+1 -1-1-1 -1-1-1 -1-1-1}}^{\text{Position}}\ \overbrace{\text{+1+1+1 -1-1-1 -1-1-1}}^{\text{Velocity}}}_{\text{CAUSE}}}\ \underbrace{\overbrace{\text{000 000 000 000 000}}^{\text{Force}}}_{\text{EFFECT}}\ .$$

With this input, the recovered pattern must be:

$$\underbrace{\overbrace{\text{-1-1-1 +1+1+1 -1-1-1 -1-1-1 -1-1-1}}^{\text{Position}}\ \overbrace{\text{+1+1+1 -1-1-1 -1-1-1}}^{\text{Velocity}}}_{\text{CAUSE}}$$

$$\underbrace{\overbrace{\text{-1-1-1 -1-1-1 -1-1-1 +1+1+1 -1-1-1}}^{\text{Force}}}_{\text{EFFECT}}\ .$$

Using a single bit per label, the resulting coding of the rules by Incremental Discretization is shown in the Table 8. As the CLAM does not impose any

Table 8. Coding of the rules with 1 bit of precision

	POSITION					VELOCITY			FORCE				
	MN	LN	ZE	LP	MP	LN	ZE	LP	MN	LN	ZE	LP	MP
1	-1	-1	-1	-1	+1	-1	+1	-1	+1	-1	-1	-1	-1
2	-1	-1	-1	+1	-1	-1	-1	+1	-1	+1	-1	-1	-1
3	-1	-1	-1	+1	-1	+1	-1	-1	-1	-1	+1	-1	-1
4	+1	-1	-1	-1	-1	-1	+1	-1	-1	-1	-1	-1	+1
5	-1	+1	-1	-1	-1	+1	-1	-1	-1	-1	-1	+1	-1
6	-1	+1	-1	-1	-1	-1	-1	+1	-1	-1	+1	-1	-1
7	-1	-1	+1	-1	-1	-1	+1	-1	-1	-1	+1	-1	-1
	CAUSE								EFFECT				

conditions on the patterns to store, preprocessing is not necessary as would occur in the case of the LAM.

The CLAM behaves like a classifier by the closest neighbour. For this reason, given an input which does not correspond to any of the stored patterns, it searches for the closest memorized pattern by understanding that the differences are due to the memory. This is desirable in the discrete model but in the linguistic case it causes problems.

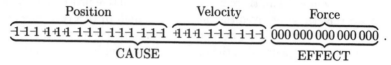

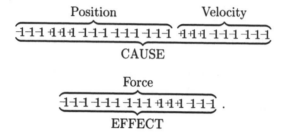

For example, if the input to the system corresponds to a situation for which there is no applicable rule, we can expect the memory not to recover any rule. Let us suppose that the position is low negative and the velocity is zero. The pattern which is presented in the memory input is:

$$\underbrace{\overbrace{-1\,+1\,-1\,-1\,-1}^{\text{Pos.}}\;\overbrace{-1\,+1\,-1}^{\text{Vel.}}}_{\text{CAUSE}}\;\underbrace{\overbrace{0\,0\,0\,0\,0}^{\text{Force}}}_{\text{EFFECT}}\;.$$

As no pattern which codes this situation has been memorized, the CLAM finds several which are just as close to the input pattern, to be more specific, those corresponding to the rules 1, 4, 5, 6 and 7 (Fig. 8) and it recovers one of them following some established criteria. This is the desired behavior.

We don't want any patterns to be recovered when there are no applicable rules. For this, we need to tell the CLAM not to recover patterns in which the components corresponding to the cause are different from those indicated in the input. In other words, a rule will only be recovered if it is applicable.

The differences between patterns are measured according to the Hamming distance. The distance between two patterns increase by one unit for each component in which both differ (Table 9).

Table 9. Examples of distances between patterns

Pairs of patterns	distance
−1 +1 +1 +1 +1 −1	2
−1 +1 +1 −1 +1 +1	
+1 −1 −1 −1 +1 −1	0
+1 −1 −1 −1 +1 −1	
+1 +1 +1 +1 −1 +1	6
−1 −1 −1 −1 +1 −1	

On the other hand, at times we leave some components at zero to indicate that they are unknown. This introduces a difference between the input pattern and the stored ones. The CLAM does not consider the distances existing between two components to be the same when one of them has a value of zero. In this case, the distance which exists is 1/2 (Table 10).

Table 10. Distances between components

Comp.	Comp.	distance
1	-1	1
1	0	1/2
-1	0	1/2

Therefore, each unknown component in a pattern increases the distance in 1/2 (Table 11).

Table 11. Distance between patterns with unknown components

Pairs of patterns	distance
+1 +1 −1 +1 −1 +1 +1 +1 −1 0 0 0	3/2
−1 +1 −1 −1 +1 +1 −1 −1 0 0 0 0	3
−1 +1 +1 +1 −1 +1 +1 −1 −1 −1 +1 0	11/2

Our intention is that the memory should complete the patterns in which we put the components corresponding to the effect as unknown. In order to do so, we must allow patterns to be recovered which are at a maximum distance which is the same as the distance entered for this unknown element. That is to say, if our patterns have 20 components corresponding to the effect, the distance entered for the unknown value is $20 \times 1/2 = 10$.

If, having 20 components to indicate the effect, in the recovery we find a pattern at a greater distance than 10, we can conclude that some of the components corresponding to the cause are different. In this case, this pattern would not be recovered. If the input pattern is not at a distance smaller than or equal to 10, none of the stored patterns will be recovered. This is the expected behavior when no rule is applicable.

In the case that concerns us, as we have 5 components corresponding to the effect, we can mark a threshold distance with the value 5/2 in such a way that any pattern which is to be found at a greater distance than this threshold will not be recovered (Table 12).

Table 12. Examples of recovered and unrecovered patterns

Input pattern	Recovered pattern	distance
-1-1-1+1-1 +1-1-1 0 0 0 0 0	-1-1-1+1-1 +1-1-1 -1-1+1-1-1	5/2
+1-1-1-1-1 -1-1+1 0 0 0 0 0	0 0 0 0 0 0 0 0 0 0 0 0 0 There is no applicable rule	> 5/2

With this we have managed to get the memory to present the desired behavior when we code the linguistic labels using 1 bit of precision. However, by using a single bit, it is not possible to use incremental degrees of measurement on the labels. For example, we cannot specify a position 1/4 low positive, 3/4 zero.

If we want to use incremental degrees of measurement on the labels we need to use more than one bit of precision. If we use 4 bits for the coding, the rules will be expressed as shown in the Table 13.

In order that only those patterns corresponding to applicable rules are recovered, the threshold is fixed at a suitable value. In this case, as we have 20 components corresponding to the effect, we can fix the threshold at $20 \times 1/2 = 10$.

The system still does not behave as well as we might expect. In order to show the problems it presents, let us look at an example.

Let us suppose that we want to discover the force that must be applied given a position 1/4 medium negative, 3/4 low negative and a velocity 1/2 low negative, 1/2 zero. The Incremental Discretization method encodes this situation as:

Position +1-1-1-1 +1+1+1-1 -1-1-1-1 -1-1-1-1 -1-1-1-1
Velocity +1+1-1-1 +1+1-1-1 -1-1-1-1 .

When we provide the system with the indicated input, on output we obtain the pattern

Position -1-1-1-1 -1-1-1-1 -1-1-1-1 -1-1-1-1 -1-1-1-1
Velocity -1-1-1-1 -1-1-1-1 -1-1-1-1
Force -1-1-1-1 -1-1-1-1 -1-1-1-1 -1-1-1-1 -1-1-1-1

This happens because the distance between the input pattern and those stored in the memory is greater than the threshold.

Anyway if we don't use the threshold we obtain a pattern which in the corresponding components shows the force with the following values:

Force -1-1-1-1 -1-1-1-1 -1-1-1-1 +1+1+1+1 -1-1-1-1 .

Table 13. Coding of the rules with 4 bits of precision

	POSITION					VELOCITY		
	MN	LN	ZE	LP	MP	LN	ZE	LP
1	-1-1-1-1	-1-1-1-1	-1-1-1-1	-1-1-1-1	+1+1+1+1	-1-1-1-1	+1+1+1+1	-1-1-1-1
2	-1-1-1-1	-1-1-1-1	-1-1-1-1	+1+1+1+1	-1-1-1-1	-1-1-1-1	-1-1-1-1	+1+1+1+1
3	-1-1-1-1	-1-1-1-1	-1-1-1-1	+1+1+1+1	-1-1-1-1	+1+1+1+1	-1-1-1-1	-1-1-1-1
4	+1+1+1+1	-1-1-1-1	-1-1-1-1	-1-1-1-1	-1-1-1-1	-1-1-1-1	+1+1+1+1	-1-1-1-1
5	-1-1-1-1	+1+1+1+1	-1-1-1-1	-1-1-1-1	-1-1-1-1	+1+1+1+1	-1-1-1-1	-1-1-1-1
6	-1-1-1-1	+1+1+1+1	-1-1-1-1	-1-1-1-1	-1-1-1-1	-1-1-1-1	-1-1-1-1	+1+1+1+1
7	-1-1-1-1	-1-1-1-1	+1+1+1+1	-1-1-1-1	-1-1-1-1	-1-1-1-1	+1+1+1+1	-1-1-1-1

	FORCE				
	MN	LN	ZE	LP	MP
1	+1+1+1+1	-1-1-1-1	-1-1-1-1	-1-1-1-1	-1-1-1-1
2	-1-1-1-1	+1+1+1+1	-1-1-1-1	-1-1-1-1	-1-1-1-1
3	-1-1-1-1	-1-1-1-1	+1+1+1+1	-1-1-1-1	-1-1-1-1
4	-1-1-1-1	-1-1-1-1	-1-1-1-1	-1-1-1-1	+1+1+1+1
5	-1-1-1-1	-1-1-1-1	-1-1-1-1	+1+1+1+1	-1-1-1-1
6	-1-1-1-1	-1-1-1-1	+1+1+1+1	-1-1-1-1	-1-1-1-1
7	-1-1-1-1	-1-1-1-1	+1+1+1+1	-1-1-1-1	-1-1-1-1

That is to say, it returns the coding of a low positive force. In its place, we can expect to obtain as an answer a force 1/4 medium positive, 1/2 low positive.

Force -1-1-1-1 -1-1-1-1 -1-1-1-1 +1+1-1-1 +1-1-1-1 .

This difference is due to the fact that the incremental degrees of measurement which are different from -1-1-1-1 y +1+1+1+1 are interpreted as erroneous patterns which must be corrected. We must bear in mind that the CLAM only recovers those patterns which were previously memorized.

If we want the system to respond correctly, we must carry out a training according to the degrees of measurement expected. For example, given a position 1/4 medium negative, 3/4 low negative and a velocity 1/2 low negative, 1/2 zero, if we want the system to respond with a force 1/4 medium positive, 1/2 low positive, we must carry out the training with the pattern:

Position: +1-1-1-1 +1+1+1-1 -1-1-1-1 -1-1-1-1 -1-1-1-1 Velocity: +1+1-1-1 +1+1-1-1 -1-1-1-1

CAUSA

Force: -1-1-1-1 -1-1-1-1 -1-1-1-1 +1+1-1-1 +1-1-1-1 .

EFFECT

This implies that each rule will give rise to a set of patterns which represent all the possibilities that can be presented as input to the system in which this rule is applicable.

For example, let us take the rule "If low positive position and positive low velocity then low negative force". In order for this rule to be applicable it is necessary for the input position to be low positive to a greater degree than zero. Likewise, the velocity must be low positive to a higher degree than zero. The different situations in which those requirements for the position and velocity are fulfilled are shown in the Tables 14 and 15. Therefore the

Table 14. Position LP

POSITION				
MN	LN	ZE	LP	MP
-1-1-1-1	-1-1-1-1	+1+1+1-1	+1-1-1-1	-1-1-1-1
-1-1-1-1	-1-1-1-1	+1+1-1-1	+1+1-1-1	-1-1-1-1
-1-1-1-1	-1-1-1-1	+1-1-1-1	+1+1+1-1	-1-1-1-1
-1-1-1-1	-1-1-1-1	-1-1-1-1	+1+1+1+1	-1-1-1-1
-1-1-1-1	-1-1-1-1	-1-1-1-1	+1+1+1-1	+1-1-1-1
-1-1-1-1	-1-1-1-1	-1-1-1-1	+1+1-1-1	+1+1-1-1
-1-1-1-1	-1-1-1-1	-1-1-1-1	+1-1-1-1	+1+1+1-1

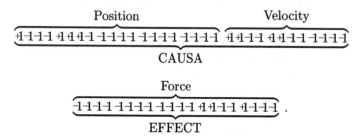

Table 15. Velocity LP

VELOCITY		
LN	ZE	LP
-1-1-1-1	+1+1+1-1	+1-1-1-1
-1-1-1-1	+1+1-1-1	+1+1-1-1
-1-1-1-1	+1-1-1-1	+1+1+1-1
-1-1-1-1	-1-1-1-1	+1+1+1+1

rule will be applicable in each one of the 28 resulting situations when the 7 position possibilities are combined with each of the 4 velocity possibilities.

In the 28 patterns resulting from this rule, it is necessary to code the resulting force. For example, for the pattern which codes the position 3/4 low positive, 1/4 medium positive and velocity 1/2 zero, 1/2 low positive, it is necessary to code a force 1/4 medium negative, 1/2 low negative.

$$\underbrace{\overbrace{-1-1-1-1\ -1-1-1-1\ -1-1-1-1\ +1+1+1-1\ +1-1-1-1}^{\text{Position}}\ \overbrace{-1-1-1-1\ +1+1-1-1\ +1+1-1-1}^{\text{Velocity}}}_{\text{CAUSE}}$$

$$\underbrace{\overbrace{+1-1-1-1\ +1+1+1-1\ -1-1-1-1\ -1-1-1-1\ -1-1-1-1}^{\text{Force}}}_{\text{EFFECT}}\ .$$

Once all the patterns have been stored, we obtain a memory which provides a correct response given the inputs presented to the system.

The method of coding linguistic variables by means of Incremental Discretization has allowed us to convert a CLAM which works with precise information into a LCLAM which works with imprecise information.

We have seen that, without much effort, a system can be obtained which is capable of working with information expressed in linguistic terms from a classic model for processing discrete information. This was achieved by applying the method of coding linguistic labels by means of Incremental Discretization and combining it with the use of the classic model.

By working as a filter, the method allows discrete, well-studied models to be used without the need for designing and studying new models.

9 Conclusions

- We have developed a method which enables us to represent information expressed in linguistic terms in binary or bipolar.
- This coding allows us to work on linguistic problems using techniques designed for crisp problems.
- By developing this method as a filter, it can be applied to any crisp model, inheriting all its characteristics.

- This method can be used in linguistic problems of well-studied crisp models without needing to design and study other new ones.
- The coding is easy to implement and with only a small computational cost.
- We can simultaneously use multiple mechanisms as a measure of uncertainty.
- The applicability of the method has been tested on two discrete associative memory models.

References

1. Anderson, J. (1970) Two models for memory organization using interacting traces, Math. Biosci., **8**, 137–160.
2. Bailón, A., Delgado, M., Fajardo, W. (1997) Clasificación de Información con Memorias Asociativas, CAEPIA'97
3. Blanco, A., Delgado, M., Fajardo, W. (1998) Extension from a Linear Associative Memory to a Linguistic Linear Associative Memory, International Journal of Intelligent Systems, Vol. 13, **1**, 41–59
4. Blanco, A., Delgado, M., Fajardo, W. (1999) Representation model of information in linguistic terms, Fuzzy Sets and Systems, 107, 277–287
5. Buckley, J., Hayashi, I., Czogala, E., (1993) On the equivalence of neural nets and fuzzy expert systems. Fuzzy Sets and Systems, **53**, 129–134.
6. Delgado, M., Verdegay, J.L., Vila, M.A. (1992) A linguistic version of the compositional rule of inference, preprints Sicica'92, Symposium of Intelligent Components and Instruments for Control Applications, 141–148.
7. Pedrycz, W. (1985) Applications of fuzzy relational equations for method of reasoning in presence of fuzzy data, Fuzzy Sets and Systems **16**, 163–175.
8. Pedrycz, W. (1995) Fuzzy Sets Engineering, CRC Press.
9. Zadeh, L.A. (1975) The concept of a linguistic variable and its applications to approximate reasoning, Part I, Inform. Sci. **8**, 199-248, Part II, Inform. Sci. **8**, 301-357, Part III, Inform. Sci. **9**, 43–80.
10. Zadeh, L.A. (1979) A theory of approximate reasoning, J.E. Hayes, D. Michie, L.I. Mikulich (Eds.), Machine Intelligence **9**, Elsevier, Amsterdam 149–194.
11. Zadeh, L.A. (1984) A theory of Commonsense knowledge, Aspects of vagueness (Skala, Termini, Trillas eds.) Reidel, Dordrecht, 257–295

A Formal Theory of Fuzzy Natural Language Quantification and its Role in Granular Computing

Ingo Glöckner and Alois Knoll

University of Bielefeld, Faculty of Technology, 33501 Bielefeld, Germany

Abstract. Fuzzy quantification is a linguistic granulation technique capable of expressing the global characteristics of a collection of individuals, or a relation between individuals, through meaningful linguistic summaries. However, existing approaches to fuzzy quantification fail to provide convincing results in the important case of two-place quantification (e.g. "many blondes are tall"). We develop an axiomatic framework for fuzzy quantification which complies with a large number of linguistically motivated adequacy criteria. In particular, we present the first models of fuzzy quantification which provide an adequate account of the "hard" cases of multiplace quantifiers, non-monotonic quantifiers, and non-quantitative quantifiers, and we show how the resulting operators can be efficiently implemented based on histogram computations.[1]

Keywords. Data summarization, fuzzy quantifiers, fuzzy set theory, human granulation techniques, natural language processing, theory of generalized quantifiers.

1 Introduction

Granular computing attempts to manage complex, large-scale problems by organizing these into different levels of detail. It is understood that each subproblem should be solved at its appropriate level of granularity, and that there are effective transformations which mediate between these levels. One essential process in this framework is that of *granulation*, i.e. the formation of granules, which permits the system to operate efficiently by reducing the amount of information, and by enhancing those aspects of the data which are of relevance to the particular subproblems. A certain degree of granulation can be achieved by abstracting from low-level features (e.g. pixel intensities in satellite images) to meaningful features (e.g. pixels classified as "cloudy"). However, this process of enhancing the description of individual objects cannot reduce their sheer number. It is hence essential for granular computing to develop granulation mechanisms which group individuals into larger meaningful units (clustering, segmentation); and to develop methods which can express accumulative properties of the resulting collections. For example, we might be interested in properties of certain regions of an image, rather than intensities of individual pixels. In a more general setting, we would like to abstract from individual observations or instances of a phenomenon to get the global picture. This particular process of granulating information by utilizing second-order properties, i.e. properties of collections of individuals, or properties of relations between individuals, will be called *summarisation*.

[1] The proofs of all theorems cited in this work have been presented in a sequence of reports [10,8,9].

The need for such transformation processes not only arises in the technical problem areas tackled by computers; every one of us needs to granulate and condense information in order to perform everyday tasks and in order to communicate efficiently with others. It is hence not surprising that natural language (NL) provides a class of expressions specifically designed to express accumulative properties and to summarise information: *natural language quantifiers*, a human granulation technique which has passed the test of many thousand years of practical language use. NL quantifiers, and in particular their approximate variety ("almost all", "a few" etc.), provide flexible means for expressing accumulative properties of collections and can also describe global (e.g., quantitative) aspects of relationships between individuals. In addition, aggregational modes of temporal or local description such as "almost always", "everywhere" are naturally modelled through quantification. Because of their suitability to describe a view of the phenomenon *as a whole*, the modelling of NL quantifiers has the promise of evolving into one of the key techniques for granular computing.

Following Zadeh [30,31], fuzzy set theory attempts to model NL quantifiers by operators called *fuzzy quantifiers*. We can discern the following main issues in fuzzy quantification:

- *Interpretation:* the development of methods for *evaluating* quantifying expressions which capture the meaning of NL quantifiers, e.g. [31,17,25,28,10];
- *Summarisation:* the development of processes for *constructing* quantifying statements ("linguistic summaries"), which succintly describe a collection of observations and/or relationships between a large number of observations (find domain concepts X and Y and a quantifier Q such that "Q X's are Y's" is true), e.g. [19];
- *Reasoning:* The development of methods which deduce further knowledge from a set of rules and/or facts involving fuzzy quantifiers, e.g. [32].

In the following, we shall focus on the interpretation task. This seems to be methodically preferable because both a convincing summarisation and appropriate rules for approximate reasoning, can only be established once the semantics of fuzzy quantifiers are better understood.

2 Existing approaches to fuzzy quantification

Several classes of operators have been proposed as properly representing the phenomenon of approximate or fuzzy NL quantification (a survey is provided in [15]), but there is no consensus about the proper choice, and notes on implausible behavior of these approaches are scattered over the literature [17,18,27,10]. In particular, it has been shown in [11] that none of these approaches provides acceptable results in the important case of two-place quantification, i.e. quantification restricted by a fuzzy predicate, as in "almost all blondes are lucky". In the following, we shall briefly review some of these counterexamples which enforced our decision to abandon these approaches, and to develop a fundamentally different approach to fuzzy quantification.[2]

[2] The exact formulas and quantifiers used in these counterexamples are explained in [11].

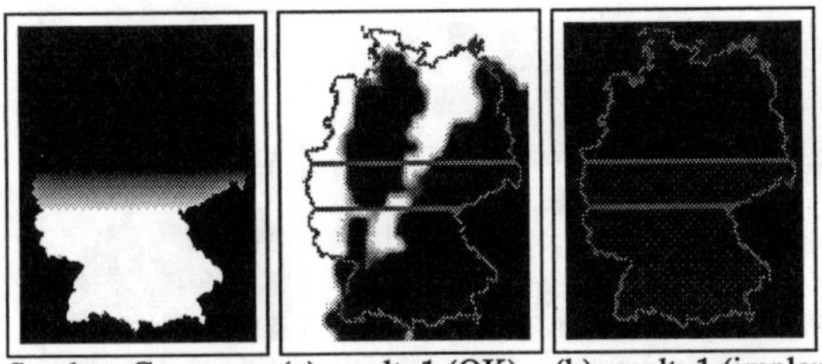

Figure1. About 10 percent of Southern Germany are cloudy (Σ-count approach)

Let us firstly consider the Σ-count approach [31]. This approach is known to accumulate "small" membership grades in an undesirable way. In the situation depicted in Fig. 1, for example, all of Southern Germany is 10 % cloudy, and the condition "about ten percent of Southern Germany are cloudy" is hence considered fully true by the Σ-count approach.[3] This is clearly implausible (to see this, consider the question *which* 10 percent are cloudy). In addition, the Σ-count approach produces discontinuous operators in the case of two-valued quantifiers like "more than 30 percent". This is inacceptable in practical applications because there is almost always some amount of noise, which can have drastic effects on the quantification results of the Σ-count approach. Because these effects also occur in the simple case of one-place quantification, we shall not consider the Σ-count approach further.

Yager [25,26] proposes an approach to fuzzy quantification based on OWA (ordered weighted averaging) operators. These perform well in the case of one-place quantification with monotonic quantifiers. However, the formula for two-place quantification with OWA operators (proposed in [25, p. 190]) exhibits unacceptable behaviour. It can be proven that from a linguistic standpoint, the *only* quantifiers which it models adequately are the existential and universal quantifier, see [11]. To provide an example, the results of the OWA-approach in Fig. 2 reveal a undesirable dependency on cloudiness grades in regions III and IV, which do not belong to Southern Germany at all.

The FG-count approach [31,24] utilizes a fuzzy measure of the cardinality of fuzzy sets to evaluate fuzzy quantifiers. This basic approach is well-behaved in the case of one-place quantifiers. Yager [26, p.72] proposes a weighting formula, which provides a definition of two-place quantification in conformance with the FG-count framework. Again, it is this formula for two-place quantification which yields implausible results. Consider the situation depicted in Fig. 3. There are no clouds at all in (the support of) SouthernGermany-1, hence we expect that "at least 5% of Southern Germany are cloudy" is false. The result of Yager's formula, however, is 0.55. The example also demonstrates that the resulting operators can be discontinuous: if we replace

[3] In the image "SouthernGermany", pixels which fully belong to Southern Germany are depicted white. In the cloudiness image, pixels classified as cloudy depicted white. The contours of Germany, split in southern, intermediate and northern part, have been added to facilitate interpretation.

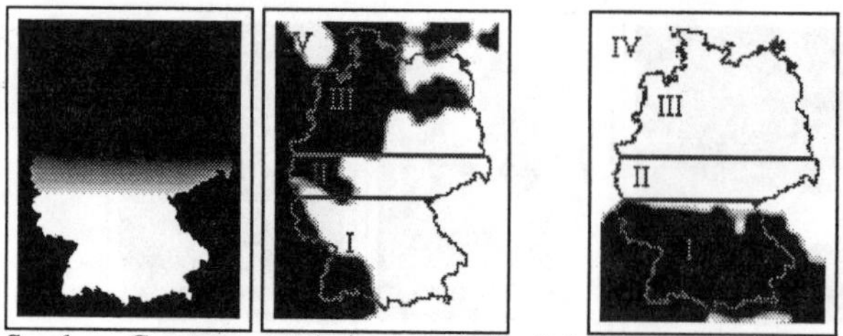

Southern Germany (a) desired: 1, OWA: 0.1 (b) desired: 0, OWA: 0.6

Figure2. At least 60 percent of Southern Germany are cloudy (OWA-approach)

SouthernGermany-1 with the slightly different SouthernGermany-2, the result jumps to 0.95, although there are still no clouds in the region of interest.

Ralescu [17] has proposed the use of the FE-count[4] for purposes of fuzzy quantification. Now consider Fig. 4, which illustrates a serious drawback of the FE-count approach. The computed "nonemptiness grade" of 1 in case (a) is adequate because (a) is a crisp nonempty region. The fuzzy image region in (b) is certainly fully nonempty to a degree of one, too, because it contains the crisp nonempty image region (a). The result of the FE-count approach, however, is 0.5. The FE-count approach hence produces counterintuitive results even in the simple case of absolute one-place quantification with a monotonic quantifier.

The above examples clearly show the drawbacks of existing approaches with respect to linguistic adequacy. In particular, the formulas proposed for the important case of two-place quantification apparently fail to grasp the intuitive meaning of NL quantifiers. In the following, we shall present a theory of fuzzy quantification built on an axiomatic foundation. We will formalize a number of linguistic adequacy conditions on approaches to fuzzy quantification and show that the axioms of our theory incorporate

[4] the FE-count is a fuzzy measure of the cardinality of fuzzy sets, cf. [31]

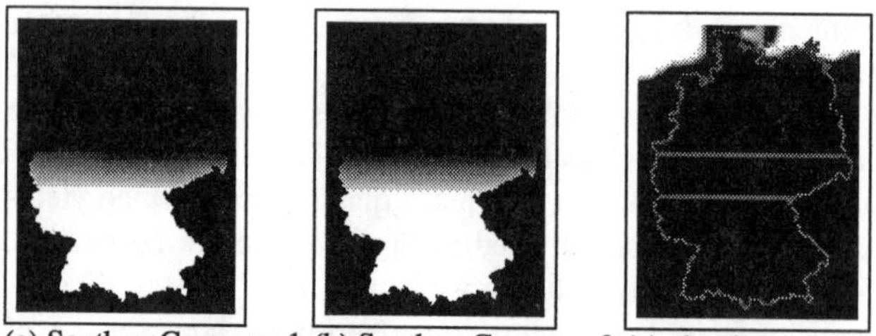

(a) SouthernGermany-1 (b) SouthernGermany-2 (c) cloudy

Result for SouthernGermany-1: 0.55
Result for SouthernGermany-2: 0.95
Desired result: 0

Figure3. At least 5 percent of Southern Germany are cloudy (FG-count approach)

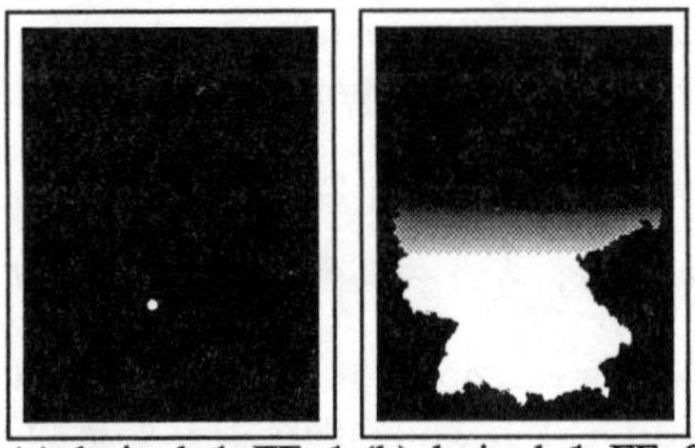

(a) desired: 1, FE: 1 (b) desired: 1, FE: 0.5

Figure4. The image region X is nonempty (FE-count approach)

(or entail) the essential conditions. A model of these axioms, called a determiner fuzzification scheme (DFS), is hence immune against the pitfalls of existing approaches.

3 The axiomatic framework of DFS theory

Because of the inadequate behaviour of all existing approaches in the case of two-place quantification, we have decided to abandon both their representation of fuzzy quantifiers by fuzzy numbers, and the idea of solving the problems of fuzzy quantification by introducing an appropriate measure for the cardinality of fuzzy sets.[5] Instead, we will assume the framework provided by the current *linguistic* theory of NL quantifcation, the theory of generalized quantifiers (TGQ [1,2]), which has been developed independently of the treatment of fuzzy quantifiers in fuzzy set theory, and provides a conceptually rather different view of natural language quantifiers. We shall introduce two-valued quantifiers in concordance with TGQ:

Definition 1. *An* n-ary generalized quantifier *on a base set* $E \neq \varnothing$ *is a mapping* $Q : \mathcal{P}(E)^n \longrightarrow \mathbf{2} = \{0, 1\}$.

A two-valued quantifier hence assigns to each n-tuple of crisp subsets $X_1, \dots, X_n \in \mathcal{P}(E)$ a two-valued quantification result $Q(X_1, \dots, X_n) \in \mathbf{2}$. Well-known examples are

$$\begin{aligned}
&\forall_E(X) = 1 \Leftrightarrow X = E \\
&\exists_E(X) = 1 \Leftrightarrow X \neq \varnothing \\
&\mathbf{all}_E(X_1, X_2) = 1 \Leftrightarrow X_1 \subseteq X_2 \\
&\mathbf{some}_E(X_1, X_2) = 1 \Leftrightarrow X_1 \cap X_2 \neq \varnothing \\
&\mathbf{at\ least\ k}_E(X_1, X_2) = 1 \Leftrightarrow |X_1 \cap X_2| \geq k\,.
\end{aligned}$$

[5] As we shall see later, it is possible to recover the cardinality-based approach to fuzzy quantification in the case of quantitative one-place quantifiers, see (Th-57).

Whenever the base set is clear from the context, we drop the subscript E; $|\bullet|$ denotes cardinality. For finite E, we can define proportional quantifiers like

$$[\mathbf{rate} \geq r](X_1, X_2) = 1 \Leftrightarrow |X_1 \cap X_2| \geq r\,|X_1|$$
$$[\mathbf{rate} > r](X_1, X_2) = 1 \Leftrightarrow |X_1 \cap X_2| > r\,|X_1|$$

for $r \in \mathbf{I}$, $X_1, X_2 \in \mathcal{P}(E)$. For example, "at least 30 percent of the X's are Y's" can be expressed as $[\mathbf{rate} \geq 0.3](X, Y)$, while $[\mathbf{rate} > 0.4]$ is suited to model "more than 40 percent". By the *scope* of an NL quantifier we denote the argument occupied by the verbal phrase (e.g. "sleep" in "all men sleep"); by convention, the scope is the last argument of a quantifier. The first argument of a two-place quantifier is its *restriction*. The two-place use of a two-place quantifier, like in "most X's are Y's" is called its *restricted use*, while its one-place use (relative to the whole domain E, like in "most elements of the domain are Y") is its *unrestricted use*. For example, the unrestricted use of $\mathbf{all} : \mathcal{P}(E)^2 \longrightarrow \mathbf{2}$ is modelled by $\forall : \mathcal{P}(E) \longrightarrow \mathbf{2}$, which has $\forall(X) = \mathbf{all}(E, X)$. TGQ has classified the wealth of quantificational phenomena in natural languages in order to unveil universal properties shared by quantifiers in all natural languages, or to single out classes of quantifiers with specific properties (we shall describe some of these properties below). However, an extension to the continuous-valued case, in order to better capture the meaning of approximate quantifiers like "many" or "about ten", has not been an issue for TGQ. In addition, TGQ has ignored the problem of providing a convincing interpretation for quantifying statements in the presence of fuzziness, i.e. in the frequent case that the arguments of the quantifier are occupied by concepts like "tall" or "cloudy" which do not possess sharply defined boundaries.
Hence let us introduce the fuzzy framework. Suppose E is a given set. A fuzzy subset $X \in \widetilde{\mathcal{P}}(E)$ of a set E assigns to each $e \in E$ a membership degree $\mu_X(e) \in \mathbf{I} = [0, 1]$; we denote by $\widetilde{\mathcal{P}}(E)$ the set of all fuzzy subsets (fuzzy powerset) of E.

Definition 2. *An n-ary* fuzzy quantifier *$\widetilde{Q}$ on a base set $E \neq \varnothing$ is a mapping $\widetilde{Q} : \widetilde{\mathcal{P}}(E)^n \longrightarrow \mathbf{I}$ which to each choice of $X_1, \ldots, X_n \in \widetilde{\mathcal{P}}(E)$ assigns a gradual result $\widetilde{Q}(X_1, \ldots, X_n) \in \mathbf{I}$.*[6]

An example is $\widetilde{\mathbf{some}}(X_1, X_2) = \sup\{\min(\mu_{X_1}(e), \mu_{X_2}(e)) : e \in E\}$, for all $X \in \widetilde{\mathcal{P}}(E)$. How can we justify that this operator is a good model of the NL quantifier "some"? How can we describe characteristics of fuzzy quantifiers and how can we locate a fuzzy quantifier based on a description of desired properties? Fuzzy quantifiers are possibly too rich a set of operators to investigate this question directly. Few intuitions apply to the behaviour of quantifiers in the case that the arguments are fuzzy, and the familiar concept of *cardinality* of crisp sets, which makes it easy to define quantifiers on crisp arguments, is no longer available. We therefore have to introduce some kind of *simplified description* of the essential aspects of a fuzzy quantifier. In order to comply with linguistic theory, this representation should be rich enough to embed all two-valued quantifiers of TGQ.

[6] This definition closely resembles Zadeh's [32, pp.756] alternative view of fuzzy quantifiers as fuzzy second-order predicates, but models these as mappings in order to simplify notation. In addition, we permit for arbitrary $n \in \mathbb{N}$.

Definition 3. *An* n-ary semi-fuzzy quantifier *on a base set* $E \neq \varnothing$ *is a mapping* $Q : \mathcal{P}(E)^n \longrightarrow \mathbf{I}$ *which to each choice of* crisp $X_1, \ldots, X_n \in \mathcal{P}(E)$ *assigns a gradual result* $Q(X_1, \ldots, X_n) \in \mathbf{I}$.

Semi-fuzzy quantifiers are half-way between two-valued quantifiers and fuzzy quantifiers because they have crisp input and fuzzy output. In particular, every two-valued quantifier of TGQ is a semi-fuzzy quantifier by definition. To give an example, a possible definition of the semi-fuzzy quantifier **almost all** : $\mathcal{P}(E)^2 \longrightarrow \mathbf{I}$ is

$$\textbf{almost all}(X_1, X_2) = \begin{cases} f_{\text{almost all}}(\frac{|X_1 \cap X_2|}{|X_1|}) & : \quad X_1 \neq \varnothing \\ 1 & : \quad \text{else} \end{cases} \qquad (1)$$

where $f_{\text{almost all}}(z) = S(z, 0.7, 0.9)$, using Zadeh's S-function (see Fig. 5). Unlike

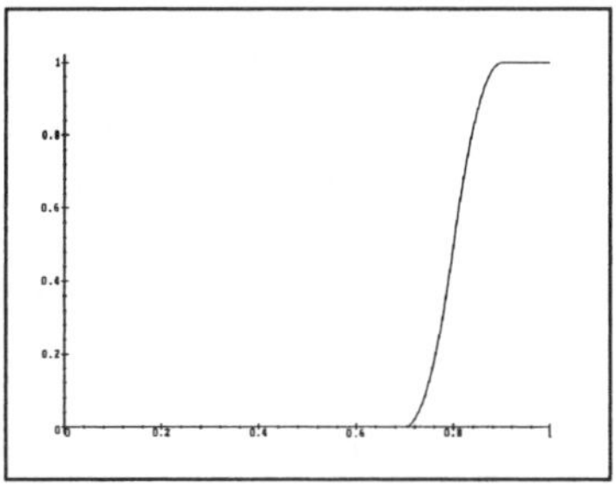

Figure5. A possible definition of $f_{\text{almost all}}$

the representations chosen by existing approaches to fuzzy quantification, semi-fuzzy quantifiers can express genuine multiplace quantification (arbitrary n); they are not restricted to the absolute and proportional types; they are not necessarily quantitative (in the sense of automorphism-invariance); and there is no a priori restriction to finite domains. Compared to fuzzy quantifiers, the main benefit of introducing semi-fuzzy quantifiers is conceptual simplicity due to the restriction to crisp argument sets, which usually makes it easy to understand the input-output behavior of a semi-fuzzy quantifier. Most importantly, we have the familiar concept of crisp cardinality available, which is of invaluable help in defining the quantifiers of interest. Being half-way between two-valued generalized quantifiers and fuzzy quantifiers, semi-fuzzy quantifiers do not accept fuzzy input, and we have to make use of a fuzzification mechanism which transports these to fuzzy quantifiers.

Definition 4. *A* quantifier fuzzification mechanism $\mathcal{F}$ *assigns to each semi-fuzzy quantifier* $Q : \mathcal{P}(E)^n \longrightarrow \mathbf{I}$ *a corresponding fuzzy quantifier* $\mathcal{F}(Q) : \widetilde{\mathcal{P}}(E)^n \longrightarrow \mathbf{I}$ *of the same arity* n *and on the same base set* E.

By viewing approaches to fuzzy quantification as instances of quantifier fuzzification mechanisms (QFM), we are able to explore the linguistic adequacy of these approaches by investigating preservation and homomorphism properties of the corresponding fuzzification mappings [10,11,8]. To this end, we first need to introduce several concepts related to (semi-) fuzzy quantifiers.

Definition 5. *Suppose* $\widetilde{Q} : \widetilde{\mathcal{P}}(E)^n \longrightarrow \mathbf{I}$ *is a fuzzy quantifier. By* $\mathcal{U}(\widetilde{Q}) : \mathcal{P}(E)^n \longrightarrow \mathbf{I}$ *we denote the* underlying semi-fuzzy quantifier, *viz.*

$$\mathcal{U}(\widetilde{Q})(Y_1,\dots,Y_n) = \widetilde{Q}(Y_1,\dots,Y_n) \tag{2}$$

for all crisp *subsets* $Y_1,\dots,Y_n \in \mathcal{P}(E)$.

Every reasonable QFM $\mathcal{F}$ should correctly generalise the semi-fuzzy quantifiers to which it is applied, i.e. for all $Q : \mathcal{P}(E)^n \longrightarrow \mathbf{I}$, we should have

$$\mathcal{F}(Q)(Y_1,\dots,Y_n) = Q(Y_1,\dots,Y_n) \tag{3}$$

for all *crisp* arguments $Y_1,\dots,Y_n \in \mathcal{P}(E)$, or equivalently: $\mathcal{U}(\mathcal{F}(Q)) = Q$.

Let us now consider a special case of quantifiers, called projection quantifiers. Suppose E is a set of persons and John $\in E$. We can then express the membership assessment "Is John contained in Y?", where Y is a crisp subset $Y \in \mathcal{P}(E)$, by computing $\chi_Y(\text{John})$, where $\chi_Y : \mathcal{P}(E) \longrightarrow \mathbf{I}$ is the characteristic function

$$\chi_Y(e) = \begin{cases} 1 & : \; e \in Y \\ 0 & : \; e \notin Y\,. \end{cases} \tag{4}$$

for all $Y \in \mathcal{P}(E)$, $e \in E$. Similarly, we can evaluate the fuzzy membership assessment "To which grade is John contained in X", where $X \in \widetilde{\mathcal{P}}(E)$ is a fuzzy subset of E, by computing $\mu_X(\text{John})$. Abstracting from argument sets, we obtain the following definitions of projection quantifiers:

Definition 6. *Suppose* $E \neq \varnothing$ *is given and* $e \in E$. *The* projection quantifier $\pi_e : \mathcal{P}(E) \longrightarrow \mathbf{2}$ *is defined by* $\pi_e(Y) = \chi_Y(e)$, *for all* $Y \in \mathcal{P}(E)$.
Similarly, the fuzzy projection quantifier $\widetilde{\pi}_e : \widetilde{\mathcal{P}}(E) \longrightarrow \mathbf{I}$ *is defined by* $\widetilde{\pi}_e(X) = \mu_X(e)$, *for all* $X \in \widetilde{\mathcal{P}}(E)$.

It is apparent from the relationship of these quantifiers with crisp / fuzzy membership assessments that $\widetilde{\pi}_e$ is the proper fuzzy counterpart of π_e, and we should have $\mathcal{F}(\pi_e) = \widetilde{\pi}_e$ in every reasonable QFM. Hence for the crisp subset **married** $\in \mathcal{P}(E)$,

$$\pi_{\mathbf{John}}(\mathbf{married}) = \begin{cases} 1 & : \; \text{John} \in \mathbf{married} \\ 0 & : \; \text{else} \end{cases}$$

and we should also have that $\mathcal{F}(\pi_{\mathbf{John}})(\mathbf{lucky}) = \widetilde{\pi}_{\mathbf{John}}(\mathbf{lucky}) = \mu_{\mathbf{lucky}}(\text{John})$, where **lucky** $\in \widetilde{\mathcal{P}}(E)$ is the fuzzy subset of lucky people.

We expect that our framework not only provides an interpretation for quantifiers, but also for the propositional part of the logic. We therefore need to associate a suitable choice of fuzzy conjunction, fuzzy disjunction etc. with a given QFM $\mathcal{F}$. By a canonical construction, $\mathcal{F}$ induces a unique fuzzy operator for each of the propositional connectives. As the starting point for the construction of induced connectives, let us observe that $\mathbf{2}^n \cong \mathcal{P}(\{1, \ldots, n\})$, using the bijection $\eta : \mathbf{2}^n \longrightarrow \mathcal{P}(\{1, \ldots, n\})$ defined by $\eta(x_1, \ldots, x_n) = \{k \in \{1, \ldots, n\} : x_k = 1\}$, for all $x_1, \ldots, x_n \in \mathbf{2}$. An analogous construction is possible in the fuzzy case, where we have $\mathbf{I}^n \cong \widetilde{\mathcal{P}}(\{1, \ldots, n\})$, using the bijection $\widetilde{\eta} : \mathbf{I}^n \longrightarrow \widetilde{\mathcal{P}}(\{1, \ldots, n\})$ defined by $\mu_{\widetilde{\eta}(x_1,\ldots,x_n)}(k) = x_k$ for all $x_1, \ldots, x_n \in \mathbf{I}$ and $k \in \{1, \ldots, n\}$.

Definition 7 (Induced fuzzy truth functions). *Suppose $\mathcal{F}$ is a QFM and f is a semi-fuzzy truth function (mapping $f : \mathbf{2}^n \longrightarrow \mathbf{I}$) of arity $n > 0$. The semi-fuzzy quantifier $Q_f : \mathcal{P}(\{1, \ldots, n\}) \longrightarrow \mathbf{I}$ is defined by*

$$Q_f(X) = f(\eta^{-1}(X))$$

for all $X \in \mathcal{P}(\{1, \ldots, n\})$. The induced fuzzy truth function $\widetilde{\mathcal{F}}(f) : \mathbf{I}^n \longrightarrow \mathbf{I}$ is

$$\widetilde{\mathcal{F}}(f)(x_1, \ldots, x_n) = \mathcal{F}(Q_f)(\widetilde{\eta}(x_1, \ldots, x_n)),$$

for all $x_1, \ldots, x_n \in \mathbf{I}$. If $f : \mathbf{2}^0 \longrightarrow \mathbf{I}$ is a nullary semi-fuzzy truth function (i.e., a constant), we shall define $\widetilde{\mathcal{F}}(f) : \mathbf{I}^0 \longrightarrow \mathbf{I}$ by $\widetilde{\mathcal{F}}(f)(\varnothing) = \mathcal{F}(c)(\varnothing)$, where $c : \mathcal{P}(\{\varnothing\})^0 \longrightarrow \mathbf{I}$ is the constant $c(\varnothing) = f(\varnothing)$.[7]

Whenever $\mathcal{F}$ is understood from context, we shall abbreviate $\widetilde{\mathcal{F}}(f)$ as $\widetilde{f}$. By pointwise application of the induced negation $\widetilde{\neg} = \widetilde{\mathcal{F}}(\neg)$, conjunction $\widetilde{\wedge} = \widetilde{\mathcal{F}}(\wedge)$, and disjunction $\widetilde{\vee} = \widetilde{\mathcal{F}}(\vee)$, $\mathcal{F}$ also induces a unique choice of fuzzy complement $\widetilde{\neg}$, fuzzy intersection $\widetilde{\cap}$, and fuzzy union $\widetilde{\cup}$. In the following, we will assume that an arbitrary but fixed choice of these connectives is given.

Definition 8. *Suppose a semi-fuzzy quantifier $Q : \mathcal{P}(E)^n \longrightarrow \mathbf{I}$ is given. The* external negation *$\widetilde{\neg} Q : \mathcal{P}(E)^n \longrightarrow \mathbf{I}$ and the* antonym *$Q\neg : \mathcal{P}(E)^n \longrightarrow \mathbf{I}$ are defined by*

$$(\widetilde{\neg} Q)(X_1, \ldots, X_n) = \widetilde{\neg}(Q(X_1, \ldots, X_n)) \tag{5}$$

$$Q\neg(X_1, \ldots, X_n) = Q(X_1, \ldots, X_{n-1}, \neg X_n) \qquad n > 0 \tag{6}$$

for all $X_1, \ldots, X_n \in \mathcal{P}(E)$, where $\neg X_n$ denotes complementation. By the dual *$Q\widetilde{\Box} : \mathcal{P}(E)^n \longrightarrow \mathbf{I}$ of Q we denote $Q\widetilde{\Box} = \widetilde{\neg} Q\neg$, $n > 0$. The definitions of negation $\widetilde{\neg}\widetilde{Q}$, antonym $\widetilde{Q}\widetilde{\neg}$ and dual $\widetilde{Q}\widetilde{\Box}$ of a fuzzy quantifier are analogous.*

For example, **less than n** is the negation of **at least n, no** is the antonym of **all,** and **some** is the dual of **all**. It is straightforward to require that a QFM be compatible with

[7] The special treatment of nullary semi-fuzzy truth functions is necessary because in this case, we would have $Q_f : \mathcal{P}(\varnothing) \longrightarrow \mathbf{I}$, which does not conform to our definition of semi-fuzzy quantifiers and fuzzy quantifiers based on nonempty base-sets. Rather than adapting the definition of semi-fuzzy and fuzzy quantifiers in such a way as to allow for empty base sets, and hence cover Q_f for nullary f, too, we prefer to treat the case of nullary truth functions by a different construction, see [8].

these constructions, i.e. we desire that $\mathcal{F}(\textbf{less than n})$ be the negation of $\mathcal{F}(\textbf{at least n})$, $\mathcal{F}(\textbf{no})$ be the antonym of $\mathcal{F}(\textbf{all})$, and $\mathcal{F}(\textbf{some})$ be the dual of $\mathcal{F}(\textbf{all})$. We hence say that $\mathcal{F}$ preserves negation, antonymy, and dualisation, if $\mathcal{F}(\neg Q) = \widetilde{\neg}\mathcal{F}(Q)$, $\mathcal{F}(Q\neg) = \mathcal{F}(Q)\widetilde{\neg}$ and $\mathcal{F}(Q\widetilde{\square}) = \mathcal{F}(Q)\widetilde{\square}$, resp.

Definition 9. *Suppose $Q : \mathcal{P}(E)^n \longrightarrow \mathbf{I}$ is a semi-fuzzy quantifier of arity $n > 0$. The semi-fuzzy quantifier $Q\cup \in: \mathcal{P}(E)^{n+1} \longrightarrow \mathbf{I}$ is defined by*

$$Q\cup(X_1, \ldots, X_{n+1}) = Q(X_1, \ldots, X_{n-1}, X_n \cup X_{n+1}),$$

for all $X_1, \ldots, X_{n+1} \in \mathcal{P}(E)$. $\widetilde{Q}\widetilde{\cup} : \widetilde{\mathcal{P}}(E)^{n+1} \longrightarrow \mathbf{I}$ is defined analogously.

In order to allow for a compositional interpretation of composite quantifiers like "all X's are Y's or $Z's$", we require that a QFM $\mathcal{F}$ be compatible with unions of the argument sets. For example, the semi-fuzzy quantifier **all**$\cup$, which has $\textbf{all}\cup(X, Y, Z) = \textbf{all}(X, Y \cup Z)$ for crisp $X, Y, Z \in \mathcal{P}(E)$, should be mapped to $\mathcal{F}(\textbf{all})\widetilde{\cup}$, i.e. for all fuzzy subsets $X, Y, Z \in \widetilde{\mathcal{P}}(E)$, $\mathcal{F}(\textbf{all}\cup)(X, Y, Z) = \mathcal{F}(\textbf{all})(X, Y \widetilde{\cup} Z)$.

Definition 10. *A semi-fuzzy quantifier $Q : \mathcal{P}(E)^n \longrightarrow \mathbf{I}$ is said to be* nonincreasing *in its i-th argument ($i \in \{1, \ldots, n\}$, $n > 0$) iff for all $X_1, \ldots, X_n, X_i' \in \mathcal{P}(E)$ such that $X_i \subseteq X_i'$,*

$$Q(X_1, \ldots, X_n) \geq Q(X_1, \ldots, X_{i-1}, X_i', X_{i+1}, \ldots, X_n).$$

Nondecreasing monotonicity of Q is defined by changing '$\geq$' to '$\leq$' in the above inequation. On fuzzy quantifiers $\widetilde{Q} : \widetilde{\mathcal{P}}(E)^n \longrightarrow \mathbf{I}$, we use an analog definition, where $X_1, \ldots, X_n, X_i' \in \widetilde{\mathcal{P}}(E)$, and "$\subseteq$" is the fuzzy inclusion relation.

For example, **all** is nonincreasing in the first and nondecreasing in the second argument. **most** is nondecreasing in its second argument, etc. It is natural to require that these monotonicity properties be preserved when applying a QFM $\mathcal{F}$; e.g. we expect that $\mathcal{F}(\textbf{all})$ is nonincreasing in the first and nondecreasing in the second argument.

Every mapping $f : E \longrightarrow E'$ uniquely determines a powerset function $\widehat{f} : \mathcal{P}(E) \longrightarrow \mathcal{P}(E')$ defined by $\widehat{f}(X) = \{f(e) : e \in X\}$, for all $X \in \mathcal{P}(E)$. The underlying mechanism which transports f to $\widehat{f}$ can be generalized to fuzzy sets.

Definition 11 (Induced extension principle). *Suppose $\mathcal{F}$ is a QFM. $\mathcal{F}$ induces an extension principle $\widehat{\mathcal{F}}$, i.e. a mechanism which to each $f : E \longrightarrow E'$ (where $E, E' \neq \varnothing$) assigns some $\widehat{\mathcal{F}}(f) : \widetilde{\mathcal{P}}(E) \longrightarrow \widetilde{\mathcal{P}}(E')$. The 'extended' mapping $\widehat{\mathcal{F}}(f)$ is defined by*

$$\mu_{\widehat{\mathcal{F}}(f)(X)}(e') = \mathcal{F}(\chi_{\widehat{f}(\bullet)}(e'))(X), \qquad \textit{for all } X \in \widetilde{\mathcal{P}}(E),\ e' \in E'.$$

The *standard extension principle* $(\hat{\hat{\bullet}})$ is defined by $\mu_{\hat{\hat{(f)}}(X)}(e') = \sup\{\mu_X(e) : e \in f^{-1}(e')\}$, cf. [29]. Now suppose $Q : \mathcal{P}(E)^n \longrightarrow \mathbf{I}$ and $f_1, \ldots, f_n : E' \longrightarrow E$ are given ($E' \neq \varnothing$). We can define a semi-fuzzy quantifier

$$Q' = Q \circ \overset{n}{\underset{i=1}{\times}} \widehat{f}_i : \mathcal{P}(E')^n \longrightarrow \mathbf{I},$$
$$Q'(Y_1, \ldots, Y_n) = Q(\widehat{f}_1(Y_1), \ldots, \widehat{f}_n(Y_n)),$$

for $Y_1, \ldots, Y_n \in \mathcal{P}(E')$, and similarly a fuzzy quantifier $\widetilde{Q}' : \mathcal{P}(E')^n \longrightarrow \mathbf{I}$ by

$$\widetilde{Q}' = \mathcal{F}(Q) \circ \overset{n}{\underset{i=1}{\times}} \widehat{\mathcal{F}}(f_i) : \mathcal{P}(E')^n \longrightarrow \mathbf{I},$$
$$\widetilde{Q}'(X_1, \ldots, X_n) = \mathcal{F}(Q)(\widehat{\mathcal{F}}(f_1)(Y_1), \ldots, \widehat{\mathcal{F}}(f_n)(X_n)),$$

for all $X_1, \ldots, X_n \in \widetilde{\mathcal{P}}(E)$. It is natural to require that a QFM $\mathcal{F}$ be compatible with its induced extension principle, i.e. $\mathcal{F}(Q') = \widetilde{Q}'$, or equivalently

$$\mathcal{F}(Q \circ \overset{n}{\underset{i=1}{\times}} \widehat{f}_i) = \mathcal{F}(Q) \circ \overset{n}{\underset{i=1}{\times}} \widehat{\mathcal{F}}(f_i)\,. \tag{7}$$

Equation (7) hence establishes a relation between powerset functions and the induced extension principle $\widehat{\mathcal{F}}$. It is of particular importance to DFS theory because it is the only axiom which relates the behaviour of $\mathcal{F}$ on different domains E, E'.

The following definition of the DFS axioms summarises our above considerations on reasonable QFMs.

Definition 12 (DFS: Determiner Fuzzification Scheme). *A QFM $\mathcal{F}$ is called a* determiner fuzzification scheme (DFS) *iff the following axioms are satisfied for every semi-fuzzy quantifier $Q : \mathcal{P}(E)^n \longrightarrow \mathbf{I}$:*

Correct generalisation	$\mathcal{U}(\mathcal{F}(Q)) = Q$ *if* $n \leq 1$	(Z-1)
Projection quantifiers	$\mathcal{F}(Q) = \widetilde{\pi}_e$ *if there exists* $e \in E$ *s.th.* $Q = \pi_e$	(Z-2)
Dualisation	$\mathcal{F}(Q\widetilde{\square}) = \mathcal{F}(Q)\widetilde{\square} \quad n > 0$	(Z-3)
Internal joins	$\mathcal{F}(Q\cup) = \mathcal{F}(Q)\widetilde{\cup} \quad n > 0$	(Z-4)
Preservation of monotonicity	Q *noninc.* n*-th arg* $\Rightarrow \mathcal{F}(Q)$ *noninc.* n*-th arg,* $n > 0$	(Z-5)
Functional application	$\mathcal{F}(Q \circ \overset{n}{\underset{i=1}{\times}} \widehat{f}_i) = \mathcal{F}(Q) \circ \overset{n}{\underset{i=1}{\times}} \widehat{\mathcal{F}}(f_i)$ *where* $f_1, \ldots, f_n : E' \longrightarrow E$, $E' \neq \varnothing$.	(Z-6)

As has been shown in [8], the axioms (Z-1) to (Z-6) form an independent axiom set.

4 Properties of DFSes

4.1 Correct generalisation

Let us firstly establish that $\mathcal{F}(Q)$ coincides with the original semi-fuzzy quantifier Q when all arguments are crisp sets, i.e. that $\mathcal{F}(Q)$ consistently extends Q.

Theorem 1. *Suppose $\mathcal{F}$ is a DFS and $Q : \mathcal{P}(E)^n \longrightarrow \mathbf{I}$ is an n-ary semi-fuzzy quantifier. Then $\mathcal{U}(\mathcal{F}(Q)) = Q$, i.e. for all* crisp *subsets $Y_1, \ldots, Y_n \in \mathcal{P}(E)$,*

$$\mathcal{F}(Q)(Y_1, \ldots, Y_n) = Q(Y_1, \ldots, Y_n) \,.$$

For example, if E is a set of persons, and **women, married** $\in \mathcal{P}(E)$ are the crisp sets of "women" and "married persons" in E, then

$$\mathcal{F}(\mathbf{some})(\mathbf{women},\ \mathbf{married}) = \mathbf{some}(\mathbf{women},\ \mathbf{married})\,,$$

i.e. the "fuzzy some" obtained by applying $\mathcal{F}$ coincides with the (original) "crisp some" whenever the latter is defined, which is of course highly desirable.

4.2 Properties of the induced truth functions

Let us now turn to the fuzzy truth functions induced by a DFS. As for negation, the standard choice in fuzzy logic is certainly $\neg : \mathbf{I} \longrightarrow \mathbf{I}$, defined by $\neg x = 1 - x$ for all $x \in \mathbf{I}$. The essential properties of this and other reasonable negation operators are captured by the following definition.

Definition 13 (Strong negation). $\tilde{\neg} : \mathbf{I} \longrightarrow \mathbf{I}$ *is called a* strong negation operator *iff it satisfies*

a. $\tilde{\neg} 0 = 1$ *(boundary condition)*
b. $\tilde{\neg} x_1 \geq \tilde{\neg} x_2$ *for all $x_1, x_2 \in \mathbf{I}$ such that $x_1 < x_2$ (i.e. $\tilde{\neg}$ is monotonically decreasing)*
c. $\tilde{\neg} \circ \tilde{\neg} = \mathrm{id}_{\mathbf{I}}$ *(i.e. $\tilde{\neg}$ is involutive).*

With conjunction, there are several common choices in fuzzy logic (although the standard is certainly $\wedge = \min$). All of these belong to the class of t-norms, which seems to capture what one would expect of a reasonable conjunction operator. The dual concept of t-norm is that of an s-norm, which expresses the essential properties of fuzzy disjunction operators (cf. Schweizer & Sklar [20]).

Theorem 2. *In every DFS $\mathcal{F}$,*

a. $\tilde{\mathcal{F}}(\mathrm{id}_2) = \mathrm{id}_{\mathbf{I}}$ *is the identity truth function;*
b. $\tilde{\neg} = \tilde{\mathcal{F}}(\neg)$ *is a strong negation operator;*
c. $\tilde{\wedge} = \tilde{\mathcal{F}}(\wedge)$ *is a t-norm;*
d. $x_1 \tilde{\vee} x_2 = \tilde{\neg}(\tilde{\neg} x_1 \tilde{\wedge} \tilde{\neg} x_2)$, *i.e. $\tilde{\vee}$ is the dual s-norm of $\tilde{\wedge}$ under $\tilde{\neg}$,*
e. $x_1 \tilde{\rightarrow} x_2 = \tilde{\neg} x_1 \tilde{\vee} x_2$

The fuzzy disjunction induced by $\mathcal{F}$ is therefore definable in terms of $\tilde{\wedge}$ and $\tilde{\neg}$, and the fuzzy implication induced by $\mathcal{F}$ is definable in terms of $\tilde{\vee}$ and $\tilde{\neg}$.

4.3 Preservation of argument structure

We now discuss homomorphism properties with respect to operations on arguments.

Definition 14 (Argument transpositions). *Suppose $Q : \mathcal{P}(E)^n \longrightarrow \mathbf{I}$ is a semi-fuzzy quantifier, $n > 0$ and $i \in \{0, \ldots, n\}$. By $Q\tau_i : \mathcal{P}(E)^n \longrightarrow \mathbf{I}$ we denote the semi-fuzzy quantifier defined by*

$$Q\tau_i(X_1, \ldots, X_n) = Q(X_1, \ldots, X_{i-1}, X_n, X_{i+1}, \ldots, X_{n-1}, X_i),$$

for all $(X_1, \ldots, X_n) \in \mathcal{P}(E)^n$. In the case of fuzzy quantifiers $\widetilde{Q} : \widetilde{\mathcal{P}}(E)^n \longrightarrow \mathbf{I}$, we define $\widetilde{Q}\tau_i : \widetilde{\mathcal{P}}(E)^n \longrightarrow \mathbf{I}$ analogously.

Theorem 3. *Every DFS $\mathcal{F}$ is compatible with argument transpositions, i.e. whenever $Q : \mathcal{P}(E)^n \longrightarrow \mathbf{I}$ and $i \in \{1, \ldots, n\}$ are given, then $\mathcal{F}(Q\tau_i) = \mathcal{F}(Q)\tau_i$.*

Because all permutations can be expressed as a sequence of transpositions, (Th-3) ensures that $\mathcal{F}$ commutes with permutations of the arguments of a quantifier. In particular, it guarantees that symmetry properties of a quantifier Q transfer to $\mathcal{F}(Q)$. Hence $\mathcal{F}(\mathbf{some})(\mathbf{rich}, \mathbf{young}) = \mathcal{F}(\mathbf{some})(\mathbf{young}, \mathbf{rich})$, i.e. the meanings of "some rich people are young" and "some young people are rich" coincide.

Theorem 4. *Every DFS $\mathcal{F}$ is compatible with the formation of antonyms, i.e. whenever $Q : \mathcal{P}(E)^n \longrightarrow \mathbf{I}$ is a semi-fuzzy quantifier of arity $n > 0$, then $\mathcal{F}(Q\neg) = \mathcal{F}(Q)\widetilde{\neg}$.*

The theorem guarantees e.g. that $\mathcal{F}(\mathbf{all})(\mathbf{rich}, \widetilde{\neg}\,\mathbf{lucky}) = \mathcal{F}(\mathbf{no})(\mathbf{rich}, \mathbf{lucky})$. Let us note that by (Th-3), the theorem generalises to arbitrary argument positions.

Theorem 5. *Every DFS $\mathcal{F}$ is compatible with the negation of quantifiers, i.e. whenever $Q : \mathcal{P}(E)^n \longrightarrow \mathbf{I}$ is a semi-fuzzy quantifier, then $\mathcal{F}(\widetilde{\neg}\, Q) = \widetilde{\neg}\, \mathcal{F}(Q)$.*

Hence $\mathcal{F}(\mathbf{at\ most\ 10})(\mathbf{young}, \mathbf{rich}) = \widetilde{\neg}\, \mathcal{F}(\mathbf{more\ than\ 10})(\mathbf{young}, \mathbf{rich})$.

Definition 15 (Internal meets). *Suppose $Q : \mathcal{P}(E)^n \longrightarrow \mathbf{I}$ is a semi-fuzzy quantifier, $n > 0$. The semi-fuzzy quantifier $Q\cap : \mathcal{P}(E)^{n+1} \longrightarrow \mathbf{I}$ is defined by*

$$Q\cap(X_1, \ldots, X_{n+1}) = Q(X_1, \ldots, X_{n-1}, X_n \cap X_{n+1}),$$

for $X_1, \ldots, X_{n+1} \in \mathcal{P}(E)$. In the case of a fuzzy quantifiers $\widetilde{Q} : \widetilde{\mathcal{P}}(E)^n \longrightarrow \mathbf{I}$, $\widetilde{Q}\widetilde{\cap} : \widetilde{\mathcal{P}}(E)^{n+1} \longrightarrow \mathbf{I}$ is defined analogously.

Theorem 6. *Every DFS $\mathcal{F}$ is compatible with intersections of arguments. Hence if $Q : \mathcal{P}(E)^n \longrightarrow \mathbf{I}$ is a semi-fuzzy quantifier of arity $n > 0$, then $\mathcal{F}(Q\cap) = \mathcal{F}(Q)\widetilde{\cap}$.*

For example, $\mathcal{F}(\mathbf{some}) = \mathcal{F}(\exists)\widetilde{\cap}$, because the two-place quantifier **some** can be expressed as $\mathbf{some} = \exists\cap$. Let us also remark that by (Th-3), this property generalises to intersections in arbitrary argument positions.

Definition 16. *Suppose $\mathcal{F}$ is some QFM. We say that $\mathcal{F}$ is* compatible with cylindrical extensions *iff the following condition holds for every semi-fuzzy quantifier $Q : \mathcal{P}(E)^n \longrightarrow \mathbf{I}$. Whenever $n' \in \mathbb{N}$, $n' \geq n$; $i_1, \ldots, i_n \in \{1, \ldots, n'\}$ such that $1 \leq i_1 < i_2 < \cdots < i_n \leq n'$, and $Q' : \mathcal{P}(E)^{n'} \longrightarrow \mathbf{I}$ is defined by*

$$Q'(Y_1, \ldots, Y_{n'}) = Q(Y_{i_1}, \ldots, Y_{i_n}), \qquad \textit{for all } Y_1, \ldots, Y_n \in \mathcal{P}(E),$$

then

$$\mathcal{F}(Q')(X_1, \ldots, X_{n'}) = \mathcal{F}(Q)(X_{i_1}, \ldots, X_{i_n}), \quad \textit{for all } X_1, \ldots, X_n \in \widetilde{\mathcal{P}}(E).$$

Being compatible with cylindrical extensions means that vacuous argument positions of a quantifier can be eliminated. For example, if $Q' : \mathcal{P}(E)^4 \longrightarrow \mathbf{I}$ is a semi-fuzzy quantifier and if there exists $Q : \mathcal{P}(E) \longrightarrow \mathbf{I}$ such that $Q'(Y_1, Y_2, Y_3, Y_4) = Q(Y_3)$ for all $Y_1, \ldots, Y_4 \in \mathcal{P}(E)$, then we know that Q' does not really depend on all arguments; it is apparent that the choice of Y_1, Y_2 and Y_4 has no effect on the quantification result. It is straightforward to require that $\mathcal{F}(Q')(X_1, X_2, X_3, X_4) = \mathcal{F}(Q)(X_3)$ for $X_1, \ldots, X_4 \in \widetilde{\mathcal{P}}(E)$, i.e. $\mathcal{F}(Q')$ is also independent of X_1, X_2, X_4, and can be computed from $\mathcal{F}(Q)$.

Theorem 7. *Every DFS $\mathcal{F}$ is compatible with cylindrical extensions.*

Let us now introduce another very fundamental adequacy condition on QFMs. Suppose $X \in \widetilde{\mathcal{P}}(E)$ is a fuzzy subset. The *support* $\text{spp}(X) \in \mathcal{P}(E)$ and the *core*, $\text{core}(X) \in \mathcal{P}(E)$ are defined by

$$\text{spp}(X) = \{e \in E : \mu_X(e) > 0\} \tag{8}$$
$$\text{core}(X) = \{e \in E : \mu_X(e) = 1\}. \tag{9}$$

$\text{spp}(X)$ contains all elements which potentially belong to X and $\text{core}(X)$ contains all elements which fully belong to X. The interpretation of a fuzzy subset X is hence ambiguous only with respect to crisp subsets Y in the context range

$$\text{cxt}(X) = \{Y \in \mathcal{P}(E) : \text{core}(X) \subseteq Y \subseteq \text{spp}(Y)\}. \tag{10}$$

For example, let $E = \{a, b, c\}$ and suppose $X \in \widetilde{\mathcal{P}}(E)$ is the fuzzy subset

$$\mu_X(e) = \begin{cases} 1 & : \ x = a \text{ or } x = b \\ \frac{1}{2} & : \ x = c \end{cases} \tag{11}$$

The corresponding context range is

$$\text{cxt}(X) = \{Y : \{a, b\} \subseteq Y \subseteq \{a, b, c\}\} = \{\{a, b\}, \{a, b, c\}\}.$$

Now let us consider $\exists : \mathcal{P}(E) \longrightarrow \mathbf{2}$. Because $\exists(\{a, b\}) = \exists(\{a, b, c\}) = 1, \exists(Y) = 1$ for all crisp subsets in the context of X. We hence expect that $\mathcal{F}(\exists)(X) = 1$: regardless of whether we assume that $c \in X$ or $c \notin X$, the quantification result is always equal to one.

Definition 17. *Assume that* $Q, Q' : \mathcal{P}(E)^n \longrightarrow \mathbf{I}$ *and* $X_1, \ldots, X_n \in \widetilde{\mathcal{P}}(E)$ *are given. We say that* Q *and* Q' *are contextually equal relative to* $(X_1, \ldots, X_n)$*, in symbols:* $Q \sim_{(X_1,\ldots,X_n)} Q'$*, if and only if* $Q|_{\mathrm{cxt}(X_1)\times\cdots\times\mathrm{cxt}(X_n)} = Q'|_{\mathrm{cxt}(X_1)\times\cdots\times\mathrm{cxt}(X_n)}$*, i.e.* $Q(Y_1, \ldots, Y_n) = Q'(Y_1, \ldots, Y_n)$ *for all* $Y_1 \in \mathrm{cxt}(X_1), \ldots, Y_n \in \mathrm{cxt}(X_n)$*.*

It is apparent that for each $E \neq \varnothing$, $n \in \mathbb{N}$ and $X_1, \ldots, X_n \in \widetilde{\mathcal{P}}(E)$, $\sim_{(X_1,\ldots,X_n)}$ is an equivalence relation on the set of all semi-fuzzy quantifiers $Q : \mathcal{P}(E)^n \longrightarrow \mathbf{I}$.

Definition 18. *A QFM* $\mathcal{F}$ *is said to be* contextual *iff for all* $Q, Q' : \mathcal{P}(E)^n \longrightarrow \mathbf{I}$ *and every choice of fuzzy argument sets* $X_1, \ldots, X_n \in \widetilde{\mathcal{P}}(E)$*:*

$$Q \sim_{(X_1,\ldots,X_n)} Q' \quad \Rightarrow \quad \mathcal{F}(Q)(X_1, \ldots, X_n) = \mathcal{F}(Q')(X_1, \ldots, X_n)\,.$$

As illustrated by our motivating example, it is highly desirable that a QFM satisfies this very elementary and fundamental adequacy condition.

Theorem 8. *Every DFS* $\mathcal{F}$ *is contextual.*

Definition 19 (Argument insertion). *Suppose* $Q : \mathcal{P}(E)^n \longrightarrow \mathbf{I}$ *is a semi-fuzzy quantifier,* $n > 0$*, and* $A \in \mathcal{P}(E)$*. By* $Q \triangleleft A : \mathcal{P}(E)^{n-1} \longrightarrow \mathbf{I}$ *we denote the semi-fuzzy quantifier defined by*

$$Q \triangleleft A(X_1, \ldots, X_{n-1}) = Q(X_1, \ldots, X_{n-1}, A), \quad \textit{for } X_1, \ldots, X_{n-1} \in \mathcal{P}(E).$$

(Analogous definition of $\widetilde{Q} \triangleleft A$ *for fuzzy quantifiers).*

Theorem 9. *Suppose* $\mathcal{F}$ *is a contextual QFM which is compatible with cylindrical extensions. Then* $\mathcal{F}$ *is compatible with argument insertion, i.e.* $\mathcal{F}(Q \triangleleft A) = \mathcal{F}(Q) \triangleleft A$ *for all semi-fuzzy quantifiers* $Q : \mathcal{P}(E)^n \longrightarrow \mathbf{I}$ *of arity* $n > 0$ *and crisp* $A \in \mathcal{P}(E)$*. In particular, every DFS is compatible with argument insertion.*

The main application is that of modelling *adjectival restriction* by a crisp adjective. For example, if $\mathbf{married} \in \mathcal{P}(E)$ is extension of the crisp adjective "married", then $Q' = \mathbf{many}\tau_1 \cap \triangleleft \mathbf{married}\tau_1$, i.e. $Q'(X, Y) = \mathbf{many}(\mathbf{married} \cap X, Y)$, models the composite quantifier "many married X's are Y's". By the above theorem, we then have $\mathcal{F}(Q')(\mathbf{rich}, \mathbf{lucky}) = \mathcal{F}(\mathbf{many})(\mathbf{married} \mathbin{\widetilde{\cap}} \mathbf{rich}, \mathbf{lucky})$, as desired. Let us remark that adjectival restriction with a *fuzzy* adjective cannot be modelled directly. This is because we cannot insert a fuzzy argument A into a semi-fuzzy quantifier.

4.4 Monotonicity properties

Theorem 10. *Suppose* $\mathcal{F}$ *is a DFS and* $Q : \mathcal{P}(E)^n \longrightarrow \mathbf{I}$*. Then* Q *is monotonically nondecreasing (nonincreasing) in its* i*-th argument* $(i \leq n)$ *if and only if* $\mathcal{F}(Q)$ *is monotonically nondecreasing (nonincreasing) in its* i*-th argument.*

For example, **some** : $\mathcal{P}(E)^2 \longrightarrow \mathbf{2}$ is monotonically nondecreasing in both arguments. By the theorem, then, $\mathcal{F}(\mathbf{some}) : \widetilde{\mathcal{P}}(E) \times \widetilde{\mathcal{P}}(E) \longrightarrow \mathbf{I}$ is nondecreasing in both arguments also. In particular,

$$\mathcal{F}(\mathbf{some})(\mathbf{young\_men}, \mathbf{very\_tall}) \leq \mathcal{F}(\mathbf{some})(\mathbf{men}, \mathbf{tall}),$$

i.e. "some young men are very tall" entails "some men are tall", if **young_men** $\subseteq$ **men** and **very_tall** $\subseteq$ **tall**.

Let us now state that every DFS preserves monotonicity properties of semi-fuzzy quantifiers even if these hold only locally.

Definition 20. *Suppose* $Q : \mathcal{P}(E)^n \longrightarrow \mathbf{I}$ *and* $U, V \in \mathcal{P}(E)^n$ *are given. We say that* Q is locally nondecreasing in the range (U, V) *iff for all* $X_1, \ldots, X_n, X_1', \ldots, X_n' \in \mathcal{P}(E)$ *such that* $U_i \subseteq X_i \subseteq X_i' \subseteq V_i$ $(i = 1, \ldots, n)$*, we have* $Q(X_1, \ldots, X_n) \leq Q(X_1', \ldots, X_n')$*. We will say that Q is locally nonincreasing in the range* (U, V) *if under the same conditions,* $Q(X_1, \ldots, X_n) \leq Q(X_1', \ldots, X_n')$*. On fuzzy quantifiers, local monotonicity is defined analogously, but* $X_1, \ldots, X_n, X_1', \ldots, X_n'$ *are taken from* $\widetilde{\mathcal{P}}(E)$*, and '*$\subseteq$*' is the fuzzy inclusion relation.*

Theorem 11. *Suppose* $\mathcal{F}$ *is a DFS,* $Q : \mathcal{P}(E)^n \longrightarrow \mathbf{I}$ *a semi-fuzzy quantifier and* $U, V \in \mathcal{P}(E)^n$*. Then Q is locally nondecreasing (nonincreasing) in the range* (U, V) *iff* $\mathcal{F}(Q)$ *is locally nondecreasing (nonincreasing) in the range* (U, V)*.*

DFSes can also be shown to be *monotonic* in the sense of preserving inequations between quantifiers. Let us firstly define a partial order $\leq$ on (semi-)fuzzy quantifiers.

Definition 21. *Suppose* $Q, Q' : \mathcal{P}(E)^n \longrightarrow \mathbf{I}$ *are semi-fuzzy quantifiers. Let us write* $Q \leq Q'$ *iff for all* $X_1, \ldots, X_n \in \mathcal{P}(E)$, $Q(X_1, \ldots, X_n) \leq Q'(X_1, \ldots, X_n)$*. On fuzzy quantifiers, we define* $\leq$ *analogously, where* $X_1, \ldots, X_n \in \widetilde{\mathcal{P}}(E)$*.*

Theorem 12. *Suppose* $\mathcal{F}$ *is a DFS and* $Q, Q' : \mathcal{P}(E)^n \longrightarrow \mathbf{I}$ *are semi-fuzzy quantifiers. Then* $Q \leq Q'$ *if and only if* $\mathcal{F}(Q) \leq \mathcal{F}(Q')$*.*

Definition 22. *Suppose* $Q_1, Q_2 : \mathcal{P}(E)^n \longrightarrow \mathbf{I}$ *are semi-fuzzy quantifiers and* $U, V \in \mathcal{P}(E)^n$*. We say that* Q_1 *is (not necessarily strictly) smaller than* Q_2 *in the range* (U, V)*, in symbols:* $Q_1 \leq_{(U,V)} Q_2$*, iff for all* $X_1, \ldots, X_n \in \mathcal{P}(E)$ *such that* $U_1 \subseteq X_1 \subseteq V_1, \ldots, U_n \subseteq X_n \subseteq V_n$*,*

$$Q_1(X_1, \ldots, X_n) \leq Q_2(X_1, \ldots, X_n).$$

$\widetilde{Q}_1 \leq_{(U,V)} \widetilde{Q}_2$ *is defined analogously, based on* $X_i \in \widetilde{\mathcal{P}}(E)$ *and fuzzy inclusion.*

Every DFS preserves inequations between quantifiers even if these hold only locally.

Theorem 13. *Suppose* $\mathcal{F}$ *is a DFS,* $Q_1, Q_2 : \mathcal{P}(E)^n \longrightarrow \mathbf{I}$ *and* $U, V \in \mathcal{P}(E)^n$*. Then* $Q_1 \leq_{(U,V)} Q_2$ *if and only if* $\mathcal{F}(Q_1) \leq_{(U,V)} \mathcal{F}(Q_2)$*.*

4.5 Properties of the induced extension principle

Theorem 14. *Suppose $\mathcal{F}$ is a DFS and $\widehat{\mathcal{F}}$ the extension principle induced by $\mathcal{F}$. Then for all $f : E \longrightarrow E'$, $g : E' \longrightarrow E''$ (where $E \neq \varnothing$, $E' \neq \varnothing$, $E'' \neq \varnothing$),*

a. $\widehat{\mathcal{F}}(g \circ f) = \widehat{\mathcal{F}}(g) \circ \widehat{\mathcal{F}}(f)$
b. $\widehat{\mathcal{F}}(\mathrm{id}_E) = \mathrm{id}_{\widetilde{\mathcal{P}}(E)}$

The induced extension principles of all DFSes coincide on injective mappings.

Theorem 15. *Suppose $\mathcal{F}$ is a DFS and $f : E \longrightarrow E'$ is an injection. Then for all $X \in \widetilde{\mathcal{P}}(E)$, $v \in E'$,*

$$\mu_{\widehat{\mathcal{F}}(f)(X)}(v) = \begin{cases} \mu_X(f^{-1}(v)) & : \quad v \in \operatorname{Im} f \\ 0 & : \quad v \notin \operatorname{Im} f \end{cases} = \begin{cases} \widetilde{\pi}_{f^{-1}(v)}(X) & : \quad v \in \operatorname{Im} f \\ 0 & : \quad v \notin \operatorname{Im} f \end{cases}$$

We shall now introduce an important property of NL quantifiers known as *having extension:* if $E \subseteq E'$ are base sets, the interpretation of the quantifier of interest in E is $Q_E : \mathcal{P}(E)^n \longrightarrow \mathbf{I}$, and its interpretation in E' is $Q_{E'} : \mathcal{P}(E')^n \longrightarrow \mathbf{I}$, then

$$Q_E(X_1, \ldots, X_n) = Q_{E'}(X_1, \ldots, X_n)\,, \tag{12}$$

for all $X_1 \ldots X_n \in \mathcal{P}(E)$. For example, suppose E is a set of men and that **married**, **have_children** $\in \mathcal{P}(E)$ are subsets of E. Further suppose we extend E to a larger base set E' which, in addition to men, also contains, say, their shoes. We should expect that $\mathbf{most}_E(\mathbf{married}, \mathbf{have\_children}) = \mathbf{most}_{E'}(\mathbf{married}, \mathbf{have\_children})$, because the shoes we have added to E are neither men, nor do they have children. The cross-domain property of having extension expresses some kind of context insensitivity: given $X_1, \ldots, X_n \in \mathcal{P}(E)$, we can add an arbitrary number of objects to our original domain E without altering the quantification result. We can also drop elements of E which are irrelevant to all arguments, i.e. not contained in the union of $X_1, \ldots, X_n$. Having extension is hence an insensitivity property of NL quantifiers against the choice of the full domain E, which is often to some degree arbitrary. An analogous definition for fuzzy quantifiers is easily obtained from (12); in this case, the property must hold for $X_1, \ldots, X_n \in \widetilde{\mathcal{P}}(E)$. We require that fuzzy quantifiers corresponding to given semi-fuzzy quantifiers which have extension also possess this property.

Definition 23. *A QFM $\mathcal{F}$ is said to* preserve extension *if each pair of semi-fuzzy quantifiers $Q : \mathcal{P}(E)^n \longrightarrow \mathbf{I}$, $Q' : \mathcal{P}(E')^n \longrightarrow \mathbf{I}$ such that $E \subseteq E'$ and $Q'|_{\mathcal{P}(E)^n} = Q$, i.e. $Q(X_1, \ldots, X_n) = Q'(X_1, \ldots, X_n)$ for $X_1, \ldots, X_n \in \mathcal{P}(E)$, is mapped to $\mathcal{F}(Q) : \widetilde{\mathcal{P}}(E)^n \longrightarrow \mathbf{I}$, $\mathcal{F}(Q') : \widetilde{\mathcal{P}}(E')^n \longrightarrow \mathbf{I}$ with $\mathcal{F}(Q')|_{\widetilde{\mathcal{P}}(E)^n} = \mathcal{F}(Q)$, i.e. $\mathcal{F}(Q)(X_1, \ldots, X_n) = \mathcal{F}(Q')(X_1, \ldots, X_n)$, for all $X_1, \ldots, X_n \in \widetilde{\mathcal{P}}(E)$.*

Theorem 16. *Every DFS $\mathcal{F}$ preserves extension.*

This is apparent from (Z-6) and (Th-15).

Definition 24. *A semi-fuzzy quantifier* $Q : \mathcal{P}(E)^n \longrightarrow \mathbf{I}$ *is called* quantitative *iff for all automorphisms*[8] $\beta : E \longrightarrow E$ *and all* $Y_1, \ldots, Y_n \in \mathcal{P}(E)$,

$$Q(Y_1, \ldots, Y_n) = Q(\widehat{\beta}(Y_1), \ \ldots, \ \widehat{\beta}(Y_n)) \,.$$

Similarly, a fuzzy quantifier $\widetilde{Q} : \widetilde{\mathcal{P}}(E)^n \longrightarrow \mathbf{I}$ *is said to be* quantitative *iff for all automorphisms* $\beta : E \longrightarrow E$ *and all* $X_1, \ldots, X_n \in \widetilde{\mathcal{P}}(E)$,

$$\widetilde{Q}(X_1, \ldots, X_n) = \widetilde{Q}(\hat{\hat{\beta}}(X_1), \ \ldots, \ \hat{\hat{\beta}}(X_n)) \,,$$

where $\hat{\hat{\beta}} : \widetilde{\mathcal{P}}(E) \longrightarrow \widetilde{\mathcal{P}}(E)$ *is obtained from the standard extension principle.*

By (Th-15), the extension principles of DFSes coincide on injective mappings. Hence the explicit mention of the standard extension principle in the above definition does *not* limit its applicability to any particular choice of extension principle.

Theorem 17. *If* $\mathcal{F}$ *is a DFS, then* $\mathcal{F}(Q)$ *is quantitative iff* Q *is quantitative.*

Hence the quantitative quantifiers **all, some** and **at least n** are mapped to quantitative fuzzy quantifiers $\mathcal{F}(\mathbf{all})$, $\mathcal{F}(\mathbf{some})$ and $\mathcal{F}(\mathbf{at\ least\ n})$. The non-quantitative **john** $= \pi_{\mathbf{John}}$ is mapped to $\mathcal{F}(\mathbf{john}) = \widetilde{\pi}_{\mathbf{John}}$, which is also non-quantitative.
The extension principle $\widehat{\mathcal{F}}$ of a DFS $\mathcal{F}$ is uniquely determined by the fuzzy existential quantifiers $\mathcal{F}(\exists) = \mathcal{F}(\exists_E) : \widetilde{\mathcal{P}}(E) \longrightarrow \mathbf{I}$. Conversely, the fuzzy existential quantifiers obtained from $\mathcal{F}$ are uniquely determined by its extension principle $\widehat{\mathcal{F}}$.

Theorem 18. *Suppose* $\mathcal{F}$ *is a DFS.*

a. *For every mapping* $f : E \longrightarrow E'$ *and all* $e' \in E'$, $\mu_{\widehat{\mathcal{F}}(f)(\bullet)}(e') = \mathcal{F}(\exists)\widetilde{\cap}\triangleleft f^{-1}(e')$.
b. *If* $E \neq \varnothing$ *and* $\exists = \exists_E : \mathcal{P}(E) \longrightarrow \mathbf{2}$, *then* $\mathcal{F}(\exists) = \widetilde{\pi}_\varnothing \circ \widehat{\mathcal{F}}(!)$, *where* $! : E \longrightarrow \{\varnothing\}$ *is the mapping defined by* $!(x) = \varnothing$ *for all* $x \in E$.

A notion related to extension principles is that of *fuzzy inverse images*. In the case of crisp sets, inverse images $f^{-1} : \mathcal{P}(E') \longrightarrow \mathcal{P}(E)$ of a given $f : E \longrightarrow E'$ are defined by $f^{-1}(V) = \{e \in E : f(e) \in V\}$, all $V \in \mathcal{P}(E')$. Generalising from this, a QFM $\mathcal{F}$ induces fuzzy inverse images by means of the following construction.

Definition 25. *Suppose* $\mathcal{F}$ *is a quantifier fuzzification mechanism and* $f : E \longrightarrow E'$ *is some mapping.* $\mathcal{F}$ *induces a* fuzzy inverse image mapping $\widehat{\mathcal{F}}^{-1}(f) : \widetilde{\mathcal{P}}(E') \longrightarrow \widetilde{\mathcal{P}}(E)$ *which to each* $Y \in \widetilde{\mathcal{P}}(E')$ *assigns the fuzzy subset* $\widehat{\mathcal{F}}^{-1}(f)$ *defined by*

$$\mu_{\widehat{\mathcal{F}}^{-1}(f)(Y)}(e) = \mathcal{F}(\chi_{f^{-1}(\bullet)}(e))(Y) \,.$$

For DFSes, these inverse images coincide with the apparent "reasonable" definition:

Theorem 19. *Suppose* $\mathcal{F}$ *is a DFS,* $f : E \longrightarrow E'$ *is a mapping and* $Y \in \widetilde{\mathcal{P}}(E')$. *Then for all* $e \in E$, $\mu_{\widehat{\mathcal{F}}^{-1}(f)(Y)}(e) = \mu_Y(f(e))$.

[8] i.e. bijections of E into itself

4.6 Properties with respect to the standard quantifiers

Theorem 20. *Suppose $\mathcal{F}$ is a DFS and $E \neq \varnothing$ is some base set. Then*

$$\mathcal{F}(\forall)(X) = \inf\left\{ \mathop{\widetilde{\wedge}}_{i=1}^{m} \mu_X(a_i) : A = \{a_1, \ldots, a_m\} \in \mathcal{P}(E) \text{ finite, } a_i \neq a_j \text{ if } i \neq j \right\},$$

$$\mathcal{F}(\exists)(X) = \sup\left\{ \mathop{\widetilde{\vee}}_{i=1}^{m} \mu_X(a_i) : A = \{a_1, \ldots, a_m\} \in \mathcal{P}(E) \text{ finite, } a_i \neq a_j \text{ if } i \neq j \right\}.$$

In particular, $\mathcal{F}(\forall)$ is a T-quantifier and $\mathcal{F}(\exists)$ is an S-quantifier in the sense of Thiele [22]. If E is finite, i.e. $E = \{e_1, \ldots, e_m\}$ where the e_i are pairwise distinct, then $\mathcal{F}(\forall)(X) = \mathop{\widetilde{\wedge}}_{i=1}^{m} \mu_X(e_i)$ and $\mathcal{F}(\exists)(X) = \mathop{\widetilde{\vee}}_{i=1}^{m} \mu_X(e_i)$. Hence the fuzzy universal (existential) quantifiers are reasonable in the sense that the important relationship between $\forall$ and $\wedge$ ($\exists$ and $\vee$, resp.), which holds in the finite case, is preserved by a DFS. The theorem also shows that in every DFS, the fuzzy existential and universal quantifiers are uniquely determined by the induced disjunction and conjunction.

Theorem 21. *Suppose $\mathcal{F}$ is a DFS, $\widehat{\mathcal{F}}$ its extension principle and $\widetilde{\vee} = \widetilde{\mathcal{F}}(\vee)$.*

a. *$\widehat{\mathcal{F}}$ is uniquely determined by $\widetilde{\vee}$ (combine equations in (Th-18) and (Th-20));*
b. *$\widetilde{\vee}$ is uniquely determined by $\widehat{\mathcal{F}}$, viz. $x_1 \mathbin{\widetilde{\vee}} x_2 = (\widetilde{\pi}_\varnothing \circ \widehat{\mathcal{F}}(!))(X)$ for all $x_1, x_2 \in \mathbf{I}$, where $X \in \widetilde{\mathcal{P}}(\{1, 2\})$ is defined by $\mu_X(1) = x_1$ and $\mu_X(2) = x_2$, and $!$ is the unique mapping $! : \{1, 2\} \longrightarrow \{\varnothing\}$.*

In particular, if $\widehat{\mathcal{F}} = (\hat{\bullet})$ is the standard extension principle, then $\widetilde{\vee} = \max$. Because we did not want QFMs in which $\widetilde{\vee} \neq \max$ to be a priori excluded from consideration, we have stated axiom (Z-6) in terms of the extension principle $\widehat{\mathcal{F}}$ induced by $\mathcal{F}$, rather than requiring the compatibility of $\mathcal{F}$ to the standard extension principle.

4.7 Special subclasses of DFSes

We now turn to subclasses of models which satisfy some additional requirements.

Definition 26. *Let $\widetilde{\neg} : \mathbf{I} \longrightarrow \mathbf{I}$ be a strong negation operator. A DFS $\mathcal{F}$ is called a $\widetilde{\neg}$-DFS if its induced negation coincides with $\widetilde{\neg}$, i.e. $\widetilde{\mathcal{F}}(\neg) = \widetilde{\neg}$. In particular, we will call $\mathcal{F}$ a $\neg$-DFS if it induces the standard negation $\neg x = 1 - x$.*

Definition 27. *Suppose $\mathcal{F}$ is a DFS and $\sigma : \mathbf{I} \longrightarrow \mathbf{I}$ a bijection. For every semi-fuzzy quantifier $Q : \mathcal{P}(E)^n \longrightarrow \mathbf{I}$ and all $X_1, \ldots, X_n \in \widetilde{\mathcal{P}}(E)$, we define*

$$\mathcal{F}^\sigma(Q)(X_1, \ldots, X_n) = \sigma^{-1}\mathcal{F}(\sigma Q)(\sigma X_1, \ldots, \sigma X_n),$$

where σQ abbreviates $\sigma \circ Q$, and $\sigma X_i \in \widetilde{\mathcal{P}}(E)$ is defined by $\mu_{\sigma X_i} = \sigma \circ \mu_{X_i}$.

Theorem 22. *If $\mathcal{F}$ is a DFS and σ an increasing bijection, then $\mathcal{F}^\sigma$ is a DFS.*

It is well-known [14, Th-3.7] that for every strong negation $\widetilde{\neg} : \mathbf{I} \longrightarrow \mathbf{I}$ there is a monotonically increasing bijection $\sigma : \mathbf{I} \longrightarrow \mathbf{I}$ such that $\widetilde{\neg} x = \sigma^{-1}(1 - \sigma(x))$ for all $x \in \mathbf{I}$. The mapping σ is called the *generator* of $\widetilde{\neg}$.

Theorem 23. *Suppose $\mathcal{F}$ is a $\widetilde{\neg}$-DFS and $\sigma : \mathbf{I} \longrightarrow \mathbf{I}$ the generator of $\widetilde{\neg}$. Then $\mathcal{F}' = \mathcal{F}^{\sigma^{-1}}$ is a $\neg$-DFS and $\mathcal{F} = \mathcal{F}'^{\sigma}$.*

This means that we can freely move from an arbitrary $\widetilde{\neg}$-DFS to a corresponding $\neg$-DFS and vice versa: we can hence restrict attention to $\neg$-DFSes.

Definition 28. *A $\neg$-DFS $\mathcal{F}$ which induces a fuzzy disjunction $\widetilde{\vee}$ is called a $\widetilde{\vee}$-DFS.*

Theorem 24. *Suppose $\mathcal{J}$ is a non-empty index set and $(\mathcal{F}_j)_{j\in\mathcal{J}}$ is a $\mathcal{J}$-indexed collection of $\widetilde{\vee}$-DFSes. Further suppose $\Psi : \mathbf{I}^{\mathcal{J}} \longrightarrow \mathbf{I}$ satisfies these conditions:*

a. If $f \in \mathbf{I}^{\mathcal{J}}$ is constant, i.e. if there is a $c \in \mathbf{I}$ such that $f(j) = c$ for all $j \in \mathcal{J}$, then $\Psi(f) = c$.
b. $\Psi(1 - f) = 1 - \Psi(f)$, where $1 - f \in \mathbf{I}^{\mathcal{J}}$ is point-wise defined by $(1 - f)(j) = 1 - f(j)$, for all $j \in \mathcal{J}$.
c. Ψ is monotonically increasing, i.e. if $f(j) \le g(j)$ for all $j \in \mathcal{J}$, then $\Psi(f) \le \Psi(g)$.

If we define $\Psi[(\mathcal{F}_j)_{j\in\mathcal{J}}]$ by

$$\Psi[(\mathcal{F}_j)_{j\in\mathcal{J}}](Q)(X_1, \ldots, X_n) = \Psi((\mathcal{F}_j(Q)(X_1, \ldots, X_n))_{j\in\mathcal{J}})$$

for all $Q : \mathcal{P}(E)^n \longrightarrow \mathbf{I}$ and $X_1, \ldots, X_n \in \widetilde{\mathcal{P}}(E)$, then $\Psi[(\mathcal{F}_j)_{j\in\mathcal{J}}]$ is a $\widetilde{\vee}$-DFS.

In particular, convex combinations (e.g., arithmetic mean) and stable symmetric sums [21] of $\widetilde{\vee}$-DFSes are again $\widetilde{\vee}$-DFSes. The $\neg$-DFSes can be ordered by "specificity" or "cautiousness", i.e. closeness to $\frac{1}{2}$. We define a partial order[9] $\preceq_c \subseteq \mathbf{I} \times \mathbf{I}$,

$$x \preceq_c y \Leftrightarrow y \le x \le \tfrac{1}{2} \text{ or } \tfrac{1}{2} \le x \le y\,, \qquad \text{for all } x, y \in \mathbf{I}. \tag{13}$$

Definition 29. *Suppose $\mathcal{F}$, $\mathcal{F}'$ are $\neg$-DFSes. We say that $\mathcal{F}$ is* consistently less specific *than $\mathcal{F}'$, in symbols: $\mathcal{F} \preceq_c \mathcal{F}'$, iff for all semi-fuzzy quantifiers $Q : \mathcal{P}(E)^n \longrightarrow \mathbf{I}$ and all $X_1, \ldots, X_n \in \widetilde{\mathcal{P}}(E)$, $\mathcal{F}(Q)(X_1, \ldots, X_n) \preceq_c \mathcal{F}'(Q)(X_1, \ldots, X_n)$.*

We now wish to establish the existence of consistently least specific $\widetilde{\vee}$-DFSes. To be able to state the theorem, we firstly need to introduce the *fuzzy median* $\mathrm{med}_{\frac{1}{2}}$.

[9] $\preceq_c$ is Mukaidono's ambiguity relation, see [16].

Definition 30. *The* fuzzy median $\mathrm{med}_{\frac{1}{2}} : \mathbf{I} \times \mathbf{I} \longrightarrow \mathbf{I}$ *is defined by*

$$\mathrm{med}_{\frac{1}{2}}(u_1, u_2) = \begin{cases} \min(u_1, u_2) & : \min(u_1, u_2) > \frac{1}{2} \\ \max(u_1, u_2) & : \max(u_1, u_2) < \frac{1}{2} \\ \frac{1}{2} & : \text{else} \end{cases}$$

The fuzzy median is an associative mean operator [3] and the only stable (i.e. idempotent) associative symmetric sum [21]. It can be generalised to an operator $\mathcal{P}(\mathbf{I}) \longrightarrow \mathbf{I}$ which accepts arbitrary subsets of $\mathbf{I}$ as its arguments.

Definition 31. *The generalised fuzzy median* $\mathrm{m}_{\frac{1}{2}} : \mathcal{P}(\mathbf{I}) \longrightarrow \mathbf{I}$ *is defined by*

$$\mathrm{m}_{\frac{1}{2}} X = \mathrm{med}_{\frac{1}{2}}(\inf X, \sup X), \qquad \textit{for all } X \in \mathcal{P}(\mathbf{I}).$$

Theorem 25. *Suppose $\widetilde{\vee}$ is an s-norm induced by some $\neg$-DFS $\mathcal{F}$. There exists a consistently least specific $\widetilde{\vee}$-DFS, i.e. a $\widetilde{\vee}$-DFS $\mathcal{L}$ such that for every $\widetilde{\vee}$-DFS $\mathcal{F}'$, $\mathcal{L} \preceq_c \mathcal{F}'$. $\mathcal{L}$ is defined by $\mathcal{L}(Q)(X_1, \ldots, X_n) = \mathrm{m}_{\frac{1}{2}}\{\mathcal{F}(Q)(X_1, \ldots, X_n) : \mathcal{F} \in \widetilde{\vee}\text{-}\mathbf{DFS}\}$, for all semi-fuzzy quantifiers $Q : \mathcal{P}(E)^n \longrightarrow \mathbf{I}$ and all $X_1, \ldots, X_n \in \widetilde{\mathcal{P}}(E)$, where $\widetilde{\vee}$-$\mathbf{DFS}$ denotes the collection of all $\widetilde{\vee}$-DFSes.*

Definition 32. *A DFS $\mathcal{F}$ with $\widetilde{\neg} = \widetilde{\mathcal{F}}(\neg)$ and $\widetilde{\vee} = \widetilde{\mathcal{F}}(\vee)$ is called a $(\widetilde{\neg}, \widetilde{\vee})$-DFS.*

Definition 33. *Suppose $\widetilde{\wedge}, \widetilde{\vee} : \mathbf{I} \times \mathbf{I} \longrightarrow \mathbf{I}$ are given. For all semi-fuzzy quantifiers $Q, Q' : \mathcal{P}(E)^n \longrightarrow \mathbf{I}$, the conjunction $Q \mathbin{\widetilde{\wedge}} Q' : \mathcal{P}(E)^n \longrightarrow \mathbf{I}$ and the disjunction $Q \mathbin{\widetilde{\vee}} Q' : \mathcal{P}(E)^n \longrightarrow \mathbf{I}$ of Q and Q' are defined by*

$$(Q \mathbin{\widetilde{\wedge}} Q')(X_1, \ldots, X_n) = Q(X_1, \ldots, X_n) \mathbin{\widetilde{\wedge}} Q'(X_1, \ldots, X_n)$$
$$(Q \mathbin{\widetilde{\vee}} Q')(X_1, \ldots, X_n) = Q(X_1, \ldots, X_n) \mathbin{\widetilde{\vee}} Q'(X_1, \ldots, X_n)$$

for all $X_1, \ldots, X_n \in \mathcal{P}(E)$. $\widetilde{Q} \mathbin{\widetilde{\wedge}} \widetilde{Q}'$ and $\widetilde{Q} \mathbin{\widetilde{\vee}} \widetilde{Q}'$ are defined analogously.

In the following, we shall be concerned with $(\widetilde{\neg}, \max)$-DFSes, which induce the standard disjunction $\widetilde{\mathcal{F}}(\vee) = \vee = \max$. For $(\widetilde{\neg}, \max)$-DFSes, we can establish a theorem on conjunctions and disjunctions of (semi-)fuzzy quantifiers.

Theorem 26. *Suppose $\mathcal{F}$ is a $(\widetilde{\neg}, \max)$-DFS. Then for all $Q, Q' : \mathcal{P}(E)^n \longrightarrow \mathbf{I}$,*

a. $\mathcal{F}(Q \wedge Q') \leq \mathcal{F}(Q) \wedge \mathcal{F}(Q')$
b. $\mathcal{F}(Q \vee Q') \geq \mathcal{F}(Q) \vee \mathcal{F}(Q')$.

We have so far not made any claims about the interpretation of $\widetilde{\leftrightarrow} = \widetilde{\mathcal{F}}(\leftrightarrow)$ and $\widetilde{\mathrm{xor}} = \widetilde{\mathcal{F}}(\mathrm{xor})$ in a given DFS $\mathcal{F}$.

Theorem 27. *Suppose $\mathcal{F}$ is a $(\widetilde{\neg}, \max)$-DFS. Then for all $x_1, x_2 \in \mathbf{I}$,*

a. $x_1 \mathbin{\widetilde{\leftrightarrow}} x_2 = (x_1 \wedge x_2) \vee (\widetilde{\neg} x_1 \wedge \widetilde{\neg} x_2)$
b. $x_1 \mathbin{\widetilde{\mathrm{xor}}} x_2 = (x_1 \wedge \widetilde{\neg} x_2) \vee (\widetilde{\neg} x_1 \wedge x_2)$.

Definition 34. *By a* standard DFS *we denote a* $(\neg, \max)$-*DFS.*

Standard DFS induce the standard connectives of fuzzy logic and by (Th-21), they also induce the standard extension principle. The propositional part of a standard DFS coincides with the K-standard sequence logic of Dienes [5], see [10, p.49]. In particular, the three-valued fragment is Kleene's three-valued logic. Standard DFSes represent a "boundary" case of DFSes because they induce the smallest existential quantifiers, the smallest extension principle, and the largest universal quantifiers.

4.8 Further adequacy criteria and theoretical adequacy bounds

Definition 35. *We say that a QFM $\mathcal{F}$ is* arg-continuous *if and only if $\mathcal{F}$ maps all $Q : \mathcal{P}(E)^n \longrightarrow \mathbf{I}$ to continuous fuzzy quantifiers $\mathcal{F}(Q)$, i.e. for all $X_1, \ldots, X_n \in \widetilde{\mathcal{P}}(E)$ and $\varepsilon > 0$ there exists $\delta > 0$ with $d(\mathcal{F}(Q)(X_1, \ldots, X_n), \mathcal{F}(Q)(X'_1, \ldots, X'_n)) < \varepsilon$ for all $X'_1, \ldots, X'_n \in \widetilde{\mathcal{P}}(E)$ with $d((X_1, \ldots, X_n), (X'_1, \ldots, X'_n)) < \delta$; where*

$$d((X_1, \ldots, X_n), (X'_1, \ldots, X'_n)) = \max_{i=1}^{n} \sup\{|\mu_{X_i}(e) - \mu_{X'_i}(e)| : e \in E\}.$$

Definition 36. *We say that a QFM $\mathcal{F}$ is* Q-continuous *if and only if for each semi-fuzzy quantifier $Q : \mathcal{P}(E)^n \longrightarrow \mathbf{I}$ and all $\varepsilon > 0$, there exists $\delta > 0$ such that $d(\mathcal{F}(Q), \mathcal{F}(Q')) < \varepsilon$ whenever $Q' : \mathcal{P}(E)^n \longrightarrow \mathbf{I}$ satisfies $d(Q, Q') < \delta$; where*

$$d(Q, Q') = \sup\{|Q(Y_1, \ldots, Y_n) - Q'(Y_1, \ldots, Y_n)| : Y_1, \ldots, Y_n \in \mathcal{P}(E)\}$$
$$d(\mathcal{F}(Q), \mathcal{F}(Q')) = \sup\{|\mathcal{F}(Q)(X_1, \ldots, X_n) - \mathcal{F}(Q')(X_1, \ldots, X_n)| : X_i \in \widetilde{\mathcal{P}}(E)\}.$$

Arg-continuity means that a small change in the membership grades $\mu_{X_i}(e)$ of the argument sets does not change $\mathcal{F}(Q)(X_1, \ldots, X_n)$ drastically; it hence expresses an important robustness condition with respect to noise. Q-continuity captures an important aspect of robustness with respect to imperfect knowledge about the precise definition of a quantifier; i.e. slightly different definitions of Q will produce similar quantification results. Both condition are crucial to the utility of a DFS and should be possessed by every practical model. They are not part of the DFS axioms because we wanted DFSes for general t-norms (including the discontinuous variety).

Theorem 28. *Suppose a DFS $\mathcal{F}$ has the property that*

$$\mathcal{F}(Q \mathbin{\widetilde{\wedge}} Q') = \mathcal{F}(Q) \mathbin{\widetilde{\wedge}} \mathcal{F}(Q') \tag{14}$$

for all semi-fuzzy quantifiers Q, $Q' : \mathcal{P}(E)^n \longrightarrow \mathbf{I}$, where $\widetilde{\wedge} = \widetilde{\mathcal{F}}(\wedge)$. Then $\widetilde{\wedge}$ satisfies $x \mathbin{\widetilde{\wedge}} \widetilde{\neg} x = 0$ for all $x, y \in \mathbf{I}$.

A DFS $\mathcal{F}$ homomorphic to conjunctions (or disjunctions) of quantifiers thus induces a t-norm which respects the law of contradiction. This is unacceptable since it excludes many t-norms; in particular, the standard choice $\widetilde{\mathcal{F}}(\wedge) = \min$. We have hence *not* required that a DFS be homomorphic to conjunctions/disjunctions of quantifiers.

The reader will certainly have noticed our special treatment of the propositional connectives $\leftrightarrow$ and xor, which is necessary because the definition of $\leftrightarrow, \mathrm{xor} : \mathbf{2} \times \mathbf{2} \longrightarrow \mathbf{2}$ involves *multiple occurrences* of propositional variables.

Definition 37. *Let a semi-fuzzy quantifier* $Q : \mathcal{P}(E)^m \longrightarrow \mathbf{I}$ *and a mapping* $\xi : \{1, \ldots, n\} \longrightarrow \{1, \ldots, m\}$ *be given. By* $Q\xi : \mathcal{P}(E)^n \longrightarrow \mathbf{I}$ *we denote the semi-fuzzy quantifier defined by* $Q\xi(Y_1, \ldots, Y_n) = Q(Y_{\xi(1)}, \ldots, Y_{\xi(n)})$, *for* $Y_1, \ldots, Y_n \in \mathcal{P}(E)$. *We use an analog definition for fuzzy quantifiers.*

The interesting case is that of non-injective ξ, which inserts the same variable in two (or more) argument positions of the original quantifier Q.

Theorem 29. *Suppose the DFS* $\mathcal{F}$ *is compatible with the duplication of variables, i.e. whenever* $Q : \mathcal{P}(E)^m \longrightarrow \mathbf{I}$ *and* $\xi : \{1, \ldots, n\} \longrightarrow \{1, \ldots, m\}$ *for some* $n \in \mathbb{N}$, *then* $\mathcal{F}(Q\xi) = \mathcal{F}(Q)\xi$. *Then* $x \mathbin{\widetilde{\wedge}} \widetilde{\neg} x = 0$ *for all* $x, y \in \mathbf{I}$, *where* $\widetilde{\wedge} = \widetilde{\mathcal{F}}(\wedge)$.

Again, we find this too restrictive and therefore have *not* required that $\mathcal{F}$ be homomorphic with respect to the duplication of variables.

Definition 38. *Suppose* $Q : \mathcal{P}(E)^n \longrightarrow \mathbf{I}$ *is an* n*-ary semi-fuzzy quantifier such that* $n > 0$. Q *is said to be* convex[10] in its i-th argument, *where* $i \in \{1, \ldots, n\}$, *iff*

$$\begin{aligned} Q(X_1, \ldots, X_n) \geq \min(& Q(X_1, \ldots, X_{i-1}, X_i', X_{i+1}, \ldots, X_n), \\ & Q(X_1, \ldots, X_{i-1}, X_i'', X_{i+1}, \ldots, X_n)) \end{aligned}$$

whenever $X_1, \ldots, X_n, X_i', X_i'' \in \mathcal{P}(E)$ *and* $X_i' \subseteq X_i \subseteq X_i''$.
Convexity of a fuzzy quantifier $\widetilde{Q} : \widetilde{\mathcal{P}}(E)^n \longrightarrow \mathbf{I}$ *in the* i*-th argument is defined analogously, where* $X_1, \ldots, X_n, X_i', X_i'' \in \widetilde{\mathcal{P}}(E)$, *and '*$\subseteq$*' is the fuzzy inclusion.*

For example, **between 10 and 20** is convex in both arguments, and **about 30 percent** is convex in the second argument.

Definition 39. *A QFM* $\mathcal{F}$ *is said to* preserve convexity of n-ary quantifiers, *where* $n \in \mathbb{N} \setminus \{0\}$, *iff every* n*-ary* $Q : \mathcal{P}(E)^n \longrightarrow \mathbf{I}$ *which is convex in its* i*-th argument is mapped to an* $\mathcal{F}(Q)$ *which is also convex in its* i*-th argument.* $\mathcal{F}$ preserves convexity *if* $\mathcal{F}$ *preserves the convexity of* n*-ary quantifiers for all* $n > 0$.

Theorem 30. *Suppose* $\mathcal{F}$ *is a contextual QFM with the following properties: for every base set* $E \neq \varnothing$,

[10] In TGQ, those quantifiers which we call 'convex' are usually dubbed 'continuous'. We have decided to change terminology because of the possible ambiguity of 'continuous', which could also mean 'smooth'.

a. *the quantifier* $\mathbb{O} : \mathcal{P}(E) \longrightarrow \mathbf{I}$, *defined by* $\mathbb{O}(Y) = 0$ *for all* $Y \in \mathcal{P}(E)$, *is mapped to the fuzzy quantifier defined by* $\mathcal{F}(\mathbb{O})(X) = 0$ *for all* $X \in \widetilde{\mathcal{P}}(E)$;
b. *If* $X \in \widetilde{\mathcal{P}}(E)$ *and there exists* $e \in E$ *with* $\mu_X(e) > 0$, *then* $\mathcal{F}(\exists)(X) > 0$;
c. *If* $X \in \widetilde{\mathcal{P}}(E)$ *and there exists* $e \in E$ *such that* $\mu_X(e) < 1$, *then* $\mathcal{F}(\sim\forall)(X) > 0$, *where* $\sim\forall : \mathcal{P}(E) \longrightarrow \mathbf{2}$ *is the quantifier defined by*

$$(\sim\forall)(Y) = \begin{cases} 1 & : \quad X \neq E \\ 0 & : \quad X = E \end{cases}$$

Then $\mathcal{F}$ *does not preserve convexity of one-place quantifiers* $Q : \mathcal{P}(E) \longrightarrow \mathbf{I}$ *on finite base sets* $E \neq \varnothing$. *In particular, no DFS preserves convexity.*

This means that even if we restrict to the simple case of one-place quantifiers on finite base sets, there is still no QFM $\mathcal{F}$ which both satisfies the very elementary adequacy conditions imposed by the theorem, and preserves convexity under the simplifying assumptions. Because contextuality is a rather fundamental condition, it seems better to weaken our requirements on the preservation of convexity, rather than compromising contextuality or the other elementary conditions.

Theorem 31. *Suppose* $\mathcal{F}$ *is a contextual QFM which is compatible with cylindrical extensions and satisfies the following properties: for all base sets* $E \neq \varnothing$,

a. *the quantifier* $\mathbb{O} : \mathcal{P}(E) \longrightarrow \mathbf{I}$, *defined by* $\mathbb{O}(Y) = 0$ *for all* $Y \in \mathcal{P}(E)$, *is mapped to the fuzzy quantifier defined by* $\mathcal{F}(\mathbb{O})(X) = 0$ *for all* $X \in \widetilde{\mathcal{P}}(E)$;
b. *If* $X \in \widetilde{\mathcal{P}}(E)$ *and there exists* $e \in E$ *with* $\mu_X(e) > 0$, *then* $\mathcal{F}(\exists)(X) > 0$;
c. *If* $X \in \widetilde{\mathcal{P}}(E)$ *and there exists and there exists some* $e \in E$ *such that* $\mu_X(e') = 0$ *for* $e' \in E \setminus \{e\}$ *and* $\mu_X(e) < 1$, *then* $\mathcal{F}(\sim\exists)(X) > 0$, *where* $\sim\exists : \mathcal{P}(E) \longrightarrow \mathbf{2}$ *is the quantifier defined by*

$$(\sim\exists)(Y) = \begin{cases} 1 & : \quad X = \varnothing \\ 0 & : \quad X \neq \varnothing \end{cases}$$

Then $\mathcal{F}$ *does not preserve the convexity of quantitative semi-fuzzy quantifiers of arity* $n > 1$ *even on finite base sets. In particular, no DFS preserves the convexity of quantitative semi-fuzzy quantifiers of arity* $n > 1$.

This leaves open the possibility that certain DFSes will preserve the convexity of quantitative semi-fuzzy quantifiers of arity $n = 1$:

Definition 40. *A QFM* $\mathcal{F}$ *is said to* weakly preserve convexity *iff* $\mathcal{F}$ *preserves the convexity of quantitative one-place quantifiers on finite domains.*

We have positive results on the existence of DFSes that weakly preserve convexity, an example will be given below. Weak preservation of convexity is strong enough to cover many NL quantifiers of interest, e.g. **between 10 and 20, about 50** etc.

One of the pervasive properties of NL quantifiers is *conservativity*. We shall call $Q : \mathcal{P}(E)^2 \longrightarrow \mathbf{I}$ *conservative* if

$$Q(X_1, X_2) = Q(X_1, X_1 \cap X_2) \tag{15}$$

for all $X_1, X_2 \in \mathcal{P}(E)$. To give an example, if E is a set of persons, **married** $\in \mathcal{P}(E)$ the subset of married persons, and **have_children** $\in \mathcal{P}(E)$ the set of persons who have children, then the conservative quantifier **almost all** : $\mathcal{P}(E)^2 \longrightarrow \mathbf{I}$ has

$$\textbf{almost all}(\textbf{married}, \textbf{have\_children}) = \textbf{almost all}(\textbf{married}, \textbf{married} \cap \textbf{have\_children})$$

i.e. the meanings of "almost all married persons have children" and "almost all married persons are married persons who have children" coincide. Like having extension, conservativity expresses an aspect of context insensitivity: if an element of the domain is irrelevant to the restriction (first argument) of a two-place quantifier, then it does not affect the quantifier at all. For example, every conservative $Q : \mathcal{P}(E)^2 \longrightarrow \mathbf{I}$ apparently has $Q(X_1, X_2 \cap X_1) = Q(X_1, X_2 \cup \neg X_1)$, i.e. if $e \notin X_1$, then it does not matter whether $e \in X_2$ or $e \notin X_2$. A corresponding $\mathcal{F}(Q) : \widetilde{\mathcal{P}}(E)^2 \longrightarrow \mathbf{I}$ should at least be *weakly conservative*, i.e.

$$\mathcal{F}(Q)(X_1, X_2) = \mathcal{F}(Q)(X_1, \mathrm{spp}(X_1) \cap X_2)\,, \tag{16}$$

for all $X_1, X_2 \in \widetilde{\mathcal{P}}(E)$, where $\mathrm{spp}(X_1)$ is the support of X_1, see (8). This definition is sufficiently strong to capture the context insensitivity aspect of conservativity: an element $e \in E$ which is irrelevant to the restriction of the quantifier, i.e. $\mu_{X_1}(e) = 0$, has no effect on the quantification result, which is independent of $\mu_{X_2}(e)$.

Theorem 32. *Every DFS $\mathcal{F}$ weakly preserves conservativity, i.e. if $Q : \mathcal{P}(E)^2 \longrightarrow \mathbf{I}$ is conservative, then $\mathcal{F}(Q)$ is weakly conservative.*

Let us say that a fuzzy quantifier $\widetilde{Q} : \widetilde{\mathcal{P}}(E)^2 \longrightarrow \mathbf{I}$ is *strongly conservative* if

$$\mathcal{F}(Q)(X_1, X_2) = \mathcal{F}(Q)(X_1, X_1 \mathbin{\widetilde{\cap}} X_2) \qquad \text{for all } X_1, X_2 \in \widetilde{\mathcal{P}}(E).$$

In addition to the context insensitivity aspect, strong conservativity also reflects the definition of crisp conservativity in terms of intersection with the first argument.

Theorem 33. *Assume the QFM $\mathcal{F}$ satisfies the following conditions: (a) $\widetilde{\mathcal{F}}(\mathrm{id}_2) = \mathrm{id}_{\mathbf{I}}$; (b) $\widetilde{\neg}$ is a strong negation; (c) $\widetilde{\wedge}$ is a t-norm; (d) $\mathcal{F}$ is compatible with internal meets, see Def. 15; (e) $\mathcal{F}$ is compatible with dualisation. Then $\mathcal{F}$ does not strongly preserve conservativity, i.e. there are conservative $Q : \mathcal{P}(E)^2 \longrightarrow \mathbf{I}$ such that $\mathcal{F}(Q)$ is not strongly conservative. In particular, no DFS strongly preserves conservativity.*

Hence strong preservation of conservativity cannot be ensured in a fuzzy framework, even under assumptions which are much weaker than the DFS axioms.

In our comments on argument insertion (see p. 15) we have remarked that adjectival restriction with fuzzy adjectives cannot be modelled directly: if $A \in \widetilde{\mathcal{P}}(E)$ is *fuzzy*, then only $\mathcal{F}(Q) \triangleleft A$ is defined, but not $Q \triangleleft A$. However, one can ask if $\mathcal{F}(Q) \triangleleft A$ can be represented by a semi-fuzzy quantifier Q', i.e. if there is a Q' with

$$\mathcal{F}(Q) \triangleleft A = \mathcal{F}(Q') . \tag{17}$$

The obvious choice for Q' is the following.

Definition 41. *Suppose $\mathcal{F}$ is a QFM, $Q : \mathcal{P}(E)^{n+1} \longrightarrow \mathbf{I}$ is a semi-fuzzy quantifier and $A \in \widetilde{\mathcal{P}}(E)$. Then $Q \,\widetilde{\triangleleft}\, A : \mathcal{P}(E)^n \longrightarrow \mathbf{I}$ is defined by*

$$Q \,\widetilde{\triangleleft}\, A = \mathcal{U}(\mathcal{F}(Q) \triangleleft A) ,$$

i.e. $Q \,\widetilde{\triangleleft}\, A(Y_1, \ldots, Y_n) = \mathcal{F}(Q)(Y_1, \ldots, Y_n, A)$ *for all crisp* $Y_1, \ldots, Y_n \in \mathcal{P}(E)$.

$Q' = Q \,\widetilde{\triangleleft}\, A$ is the only choice of Q' which possibly satisfies (17), because any Q' which satisfies $\mathcal{F}(Q') = \mathcal{F}(Q) \triangleleft A$ also satisfies

$$Q' = \mathcal{U}(\mathcal{F}(Q')) = \mathcal{U}(\mathcal{F}(Q) \triangleleft A) = Q \,\widetilde{\triangleleft}\, A ,$$

which is apparent from (Th-1). Unfortunately, $Q \,\widetilde{\triangleleft}\, A$ is not guaranteed to fulfill (17) in a QFM, not even in a DFS. We hence turn this equation into an adequacy condition which ensures that $Q \,\widetilde{\triangleleft}\, A$ convey the intended meaning in a given QFM:

Definition 42. *Suppose $\mathcal{F}$ is a QFM. We say that $\mathcal{F}$ is* compatible with fuzzy argument insertion *iff for every semi-fuzzy quantifier $Q : \mathcal{P}(E)^n \longrightarrow \mathbf{I}$ of arity $n > 0$ and every $A \in \widetilde{\mathcal{P}}(E)$, $\mathcal{F}(Q \,\widetilde{\triangleleft}\, A) = \mathcal{F}(Q) \triangleleft A$.*

The main application in natural language is that of adjectival restriction of a quantifier with a fuzzy adjective. For example, suppose E is a set of people, and **lucky** $\in \widetilde{\mathcal{P}}(E)$the fuzzy subset of people in E who are lucky. Further suppose **almost all** : $\mathcal{P}(E)^2 \longrightarrow \mathbf{I}$ is a semi-fuzzy quantifier which models "almost all". Finally, let a DFS $\mathcal{F}$ be chosen as the model of fuzzy quantification. We can then construct the semi-fuzzy quantifier $Q' =$ **almost all**$\cap \,\widetilde{\triangleleft}\,$ **lucky** . If $\mathcal{F}$ is compatible with fuzzy argument insertion, then the semi-fuzzy quantifier Q' is guaranteed to adequately model the composite expression "almost all X's are lucky Y's", because

$$\mathcal{F}(Q')(X_1, X_2) = \mathcal{F}(Q)(X, Y \,\widetilde{\cap}\, \mathbf{lucky})$$

for all fuzzy arguments $X, Y \in \widetilde{\mathcal{P}}(E)$, which (relative to $\mathcal{F}$) is the proper expression for interpreting "almost all X's are lucky Y's" in the fuzzy case. Compatibility with fuzzy argument insertion is a very restrictive adequacy condition. We shall present the unique standard DFS which fulfills this condition on p. 31.

Finally, let us recall the specificity order $\preceq_c$ defined in (13). We can extend $\preceq_c$ to fuzzy sets $X \in \widetilde{\mathcal{P}}(E)$, semi-fuzzy quantifiers $Q : \mathcal{P}(E)^n \longrightarrow \mathbf{I}$ and fuzzy quantifiers $\widetilde{Q} : \widetilde{\mathcal{P}}(E)^n \longrightarrow \mathbf{I}$ as follows:

$$\begin{aligned}
X \preceq_c X' &\iff \mu_X(e) \preceq_c \mu_{X'}(e) && \text{for all } e \in E; \\
Q \preceq_c Q' &\iff Q(Y_1, \ldots, Y_n) \preceq_c Q'(Y_1, \ldots, Y_n) && \text{for all } Y_1, \ldots, Y_n \in \mathcal{P}(E); \\
\widetilde{Q} \preceq_c \widetilde{Q}' &\iff \widetilde{Q}(X_1, \ldots, X_n) \preceq_c \widetilde{Q}'(X_1, \ldots, X_n) && \text{for all } X_1, \ldots, X_n \in \widetilde{\mathcal{P}}(E) .
\end{aligned}$$

Intuitively, we expect the results to become less specific when the quantifier or the arguments become less specific: the fuzzier the input, the fuzzier the output.

Definition 43. *We say that a QFM $\mathcal{F}$* propagates fuzziness in arguments *iff for all $Q : \mathcal{P}(E)^n \longrightarrow \mathbf{I}$ and $X_1, \ldots, X_n, X'_1, \ldots, X'_n$: whenever $X_i \preceq_c X'_i$ for all $i = 1, \ldots, n$, then $\mathcal{F}(Q)(X_1, \ldots, X_n) \preceq_c \mathcal{F}(Q)(X'_1, \ldots, X'_n)$. We say that $\mathcal{F}$* propagates fuzziness in quantifiers *iff $\mathcal{F}(Q) \preceq_c \mathcal{F}(Q')$ whenever $Q \preceq_c Q'$.*

5 A class of models of the axiomatic framework

We now address the issue of models of our axiomatic framework. We introduce a class of QFMs defined in terms of three-valued cuts of arguments and subsequent aggregation by the fuzzy median. We hence generalise the construction successfully used in [10] to define DFSes. In particular, we present a characterisation of the class of $\mathcal{M}_\mathcal{B}$-DFSes in terms of necessary and sufficient conditions on the aggregation mapping $\mathcal{B}$. To define the unrestricted class of $\mathcal{M}_\mathcal{B}$-QFMs, we recall some concepts introduced in [10]. We need the cut range $\mathcal{T}_\gamma(X) \subseteq \mathcal{P}(E)$ which represents a three-valued cut at the "cautiousness level" $\gamma \in \mathbf{I}$ by a set of alternatives $\{Y : (X)^{\min}_\gamma \subseteq Y \subseteq (X)^{\max}_\gamma\}$:

Definition 44. *Suppose E is some set, $X \in \widetilde{\mathcal{P}}(E)$ $\gamma \in \mathbf{I}$. $(X)^{\min}_\gamma, (X)^{\max}_\gamma \in \mathcal{P}(E)$ and $\mathcal{T}_\gamma(X) \subseteq \mathcal{P}(E)$ are defined by*

$$(X)^{\min}_\gamma = \begin{cases} X_{>\frac{1}{2}} & : \ \gamma = 0 \\ X_{\geq \frac{1}{2} + \frac{1}{2}\gamma} & : \ \gamma > 0 \end{cases}$$

$$(X)^{\max}_\gamma = \begin{cases} X_{\geq\frac{1}{2}} & : \ \gamma = 0 \\ X_{> \frac{1}{2} - \frac{1}{2}\gamma} & : \ \gamma > 0 \end{cases}$$

$$\mathcal{T}_\gamma(X) = \{Y : (X)^{\min}_\gamma \subseteq Y \subseteq (X)^{\max}_\gamma\},$$

where $X_{\geq\alpha} = \{e \in E : \mu_X(e) \geq \alpha\}$ is α-cut and $X_{>\alpha} = \{e \in E : \mu_X(e) > \alpha\}$ is strict α-cut.

The basic idea is to view the crisp range $\mathcal{T}_\gamma(X)$ as providing alternatives to be evaluated. For example, in order to evaluate a quantifier Q at a certain cut level γ, we have to consider all choices of $Q(Y_1, \ldots, Y_n)$, where $Y_i \in \mathcal{T}_\gamma(X_i)$. The set of results obtained in this way must then be aggregated to a single result in the unit interval. The generalised fuzzy median (see Def. 31) is particularly suited to carry out this aggregation because the resulting fuzzification mechanism will then contain Kleene's three-valued logic as (see remark on p. 22). Let us hence use the crisp ranges $\mathcal{T}_\gamma(X_i)$ of the argument sets to define a family of QFMs $(\bullet)_\gamma$, indexed by the cautiousness parameter $\gamma \in \mathbf{I}$:

Definition 45. *For every $\gamma \in \mathbf{I}$, we denote by $(\bullet)_\gamma$ the QFM defined by*

$$Q_\gamma(X_1, \ldots, X_n) = \mathrm{m}_{\frac{1}{2}}\{Q(Y_1, \ldots, Y_n) : Y_i \in \mathcal{T}_\gamma(X_i)\},$$

for all semi-fuzzy quantifiers $Q : \mathcal{P}(E)^n \longrightarrow \mathbf{I}$.

None of the QFMs $(\bullet)_\gamma$ is a DFS, the fuzzy median suppresses too much structure. Nevertheless, these QFMs prove useful in defining DFSes. The basic idea is that in order to compute $\mathcal{F}(Q)(X_1,\ldots,X_n)$, we should consider the results obtained at all levels of cautiousness γ, i.e. $(Q_\gamma(X_1,\ldots,X_n))_{\gamma\in\mathbf{I}}$. We can then apply various aggregation operators on these γ-indexed results to obtain new QFMs, which might be DFSes. Now we define the domain on which the aggregation operators will act.

Definition 46. *$\mathbb{B}^+, \mathbb{B}^{\frac{1}{2}}, \mathbb{B}^-$ and $\mathbb{B} \subseteq \mathbf{I}^\mathbf{I}$ are defined by*

$$\mathbb{B}^+ = \{f \in \mathbf{I}^\mathbf{I} : f(0) > \tfrac{1}{2} \text{ and } f(\mathbf{I}) \subseteq [\tfrac{1}{2}, 1] \text{ and } f \text{ nonincreasing}\}$$

$$\mathbb{B}^{\frac{1}{2}} = \{\mathrm{c}_{\frac{1}{2}}\}, \quad \text{where } \mathrm{c}_{\frac{1}{2}} : \mathbf{I} \longrightarrow \mathbf{I} \text{ is the constant } \mathrm{c}_{\frac{1}{2}}(x) = \tfrac{1}{2} \text{ for all } x \in \mathbf{I}$$

$$\mathbb{B}^- = \{f \in \mathbf{I}^\mathbf{I} : f(0) < \tfrac{1}{2} \text{ and } f(\mathbf{I}) \subseteq [0, \tfrac{1}{2}] \text{ and } f \text{ nondecreasing}\}$$

$$\mathbb{B} = \mathbb{B}^+ \cup \mathbb{B}^{\frac{1}{2}} \cup \mathbb{B}^- .$$

Theorem 34.

a. *Suppose $Q : \mathcal{P}(E)^n \longrightarrow \mathbf{I}$ and $X_1, \ldots, X_n \in \widetilde{\mathcal{P}}(E)$ are given. Then*

$$(Q_\gamma(X_1,\ldots,X_n))_{\gamma\in\mathbf{I}} \in \begin{cases} \mathbb{B}^+ & : \ Q_0(X_1,\ldots,X_n) > \frac{1}{2} \\ \mathbb{B}^{\frac{1}{2}} & : \ Q_0(X_1,\ldots,X_n) = \frac{1}{2} \\ \mathbb{B}^- & : \ Q_0(X_1,\ldots,X_n) < \frac{1}{2} \end{cases}$$

b. *For each $f \in \mathbb{B}$ there exists $Q : \mathcal{P}(E)^n \longrightarrow \mathbf{I}$ and $X_1, \ldots, X_n \in \widetilde{\mathcal{P}}(E)$ such that $f = (Q_\gamma(X_1,\ldots,X_n))_{\gamma\in\mathbf{I}}$.*

Given an aggregation operator $\mathcal{B} : \mathbb{B} \longrightarrow \mathbf{I}$, we define the corresponding QFM $\mathcal{M}_\mathcal{B}$ as follows.

Definition 47. *Suppose $\mathcal{B} : \mathbb{B} \longrightarrow \mathbf{I}$ is given. The QFM $\mathcal{M}_\mathcal{B}$ is defined by*

$$\mathcal{M}_\mathcal{B}(Q)(X_1,\ldots,X_n) = \mathcal{B}((Q_\gamma(X_1,\ldots,X_n))_{\gamma\in\mathbf{I}}), \tag{18}$$

for all semi-fuzzy quantifiers $Q : \mathcal{P}(E)^n \longrightarrow \mathbf{I}$ and $X_1 \ldots, X_n \in \widetilde{\mathcal{P}}(E)$.

By the class of $\mathcal{M}_\mathcal{B}$-QFMs we mean the class of all QFMs $\mathcal{M}_\mathcal{B}$ defined in this way. It is apparent that if we do not impose restrictions on admissible choices of $\mathcal{B}$, the resulting QFMs will often fail to be DFSes. We are hence interested in stating necessary and sufficient conditions on $\mathcal{B}$ for $\mathcal{M}_\mathcal{B}$ to be a DFS. In order to do so, we first need to introduce some constructions on $\mathbb{B}$.

Definition 48. *Suppose $f : \mathbf{I} \longrightarrow \mathbf{I}$ is a monotonic mapping (i.e., nondecreasing or nonincreasing). The mappings $f^\flat$, $f^\sharp : \mathbf{I} \longrightarrow \mathbf{I}$ are defined by:*

$$f^\sharp = \begin{cases} \lim_{y\to x^+} f(y) & : \ x < 1 \\ f(1) & : \ x = 1 \end{cases}$$

$$f^\flat = \begin{cases} \lim_{y\to x^-} f(y) & : \ x > 0 \\ f(0) & : \ x = 0 \end{cases} \qquad \text{for all } f \in \mathbb{B},\ x \in \mathbf{I}.$$

It is apparent that if $f \in \mathbb{B}$, then $f^\sharp \in \mathbb{B}$ and $f^\flat \in \mathbb{B}$. We shall further introduce several coefficients which describe certain aspects of a mapping $f : \mathbf{I} \longrightarrow \mathbf{I}$.

Definition 49. *For every monotonic mapping $f : \mathbf{I} \longrightarrow \mathbf{I}$ (i.e., either nondecreasing or nonincreasing), we define*

$$f_0^* = \lim_{\gamma \to 0^+} f(\gamma) \tag{19}$$

$$f_*^0 = \inf\{\gamma \in \mathbf{I} : f(\gamma) = 0\} \tag{20}$$

$$f_*^{\frac{1}{2}} = \inf\{\gamma \in \mathbf{I} : f(\gamma) = \tfrac{1}{2}\} \tag{21}$$

$$f_1^* = \lim_{\gamma \to 1^-} f(\gamma) \tag{22}$$

$$f_*^1 = \sup\{\gamma \in \mathbf{I} : f(\gamma) = 1\}\,. \tag{23}$$

Based on these concepts, we can now state a number of axioms governing the behaviour of reasonable choices of $\mathcal{B}$.

Definition 50. *Suppose $\mathcal{B} : \mathbb{B} \longrightarrow \mathbf{I}$ is given. For all $f, g \in \mathcal{B}$, we define the following conditions on $\mathcal{B}$:*

$$\mathcal{B}(f) = f(0) \quad \textit{if } f \textit{ is constant, i.e. } f(x) = f(0) \textit{ for all } x \in \mathbf{I} \tag{B-1}$$

$$\mathcal{B}(1 - f) = 1 - \mathcal{B}(f) \tag{B-2}$$

$$\textit{If } \widehat{f}(\mathbf{I}) \subseteq \{0, \tfrac{1}{2}, 1\}, \textit{ then} \tag{B-3}$$

$$\mathcal{B}(f) = \begin{cases} \frac{1}{2} + \frac{1}{2} f_*^{\frac{1}{2}} & : \; f \in \mathbb{B}^+ \\ \frac{1}{2} & : \; f \in \mathbb{B}^{\frac{1}{2}} \\ \frac{1}{2} - \frac{1}{2} f_*^{\frac{1}{2}} & : \; f \in \mathbb{B}^- \end{cases}$$

$$\mathcal{B}(f^\sharp) = \mathcal{B}(f^\flat) \tag{B-4}$$

$$\textit{If } f \le g, \textit{ then } \mathcal{B}(f) \le \mathcal{B}(g) \tag{B-5}$$

Theorem 35.

a. *The conditions* (B-1) *to* (B-5) *are sufficient for $\mathcal{M}_\mathcal{B}$ to be a standard DFS.*
b. *The conditions* (B-1) *to* (B-5) *are necessary for $\mathcal{M}_\mathcal{B}$ to be a DFS.*
c. *The conditions* (B-1) *to* (B-5) *are independent.*

In particular, $\mathcal{B}(f) = 1 - \mathcal{B}(1 - f)$ for all $f \in \mathbb{B}$, and $\mathcal{B}(f) \ge \frac{1}{2}$ whenever $f \in \mathbb{B}^+$. We can hence give a more concise description of $\mathcal{M}_\mathcal{B}$-DFSes, by considering only their behaviour on $\mathbb{B}^+$:

Definition 51. *By $\mathbb{H} \subseteq \mathbf{I}^\mathbf{I}$ we denote the set of nonincreasing $f : \mathbf{I} \longrightarrow \mathbf{I}$, $f \neq 0$,*

$$\mathbb{H} = \{f \in \mathbf{I}^\mathbf{I} : \, f \textit{ nonincreasing and } f(0) > 0\,\}\,.$$

We can associate with each $\mathcal{B}' : \mathbb{H} \longrightarrow \mathbf{I}$ a $\mathcal{B} : \mathbb{B} \longrightarrow \mathbf{I}$ as follows:

$$\mathcal{B}(f) = \begin{cases} \frac{1}{2} + \frac{1}{2}\mathcal{B}'(2f-1) & : \quad f \in \mathbb{B}^+ \\ \frac{1}{2} & : \quad f \in \mathbb{B}^{\frac{1}{2}} \\ \frac{1}{2} - \frac{1}{2}\mathcal{B}'(1-2f) & : \quad f \in \mathbb{B}^- \end{cases} \tag{24}$$

Theorem 36. *If $\mathcal{M}_\mathcal{B}$ is a DFS, then $\mathcal{B}$ can be defined in terms of a mapping $\mathcal{B}' : \mathbb{H} \longrightarrow \mathbf{I}$ according to equation* (24). *$\mathcal{B}'$ is defined by*

$$\mathcal{B}'(f) = 2\mathcal{B}(\tfrac{1}{2} + \tfrac{1}{2}f) - 1\,. \tag{25}$$

We can hence focus on mappings $\mathcal{B}' : \mathbb{H} \longrightarrow \mathbf{I}$ without loosing any desired models.

Definition 52. *Suppose $\mathcal{B}' : \mathbb{H} \longrightarrow \mathbf{I}$ is given. For all $f, g \in \mathbb{H}$, we define the following conditions on $\mathcal{B}'$:*

$$\mathcal{B}'(f) = f(0) \quad \text{if } f \text{ is constant, i.e. } f(x) = f(0) \text{ for all } x \in \mathbf{I} \tag{C-1}$$

$$\text{If } \widehat{f}(\mathbf{I}) \subseteq \{0,1\}, \text{ then } \mathcal{B}'(f) = f_*^0, \tag{C-2}$$

$$\mathcal{B}'(f^\sharp) = \mathcal{B}'(f^\flat) \quad \text{if } \widehat{f}((0,1]) \neq \{0\} \tag{C-3}$$

$$\text{If } f \leq g, \text{ then } \mathcal{B}'(f) \leq \mathcal{B}'(g) \tag{C-4}$$

A theorem analogous to (Th-35) can be proven for (C-1) to (C-4). Our introducing of $\mathcal{B}'$ is only a matter of convenience: we can now succintly define examples of $\mathcal{M}_\mathcal{B}$-QFMs.

Definition 53. *By $\mathcal{M}$ we denote the $\mathcal{M}_\mathcal{B}$-QFM defined by*

$$\mathcal{B}'_{\int}(f) = \int_0^1 f(x)\,dx\,, \qquad \text{for all } f \in \mathbb{H}$$

Theorem 37. *$\mathcal{M}$ is a standard DFS.*

$\mathcal{M}$ is Q-continuous and arg-continuous and hence a good choice for applications.

Definition 54. *By $\mathcal{M}_U$ we denote the $\mathcal{M}_\mathcal{B}$-QFM defined by*

$$\mathcal{B}'_U(f) = \max(f_*^1, f_1^*) \qquad \text{for all } f \in \mathbb{H}, \text{ see (22) and (23).}$$

Theorem 38. *Suppose $\oplus : \mathbf{I}^2 \longrightarrow \mathbf{I}$ is an s-norm and $\mathcal{B}' : \mathbb{H} \longrightarrow \mathbf{I}$ is defined by*

$$\mathcal{B}'(f) = f_*^1 \oplus f_1^*\,,$$

for all $f \in \mathbb{H}$. Further suppose that $\mathcal{M}_\mathcal{B}$ is defined in terms of $\mathcal{B}'$ according to equations (18) *and* (24). *Then $\mathcal{M}_\mathcal{B}$ is a standard DFS.*

In particular, $\mathcal{M}_U$ is a standard DFS. It is neither Q-continuous nor arg-continuous and hence not practical. However, $\mathcal{M}_U$ is of theoretical interest because it represents an extreme case of $\mathcal{M}_\mathcal{B}$-DFS in terms of specificity:

Theorem 39. *$\mathcal{M}_U$ is the least specific $\mathcal{M}_\mathcal{B}$-DFS.*

Let us now consider the issue of most specific $\mathcal{M}_\mathcal{B}$-DFSes.

Definition 55. *By $\mathcal{M}_S$ we denote the $\mathcal{M}_\mathcal{B}$-QFM defined by*

$$\mathcal{B}'_S(f) = \min(f_*^0, f_0^*) \qquad \textit{for all } f \in \mathbb{H}\textit{; see } (19) \textit{ and } (20).$$

Theorem 40. *Suppose $\mathcal{B}' : \mathbb{H} \longrightarrow \mathbf{I}$ is defined by*

$$\mathcal{B}'(f) = f_*^0 \odot f_0^*$$

for all $f \in \mathbb{H}$, where $\odot : \mathbf{I}^2 \longrightarrow \mathbf{I}$ is a t-norm. Further suppose that the QFM $\mathcal{M}_\mathcal{B}$ is defined in terms of $\mathcal{B}'$ according to (18) *and* (24). *Then $\mathcal{M}_\mathcal{B}$ is a standard DFS.*

In particular, $\mathcal{M}_S$ is a standard DFS. $\mathcal{M}_S$ fails on both continuity conditions, but:

Theorem 41. *$\mathcal{M}_S$ is the most specific $\mathcal{M}_\mathcal{B}$-DFS.*

Definition 56. *By $\mathcal{M}_{CX}$ we denote the $\mathcal{M}_\mathcal{B}$-QFM defined by*

$$\mathcal{B}'_{CX}(f) = \sup\{\min(x, f(x)) : x \in \mathbf{I}\} \qquad \textit{for all } f \in \mathbb{H}$$

Theorem 42. *Suppose $\odot : \mathbf{I}^2 \longrightarrow \mathbf{I}$ is a continuous t-norm and $\mathcal{B}' : \mathbb{H} \longrightarrow \mathbf{I}$ is defined by*

$$\mathcal{B}'(f) = \sup\{\gamma \odot f(\gamma) : \gamma \in \mathbf{I}\}$$

for all $f \in \mathbb{H}$. Further suppose that $\mathcal{M}_\mathcal{B}$ is defined in terms of $\mathcal{B}'$ according to (18) *and* (24). *Then $\mathcal{M}_\mathcal{B}$ is a standard DFS.*

Hence $\mathcal{M}_{CX}$ is a standard DFS. It is Q-continuous and arg-continuous and hence a good choice for applications. Indeed, $\mathcal{M}_{CX}$ is a DFS with unique properties.

Theorem 43.

a. $\mathcal{M}_{CX}$ weakly preserves convexity.
b. If an $\mathcal{M}_\mathcal{B}$-DFS weakly preserves convexity, then $\mathcal{M}_{CX} \preceq_c \mathcal{M}_\mathcal{B}$.

Theorem 44. *$\mathcal{M}_{CX}$ is the only standard DFS compliant to fuzzy argument insertion.*

Definition 57. *Suppose $Q : \mathcal{P}(E) \longrightarrow \mathbf{I}$ is a nondecreasing semi-fuzzy quantifier and $X \in \widetilde{\mathcal{P}}(E)$. The* Sugeno integral *$(S) \int X \, dQ$ is defined by*

$$(S) \int X \, dQ = \sup\{\min(\alpha, Q(X_{\geq\alpha})) : \alpha \in \mathbf{I}\} .$$

Theorem 45. *Suppose $Q : \mathcal{P}(E) \longrightarrow \mathbf{I}$ is nondecreasing. Then for all $X \in \widetilde{\mathcal{P}}(E)$,*

$$(S) \int X \, dQ = \mathcal{M}_{CX}(Q)(X) .$$

Hence $\mathcal{M}_{CX}$ coincides with the Sugeno integral whenever the latter is defined.

Definition 58. *Suppose $E \neq \varnothing$ is a finite base set of cardinality $|E| = m$. For a fuzzy subset $X \in \widetilde{\mathcal{P}}(E)$, let us denote by $\mu_{[j]}(X) \in \mathbf{I}$, $j = 1, \ldots, m$, the j-th largest membership value of X (including duplicates).*[11] *We also stipulate that $\mu_{[0]}(X) = 1$ and $\mu_{[j]}(X) = 0$ whenever $j > m$.*

Using this notation, we obtain the following corollary (cf. [4]):

Theorem 46. *Suppose $E \neq \varnothing$ is a finite base set, $q : \{0, \ldots, |E|\} \longrightarrow \mathbf{I}$ is a nondecreasing mapping and $Q : \mathcal{P}(E) \longrightarrow \mathbf{I}$ is defined by $Q(Y) = q(|Y|)$ for all $Y \in \mathcal{P}(E)$. Then for all $X \in \widetilde{\mathcal{P}}(E)$,*

$$\mathcal{M}_{CX}(Q)(X) = \max\{\min(q(j), \mu_{[j]}(X)) : 0 \leq j \leq |E|\} ,$$

i.e. $\mathcal{M}_{CX}$ consistently generalises the FG-count approach of [31,24].

Let us also observe that $\mathcal{M}_{CX}$ is a concrete implementation of a so-called "substitution approach" to fuzzy quantification [23], i.e. the fuzzy quantifier is modelled by constructing an equivalent logical formula as follows.[12]

Theorem 47. *For every $Q : \mathcal{P}(E)^n \longrightarrow \mathbf{I}$ and $X_1, \ldots, X_n \in \widetilde{\mathcal{P}}(E)$,*

$$\begin{aligned} &\mathcal{M}_{CX}(Q)(X_1, \ldots, X_n) \\ &= \sup\{\widetilde{Q}^L_{V,W}(X_1, \ldots, X_n) : V, W \in \mathcal{P}(E)^n, V_1 \subseteq W_1, \ldots, V_n \subseteq W_n\} \\ &= \inf\{\widetilde{Q}^U_{V,W}(X_1, \ldots, X_n) : V, W \in \mathcal{P}(E)^n, V_1 \subseteq W_1, \ldots, V_n \subseteq W_n\} \end{aligned}$$

where

$$\begin{aligned} &\widetilde{Q}^L_{V,W}(X_1, \ldots, X_n) \\ &= \min(\Xi_{V,W}(X_1, \ldots, X_n), \inf\{Q(Y_1, \ldots, Y_n) : V_i \subseteq Y_i \subseteq W_i, \text{ all } i\}) \\ &\widetilde{Q}^U_{V,W}(X_1, \ldots, X_n) \\ &= \max(1 - \Xi_{V,W}(X_1, \ldots, X_n), \sup\{Q(Y_1, \ldots, Y_n) : V_i \subseteq Y_i \subseteq W_i, \text{ all } i\}) \\ &\Xi_{V,W}(X_1, \ldots, X_n) \\ &= \min_{i=1}^{n} \min(\inf\{\mu_X(e) : e \in V_i\}, \inf\{1 - \mu_X(e) : e \notin W_i\}) . \end{aligned}$$

---

[11] More formally, we can order the elements of E such that $E = \{e_1, \ldots, e_m\}$ and $\mu_X(e_1) \geq \cdots \geq \mu_X(e_m)$ and then define $\mu_{[j]}(X) = \mu_X(e_j)$.

[12] In the finite case, inf and sup reduce to logical connectives $\wedge = \max$ and $\vee = \min$ as usual. We need to allow for occurrences of constants $Q(Y_1, \ldots, Y_n) \in \mathbf{I}$ in the resulting formula because the fuzzification mechanism is applied to semi-fuzzy quantifiers, not only to two-valued quantifiers.

Returning to $\mathcal{M}_\mathcal{B}$-DFSes in general, we can state that:

Theorem 48.

- *All $\mathcal{M}_\mathcal{B}$-DFSes coincide on three-valued arguments, i.e. if $X_1, \ldots, X_n \in \widetilde{\mathcal{P}}(E)$ satisfy $\mu_{X_i}(e) \in \{0, \frac{1}{2}, 1\}$ for all $e \in E$;*
- *all $\mathcal{M}_\mathcal{B}$-DFSes coincide on three-valued semi-fuzzy quantifiers $Q : \mathcal{P}(E)^n \longrightarrow \{0, \frac{1}{2}, 1\}$;*
- *Every $\mathcal{M}_\mathcal{B}$-DFS propagates fuzziness in quantifiers and arguments.*

In addition, every $\mathcal{M}_\mathcal{B}$-DFS is a consistent generalisation of the fuzzification mechanism proposed by Gaines [7] as a foundation of fuzzy reasoning. A broader class of DFS models is obtained if we drop the median-based aggregation mechanism of $\mathcal{M}_\mathcal{B}$-DFSes [9]. To this end, let us observe that $(\bullet)_\gamma$ can be expressed as

$$Q_\gamma(X_1, \ldots, X_n) = \mathrm{med}_{\frac{1}{2}}(Q_\gamma^{\min}(X_1, \ldots, X_n), Q_\gamma^{\max}(X_1, \ldots, X_n)), \qquad (26)$$

where we abbreviate

$$Q_\gamma^{\min}(X_1, \ldots, X_n) = \inf\{Q(Y_1, \ldots, Y_n) : Y_i \in \mathcal{T}_\gamma(X_i)\} \qquad (27)$$

$$Q_\gamma^{\max}(X_1, \ldots, X_n) = \sup\{Q(Y_1, \ldots, Y_n) : Y_i \in \mathcal{T}_\gamma(X_i)\}. \qquad (28)$$

This is apparent from Def. 31 and Def. 45. The fuzzy median can then be replaced with other connectives, e.g. the arithmetic mean $(x + y)/2$. An example of a DFS not based in the fuzzy median is the following:

Definition 59. *The QFM $\mathcal{F}_{Ch}$ is defined by*

$$\mathcal{F}_{Ch}(Q)(X_1, \ldots, X_n) = \tfrac{1}{2}\int_0^1 Q_\gamma^{\min}(X_1, \ldots, X_n)\, d\gamma + \tfrac{1}{2}\int_0^1 Q_\gamma^{\max}(X_1, \ldots, X_n)\, d\gamma,$$

for all $Q : \mathcal{P}(E)^n \longrightarrow \mathbf{I}$, $X_1, \ldots, X_n \in \widetilde{\mathcal{P}}(E)$.

Theorem 49. *$\mathcal{F}_{Ch}$ is a standard DFS.*

The DFS $\mathcal{F}_{Ch}$ is Q-continuous and arg-continuos, but it neither propagates fuzziness in quantifiers nor in arguments. This reveals that $\mathcal{F}_{Ch}$ is rather different from $\mathcal{M}_\mathcal{B}$-DFSes.

Definition 60. *Suppose $Q : \mathcal{P}(E) \longrightarrow \mathbf{I}$ is a nondecreasing semi-fuzzy quantifier and $X \in \widetilde{\mathcal{P}}(E)$. The* Choquet integral *$(Ch) \int X\, dQ$ is defined by*

$$(Ch) \int X\, dQ = \int_0^1 Q(X_{\geq \alpha})\, d\alpha.$$

Theorem 50. *Suppose $Q : \mathcal{P}(E) \longrightarrow \mathbf{I}$ is nondecreasing. Then for all $X \in \widetilde{\mathcal{P}}(E)$,*

$$(Ch) \int X\, dQ = \mathcal{F}_{Ch}(Q)(X).$$

Hence $\mathcal{F}_{Ch}$ coincides with the Choquet integral whenever the latter is defined.

Theorem 51. *Suppose $E \neq \varnothing$ is a finite base set, $q : \{0, \ldots, |E|\} \longrightarrow \mathbf{I}$ is a nondecreasing mapping such that $q(0) = 0$, $q(|E|) = 1$, and $Q : \mathcal{P}(E) \longrightarrow \mathbf{I}$ is defined by $Q(Y) = q(|Y|)$ for all $Y \in \mathcal{P}(E)$. Then for all $X \in \widetilde{\mathcal{P}}(E)$,*

$$\mathcal{F}_{Ch}(Q)(X) = \sum_{j=1}^{|E|}(q(j) - q(j-1)) \cdot \mu_{[j]}(X)\,,$$

i.e. $\mathcal{M}_{CX}$ consistently generalises Yager's OWA approach [25], see also [4].

6 Evaluation of fuzzy quantifiers in DFS

Let us now discuss the computational aspects and show how the fuzzy quantifiers $\mathcal{F}(Q)$ can be efficiently implemented.

6.1 Evaluation of "simple" quantifiers

Theorem 52. *In every standard DFS $\mathcal{F}$ and for all $E \neq \varnothing$,*

$$\begin{aligned}
&\mathcal{F}(\exists)(X) = \sup\{\mu_X(e) : e \in E\}\\
&\mathcal{F}(\forall)(X) = \inf\{\mu_X(e) : e \in E\}\\
&\mathcal{F}(\mathbf{all})(X_1, X_2) = \inf\{\max(1 - \mu_{X_1}(e), \mu_{X_2}(e)) : e \in E\}\\
&\mathcal{F}(\mathbf{some})(X_1, X_2) = \sup\{\min(\mu_{X_1}(e), \mu_{X_2}(e)) : e \in E\}\\
&\mathcal{F}(\mathbf{no})(X_1, X_2) = \inf\{\max(1 - \mu_{X_1}(e), 1 - \mu_{X_2}(e)) : e \in E\}\\
&\mathcal{F}(\mathbf{at\ least\ k})(X_1, X_2) = \sup\{\alpha \in \mathbf{I} : |(X_1 \cap X_2)_{\geq\alpha}| \geq k\}\,,
\end{aligned}$$

for all $X, X_1, X_2 \in \widetilde{\mathcal{P}}(E)$. In particular, if E is finite, then

$$\mathcal{F}(\mathbf{at\ least\ k})(X_1, X_2) = \mu_{[k]}(X_1 \cap X_2)$$

Note that **more than k** $=$ **at least k+1, less than k** $= 1 -$ **at least k** and **at most k** $=$ $1 -$ **more than k,** i.e. these quantifiers are covered by the formula for **at least k**.

6.2 Evaluation of quantitative one-place quantifiers

We first need some observations on quantitative one-place quantifiers.

Theorem 53. *A one-place semi-fuzzy quantifier $Q : \mathcal{P}(E) \longrightarrow \mathbf{I}$ on a finite base set $E \neq \varnothing$ is quantitative if and only if there exists a mapping $q : \{0, \ldots, |E|\} \longrightarrow \mathbf{I}$ such that $Q(Y) = q(|Y|)$, for all $Y \in \mathcal{P}(E)$. q is defined by*

$$q(j) = Q(Y_j) \tag{29}$$

for $j \in \{0, \ldots, |E|\}$, with $Y_j \in \mathcal{P}(E)$ an arbitrary subset of cardinality $|Y_j| = j$.

In particular, if the quantifier has extension, then there exists $\mu_Q : \mathbb{N} \longrightarrow \mathbf{I}$ such that for all finite base sets $E \neq \varnothing$, $q(j) = \mu_Q(j)$ for all $j \in \{0, \ldots, |E|\}$.

Theorem 54. *A quantitative one-place semi-fuzzy quantifier $Q : \mathcal{P}(E) \longrightarrow \mathbf{I}$ on a finite base set is convex if and only if there exists $j_{\mathrm{pk}} \in \{0, \ldots, m\}$ such that $q(\ell) \leq q(u)$ for all $\ell \leq u \leq j_{\mathrm{pk}}$, and $q(\ell) \geq q(u)$ for all $j_{\mathrm{pk}} \leq \ell \leq u$; where $q : \{0, \ldots, |E|\} \longrightarrow \mathbf{I}$ is the mapping defined by* (29).

Theorem 55. *A quantitative one-place semi-fuzzy quantifier $Q : \mathcal{P}(E) \longrightarrow \mathbf{I}$ on a finite base set is nondecreasing (nonincreasing) if and only if the mapping q defined by* (29) *is nondecreasing (nonincreasing).*

Let us now simplify the formulas for $Q_\gamma^{\min}(X_1, \ldots, X_n)$ and $Q_\gamma^{\max}(X_1, \ldots, X_n)$ in the case of a quantitative Q, cf. (26), (27), (28). Given a fuzzy subset $X \in \widetilde{\mathcal{P}}(E)$ of a finite base set $E \neq \varnothing$ and $\gamma \in \mathbf{I}$, let us abbreviate

$$|X|_\gamma^{\min} = |(X)_\gamma^{\min}| \tag{30}$$

$$|X|_\gamma^{\max} = |(X)_\gamma^{\max}| \tag{31}$$

For all $0 \leq \ell \leq u \leq |E|$, we further define

$$q^{\min}(\ell, u) = \min\{q(k) : \ell \leq k \leq u\} \tag{32}$$

$$q^{\max}(\ell, u) = \max\{q(k) : \ell \leq k \leq u\}\,. \tag{33}$$

Theorem 56. *For every quantitative one-place semi-fuzzy quantifier $Q : \mathcal{P}(E) \longrightarrow \mathbf{I}$ on a finite base set, all $X \in \widetilde{\mathcal{P}}(E)$ and $\gamma \in \mathbf{I}$,*

$$\begin{aligned} Q_\gamma^{\min}(X) &= q^{\min}(\ell, u) \\ Q_\gamma^{\max}(X) &= q^{\max}(\ell, u) \\ Q_\gamma(X) &= \mathrm{med}_{\frac{1}{2}}(q^{\min}(\ell, u), q^{\max}(\ell, u))\,, \end{aligned}$$

abbreviating $\ell = |X|_\gamma^{\min}$ and $u = |X|_\gamma^{\max}$.

It is then apparent from (Th-54) and (Th-55) that whenever Q is *convex*, then

$$q^{\min}(\ell, u) = \min(q(\ell), q(u))$$

$$q^{\max}(\ell, u) = \begin{cases} q(\ell) & : \ \ell > j_{\mathrm{pk}} \\ q(u) & : \ u < j_{\mathrm{pk}} \\ q(j_{\mathrm{pk}}) & : \ \ell \leq j_{\mathrm{pk}} \leq u \end{cases}$$

and if Q is *monotonic*, then

$$q^{\min}(\ell, u) = q(\ell), \quad q^{\max}(\ell, u) = q(u) \qquad \text{if } Q \text{ nondecreasing}$$

$$q^{\min}(\ell, u) = q(u), \quad q^{\max}(\ell, u) = q(\ell) \qquad \text{if } Q \text{ nonincreasing.}$$

In the case of the DFS $\mathcal{M}_{CX}$, we can use the following *fuzzy interval cardinality* to evaluate quantitative one-place quantifiers.

Definition 61. *For every fuzzy subset* $X \in \widetilde{\mathcal{P}}(E)$, *the* fuzzy interval cardinality $\|X\|_{\mathrm{iv}} \in \widetilde{\mathcal{P}}(\mathbb{N} \times \mathbb{N})$ *is defined by*

$$\mu_{\|X\|_{\mathrm{iv}}}(\ell, u) = \begin{cases} \min(\mu_{[\ell]}(X), 1 - \mu_{[u+1]}(X)) & : \ \ell \leq u \\ 0 & : \ else \end{cases} \quad \textit{for all } \ell, u \in \mathbb{N}. \tag{34}$$

Theorem 57. *For every quantitative one-place quantifier* $Q : \mathcal{P}(E) \longrightarrow \mathbf{I}$ *on a finite base set and all* $X \in \widetilde{\mathcal{P}}(E)$,

$$\begin{aligned} \mathcal{M}_{CX}(Q)(X) &= \max\{\min(\mu_{\|X\|_{\mathrm{iv}}}(\ell, u), q^{\min}(\ell, u)) : 0 \leq \ell \leq u \leq |E|\} \\ &= \min\{\max(1 - \mu_{\|X\|_{\mathrm{iv}}}(\ell, u), q^{\max}(\ell, u)) : 0 \leq \ell \leq u \leq |E|\}. \end{aligned}$$

For other $\mathcal{M}_{\mathcal{B}}$-DFSes and $\mathcal{F}_{Ch}$, a histogram-based approach can be used to efficiently implement the resulting quantifiers. For simplicity of presentation, we will describe a computation procedure suited to integer-arithmetics. We hence assume that, for a fixed $m' \in \mathbb{N} \setminus \{0\}$, all membership values of fuzzy argument sets $X_1, \ldots, X_n$ satisfy

$$\mu_{X_i}(e) \in \left\{0, \frac{1}{m'}, \ldots, \frac{m'-1}{m'}, 1\right\} \tag{35}$$

for all $e \in E$. If $X \in \widetilde{\mathcal{P}}(E)$ satisfies (35), we can represent the required histogram of X as an $(m'+1)$-dimensional array $\mathrm{Hist}_X : \{0, \ldots, m'\} \longrightarrow \mathbb{N}$, defined by

$$\mathrm{Hist}_X[j] = \left|\{e \in E : \mu_X(e) = \frac{j}{m'}\}\right|$$

for all $j = 0, \ldots, m'$. We further assume that m' is even, (i.e. $m' = 2m$ for a given $m \in \mathbb{N} \setminus \{0\}$). The computation procedures for the DFSes $\mathcal{M}$ and $\mathcal{F}_{Ch}$ are presented in Table 1. In the algorithm for $\mathcal{M}$, we have utilized that $Q_\gamma(X) = \max(\frac{1}{2}, q^{\min}(\ell, u))$ if $Q_0(X) > \frac{1}{2}$ and $Q_\gamma(X) = \min(\frac{1}{2}, q^{\max}(\ell, u))$ otherwise. A further simplification is possible if Q is monotonic. For example, if Q is nondecreasing, then $q^{\min}(\ell, u) = q(\ell)$ and $q^{\max}(\ell, u) = q(u)$, i.e. we can omit the updating of u in the first for-loop and likewise omit ℓ in the second for-loop.

6.3 Evaluation of absolute quantifiers and quantifiers of exception

Definition 62. *For every two-place semi-fuzzy quantifier* $Q : \mathcal{P}(E)^2 \longrightarrow \mathbf{I}$,

- Q *is* absolute *iff there exists a quantitative one-place quantifier* $Q' : \mathcal{P}(E) \longrightarrow \mathbf{I}$ *such that* $Q = Q'\cap$, *i.e.* $Q(Y_1, Y_2) = Q'(X_1 \cap X_2)$ *for all* $Y_1, Y_2 \in \mathcal{P}(E)$.
- Q *is called a* quantifier of exception *iff there exists an absolute quantifier* Q'' : $\mathcal{P}(E)^2 \longrightarrow \mathbf{I}$ *with* $Q = Q''\neg$, *i.e.* $Q(Y_1, Y_2) = Q''(X_1, \neg X_2)$ *for* $Y_1, Y_2 \in \mathcal{P}(E)$.

For example, the two-place quantifier **about 50** is an absolute quantifier. Some examples of quantifiers of exception are presented in Table 2. The DFS axioms ensure that

Algorithm for computing $\mathcal{M}(Q)(X)$	Algorithm for computing $\mathcal{F}_{Ch}(Q)(X)$

Algorithm for computing $\mathcal{M}(Q)(X)$

```
INPUT: X
// initialise H, ℓ, u
H := Hist_X;
ℓ := Σ_{j=1}^{m} H[m+j];
u := ℓ + H[m];
cq := med_{1/2}(q^min(ℓ,u), q^max(ℓ,u));
if( cq == 1/2 ) return 1/2;
sum := cq;
if( cq > 1/2 )
    for( j := 1; j<m; j := j + 1 ) {
        nc:= true; // "no change"
        // update clauses for ℓ and u
        if( H[m+j] ≠ 0 )
            { ℓ := ℓ - H[m+j]; nc:= false; }
        if( H[m-j] ≠ 0 )
            { u := u + H[m-j]; nc:= false; }
        if( nc)
            { sum := sum + cq; continue; }
        // one of ℓ or u has changed
        cq := q^min(ℓ,u);
        if( cq ≤ 1/2 ) break;
        sum := sum + cq;
        }
else
    for( j := 1; j<m; j := j + 1 ) {
        nc:= true;
        .
        : // update clauses etc. as above
        // one of ℓ or u has changed
        cq := q^max(ℓ,u);
        if( cq ≥ 1/2 ) break;
        sum := sum + cq;
        }
return (sum + 1/2*(m-j)) / m;
END
```

Algorithm for computing $\mathcal{F}_{Ch}(Q)(X)$

```
INPUT: X
// initialise H, ℓ, u
H := Hist_X;
ℓ := Σ_{j=1}^{m} H[m+j];
u := ℓ + H[m];
cq := q^min(ℓ,u) + q^max(ℓ,u);
sum := cq;
for( j := 1; j<m; j := j + 1 ) {
    ch := false; // "change"
    // update clauses for ℓ and u
    if( H[m+j] ≠ 0 )
        { ℓ := ℓ - H[m+j]; ch := true; }
    if( H[m-j] ≠ 0 )
        { u := u + H[m-j]; ch := true; }
    if( ch )
        // one of ℓ or u has changed
        { cq := q^min(ℓ,u) + q^max(ℓ,u); }
    sum := sum + cq;
    }
return sum / m'; // where m' = 2*m
END
```

Table1. Algorithms for evaluating quantitative one-place quantifiers

$\mathcal{F}(Q)(X_1, X_2) = \mathcal{F}(Q')(X_1 \cap X_2)$, whenever $Q = Q'\cap$ is absolute and $X_1, X_2 \in \widetilde{\mathcal{P}}(E)$. Similarly if $Q = Q'\cap\neg$ is a quantifier of exception, then $\mathcal{F}(Q)(X_1, X_2) = \mathcal{F}(Q')(X_1 \cap \neg X_2)$, for all $X_1, X_2 \in \widetilde{\mathcal{P}}(E)$, where $Q' : \mathcal{P}(E) \longrightarrow \mathbf{I}$ is quantitative. We can hence use the algorithm for $\mathcal{F}(Q')(X)$, $\mathcal{F} \in \{\mathcal{M}_{CX}, \mathcal{M}, \mathcal{F}_{Ch}\}$ to evaluate absolute quantifiers and quantifiers of exception.

6.4 Evaluation of proportional quantifiers

Definition 63. *A two-place semi-fuzzy quantifier* $Q : \mathcal{P}(E)^2 \longrightarrow \mathbf{I}$ *on a finite base set is called* proportional *if there exist* $v_0 \in \mathbf{I}$, $f : \mathbf{I} \longrightarrow \mathbf{I}$ *such that*

$$Q(Y_1, Y_2) = \begin{cases} f(|Y_1 \cap Y_2|/|Y_1|) & : \; Y_1 \neq \varnothing \\ v_0 & : \; else \end{cases} \qquad \textit{for all } Y_1, Y_2 \in \mathcal{P}(E).$$

For example, we have provided a definition of **almost all** where $f(z) = S(x, 0.7, 0.9)$ and $v_0 = 1$, see equation (1). Usually f and v_0 can be chosen independently of E, i.e. Q has extension. We shall restrict our attention to those proportional quantifiers where

Quantifier	**Antonym (absolute)**
all	**no**
all except exactly k	**exactly k**
all except about k	**about k**
all except at most k	**at most k**

Table2. Examples of quantifiers of exception

$f : \mathbf{I} \longrightarrow \mathbf{I}$ is nondecreasing.[13] Suppose Q is such a quantifier and $X_1, X_2 \in \widetilde{\mathcal{P}}(E)$. We are using abbreviations $Z_1 = X_1$, $Z_2 = X_1 \cap X_2$ and $Z_3 = X_1 \cap \neg X_2$; further let $\ell_k = |Z_k|_\gamma^{\min}$ and $u_k = |Z_k|_\gamma^{\max}$, $k \in \{1, 2, 3\}$, $f^{\min} = f(\ell_2/(\ell_2 + u_3))$ and $f^{\max} = f(u_2/(u_2 + \ell_3))$. Then

$$Q_\gamma(X_1, X_2) = \mathrm{med}_{\frac{1}{2}}(q^{\min}(\ell_1, \ell_2, u_1, u_3), q^{\max}(\ell_1, \ell_3, u_1, u_2)).$$

For the definitions of $q^{\min}, q^{\max} : \{0, \dots, |E|\}^4 \longrightarrow \mathbf{I}$ and the actual algorithms for evaluating proportional quantifiers, see Table 3 and 4 at the end of the paper. As shown in Table 4.a, a slight modification of the algorithm for $\mathcal{M}(Q)(X)$ in the case of one-place quantitative quantifiers is sufficient to compute $\mathcal{M}(Q)(X_1, X_2)$ for proportional quantifiers. 4.b depicts the algorithm for evaluating $\mathcal{M}_{CX}(Q)(X_1, X_2)$ in proportional case. The algorithm for computing $\mathcal{F}_{Ch}(Q)(X)$ can be adapted in a similar fashion to implement $\mathcal{F}_{Ch}(Q)(X_1, X_2)$ for proportional Q.

7 Conclusion

In this work, we have argued that fuzzy quantification is one of the enabling techniques for granular computing. We have focused on the task of developing an axiomatic theory in order to guarantee predictable and linguistically well-motivated results. To this end, we have introduced the framework of DFS theory by presenting an independent axiom system for "reasonable" approaches to fuzzy quantification, that are consistent with the use of quantifiers in NL. We have formalized a number of linguistic adequacy criteria and shown that every model of our axioms exhibits these essential properties. However, some principled adequacy bounds for approaches to fuzzy quantification have also been established, which in most cases result from the known conflict between idempotence/distributivity and the law of contradiction in the presence of fuzziness. We have also presented a broad class of models of the axiomatic framework. In particular, the model $\mathcal{M}_{CX}$, which consistently generalises the Sugeno integral (and hence the FG-count approach) to non-monotonic and multiplace quantifiers, can be shown to exhibit unique adequacy properties. Our analysis of $\mathcal{M}_{CX}$ has unvealed the fuzzy interval cardinality $\|\bullet\|_{\mathrm{iv}}$, which is the first definition of fuzzy cardinality to yield adequate results with arbitrary quantitative one-place quantifiers. In addition, we have defined a model which generalises the Choquet integral (and hence the OWA approach)

[13] if f is nonincreasing, we can compute $\mathcal{F}(Q) = \neg\mathcal{F}(\neg Q)$, noting that the negation $\neg Q$ is proportional and nondecreasing.

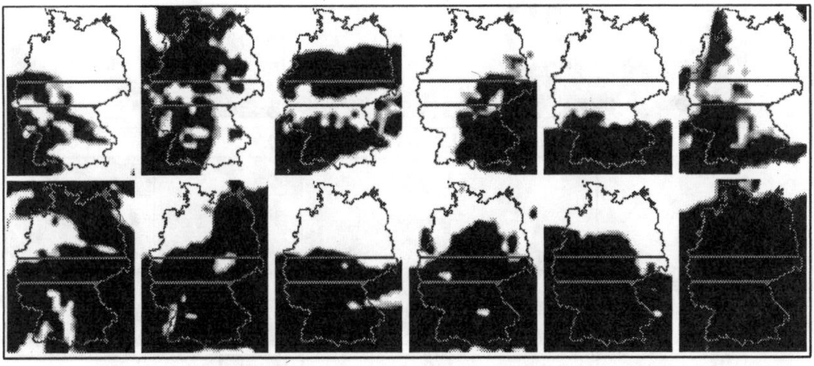

Figure6. Cloudy in Southern Germany (relevant images on top)

to non-monotonic and multiplace quantifiers. Finally, algorithms for evaluating quantifying expressions in these models have been presented.

The model denoted $\mathcal{M}$ is being applied in a multimedia retrieval system for meteorological documents [12]. Fig. 6 depicts a ranking of cloudiness situations according to the criterion "cloudy in Southern Germany". The ranking has been computed by evaluating $\mathcal{M}(\mathbf{prop})(\mathbf{SouthernGermany}, \mathbf{cloudy})$, where $\mathbf{prop}(Y_1, Y_2) = |Y_1 \cap Y_2|/|Y_1|$. In order to determine the ranking, a quantification result had to be computed for each image, based on the domain of 219063 (411x533) pixel coordinates. This took approx. 0.03 seconds per image on an AMD Athlon 600 MHz PC running Linux (using 8 bit arithmetics). Fig. 7 presents the results of evaluating several accumulative conditions on a set of cloudiness images. These conditions are of the type "Q-times cloudy in the last days". A trapezoidal proportional quantifier $\mathbf{trp}_{a,b,c} : \mathcal{P}(E)^n \longrightarrow \mathbf{I}$, defined by

$$\mathbf{trp}_{a,b,c}(Y_1, Y_2) = \begin{cases} t_{a,b}(|Y_1 \cap Y_2|/|Y_1|) & : \quad Y_1 \neq \varnothing \\ c & : \quad Y_1 = \varnothing \end{cases} \qquad t_{a,b}(z) = \begin{cases} 0 & : \quad z < a \\ \frac{z-a}{b-a} & : \quad a \leq z \leq b \\ 1 & : \quad z > b \end{cases}$$

has been used to model some of the quantifiers. The resulting image R has pixel intensities $R_p = \mathcal{M}(Q)(X_1, X_{2,p})$, for all pixels p where E is the set of images, $\mu_{X_1}(e)$ expresses the degree to which an image $e \in E$ belongs to the fuzzy time interval "in the last days", and $\mu_{X_{2,p}}(e)$ is the degree to which pixel p is cloudy in image e. In this case, 219063 quantification results involving a domain of 8 images had to be computed for each result image (i.e. one quantifying expression for each of the 411x533 pixels). This took a total of about 5 seconds per result image on the same computer platform as above.

The result images in Fig. 7 can be considered as a description of the same underlying phenomenon (the given time sequence of images) from different perspectives, as expressed by distinct fuzzy quantifiers. When the quantifiers which measure the accumulative properties are carefully chosen, it should be possible to capture the important characteristics of a complex sequence in a small set of summarizing images of even of scalar evaluations. In order to achieve this, the users of a data summarization system

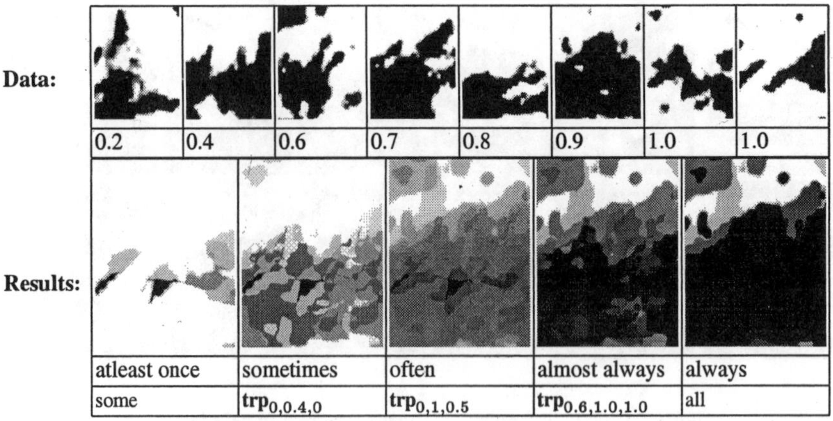

Figure7. Cloudy in the last days: Summarisation results for some choices of the quantifier Q.

should provide feedback and adjust the operators by manipulating the graphical representation of parametrized fuzzy quantifiers (like in Fig. 5). This will allow the users to enter the process of data summarization and to optimize the system-generated descriptions of granules (images regions etc.) in terms of their accumulative properties.

In addition, it would be instructive to evaluate the present approach from the perspective of real users. A bibliographic retrieval system which utilizes DFS theory to evaluate weighted quantifying search criteria ("many important search terms apply to the document") is being constructed which serves to carry out such an experiment [13]. The retrieval system collects relevance judgements from its users which will be used for an empirical comparison of rivaling approaches to fuzzy quantification and of different DFS models, in particular $\mathcal{M}_{CX}$ and $\mathcal{F}_{Ch}$.

References

1. J. Barwise and R. Cooper. Generalized quantifiers and natural language. *Linguistics and Philosophy*, 4:159–219, 1981.
2. J. van Benthem. Questions about quantifiers. *J. of Symb. Logic*, 49, 1984.
3. I. Bloch. Information combination operators for data fusion: a comparative review with classification. *IEEE Transactions on Systems, Man, and Cybernetics*, 26(1):52–67, 1996.
4. P. Bosc and L. Lietard. Monotonic quantified statements and fuzzy integrals. In *Proc. of the NAFIPS/IFI/NASA '94 Joint Conference*, pages 8–12, San Antonio, Texas, 1994.
5. Z.P. Dienes. On an implication function in many-valued systems of logic. *Journal of Symbolic Logic*, 14:95–97, 1949.
6. D. Dubois, H. Prade, and R.R. Yager, editors. *Fuzzy Information Engineering*. Wiley, New York, 1997.
7. B.R. Gaines. Foundations of fuzzy reasoning. *Int. J. Man-Machine Studies*, 8:623–668, 1978.
8. I. Glöckner. Advances in DFS theory. TR2000-01, Technical Faculty, University Bielefeld, 33501 Bielefeld, Germany, 2000.

9. I. Glöckner. A broad class of standard DFSes. TR2000-02, Technical Faculty, University Bielefeld, 33501 Bielefeld, Germany, to appear.
10. I. Glöckner. DFS – an axiomatic approach to fuzzy quantification. TR97-06, Technical Faculty, University Bielefeld, 33501 Bielefeld, Germany, 1997.
11. I. Glöckner. A framework for evaluating approaches to fuzzy quantification. TR99-03, Technical Faculty, University Bielefeld, 33501 Bielefeld, Germany, 1999.
12. I. Glöckner and A. Knoll. Query evaluation and information fusion in a retrieval system for multimedia documents. In *Proceedings of Fusion'99*, pages 529–536, Sunnyvale, CA, 1999.
13. I. Glöckner and A. Knoll. Architecture and retrieval methods of a search assistant for scientific libraries. In W. Gaul and R. Decker, editors, *Classification and Information Processing at the Turn of the Millennium*. Springer, Heidelberg, 2000.
14. G.J. Klir and B. Yuan. *Fuzzy Sets and Fuzzy Logic: Theory and Applications*. Prentice Hall, Upper Saddle River, NJ, 1995.
15. Y. Liu and E.E. Kerre. An overview of fuzzy quantifiers. (I). interpretations. *Fuzzy Sets and Systems*, 95:1–21, 1998.
16. M. Mukaidono. On some properties of fuzzy logic. *Syst.—Comput.—Control*, 6(2):36–43, 1975.
17. A.L. Ralescu. A note on rule representation in expert systems. *Information Sciences*, 38:193–203, 1986.
18. D. Ralescu. Cardinality, quantifiers, and the aggregation of fuzzy criteria. *Fuzzy Sets and Systems*, 69:355–365, 1995.
19. D. Rasmussen and R. Yager. A fuzzy SQL summary language for data discovery. In Dubois et al. [6], pages 253–264.
20. B. Schweizer and A. Sklar. *Probabilistic metric spaces*. North-Holland, Amsterdam, 1983.
21. W. Silvert. Symmetric summation: A class of operations on fuzzy sets. *IEEE Transactions on Systems, Man, and Cybernetics*, 9:657–659, 1979.
22. H. Thiele. On T-quantifiers and S-quantifiers. In *The Twenty-Fourth International Symposium on Multiple-Valued Logic*, pages 264–269, Boston, MA, 1994.
23. R.R. Yager. Quantified propositions in a linguistic logic. *Int. J. Man-Machine Studies*, 19:195–227, 1983.
24. R.R. Yager. Approximate reasoning as a basis for rule-based expert systems. *IEEE Trans. on Systems, Man, and Cybernetics*, 14(4):636–643, Jul./Aug. 1984.
25. R.R. Yager. On ordered weighted averaging aggregation operators in multicriteria decisionmaking. *IEEE Trans. on Systems, Man, and Cybernetics*, 18(1):183–190, 1988.
26. R.R. Yager. Connectives and quantifiers in fuzzy sets. *Fuzzy Sets and Systems*, 40:39–75, 1991.
27. R.R. Yager. Counting the number of classes in a fuzzy set. *IEEE Trans. on Systems, Man, and Cybernetics*, 23(1):257–264, 1993.
28. R.R. Yager. Families of OWA operators. *Fuzzy Sets and Systems*, 59:125–148, 1993.
29. L.A. Zadeh. The concept of a linguistic variable and its application to approximate reasoning. *Information Sciences*, 8,9:199–249,301–357, 1975.
30. L.A. Zadeh. A theory of approximate reasoning. In J. Hayes, D. Michie, and L. Mikulich, editors, *Mach. Intelligence*, volume 9, pages 149–194. Halstead, New York, 1979.
31. L.A. Zadeh. A computational approach to fuzzy quantifiers in natural languages. *Computers and Math. with Appl.*, 9:149–184, 1983.
32. L.A. Zadeh. Syllogistic reasoning in fuzzy logic and its application to usuality and reasoning with dispositions. *IEEE Trans. on Systems, Man, and Cybernetics*, 15(6):754–763, 1985.

1. $\ell_1 > 0$. Then $q^{\min} = f^{\min}$. 2. $\ell_1 = 0$. a. $\ell_2 + u_3 > 0$. Then $q^{\min} = \min(v_0, f^{\min})$. b. $\ell_2 + u_3 = 0$. i. $u_1 > 0$. Then $q^{\min} = \min(v_0, f(1))$. ii. $u_1 = 0$. Then $q^{\min} = v_0$. *Note.* If $v_0 \leq f(1)$, then $\min(v_0, f(1)) = v_0$, i.e. we need not distinguish 2.b.i and 2.b.ii.	For $q^{\max}(\ell_1, \ell_3, u_1, u_2)$, we have: 1. $\ell_1 > 0$. Then $q^{\max} = f^{\max}$. 2. $\ell_1 = 0$. a. $u_2 + \ell_3 > 0$. Then $q^{\max} = \max(v_0, f^{\max})$. b. $u_2 + \ell_3 = 0$. i. $u_1 > 0$. Then $q^{\max} = \max(v_0, f(0))$. ii. $u_1 = 0$. Then $q^{\max} = v_0$. *Note.* If $f(0) \leq v_0$, then 2.b.i and 2.b.ii need not be distinguished.

Table3. Computation of $q^{\min}$ and $q^{\max}$ for proportional quantifiers

a. Algorithm for computing $\mathcal{M}(Q)(X_1, X_2)$

```
INPUT: X_1, X_2
// initialise H_k, l, u
H_1 := Hist_{X_1};
H_2 := Hist_{X_1 ∩ X_2};
H_3 := Hist_{X_1 ∩ ¬X_2};
for( k := 1; k ≤ 3; k := k+1 ) {
    l_k := Σ_{j=1}^{m} H_k[m+j]; u_k := l_k + H_k[m];
    }
cq := med_{1/2}(q^min(l_1, l_2, u_1, u_3), q^max(l_1, l_3, u_1, u_2))
if( cq == 1/2 ) return 1/2;
sum := cq;
if( cq > 1/2 )
    for( j := 1; j<m; j := j + 1 ) {
        nc:= true; // "no change"
        // update clauses for l_1, l_2, u_1, u_3
        if( H_1[m+j] ≠ 0 )
            { l_1 := l_1 - H_1[m+j]; nc:= false; }
        if( H_2[m+j] ≠ 0 )
            { l_2 := l_2 - H_2[m+j]; nc:= false; }
        if( H_1[m-j] ≠ 0 )
            { u_1 := u_1 + H_1[m-j]; nc:= false; }
        if( H_3[m-j] ≠ 0 )
            { u_3 := u_3 + H_3[m-j]; nc:= false; }
        if( nc)
            { sum := sum + cq; continue; }
        // one of l_1, l_2, u_1, u_3 has changed
        cq := q^min(l_1, l_2, u_1, u_3);
        if( cq ≤ 1/2 ) break;
        sum := sum + cq;
        }
else
    for( j := 1; j<m; j := j + 1 ) {
        nc:= true;
        // update clauses for l_1, l_3, u_1, u_2
        if( H_1[m+j] ≠ 0 )
            { l_1 := l_1 - H_1[m+j]; nc:= false; }
        if( H_3[m+j] ≠ 0 )
            { l_3 := l_3 - H_3[m+j]; nc:= false; }
        if( H_1[m-j] ≠ 0 )
            { u_1 := u_1 + H_1[m-j]; nc:= false; }
        if( H_2[m-j] ≠ 0 )
            { u_2 := u_2 + H_2[m-j]; nc:= false; }
        if( nc)
            { sum := sum + cq; continue; }
        // one of l_1, l_3, u_1, u_2 has changed
        cq := q^max(l_1, l_3, u_1, u_2);
        if( cq ≥ 1/2 ) break;
        sum := sum + cq;
        }
return (sum + 1/2*(m-j)) / m;
END
```

b. Algorithm for computing $\mathcal{M}_{CX}(Q)(X_1, X_2)$

```
INPUT: X_1, X_2
// initialise H_k, l, u
H_1 := Hist_{X_1};
H_2 := Hist_{X_1 ∩ X_2};
H_3 := Hist_{X_1 ∩ ¬X_2};
for( k := 1; k ≤ 3; k := k+1 ) {
    l_k := Σ_{j=1}^{m} H_k[m+j]; u_k := l_k + H_k[m];
    }
cq := med_{1/2}(q^min(l_1, l_2, u_1, u_3), q^max(l_1, l_3, u_1, u_2))
if( cq == 1/2 ) return 1/2;
sum := cq;
if( cq > 1/2 )
{
    for( j := 1; j<m; j := j + 1 ) {
        ch := false; // "change"
        // update clauses for l_1, l_2, u_1, u_3
        if( H_1[m+j] ≠ 0 )
            { l_1 := l_1 - H_1[m+j]; ch := true; }
        if( H_2[m+j] ≠ 0 )
            { l_2 := l_2 - H_2[m+j]; ch := true; }
        if( H_1[m-j] ≠ 0 )
            { u_1 := u_1 + H_1[m-j]; ch := true; }
        if( H_3[m-j] ≠ 0 )
            { u_3 := u_3 + H_3[m-j]; ch := true; }
        if( ch )
            // one of l_1, l_2, u_1, u_3 has changed
            { cq := q^min(l_1, l_2, u_1, u_3); }
        if( cq ≤ m+j )
            { return (m+j)/m' }
        }
    return 1;
}
else
    for( j := 1; j<m; j := j + 1 ) {
        ch := false;
        // update clauses for l_1, l_3, u_1, u_2
        if( H_1[m+j] ≠ 0 )
            { l_1 := l_1 - H_1[m+j]; ch := true; }
        if( H_3[m+j] ≠ 0 )
            { l_3 := l_3 - H_3[m+j]; ch := true; }
        if( H_1[m-j] ≠ 0 )
            { u_1 := u_1 + H_1[m-j]; ch := true; }
        if( H_2[m-j] ≠ 0 )
            { u_2 := u_2 + H_2[m-j]; ch := true; }
        if( ch )
            // one of l_1, l_3, u_1, u_2 has changed
            { cq := q^max(l_1, l_3, u_1, u_2); }
        if( cq ≥ m-j )
            { return (m-j)/m'; }
        }
return 0;
END
```

Table4. Algorithms for evaluating two-place proportional quantifiers

Granularity and Specificity in Fuzzy Rule-Based Systems

Thomas Sudkamp

Department of Computer Science, Wright State University, Dayton OH 45435, USA

Abstract. The structure of fuzzy models produced by a heursitic analysis of the problem domain is compared with that of models algorithmically generated from training data. The trade-offs between granularity, specificity, interpretability, and efficiency are examined for rule-bases produced in each of these manners. An algorithm that combines rule learning with region merging is introduced to incorporate beneficial features of both the heuristic and learning approaches to producing fuzzy models.

1 Introduction

Fuzzy set theoretic techniques have been successfully employed to construct rule-based systems for classification, diagnostic reasoning, function approximation, and automatic control applications. A rule encapsulates relationships among the components of the underlying system and the combination of the rule-base and the inference procedure produces a model of the system. The resulting model can then be used for prediction, explanation, or decision making. The utility of a model is determined by several sometimes competing factors: its granularity, its specificity, its interpretability, it efficiency, and its ability to generalize information.

Two strategies are commonly employed for constructing fuzzy models: obtaining rules heuristically from experts in the application area and generating rules from training data. Experts use linguistic terms common to the problem domain to describe states of the system and actions that are suitable responses for particular system configurations. A rule-base produced in this manner generally contains a small number of highly granular rules with each rule covering a large number of potential situations.

When learning algorithms are used to generate a rule-base from a set of known instances, no semantic interpretation of the input or output domains is guaranteed. The fuzzy sets are often selected based on the distribution of the training data or for the efficiency of the calculations required to evaluate input and produce a response. When sufficient data are available, learning algorithms may construct precise rules with a limited range of applicability.

This paper begins by describing characteristics of fuzzy rule-bases designed for system modeling and function approximation. In particular, we consider the granularity, specificity, and the run-time efficiency of fuzzy models. Based on these characteristics, we will compare the structure of models produced via heursitic analysis of the application with that of models algorithmically generated from training data. Finally we examine two rule learning strategies: the first exhibits the role of granularity in learning rules from training data and the second combines rule learning with region merging to incorporate beneficial features of both the heuristic and learning approaches to producing fuzzy models.

2 Characteristics of rule-bases

A fuzzy rule expresses an approximate relationship between the input domains and the output domain. The antecedent of a rule defines conditions under which the rule is applicable while the consequent gives an approximate response for these conditions. The two predominant types of fuzzy rules differ in the form of the consequent. In Mamdani style rules [1] the consequent is a fuzzy subset of the output domain while the consequent of a Takagi-Sugeno-Kang (TSK) style rule [2] is a function of the input values.

Granularity, specificity, interpretability, and efficiency are features that affect the utility of a fuzzy model. Each of these properties is determined by the constitutents of the antecedents and consequents of the rules. The analysis of properties of models in this section and algorithms presented in next section will consider models defined using Mamdani style rules. The number of input variables is referred to as the dimension of the model. Throughout the paper we will consider models with one or two dimensions and all domains will be normalized to the interval $[-1, 1]$. The definitions, relationships, and algorithms described are immediately extendable to higher dimensional models.

2.1 Term sets, decompositions, and rules

An initial step in the construction of a fuzzy model is the selection of a language to describe properties or features of the input and output domains. The elements of this language are then used in the antecedents and consequents of the rules. The terms in the language define the granularity of the model. Zadeh [3] describes a granule as "points drawn together by indistinguishability, similarity, proximity or functionality." In fuzzy modeling, a subset of the input domain is considered to be a granule when the elements in the set represent similar input conditions and form the support of the antecedent of a rule.

The heuristic approach to rule-base construction uses similarity within a domain to produce a reasonably small number of categories that cover the entire domain. This is accomplished by choosing a set of linguistic terms [4] that define basic characteristics of the domain. For example, a linguistic variable *speed* may be used to describe speed of an automobile. The universe of this domain consists of possible automobile speeds. A term set for the

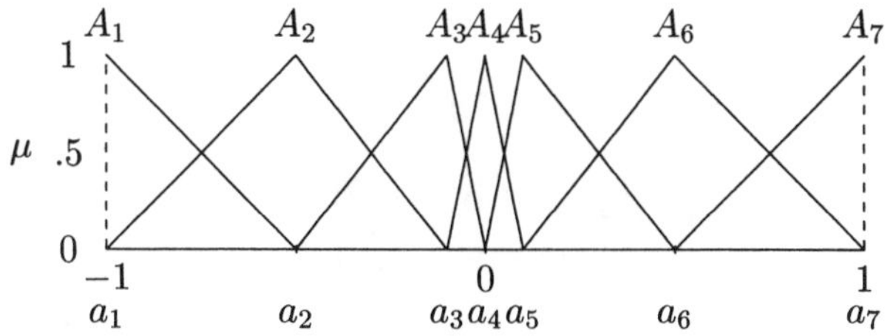

Figure 1: Domain decomposition

variable *speed* may contain general descriptions such as *fast*, *average*, and *slow* as well as more specific descriptions such as *about-50-mph*. Associated with each linguistic term is a fuzzy set that defines the concept denoted by the term. The language of an application consists of all the linguistic terms used to describe regions of the input and output universes.

Obtaining a domain decomposition via the heuristic selection of linguistic terms was popularized by the success of fuzzy rule-based systems for automatic control applications [1, 5, 6]. In fuzzy control systems, as in fuzzy modeling in general, the additional condition that a domain decomposition comprise a fuzzy partition with .5 coverage is frequently required [7]. A fuzzy partition of a domain U consists of a sequence of fuzzy sets $A_1, \dots, A_n$ that satisfies

$$\sum_{i=1}^{n} \mu_{A_i}(u) = 1$$

for every $u \in U$. A decomposition of U has .5 coverage if, for each $u \in U$, there is at least one fuzzy set A_i in the decomposition with $\mu_{A_i}(u) \geq .5$. In rule-based inference, this ensures that there is at least one rule that significantly matches each possible input.

Combining fuzzy partitioning and .5 coverage with the simplicity of triangular or trapezodial membership functions produces the type of domain decomposition routinely employed in control applications (Figure 1). The examples throughout this paper will use triangular decompositions as shown in Figure 1. The peak point of a triangular fuzzy set A_i is the unique point a_i for which $\mu_{A_i}(a_i) = 1$. In a triangular decomposition, the peak points $a_1, \dots, a_n$ completely determine the membership functions of the fuzzy sets in the decomposition.

2.2 Rule-bases and models

The role of rules in fuzzy modeling is demonstrated by considering a system with input domain U and output domain W. A Mamdani style rule for such a system has the form

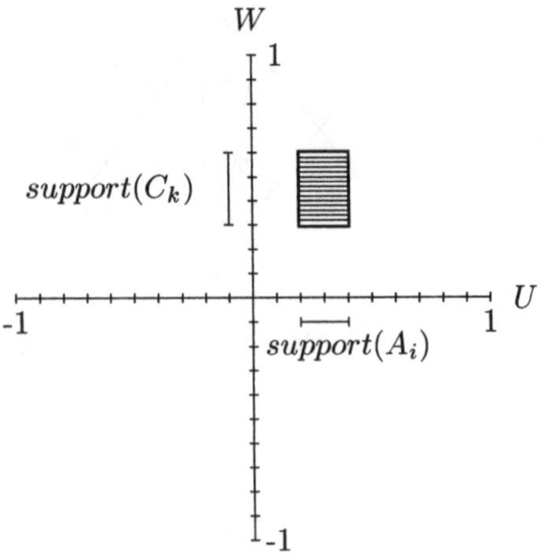

Figure 2: Relation defined by $A_i \times C_k$.

$$\text{`if } X \text{ is } A_i \text{ then } Z \text{ is } C_k\text{'},$$

where A_i and C_k are fuzzy sets from the decompositions of the input and output domains, respectively. The rule defines a fuzzy relation $A_i \times C_k$ over $U \times W$ where $\mu_{A_i \times C_k}(x, y) = \min\{\mu_{A_i}(x), \mu_{C_k}(y)\}$. The antecedent of a rule defines its region of applicability in the input domain. The relation generated by the antecedent and the consequent defines a subspace of the product space $U \times W$ (Figure 2).

A system model is obtained by combining the relations generated by the rules. Requiring the decomposition to partition the input domain ensures that the union of the relations covers the entire input domain. Figure 3 depicts a model defined by the union of relations, where the rectangles represent the relations defined by the individual rules. Continuity is frequently required in the construction of rule bases; this condition is satisfied when the adjacent relations overlap as in Figure 3.

In modeling with fuzzy rules, the rule-base defines a mapping from the input domain to fuzzy sets over the output domain. That is, the model is a function $\tilde{f}$ from U to $\mathcal{F}(W)$, where $\mathcal{F}(W)$ is the set of fuzzy sets over W. For an input value u, the output is determined by the set of rules whose antecedents match u to some nonzero degree. Assume that the rules are written 'if X is A_i then Z is C_{k_i}', for $i = 1, \ldots, n$. For an input value u, let $R = \{i \mid u \in \mathit{support}(A_i)\}$ be the set of indices of the applicable rules. The rules indicate that the response for input u should be within the range

$$\bigcup_{i \in R} \mathit{support}(C_{k_i}).$$

The output of the model is a fuzzy set C whose membership values are determined by a disjunction of the consequents of the applicable rules.

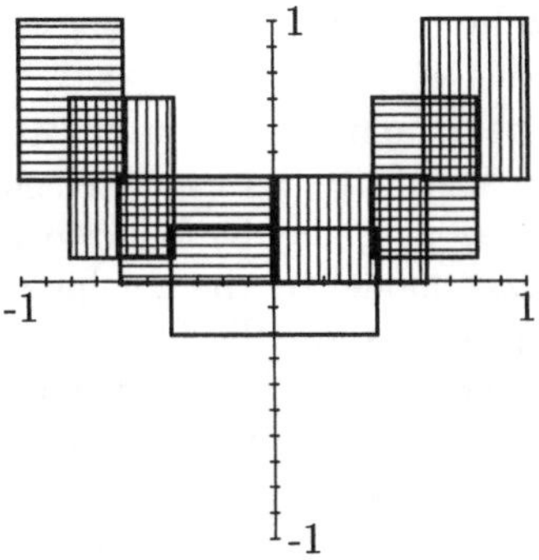

Figure 3: Fuzzy model from relations.

A rule for two input system is written

'if X is A_i and Y is B_j then Z is C_k'

where A_i and B_j are fuzzy sets describing the state of the input domains U and V and C_k is a fuzzy set over the output domain W. The support of $A_i \times B_j$ is the region of applicability of the rule and the rule generates the relation $A_i \times B_j \times C_k$ in $U \times V \times W$. In a like manner, the region of applicability of an n-dimensional rule is obtained from the n conjuncts in the antecedent of the rule.

2.3 Granularity and specificity

The fuzzy sets used in the antecedent and consequent of a rule are obtained from the decompositions of the input and output domains. The region of applicability of a rule determines its *granularity*. The larger the support of the antecedent, the larger the granularity of the rule. As illustrated in Figure 1, the fuzzy sets in a domain decomposition may have varying sizes of support and, as a consequence, a rule-base may contain rules of greatly varying granularity.

The granularity of a rule-base is determined by the number and the size of the fuzzy sets in the decomposition of the input domains. In this presentation we will consider one rule-base to be of higher granularity than another if the average rule granularity of the first is larger than that of the second (rule of thumb, fewer rules–higher granularity). Although we will not do so, it is reasonable to consider granularity as a local property of a model and focus granularity analysis on subregions of the universe.

The *specificity* of a fuzzy set measures the degree to which the set designates a unique item. For simplicity, we will consider specificity to be the size of the support of the fuzzy set. The specificity of a rule is the specificity of the fuzzy set in the consequent of the rule. A lower specificity indicates that the consequent provides a wider range of possible values. When the output domain is partitioned by triangular fuzzy sets with peak points $\{c_1, \ldots, c_t\}$, the specificity of the rule 'if X is A_i then Z is C_k' is the distance $|c_{k-1} - c_{k+1}|$ between neighboring peak points in the output domain decomposition. A detailed introduction to specificity, its measurement, and its role in assessing uncertainty in fuzzy reasoning may be found in Klir and Folger [8].

Constructing a fuzzy model by the combination of fuzzy relations generated by Cartesian products greatly limits the ability to produce a fuzzy model with both high granularity and high specificity. To illustrate this we consider a one-input system whose input and output domain decompositions are $A_1, \ldots, A_n$ and $C_1, \ldots, C_t$ with peak points $a_1, \ldots, a_n$ and $c_1, \ldots, c_t$, respectively. We will examine the constraint on the specificity of fuzzy sets in the output decomposition imposed by the desire to maintain continuity.

Assume that the points $(a_i, f(a_i))$ and $(a_{i+1}, f(a_{i+1}))$ represent actual input and output points for the system. The model should have $(a_i, f(a_i))$ within the fuzzy relation defined by the rule with antecedent 'X is A_i' and $(a_{i+1}, f(a_{i+1}))$ within the relation defined by the rule with antecedent 'X is A_{i+1}'. Let c_k and $c_{k'}$ be the peak points in the decomposition of the output domain that are closest to $f(a_i)$ and $f(a_{i+1})$ respectively. Thus, the rules applicable for inputs near a_i and a_{i+1} are

'if X is A_i then Z is C_k'
'if X is A_{i+1} then Z is $C_{k'}$'.

Figure 4 shows the relations in the product space corresponding to these rules with the peak points of the antecedent fuzzy sets on the horizontal axis and consecutive peak points of output fuzzy sets along the vertical axis. If there are any peak points of the output domain decomposition between c_k and $c_{k'}$, the resulting fuzzy model $\tilde{f}$ is discontinuous; an input value within (a_i, a_{i+1}) would produce a disconnected fuzzy set as output.

Continuity is achieved by requiring that the output fuzzy sets for consecutive rules differ by at most one location in the output domain decomposition. Consider a sequence of rules

'if X is A_i then Z is C_k'
'if X is A_{i+1} then Z is C_{k+1}'

which indicates that the output values are increasing with an increase of the input values, as on the right-hand side of Figure 3. If the system values for peak points a_i and a_{i+1} are $f(a_i)$ and $f(a_{i+1})$ respectively, the projection of the relations defined of the rules onto the W axis must contain the interval $[f(a_i), f(a_{i+1})]$. That is, the minimal specificity is determined the size of the input domain decompositions and the change in the system output over that region. In a region where the system values are changing, increasing the specificity requires lowering the granularity (dividing the input domains into smaller regions).

The inability to construct n-dimensional models with both high granularity and high specificity from the combination of $n+1$-dimensional rectangles

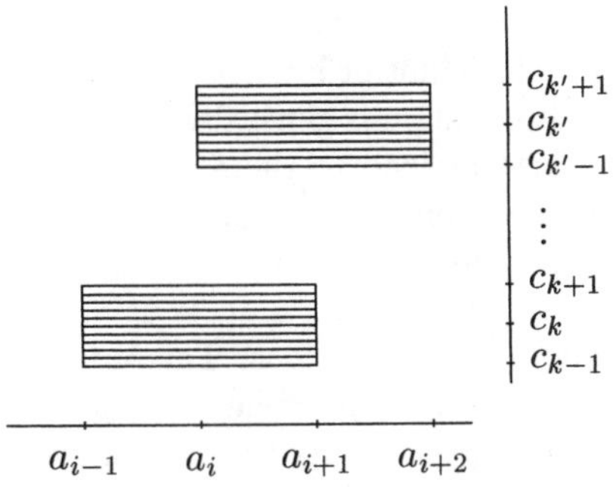

Figure 4: Specificity and continuity

has motivated research into the feasibility of employing alternative geometric shapes to represent rules. Ellipsoids have been used in classification by Abe and Thawonmas [9, 10] and in function approximation by Dickerson and Kosko [11]. A fuzzification of TSK type rules has been proposed by Sudkamp and Hammell [12]. The results of an extensive experimental analysis of geometric forms for function approximation performed by Mitaim and Kosko are reported in [13].

Regardless of the type of fuzzy sets used to partition the input domain, the relationship between the number of rules and the granularity remains the same: increasing the rule set decreases the granularity. However, the use of Gaussian or trapezoidal fuzzy sets in the antecedents of rules or alternative geometric shapes to represent rules may reduce the overall number of rules needed to produce a model.

2.4 Precision and universal approximation

As seen in Figure 3, a rule-base defines a 'fuzzy' function $\tilde{f}$ from the input domain U to the set of fuzzy subsets $\mathcal{F}(W)$ of the output universe. For applications in which a precise response is required, defuzzification is added to the model to produce a crisp approximating function $\hat{f} : U \to W$ from the model. For an input $u \in [a_i, a_{i+1}]$, the consecutive rules 'if X is A_i then Z is C_r' and 'if X is A_{i+1} then Z is C_s' determine the value of $\hat{f}(u)$. Using weighted average defuzzification, the approximating function $\hat{f}$ is

$$\hat{f}(u) \quad = \quad \frac{\mu_{A_i}(u)c_r + \mu_{A_{i+1}}(u)c_s}{\mu_{A_i}(u) + \mu_{A_{i+1}}(u)}.$$

where a_i, a_{i+1}, c_r, and c_s are the peak points of the fuzzy sets A_i, A_{i+1}, C_r, and C_s. The derivation of this approximating function can be obtained using compositional rule of inference and weighted average defuzzification (see, for example, [7]). Approximating functions may be obtained from multidimensional models using weighted average defuzzification in a similar manner.

When defuzzification is employed, the notion of specificity is supplanted by that of precision. The precision of a model is determined by the difference between the model values and the system values. In building fuzzy models, either by learning algorithms or heuristic evaluation, a mathematical description of the underlying system is generally not known and the precision cannot be determined. This has lead to the evaluation of the potential to produce models with high precision.

Determining the potential modeling ability of fuzzy systems usually focuses on the flexibility of defining models using a well structured or parameterized set of domain decompositions. The objective is to ensure that, within the constraints imposed upon possible domain decompositions, every continuous function can be approximated to any desired degree of precision. This has lead to a number of theorems that can all be written as:

The set of functions obtained from fuzzy rules using x-type domain decompositions and y-type defuzzification are univeral approximators of continuous functions.

The import of universal approximation is that, for any continuous function f, a model can be constructed whose approximating function $\hat{f}$ satisfies $\max\{|f(u) - \hat{f}(u)| \mid u \in U\} \leq \epsilon$ for any $\epsilon > 0$. For example, universal approximation theorems have been proven when the domain decompositions are limited to Gaussian membership functions [14] or equally spaced triangular membership functions [15]. Other conditions that ensure universal approximation by fuzzy systems have been established in [16, 17, 18, 19].

Universal approximation proofs usually employ the Stone-Weierstrass Theorem or analyze the change in the variability of function values over each region of the input domain. Variability within a region requires more fuzzy rules to produce the desired degree of precision in the approximation. In the general case, arbitrary precision is obtained by reducing the granularity of the domain decomposition. That is, as ϵ approaches 0 the number of rules required to ensure the desired precision approaches infinity.

Universal approximation results describe a theoretical potential to approximate functions with fuzzy rules. In two important ways these results are irrelevant to modeling since they focus on conditions that are generally not applicable when constructing models. A goal in constructing a fuzzy model is, to the extent possible, to have a minimal number of highly granular rules. Thus results predicated on the presence of an unlimited number of rules do not address this major objective of modeling. Moreover, the systems being modeled are not represented by arbitrary continuous functions. Rather they usually have well behaved properties with naturally occurring limits on the functional variation within regions.

Some recent research has concentrated on the more practical issues of the tradeoff between precision and granularity. In [20, 21, 22], relationships were established between the form of the function being modeled and the number of rules needed to obtain a specified precision. An investigation analyzing the granularity-precision tradeoffs for TSK systems can be found in [23].

2.5 Granularity and run-time efficiency

The execution cycle of a fuzzy model consists the acquistion of input, the determination of the appropriate set of rules, the evaluation of the rules, and the aggregation of the results. The final value, either fuzzy or precise depending upon the application, is then returned as the appropriate response or action. The work required in the evaluation of the rules and the aggregation of the results is independent of the granularity of the rule-base. The form of the domain decompositions, however, directly influences the work required in the second step of the evaluation cycle.

The determination of the set of applicable rules is accomplished by comparing the information describing the state of an input domain U with the fuzzy sets in the domain decomposition of U. When a domain decomposition is constructed heuristically, the identification of the applicable rules may require a comparison of the input with the antecedent of each rule or at least a subset of the rules in the rule-base.

To facilitate the representation and evaluation of a fuzzy model, the rule-base is frequently represented in a tabular form known as a fuzzy associative memory or FAM. We will demonstrate this compact representation using a model with input domains U and V. If $A_1, \ldots, A_n$ and $B_1, \ldots, B_m$ are the decompositions of U and V, the rule-base may be represented by the matrix

	B_1	B_2	$\cdots$	B_m
A_1	$C_{1,1}$	$C_{1,2}$	$\cdots$	$C_{1,m}$
A_2	$C_{2,1}$	$C_{2,2}$	$\cdots$	$C_{2,m}$
$\vdots$	$\vdots$	$\vdots$		$\vdots$
A_n	$C_{n,1}$	$C_{n,2}$	$\cdots$	$C_{n,m}$

where the i, jth entry in the FAM represents the rule

$$\text{'if } X \text{ is } A_i \text{ and } Y \text{ is } B_j \text{ then } Z \text{ is } C_{i,j}\text{'.}$$

Each dimension of a FAM corresponds to one of the input variables of the system being modeled.

Domain decompositions employed by rule-base learning algorithms are frequently well-structured or parametrically defined. For structured domain decompositions, hash-functions may be used to compute the indices of the applicable rules in the FAM.

Consider a domain decomposition consisting of evenly spaced triangular fuzzy sets over the universe $[-1, 1]$. In this case, the hash function

$$index(u) = trunc((n-1)(u+1)/2) + 1$$

returns the value i indicating that A_i and A_{i+1} are the applicable fuzzy sets for the evaluation of input u. A similar direct calculation may be used to obtain the indices of the applicable rules for other input dimensions. Within

	Heuristic Evaluation	Learning Algorithms	Desirable Characteristic
intepretability	yes	no	yes
granularity	high	low	high
specificity	low	high	high
efficiency	low	high	high

Table 1: Properties of fuzzy models

the structure of well defined decompositions, the evaluation is independent of the number of rules and hence the granularity of the rule-base. Moreover, the work required in identifying the applicable rules grows linearly with the dimension of the input space with a single hash function required for each dimension.

2.6 Desiderata

The preceding sections introduced several important properties of fuzzy rule-bases. The two standard methods for generating rules, heuristic evaluation and learning algorithms, generally produce rule-bases with greatly varying characteristics. A summary of the properties is given in Table 1.

Heuristic evaluation produces rules of high granularity; domain decompositions generally consist of between five and nine fuzzy sets. Due to the relationship developed in Section 2.3, the rules must necessarily have a low specificity. The selection of the term set by the domain expert ensures that the rules have a readily understandable linguistic interpretation. The lack of structure in the decompositions, however, does not permit an efficient determination of the set of applicable rules.

The algorithmic generation of rule-bases from training data produces rule-bases with almost exactly the opposite characteristics. The domain decompositions consist of a predetermined parameterized sequence of fuzzy sets and do not necessarily represent concepts or linguistically recognizable subsets of the input domains. Without the interpretability restriction on the rule antecedents, multiple rules of high specificity may be generated when sufficient training data are available.

In the next section we will examine the effect of training data on the granularity of models. A combination of rule learning and region merging will be proposed to incorporate beneficial features of rule-bases produced by both techniques into the resulting model.

3 Granularity and rule learning

In this section we will analyze the relationship between granularity, training data, and precision in the algorithmic generation of fuzzy rules. In the algorithms that we examine, domain decompositions are selected without regard to the distribution of training data or any knowledge of the underlying system. Decompositions are chosen that facilitate the computations and have a granularity that permits the production of models of high precision. Once the input domain decompositions are selected, a learning algorithm determines an appropriate consequent for each combination of input fuzzy sets.

Regardless of the maner in which a rule base is constructed, it is possible for rules whose regions of applicability are adjacent in the input domain to specify identical or similar responses. This is particularly true when a rule-base contains a large number of rules like those frequently generated by rule learning algorithms. When this occurs, these rules may be merged to produce a single rule with a larger region of applicability.

There are two potential benefits for generating fewer rules with larger regions of applicability. For models produced by heuristic evaluation, the determination of the appropriate rules requires a comparison of the input with the rules. A reduction in the number of rules will improve the efficiency by reducing the number of required comparisons. When models are produced by the FLM algorithm, the input domain decompositions have low granularity. Increasing the region of applicability of a rule may provide a more readily interpretable region in the input space.

The second learning algorithm we examine combines the selection of the consequent with a region merging strategy. A rule is produced by growing its region of applicability until a precision bound is exceeded. The region merging strategy preserves sufficient structure to maintain the efficient determination of the set of applicable rules. Region merging differs from rule reduction (see [25, 26, 27]) in that the latter begins after the rule-base has been constructed while merging constructs the rules as it expands the regions of applicability.

3.1 Granularity, completion, and precision

Perhaps the simplest rule learning strategy is the FLM algorithm of Wang and Mendel [24]. The FLM algorithm employs a local analysis of the training data to produce rules. The domain decompositions are independent of the distribution of the data and are usually selected for reasons of simplicity of evaluation. With this in mind, we will assume that each input domain is decomposed into a sequence of equally spaced triangular fuzzy sets.

A two-dimensional system with input domains U and V with decompositions $A_1, \dots, A_n$ and $B_1, \dots, B_m$ and output domain W with decomposition $C_1, \dots, C_t$ will be used to describe the rule generation algorithm. A set $T = \{(u_s, v_s, z_s) \mid s = 1, \dots, k\}$ of training examples, where $u_i \in U$ and

$v_i \in V$ are input values and $z_i \in W$ is the associated response, provides the information needed for the generation of the rules. The consequent of a rule with antecedent 'X is A_i and Y is B_j' is determined by the training points (u_s, v_s, z_s) that satisfy $\min\{\mu_{A_i}(u_s), \mu_{B_j}(v_s)\} \geq .5$. We will call this set $T_{i,j}$.

The FLM algorithm uses a training example that has maximal membership in $A_i \times B_j$ to determine the consequent of the rule. The FLM rule generation strategy for a two-dimensional FAM is given below.

FLM: A training example (u_s, v_s, z_s) that has the maximal membership in $A_i \times B_j$ is selected from the training set $T_{i,j}$. If more than one example assumes the maximal membership in $A_i \times B_j$, one is selected arbitrarily. The rule 'if X is A_i and Y is B_j then Z is C_r' is constructed where the consequent C_r is the fuzzy set from the decomposition of the output domain in which z_s has maximal membership. If z_s has maximal value in two adjacent regions $(\mu_{C_r}(z_i) = \mu_{C_{r+1}}(z_i) = .5)$, then the consequent C_r is selected.

Variations of the FLM strategy using all training examples that fall within $T_{i,j}$ to determine the consequent and extending $T_{i,j}$ to the entire region of applicability of the rule have been investigated in [15, 28]. A detailed presentation of the FLM algorithm and comparisons with other fuzzy rule learning algorithms and neural network learning can be found in [15, 29].

Since the FLM algorithm requires no interpretation for the fuzzy sets in the domain decomposition, there is no inherent restriction on the granularity of the decompositions. There are t^{mn} distinct models realizable with domain decompositions of n, m, and t fuzzy sets. Increasing the number of fuzzy sets in the decompositions (decreasing the granularity), increases both the number of models and the potential to generate a highly specific or precise model. However, the achievability of this potential is dependent upon the training set.

The FLM algorithm requires at least one training instance in the set $T_{i,j}$ to produce the consequent of the rule 'if X is A_i and Y is B_j'. Decreasing the granularity increases the number of rules and the amount of training data that must be acquired and processed. This difficulty is accentuated with multidimensional systems. A system with five input domains, each of which is decomposed into five regions, produces a system with 3125 rules. If the training set is obtained from sampling successful operations, many combinations of input fuzzy sets may not contain a suitable training example.

To mitigate the need for training data, an interpolation strategy known as *completion* [15, 30] was introduced to complement the FLM rule generation algorithm. After the training data have been processed, an unfilled entry in the FAM represents a combination of inputs for which no action is specified. Region growing completion is accomplished by iteratively extending the information provided by the filled FAM locations to their unfilled neighboring locations.

Extensive experimentation with the FLM algorithm augmented with region growing completion has been reported in [15, 29]. Several relationships concerning the effect of training set size on the granularity–precision tradeoff were evident in the experimental results.

1. For fixed input domain decompositions, there is a point at at which increasing the size of the training set has little effect on the performance of the model.

2. For fixed input domain decompositions, completion is effective when more than 50% of the FAM locations are filled by the FLM algorithm.

3. For a fixed size training set, there is a point at which decreasing the granularity of the model degrades its precision.

The first observation indicates that a saturation condition occurs in the FLM algorithm. When sufficient training information is available, the precision of the model is limited by the set of functions realizable from the domain decompositions. Items 2 and 3 indicate that rule granularity imposes limitations on the effectiveness of FLM and completion. When the data is scattered but not sparse, completion enhances the ability to produce precise models.

Increasing the number of rules provides a theoretical potential for constructing models of ever greater precision. In reality, however, this potential is only achievable with the unrealistic assumption of having an unlimited supply of training data. The precision of a rule generated by completion is dependent upon its proximity in the FAM to rules produced directly from data. Consequently, decreasing the granularity beyond the limits of the training data results in models of decreased precision.

3.2 Merging in learning

The FLM algorithm produces large rule-bases of high precision when sufficient training data are available. The objective of the learning with region merging strategy is not to maximize precision, but rather to produce a model with a reasonably small set of rules within a predetermined degree of precision. Ideally, the goal would be to identify the smallest set of rules satisfying the precision bound. However, the algorithm that we present is greedy and does not ensure the generation of a minimal rule-base. The approach follows that employed in [12] for extending the scope of fuzzified TSK rules.

As with the FLM algorithm, the first step is to partition the input domain. This is accomplished by selecting evenly spaced sets of points $\{a_1, \ldots, a_n\}$ and $\{b_1, \ldots, b_m\}$ along the input axes. These points divide the input domain into $(n-1)(m-1)$ rectangular regions. Part of the learning process is to construct the fuzzy partition of the input domain. The partition will consist of fuzzy sets with both rectangular supports and cores. The core of such a fuzzy set is defined by a pair of points $[(a_i, b_j), (a_r, b_s)]$ with $i < r$ and $j < s$; (a_i, b_j) is the lower left-hand corner of the rectangle and (a_r, b_s) is the upper right-hand corner. In the general case, the support of the fuzzy set

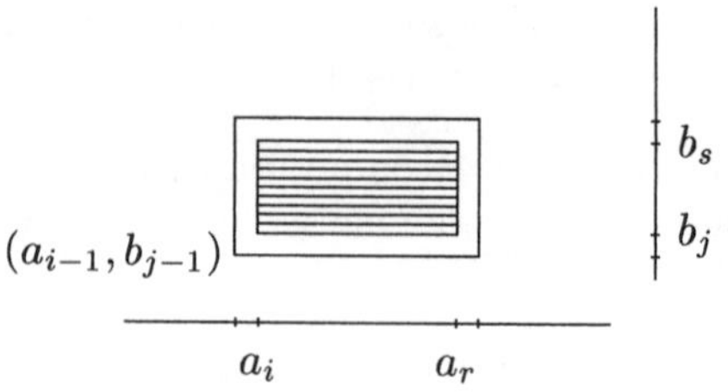

Figure 5: Rectangular fuzzy set

is the rectangle specified by the points $[(a_{i-1}, b_{j-1}), (a_{r+1}, b_{s+1})]$. The outer rectangle in Figure 5 shows the support of a rectangular fuzzy set and the shaded region is the core. If a core rectangle abuts the boundary of the input domain, the support does not extend beyond the core in that direction. The region of the support not in the core forms a buffer which provides for the smooth transition between fuzzy sets in the partition.

A rule has the form

$$\text{'if } X \times Y \text{ is } A = [(a_i, b_j), (a_r, b_s)] \text{ then } [c_{i,j}, c_{i,s}, c_{r,j}, c_{r,s}]\text{'},$$

where the antecedent describes the core of a rectangular fuzzy set in the input domain and the consequent gives output values for the corners of the core of A. The output value for any point (u, v) in the core of A may be obtained by three linear interpolations using the values at the corners of the rectangle. The output values for input in a buffer region are obtained by a weighted average of the output values of nearest point in each adjacent core regions.

For a rule with antecedent '$X \times Y$ is $[(a_i, b_j), (a_r, b_s)]$', rule learning consists of selecting the constants $[c_{i,j}, c_{i,s}, c_{r,j}, c_{r,s}]$. To simplify the selection, restrictions may be put upon the choice of the c's or they may be chosen to simply minimize the error of the training between the output values and the training examples in the region $[(a_i, b_j), (a_r, b_s)]$.

The fuzzy set generation merges adjacent regions as long as the difference in the output values and the training data in the expanded region satisfy the precision bound. The algorithm is greedy; it begins with rectangle $[(a_1, b_1), (a_2, b_2)]$ and merges with adjacent rectangles in the same row until the precision bound stops the process. The resulting rectangular core $[(a_1, b_1), (a_r, b_2)]$ is then grown vertically until halted by the precision bound. If achievable within the precision bound, the first growth in the vertical direction would produce the rectangle $[(a_1, b_1), (a_r, b_3)]$. The result is a fuzzy set with core $[(a_1, b_1), (a_r, b_s)]$ and support $[(a_1, b_1), (a_{r+1}, b_{s+1})]$, unless the core extends across the entire domain in which case the corresponding buffer is not present. The merging process is reinitiated with the first rectangle following the buffer. The process continues left-to-right, bottom-to-top until the entire domain is in a core or a buffer.

Restricting the merging algorithm to producing rectangular fuzzy sets maintains the efficient evaluation of input. An $n \times m$ matrix is constructed to represent the regions in the original division of the input domain. For a given input, hashing may be used to determine the region in which the input occurs. Rather than directly providing the consequent of a rule, as in the FAM representation, the matrix entry designates the four points used to determine the output value.

A number of preliminary experiments have been undertaken to demonstrate the ability of the merging algorithm to produce small rule sets while satisfying a specified precision bound. An experiment consists of selecting a target function f which represents the underlying system and randomly generating a test set of the form $T = \{(x_i, y_i, f(x_i, y_i))\}$. The input domain is initially divided into 10,000 square subregions. The results of a suite of experiments with target function $f(x, y) = 1 - (x^2 + y^2)$ are given in Table 2, where the errors listed are over the points in the training set. In these tests, the consequent of a rule was restricted to sequences in which $c_{i,j} = c_{i,s} = c_{r,j} = c_{r,s}$. Choosing $c_{i,j}$ as the center point of a fuzzy set C in the output domain decomposition, the rule will have the interpretable Mamdani form 'if $X \times Y$ is $A = [(a_i, b_j), (a_r, b_s)]$ then Z is C'.

As would be expected, increasing the precision bound decreases the number of rules needed. When the training data are sparse in the space, merging will be successful since extending an existing rule to include an adjacent region with no training data will not violate the precision bound. As the size of the training set increases, the merging process will encounter more regions with training data. If the selection of the consequent is unable to approximate these data within the precision bound, the extension of the current rule will halt. Consequently, additional training data may increase the number of rules.

A more extensive set of experimental results on the effectiveness of the combination of learning and merging can be found in [31]. Along with the planar fuzzy sets generated by requiring $c_{i,j} = c_{i,s} = c_{r,j} = c_{r,s}$, these experiments consider fuzzy sets obtained by independently selecting values for $c_{i,j}$, $c_{i,s}$, $c_{r,j}$, and $c_{r,s}$ and generating a nonlinear surface over the core of the rule from the values at the corners.

4 Conclusions

The characteristics of a model are influenced by the technique used to generate the rules. Models constructed by expert evaluation are highly granular with interpretable rules while those generated by learning algorithms are less granular but more specific and frequently amenable to efficient evaluation. Increasing the granularity of models generated by learning algorithms has the potential of improving their interpretability while maintaining their advantages in specificity and efficiency. The added explanatory capability

Precision Bound	Training Examples	Rules Generated	Maximum Error	Average Error
0.1	50	54	.0001	.0001
	100	65	.004	.007
	1000	253	.098	.038
0.2	50	16	.163	.066
	100	42	.200	.055
	1000	114	.119	.073
0.5	50	8	.490	.213
	100	8	.360	.169
	1000	17	.484	.184

Table 2: Rule merging approximations of $f(x, y) = 1 - (x^2 + y^2)$

associated with interpretable rules may facilitate the acceptance of models generated from training data in critical problem domains.

Acknowledgement: The author would like to thank the US Army Research Laboratory for its support of this research.

References

[1] E. H. Mamdani and S. Assilian (1975). An experiment in linguistic synthesis with a fuzzy logic controller. *International Journal of Man-Machine Studies* 7, 1–13.

[2] T. Takagi and M. Sugeno (1985). Fuzzy identification of systems and its applications to modeling and control. *IEEE Transactions on Systems, Man, and Cybernetics* 15, 329–346.

[3] L. A. Zadeh (1997). Toward a theory of fuzzy information granulation and its centrality in human reasoning and fuzzy logic. *Fuzzy Sets and Systems* 90(2), 111–127.

[4] L. A. Zadeh (1975). The concept of a linguistic variable and its application to approximate reasoning: Part 1. *Information Sciences* 8, 199–249.

[5] L. A. Zadeh (73). Outline of a new approach to the analysis of complex systems and decision processes. *IEEE Transactions on Systems, Man, and Cybernetics* 3, 28–44.

[6] E. H. Mamdani (1976). Advances in the linguistic synthesis of fuzzy controllers. *International Journal of Man-Machine Studies* 8, 669–678.

[7] C. C. Lee (1990). Fuzzy logic in control systems: Part I. *IEEE Transactions on Systems, Man, and Cybernetics* 20(2), 404–418.

[8] G. Klir and T. Folger (1988). *Fuzzy Sets, Uncertainty, and Information.* Englewood Cliffs, NJ: Prentice Hall.

[9] S. Abe and T. Thawonmas (1997). A fuzzy classifier with ellipsoid regions, *IEEE Transactions on Fuzzy Systems* 5(3), 358–368.

[10] T. Thawonmas and S. Abe (1999). Function approximation based on fuzzy rules extracted from partitioned numerical data. *IEEE Transactions on Systems, Man, and Cybernetics:B* 29(4), 525–534.

[11] J. A. Dickerson and B. A. Kosko (1996). Fuzzy function approximation with ellipsoidal rules. *IEEE Transactions on Systems, Man, and Cybernetics* 26(4), 542–560.

[12] T. A. Sudkamp and R. J. Hammell II (1998). Granularity and specificity in fuzzy function approximation. In *Proceedings of the NAFIPS-98*, Pensacola, Florida, 105–109.

[13] S. Mitaim and B. Kosko (1996). What is the best shape for a fuzzy set in function approximation? In *Proceedings of the 5th IEEE International Conference on Fuzzy Systems*, New Orleans, 1237–1243.

[14] L. X. Wang and J. M. Mendel (1992). Fuzzy basis functions, universal approximation, and orthogonal least-squares learning. *IEEE Transactions on Neural Networks* 3(5), 807–814.

[15] T. Sudkamp and R. J. Hammell II (1994). Interpolation, completion, and learning fuzzy rules. *IEEE Transactions on Systems, Man, and Cybernetics* 24(2), 332–342.

[16] J. J. Buckley (1993). Sugeno controllers are universal controllers. *Fuzzy Sets and Systems* 53(3), 299–304.

[17] B. Kosko (1994). Fuzzy systems as universal approximators. *IEEE Transactions on Computers* 43(11), 1329–1333.

[18] X. Zeng and M. G. Singh (1994). Approximation theory of fuzzy systems–SISO case. *IEEE Transactions on Fuzzy Systems* 2(2), 162–176.

[19] X. Zeng and M. G. Singh (1995). Approximation theory of fuzzy systems–MISO case. *IEEE Transactions on Fuzzy Systems* 3(2), 219–235.

[20] Y. Ding, H. Ying, and S. Shao (1997). Necessary conditions for general MISO fuzzy systems as universal approximators. In *Proceedings of the 1997 IEEE International Conference on Systems, Man, and Cybernetics*, 1153–1162.

[21] H. Ying and G. Chen (1997). Necessary conditions for some typical fuzzy systems to be universal approximators. *Automatica* 30, 1333–1338.

[22] D. Lisin and M. A. Gennert (1997). Optimal functional approximation using fuzzy rules. In *Proceedings of the Eighteenth International Conference of NAFIPS*, New York, 184–188.

[23] H. Ying, Y. Ding, S. Li, and S. Shao (1999). Comparison of necessary conditions for typical Takagi-Sugeno and Mamdani fuzzy systems as universal approximators. *IEEE Transactions on Systems, Man, and Cybernetics:A* 25(5), 508–514.

[24] L. Wang and J. M. Mendel (1991). Generating fuzzy rules from numerical data, with applications. Tech. Rep. USC-SIPI-169, Signal and Image Processing Institute, University of Southern California, Los Angeles, CA 90089, 1991.

[25] J. Yen and L. Wang (1996). An SVD-based fuzzy model reduction strategy. In *Proceedings of the Fifth IEEE International Conference on Fuzzy Systems*, New Orleans, 835–841.

[26] M. Setnes, R. Babuska, U. Kaymak, and H. R. van Nauta Lemke (1998). Similarity measures in fuzzy rule base simplification. *IEEE Transactions on Systems, Man, and Cybernetics:B* 28(3), 376–386.

[27] Y. Yam, P. Baranyi, and C. Yang (1999). Reduction of fuzzy rule base via singular value decomposition. *IEEE Transactions on Fuzzy Systems* 7(2), 120–132.

[28] B. Kosko (1992). *Neural Networks and Fuzzy Systems: A dynamical systems approach to machine intelligence.* Englewood Cliffs, NJ: Prentice Hall.

[29] R. J. Hammell II and T. Sudkamp (1995). A two-level architecture for fuzzy learning. *Journal of Intelligent and Fuzzy Systems* 3(4), 273–286.

[30] T. Sudkamp and R. J. Hammell II (1996). Rule base completion in fuzzy models. In *Fuzzy Modeling: Paradigms and Practice*, W. Pedrycz, ed., Kluwer Academic Publishers, 313–330.

[31] T. Sudkamp, A. Knapp, and J. Knapp (2000). A greedy approach to rule reduction in fuzzy models. In *Proceedings of the 2000 IEEE Conference on Systems, Man, and Cybernetics*, Nashville, Tennessee.

Granular Computing in Neural Networks

Scott Dick, Abraham Kandel

University of South Florida, Department of Computer Science and Engineering, Tampa, FL, 33620, USA

Abstract. The basic premise of granular computing is that, by reducing precision in our model of a system, we can suppress minor details and focus on the most significant relationships in the system. In this chapter, we will test this premise by defining a granular neural network and testing it on the Iris data set. Our hypothesis is that the granular neural network will be able to learn the Iris data set, but not as accurately as a standard neural network. Our network is a novel neuro-fuzzy systems architecture called the linguistic neural network. The defining characteristic of this network is that all connection weights are linguistic variables, whose values are updated by adding linguistic hedges. We define two new hedges, whose semantics require a generalization of the standard definition of linguistic variables. These generalized linguistic variables lead naturally to a linguistic arithmetic, which we prove forms a vector space. The node functions of the linguistic neural network are defined in terms of this linguistic arithmetic. The learning method used for the network is a modified Backpropagation algorithm, with the original arithmetic operations replaced by their linguistic equivalents. In a simulation experiment, this granulated version of the multilayer perceptron achieved 90% accuracy on the Iris data set, using a coarse granulation. This result supports our hypothesis.

Keywords. Granular computing, linguistic variables, linguistic arithmetic, linguistic hedges, linguistic space, machine learning, multiplayer perceptrons, Backpropagation learning, neuro-fuzzy systems

1. Introduction

Over the past several years, there has been a great deal of interest in the notion of "information granularity." Information granularity refers to methods of reasoning about aggregations of objects. Instead of reasoning about individual objects, we group objects together, and reason about the resulting groups. This grouping collects similar objects into a single group, using whatever definition of similarity is appropriate. The goal of this process is to provide an abstraction of the original problem, in which the most important relationships still hold true, but fine details are suppressed. This is a very important technique, as it appears to be a key mechanism in human cognition [1]. One key usage of information granularity is in the modeling and control of complex, ill-defined systems. A mathematical model for such a system may be difficult or impossible to build. However, it may be possible to reduce the complexity of the system by granulating it. The main premise of granular computing is that granulation removes the fine detail that creates much of the complexity of the system, while leaving the most significant relationships intact.

One very complex problem is the extraction of knowledge from a trained neural network. A trained neural network has learned how to solve a problem by repeatedly encountering examples of that problem, and learning from them. Thus, useful knowledge is stored in the network, which may well be unavailable from other sources. However, as connectionist architectures, neural networks store their knowledge as a pattern of connection weights between simple processing nodes. In order to extract information from a neural network, an analyst must interpret a distributed pattern of numeric values. Experience indicates that human beings have great difficulty in performing this task.

We will test the basic premise of granular computing by creating a neural network in which the connection weights are granular. This neural network will store knowledge and learn new knowledge in a granular form. If the granular computing premise is correct, the granular neural network will be able to learn input-output relationships just like a standard neural network, but at a lower level of precision. We will test this hypothesis using the well-known Iris data set. Standard neural networks achieve a testing accuracy of about 95% on the Iris data set. We predict that our network will have a testing accuracy that is somewhat lower, but still significant.

In this chapter, we propose a novel neuro-fuzzy system architecture called the Linguistic Neural Network (LNN), which uses information granularity to simplify the representation of knowledge in a neural network. Our architecture is a multilayer perceptron, in which the connection weights are linguistic values. The learning algorithm in this network will operate by adding linguistic hedges to these linguistic weights. We use the ideas of linguistic variables and hedges from fuzzy set theory as the basis of our method. Each linguistic term corresponds to a fuzzy set on some universe of discourse, and each hedge is a function altering these fuzzy sets. However, all processing in the LNN takes place at the symbolic level. We define two novel hedges, whose semantics require us to change the syntactic rule of a linguistic variable from a context-free grammar to a phrase-structure grammar. This leads us to the creation of a linguistic arithmetic, in which addition, subtraction and multiplication can be carried out over linguistic values. A major result of this chapter is that the linguistic arithmetic forms a vector space over the field of integers. We use the linguistic arithmetic to define the node functions of the LNN, as well as the learning rule, called Linguistic Backpropagation. Linguistic Backpropagation is a modification of the Backpropagation algorithm, in which the original arithmetic operations have been replaced by their linguistic equivalents. In a simulation experiment, the LNN achieved a 90% accuracy rate on the Iris data set, using a fairly coarse granulation of the network weights. One possible application of the LNN is in the domain of knowledge extraction form trained neural networks. A standard neural network encodes knowledge as a distributed pattern of connection weights. This distributed, numeric form of knowledge is extremely difficult for humans to understand. By granulating the connection weights, the LNN may simplify an analyst's task by permitting them to concentrate on the distributed pattern of knowledge in the network, without the added burden of interpreting numeric information.

The remainder of this chapter is organized as follows. In Section 2, we review the relevant literature on fuzzy set theory, neuro-fuzzy systems and information granularity. In Section 3, we define the linguistic variables and hedges used as connection weights in the neural network. We develop a linguistic arithmetic, and prove that it is a vector space. In Section 4, we define the architecture and learning rule of the LNN. In Section 5, we present the results of our experiment on the Iris data set, which supports the granular computing premise. In Section 6, we provide a summary and discussion of future work.

2. Review of Related Work

In this section, we review the relevant literature on fuzzy set theory, neuro-fuzzy systems, and information granularity. We will discuss fuzzy sets, linguistic variables, fuzzy arithmetic and linguistic spaces, in order to provide the theoretical background for the linguistic arithmetic used in this chapter. We will review several neuro-fuzzy architectures, including the ANFIS architecture. Finally, we review several important ideas in information granularity, such as computing with words, using mixed granularities, and qualitative reasoning.

2.1 Fuzzy Set Theory

Fuzzy sets were first proposed by Lotfi Zadeh in [5]. A fuzzy set is a generalization of a classical ("crisp") set, in which elements may have partial memberships. Unlike crisp sets, which have a sharp boundary between elements that are in the set and elements not in the set, the boundaries of a fuzzy set are imprecise. This makes fuzzy sets an excellent tool for dealing with vague or imprecise information. Formally, a fuzzy set is a collection of ordered pairs (x, $\mu(x)$), in which x is an element of some universe of discourse U, and $\mu(x)$ is a function

$$\mu: U \rightarrow [0,1]$$

that describes the degree to which x is an element of that fuzzy set. μ is often referred to as the membership of element x in the fuzzy set. Obviously, a crisp set is a special case of a fuzzy set, in which the range of μ is restricted to {0,1}. A fuzzy set termed normal if there is some $x \in U$ for which $\mu(x) = 1$. A fuzzy set A on U is termed convex iff

$\mu_A(\lambda x_1 + (1-\lambda)x_2) \geq \min(\mu_A(x_1), \mu_A(x_2))$

for any x_1, $x_2 \in U$, where min denotes the minimum operator and $\mu_A(x)$ is the membership of x in A [6]. Convex fuzzy sets are important because they have an intuitive meaning: there is a central core of elements with total membership in the fuzzy set, surrounded by a boundary layer of elements with partial membership. The further a boundary element is from the core, the lower its membership.

One of the most important operations on fuzzy sets is the α-cut. An α-cut of a fuzzy set A is a crisp set consisting of every $x \in U$ such that $\mu_A(x) \geq \alpha$. The family of all distinct α-cuts of a fuzzy set will uniquely determine that fuzzy set. α-cuts are used transform a fuzzy set into a crisp set, permitting operations on crisp sets to be extended to fuzzy sets. An important point to note is that the α-cuts of a convex fuzzy set form a family of nested sets. One special class of convex fuzzy sets is the class of *fuzzy numbers.* A fuzzy number is a convex, normal fuzzy set, whose α-cuts are compact for all $\alpha \in [0,1]$. This type of fuzzy set captures the idea of *a number that is not precisely known* [6]. A example of a fuzzy number is shown in Figure 1.

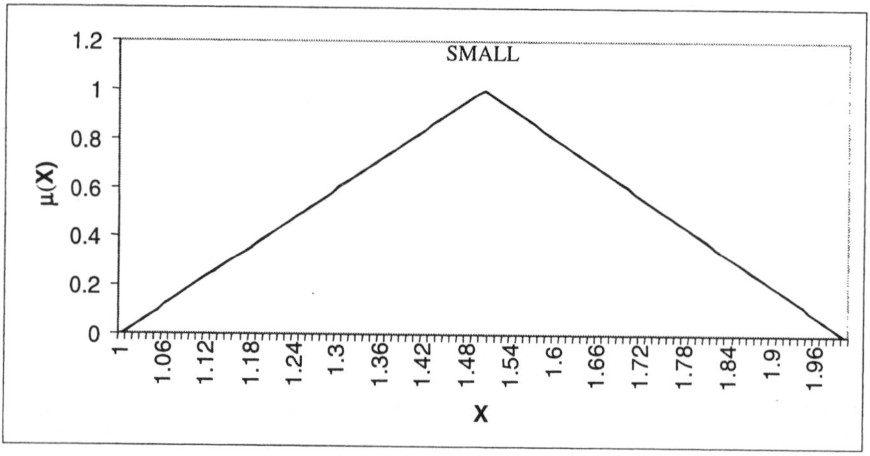

Figure 1: Fuzzy Number

2.2 Linguistic Variables

Fuzzy set theory is often said to provide a mathematical means of describing perceptions. The key to this interpretation is a construct called a linguistic variable. A linguistic variable is a variable whose possible values are drawn from a set of words in natural language. A specific meaning is attached to each word by associating it with a fuzzy set, which is often a fuzzy number. Linguistic variables were introduced by Zadeh in [7], and greatly elaborated in Zadeh's 1975 monologue [8]. Formally, a linguistic variable is a 5-tuple (X,U,T,S,M) where X is the name of the variable, U is the universe of discourse for the variable, T is a set of atomic terms and hedges for the variable, S is a syntactic rule, and M is a semantic rule. The set T contains the basic words and hedges that may be used in the variable. The syntactic rule S is a grammar for generating composite terms out of the atomic terms and hedges. In other words, S is a grammar (in general, a context-free grammar), and T is the set of terminal symbols for S. A composite term will consist of an atomic term modified by one or more hedges. The semantic rule M gives each possible term generated by S (including the atomic terms in T) a meaning by associating it with a fuzzy set in U. This is done by assigning a fuzzy set to each atomic term in T, and a function h to each hedge in T. To determine the meaning of a composite term t, we first determine the parse tree for t given S. Then, we apply hedge functions to modify the atomic term's meaning, in the order indicated by the parse tree [9].

The idea of linguistic variables has given rise to an entire industry that uses fuzzy reasoning in modeling and control problems. However, comparatively little attention has been paid to linguistic hedges. A hedge is a function

$$h: [0,1] \rightarrow [0,1]$$

associated with a linguistic modifier, such as the word "very." The function h modifies the fuzzy set associated with an atomic term, and thus modifies its meaning. Hedges were first proposed by Zadeh in [9], and further explored in [10], where the idea of treating hedges as functions was introduced. This introduced the idea of coupling a change in a term (a *syntactic* operation) with a change in the meaning of that term (a *semantic* operation). This coupling of syntax and semantics is a powerful feature of linguistic variables. Lakoff published a centerpiece of the literature on hedges in 1973 [11]. His analysis was based on psychological experiments, in which people were asked to assign objects a degree of membership to a given category. He concluded that hedges must be represented as a vector-valued function, whose components are themselves membership functions.

2.3 Fuzzy Mathematics

The mathematics based on fuzzy set theory is a very broad topic, subsuming arithmetic operations on fuzzy numbers, possibility theory, fuzzy relations, fuzzy logic, and others. One unifying work on this topic is a textbook by Kandel [12]. Our discussion in this section will be limited to arithmetic operations on fuzzy numbers and linguistic terms, which are key to the theoretical material presented in Section 3. We introduced the idea of a fuzzy number in Subsection 2.1. This is a fuzzy set that represents a number that is not precisely known. Given this basic definition, it is no surprise that a considerable amount of work has been done in defining arithmetic operations whose operands are fuzzy numbers. It is also no surprise that the traditional numeric operators are useless in this context. A statement such as "hot + cold = warm" makes perfect linguistic sense, but if the "+" operator is the numeric sum, then the statement becomes an absurdity.

There are two methods in general use for defining arithmetic operators on fuzzy numbers. The first and best-known is the extension principle, proposed by Zadeh in [8] and explored by Yager in [13]. The extension principle "fuzzifies" an operation, shifting its domain and range from the set of real numbers to the set of fuzzy numbers of the real line. Given a crisp function z=f(x,y), the extension of this function to fuzzy numbers is given by

$$\mu_{f(A,B)}(z) = \sup_{z=f(x,y)} (T(\mu_A(x), \mu_B(y))) \tag{1}$$

where x belongs to the universe of discourse X, y belongs to the universe of discourse Y, $z \in X \times Y$, A is a fuzzy number on X, B is a fuzzy number on Y, f(A,B) is a fuzzy number on X×Y, and T is any t-norm [6]. Since t-norms are associative, we can extend this definition to any number of arguments in f. The implications of this formula are currently an active area of research; see for example [14-21].

The second method of defining fuzzy arithmetic operators is restricted to a class of fuzzy numbers called LR fuzzy numbers. An LR fuzzy number is a fuzzy set A whose membership function is of the form

$$A(x) = \begin{cases} L[(a-x)/\alpha] & if\,(a-\alpha) \le x \le a \\ R[(x-a)/\beta] & if\, a \le x \le (a+\beta) \\ 0 & otherwise \end{cases} \tag{2}$$

where a is the unique value for which $\mu_A(x=a)=1$, α and β are respectively the left and right spread of the fuzzy number, and L, R are monotonic functions for which L(0) = R(0) = 1 and L(1) = R(1) = 0. Fuzzy arithmetic operators on LR fuzzy numbers can be defined using the α-cuts of A. Since the α-cuts of an LR fuzzy number are necessarily closed intervals on the real line, the definitions of classical interval analysis can be used directly. This method is more efficient than using the more general extension principle, which generally reduces to solving a nonlinear programming problem [22,23]. Dubois and Prade have produced a large body of work on LR fuzzy numbers; see for example [24-27].

It is also possible to derive mathematical operators that act only on certain classes of fuzzy numbers. For instance, Meier defines sum, difference and product operators for the class of triangular fuzzy numbers in [28]. These operators are extremely efficient, since their operands are the parameters of a triangular fuzzy number (α,β,γ), where α and γ are respectively the x-intercepts of the left and right sides of the triangular membership function, and β is the peak of the membership function. Obviously, this method cannot be extended to other classes of fuzzy numbers. In a similar vein, Giachetti and Young develop a single parameterized representation of several classes of fuzzy numbers in [23], including triangular fuzzy numbers. They then develop a single set of mathematical operators for the entire set of parameterized fuzzy numbers, and show that this set is closed under their operators.

2.4 Linguistic Space

The ideas of linguistic space, linguistic trajectories, and linguistic functions can be traced to a pair of papers by Braae and Rutherford [29,30]. In these papers, one- and two-dimensional linguistic spaces are defined. These linguistic spaces are used as system state spaces, and a linguistic trajectory is the sequence of linguistic states a system trajectory passes through. The linguistic space is defined by a partition of the system state space, which is induced by the linguistic variables representing the system. These concepts are used to determine an optimal structure for linguistic rules in a fuzzy controller for that system. A discussion of stability from a geometric viewpoint is also given. These ideas are summarized in [31], and criticized for not extending to higher-dimensional spaces. Another discussion of fuzzy system stability is given in [32], based on geometric criteria.

A somewhat different interpretation of a linguistic space is given in [33]. This paper defines a linguistic state space, and formalizes the ordering of linguistic terms on each axis of the space, by using a *standard vector*. This standard vector contains all atomic terms of a linguistic variable, arranged in a "reasonable" ordering. For an N-dimensional state space, there will thus be N standard vectors, each associated with one dimension of the state space.

2.5 Neuro-Fuzzy Systems

Neuro-fuzzy systems are hybrids of fuzzy systems and neural networks. The goal of these systems is to combine the learning capability of a neural network with the intuitive representation of knowledge found in a fuzzy system. This may be accomplished by designing a network architecture to mimic a fuzzy system, by incorporating linguistic terms into the computations performed by the network, by means of an explanation mechanism for the network, and so forth. In the past decade, a very large number of papers have been published on neuro-fuzzy systems, describing a wide variety of architectures and applications. A good introduction to neuro-fuzzy systems may be found in [34]. We will review four architectures of particular interest in this paper: the well-known ANFIS architecture developed by Jang [35], a fuzzy neural network developed by Pal and Mitra [36], an architecture for processing linguistic values developed by Park and Kandel [37], and a fuzzified multilayer perceptron developed by Hayashi et al. [38].

Jang's ANFIS system is perhaps the best-known neuro-fuzzy system, in part because an implementation of ANFIS is provided in the MATLAB® Fuzzy Logic Toolbox. ANFIS is a feedforward neural network whose structure mimics the operation of a TSK fuzzy system, whose rules are of the form

If x_1 *is* A_1 *and...and* x_n *is* A_n *Then* $y=a_0+a_1x_1+\ldots+a_nx_n$ (3)

where x_i is the i-th system input, originating from the universe of discourse U_i, A_i is a fuzzy set on U_i, and $a_j \in R$, i=(1,2,...,n), j=(0,1,...,n). ANFIS mimics the operation of a TSK fuzzy system by using a specialized network architecture. The ANFIS network is a 5-layer network, in which the layers are not fully connected, and the transfer function of a neuron is determined by what layer the neuron is in. Referring to Figure 2, layer 0 of the network is a non-computational layer for distributing the system inputs to layer 1. Layer 1 nodes each implement the membership function of a fuzzy set. Each system input is associated with a linguistic variable, having a fixed number of possible linguistic values, represented by fuzzy sets. Layer 0 distributes each system input to the subset of layer 1 nodes that represent the possible linguistic values for that input. Each layer 2 node computes the firing strength of a rule. A layer 2 node receives its inputs from only those layer 1 nodes representing the antecedents of a rule. The structure of a specific fuzzy rulebase is thus encoded in the connection pattern between layers 1 and 2. Layer 3 nodes compute the normalized firing strength of a rule, and layer 4 nodes compute the normalized output of a rule. Finally, the layer 5 node computes the weighted sum of the rule outputs. Clearly, each ANFIS network corresponds to a single fuzzy system. This correspondence means that the knowledge in an ANFIS network can be easily understood by human beings. However, ANFIS does not yield new insights into the distributed form of knowledge employed by neural networks in general.

Pal and Mitra's fuzzy neural network encodes each input as the membership grade of that input for 3 predetermined fuzzy sets. Thus, N system inputs are preprocessed into 3N inputs to a standard multilayer perceptron. The outputs of the network are treated as fuzzy classifications. While this approach is interesting, in that it is a form of symbolic processing, the internal operation of the network is unchanged; only the encoding of the inputs and outputs has been altered.

Park and Kandel have proposed another encoding scheme for linguistic symbols. This approach relies on partitioning the interval [0,1] into *k* uniform intervals, with each interval representing one of the *k* possible values of a linguistic variable. The inputs will be encoded as the midpoint of the interval representing that linguistic value, while each output o_i is interpreted as the linguistic term associated with the interval containing o_i. The authors refer to this encoding as the Multi-Level Grading Rule (MLGR). For the i-th term of a linguistic variable having *k* possible values, the input may be directly encoded using the formula

$$\frac{2(i-1)+1}{2k} \tag{4}$$

This work is significant because it uses neural networks to explicitly process linguistic symbols, whereas Pal and Mitra's network implicitly assumes that the inputs are encoded into a representation of symbolic values. However, as with Pal and Mitra's work, networks using the MLGR are merely interpreting encoded knowledge; no alteration to the network itself has been undertaken, and thus they shed no light on the knowledge stored within the network.

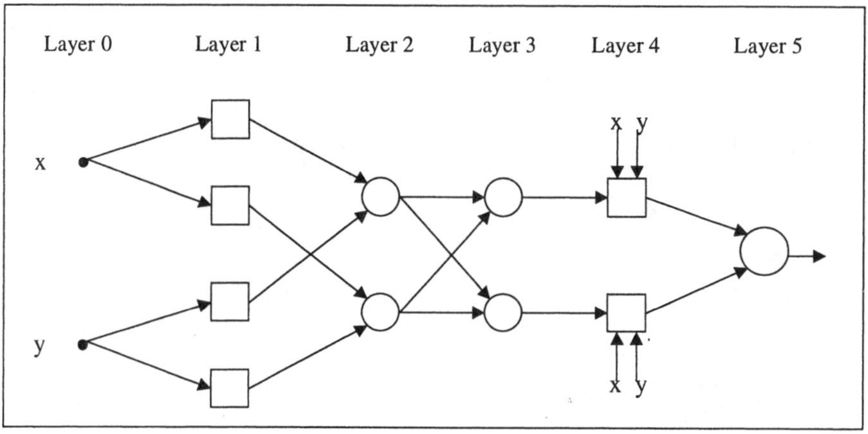

Figure 2: ANFIS Network [35]

Hayashi et al. fuzzified the multilayer perceptron using two distinct approaches. In both approaches, the inputs, connection weights, and the outputs were defined as fuzzy numbers. The first approach was to replace numeric values with fuzzy numbers in the computational formulas and the Backpropagation algorithm. This differs from performing a new derivation of the Backpropagation algorithm for fuzzy numbers, as the operations of fuzzy differential calculus are explicitly not used. This approach was tested, and produced good results. The second approach was to use α-cuts of the fuzzy numbers. The authors analytically determined that this approach would not necessarily result in proper fuzzy numbers, and so it was abandoned. In this paper, we will adopt the first approach for defining the node functions and learning rule of our network; we will replace the numeric values and operators with their linguistic equivalents.

2.6 Information Granularity

The formal investigation of granularity in fuzzy systems began with a paper by Zadeh in 1979 [39]. In this early paper, Zadeh defines a granule as a collection of indistinguishable objects, and information granularity as the grouping of objects into granules. The size of the granules is determined by the precision with which we are able to measure (or *desire* to measure) some attributes of the underlying objects. Formally, a granule in this paper is defined as a possibility distribution. In a second paper [40], Zadeh outlines a method for executing computations on purely linguistic quantities, by means of constraint propagation. Information granules are the basic units of computation in this paper, with each linguistic term being associated with a granule. Ideas from both of these papers are drawn together in [41], in which granules are treated as constraints on a variable. Possibilistic constraints are one of several classes of constraint that can define a granule. In [41], Zadeh describes fuzzy information granulation as the unifying principle in fuzzy set theory, fuzzy systems, and fuzzy logic, as well as a key mechanism in human cognition.

Hathaway et al. identified three types of data in a general sensor fusion problem: numeric data, interval-valued data, and fuzzy numbers [42]. A parametric model for fusing all three forms of data was developed. While these three forms of data are properly viewed as different granularities of information, the authors do not rank them by their coarseness. In [43], Bortolan and Pedrycz show that a reasonable ranking would be (numeric, interval, fuzzy) in order of increasing coarseness. This is determined by elucidating mechanisms to transform one kind of data to another. Numeric and interval-valued data can both be transformed into fuzzy data through possibility theory. However, while we need only compute the possibility measure of numeric data, we must compute possibility and necessity measures to transform interval-valued data into fuzzy data. These ideas are applied to creating non-parametric data fusion methods for numeric, interval-valued and fuzzy data in [44].

A recent development is the use of type-2 fuzzy sets in [45]. Type-2 fuzzy sets are fuzzy sets whose membership grades are themselves fuzzy sets. Type-2 fuzzy sets were first described by Zadeh in [8], but have received scant attention since then. Formally, a type-2 fuzzy set A is an ordered pair $(x, \mu(x))$, in which x is a element of some universe of discourse U, and $\mu(x)$ is a fuzzy set on U representing the membership of x in A. Type-2 fuzzy sets are useful when we cannot precisely determine the membership of x in A. Type-2 fuzzy sets are a means of catching types of uncertainty that, while falling under the general category of fuzziness, are not represented well by ordinary fuzzy sets. For example, if the membership values of a fuzzy system were determined from a data set, then any noise in the data set will introduce errors into the membership functions. However, if type-2 fuzzy sets are used, then the fuzzy membership value of each element can include information about the noise present in the data set.

While information granularity as a topic in itself has only recently become a major research focus for the fuzzy systems community, applications of the basic ideas have been proposed for years. Linguistic models of complex systems were an early application of fuzzy logic. As the name implies, linguistic models are models of some system whose quantities are treated as linguistic values instead of numeric values. In all of these models, the dynamics of the model are controlled by a set of fuzzy rules. Thus, linguistic models and fuzzy systems are closely related. One of the main differences is in the application areas. Fuzzy systems are usually used to model or control complex, ill-defined plants. While linguistic models have been used to model complex plants, they have also been used in the social sciences to create simulations of human behavior. Some examples of this latter field include modeling organizational behaviors [46] and modeling how people gain and exercise personal power [47]. Both of these areas are highly subjective, and difficult to model numerically. A recent paper by Pedrycz describes a fuzzy modeling algorithm in which the granularity of the model is an explicit parameter [55]. A context-based clustering algorithm is used, in which each context is a fuzzy set. The model generated by this process is encoded into a fuzzy neural network.

The human ability to operate under uncertain conditions has not escaped the notice of artificial intelligence researchers working outside the domain of fuzzy systems. A branch of AI research known as "qualitative reasoning" also deals with information granularity. In qualitative reasoning, numerical values are grouped into intervals representing qualitative values, and these intervals receive labels. Mathematical operators may then be defined for the qualitative values. The sign algebra is the most common form of qualitative reasoning. It consists of the set of values {-, 0, +, ?}, representing the intervals $(-\infty,0)$, 0, $(0,\infty)$ and unknown, respectively. Reasoning in this algebra consists of determining the sign of a quantity, if it can be determined without appealing to numerical precision. If it cannot, then the result is said to be ambiguous (?). One application of the sign algebra was in developing a qualitative physics [48]. Since any interval of the real line is by definition a convex, normal fuzzy set, there is a clear relation between qualitative reasoning and linguistic modeling. This relationship was explored in [49].

An application of qualitative reasoning that is of great significance to this paper was presented by Seix et al. in [50]. In this paper, the authors identify the main contribution of the Backpropagation algorithm to be the *direction* of the update for each parameter, rather than the exact value of the update. That being the case, the authors propose a set of modified Backpropagation algorithms in which different parameters were treated as qualitative values. The network error, the connection weights and the derivative of the activation function can all be treated as qualitative values. In total, the authors report the development of 14 qualitative versions of Backpropagation, of which three performed comparably to the standard Backpropagation algorithm over four test problems.

3. Linguistic Connection Weights

In this section, we describe the theoretical developments underlying the LNN architecture. Generalized Linguistic Variables are mathematical quantities that subsume linguistic variables as defined by Zadeh in [8]. The development of these quantities was required by the syntactic and semantic properties of linguistic connection weights. Using the values of a generalized linguistic variable as a set of vectors, we are able to define a linguistic arithmetic, which forms a vector space over the field of integers. A proof of this contention is an important result in this section. Linguistic arithmetic is used in Section 4 to perform both the forward computational pass and the backward learning pass in the LNN.

3.1 Generalized Linguistic Variables

As noted in Subsection 2.2, a linguistic variable establishes a close coupling between the syntax and semantics of a linguistic value. New linguistic values are generated by adding hedges to an existing value, following the grammar in the syntactic rule S. The meaning of these new values are determined by applying hedge functions to the fuzzy set representing the meaning of the previous value, as determined by the semantic rule M. Virtually all practical applications of fuzzy systems have the syntactic rule S equal to null for all LVs. Thus, only atomic terms exist, and the semantic rule associates each term with a fuzzy set. Fuzzy rulebases can thus be implemented as lookup tables, rather than having to compute a composition of several functions for each term. While this provides simplicity, any adaptation involving linguistic values must be performed by altering the semantic rules of an LV. This, of course, is a numeric procedure.

The goal of the LNN is to perform learning without using numeric weights. We propose that each connection weight in the network be defined as a linguistic variable. Adaptation in the neural network will consist of adding hedges to the weights. Philosophically, the hedges used to update the connection weights should increase or decrease the value of a weight. The hedges proposed by Zadeh in [10], which remains a definitive work in the area, do not have this effect on symmetric fuzzy sets. Thus, we now propose two novel hedges, "less than" and "greater than."

Definition 1: The hedge *less than* is defined by the function

$$\mu_{lessthan}(x) = \begin{cases} (\mu(x))^{0.5} & if\ x \leq x_0 \\ (\mu(x))^2 & if\ x \geq x_0 \end{cases} \tag{5}$$

Definition 2: The hedge *greater than* is defined by the function

$$\mu_{greaterthan}(x) = \begin{cases} (\mu(x))^2 & if\ x \leq x_0 \\ (\mu(x))^{0.5} & if\ x \geq x_0 \end{cases} \tag{6}$$

where x is an element of a universe of discourse U, $\mu(x)$ is the membership function of some fuzzy set A on U, x_0 is the single unique point for which $\mu(x_0) = 1$, $\mu_{\text{less than}}(x)$ is the membership function for "less than A", and $\mu_{\text{greater than}}(x)$ is the membership function for "greater than A." As can be seen in Figure 3 (where "small" is defined by Figure 1), the effect of these hedges is to skew a membership function in the positive or negative direction, which is what we desire. These hedges are different from most hedges in the current literature because they do not act uniformly on an entire membership function. They do resemble the INT() operator found in [10], which also acts non-uniformly on a membership function. However, where the INT() operator discriminates between values of the membership function, the hedges *less than* and *greater than*

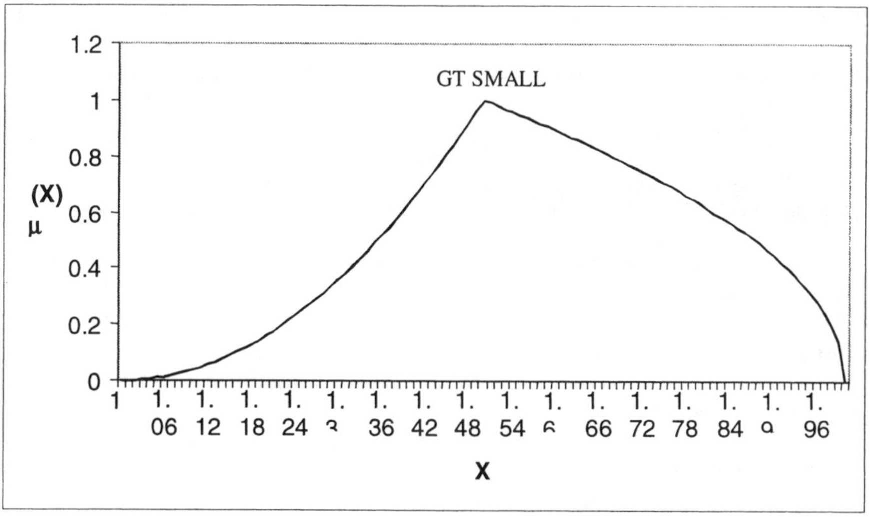

Figure 3(a): Greater Than Small (see Fig. 1)

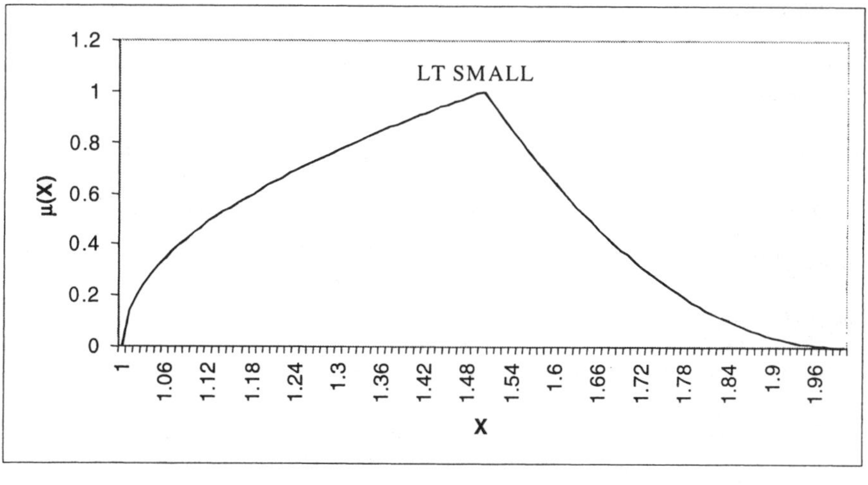

Figure 3(b): Less Than Small (see Fig. 1)

discriminate between the values of elements of the universe of discourse. Note also that Zadeh provides an alternative definition of *greater than* in [10].

Having defined our hedges, we now define the linguistic variables that will form the connection weights of the LNN. Each weight variable will be defined on the universe of discourse [-1,1], with a term set containing the atomic terms Negative Large, Negative Medium, Negative Small, Almost Zero, Positive Small, Positive

Medium, Positive Large, and the hedges Greater Than and Less Than. In the remainder of this paper, we will refer to these terms and hedges solely by their acronyms. The term set is thus {NL, NM, NS, AZ, PS, PM, PL, GT, LT}. The semantic rule of each linguistic variable will associate each atomic term with a fuzzy set as shown in Figure 4, and will associate LT and GT with the functions in Definitions 1 and 2, respectively. Note that this semantic rule uses triangular fuzzy sets with a 50% overlap, forming a uniform partition of [-1,1] (see [31]).

We now encounter a problem with the classical definition of the syntactic rule. Linguistically, there is a point at which "greater than...greater than positive small" is equivalent to "less than...less than positive medium." This is reflected in the semantic rule we have defined; as can be seen in Figure 5, applying an infinite number of GT hedges to PS yields the same fuzzy set as applying an infinite number of LT hedges to PM. Thus, the syntactic rule must rewrite GT...GT PS as LT...LT PM. Henceforth, we refer to this rewrite as the *crossover* operation. However, that is not possible under the current definition of a linguistic variable. The syntactic rule S is a context-free grammar, which can only replace a single nonterminal with a string of terminal and nonterminal symbols. In order to implement the syntactic crossover, we must redefine S to be a phrase structure grammar. Since a phrase structure grammar is more general than a context-free grammar, we refer to this new entity as a generalized linguistic variable (GLV). Zadeh briefly considered using a phrase structure grammar as a syntactic rule in [9], but decided the computational burden of computing memberships would be too great. In this paper, however, we do not need to compute membership values, and so GLVs are useful to us.

Definitions (1) and (2) clearly imply that GT and LT are semantic inverses of each other; LT GT t = GT LT t = t. We will include this behavior in the syntactic rule by adding the productions LT GT→ null and GT LT→ null to S. Since these productions remove any interleaving of hedges, we can implement the crossover operation by choosing some limit χ on the number of GT or LT hedges that may be added to an atomic term. To formalize this notion, let us first order the atomic terms of a weight variable ω based on the ordering of their fuzzy sets [58]. Adopting the convention of a standard vector from [33], we represent the ordering of atomic terms in ω as the standard vector W_ω=(NL,NM,NS,AZ,PS,PM,PL). A crossover between two consecutive elements of W_ω, t_1 and t_2, ($t_1 < t_2$) occurs when $\chi+1$ GT hedges are added to t_1. This composite term must be re-written as χ LT hedges added to t_2. Conversely, if $\chi+1$ LT hedges are added to t_2, then we re-write the term as χ GT hedges added to t_1. We accomplish this by adding 2 productions for each pair of consecutive terms in W_ω to the syntactic rule S_ω:

$$(GT)^{\chi+1}\, t_1 \rightarrow (LT)^{\chi}\, t_2 \tag{7a}$$
$$(LT)^{\chi+1}\, t_2 \rightarrow (GT)^{\chi}\, t_1 \tag{7b}$$

where $(GT)^k$ t means that k GT hedges are added to the atomic term t. We have seen that two consecutive atomic terms of W_ω will converge semantically if an infinite number of hedges are added to each (see Figure 5). However, setting χ to infinity does not granulate the network weights at all; we have merely renamed the values of [-1,1]. Instead, we desire a finite value for χ that gives an acceptable approximation to $\chi=\infty$. Our criteria for an "acceptable" approximation was that the center of gravity for $(GT)^{\chi}\ t_1$ and $(LT)^{\chi}\ t_2$ be within some small neighborhood of the center of gravity for $(GT)^{\infty}\ t_1$ (or equivalently, $(LT)^{\infty}\ t_2$). After numerical exploration, we determined that $\chi=7$ was an appropriate crossover limit. Note that, since NL is the least element in the standard vector and PL is the greatest element of the standard vector, an infinite number of LT hedges may be added to NL, and an infinite number of GT hedges may be added to PL.

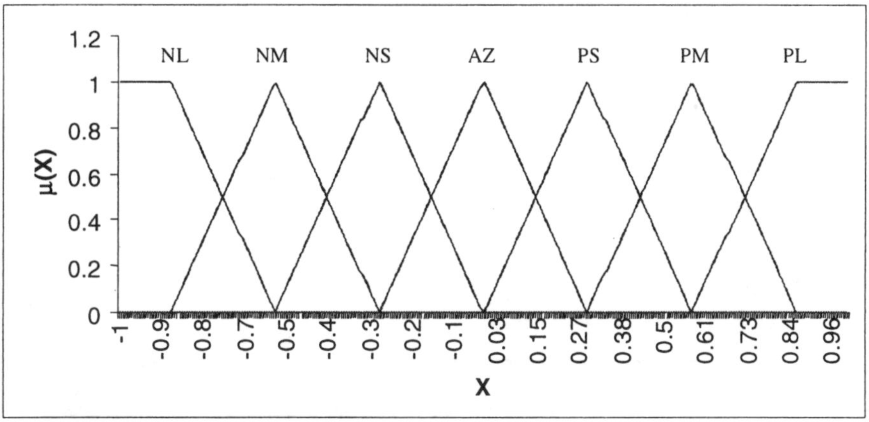

Figure 4: Semantic Rule for Linguistic Weights

3.2 Linguistic Arithmetic

In Section 4, we will find that four operations are required to implement the LNN: linguistic addition, linguistic subtraction, linguistic product, and the product of a real-valued scalar with a linguistic term (scalar product). In this section, we define these operations. The operators in linguistic arithmetic are based on two concepts: an ordering of all the terms in a weight variable ω, and a difference between two linguistic terms. We represent the ordering of terms in the set of possible values τ of ω via a standard sequence σ. This is an adaptation of the standard vector for the case of an infinite number of terms in τ. As with a standard vector, if a term t_1 precedes a term t_2 in σ, then $t_1 < t_2$. The ordering of terms in σ is based on four rules:

- For any term t, LT t < t
- For any term t, t < GT t
- For any term t, if GT t triggers a crossover, then t < crossover(GT t)
- For any term t, if LT t triggers a crossover, then crossover(LT t) < t

where crossover(t) denotes the crossover operation. The standard sequence σ is

σ = ...$(LT)^2$ NL, LT NL, **NL**, GT NL,...,$(GT)^7$ NL, $(LT)^7$ NM,...,LT NM, **NM**, GT NM,...,$(GT)^7$ NM, $(LT)^7$ NS,..., LT NS, **NS**, GT NS,...,$(GT)^7$ NS, $(LT)^7$ AZ,...,LT AZ, **AZ**, GT AZ,...,$(GT)^7$ AZ, $(LT)^7$ PS,...,LT PS, **PS**, GT PS,..., $(GT)^7$ PS, $(LT)^7$ PM,...,LT PM, **PM**, GT PM,...,$(GT)^7$ PM, $(LT)^7$ PL,...,LT PL, **PL**, GT PL, $(GT)^2$ PL,... (8)

where $(GT)^k$ A means the term formed by adding k GT hedges to the atomic term A, and $(LT)^k$ A means the term formed by adding k LT hedges to A. Note that this definition of σ is specific to the term set τ; in general, it will be necessary to construct a separate standard sequence for each GLV.

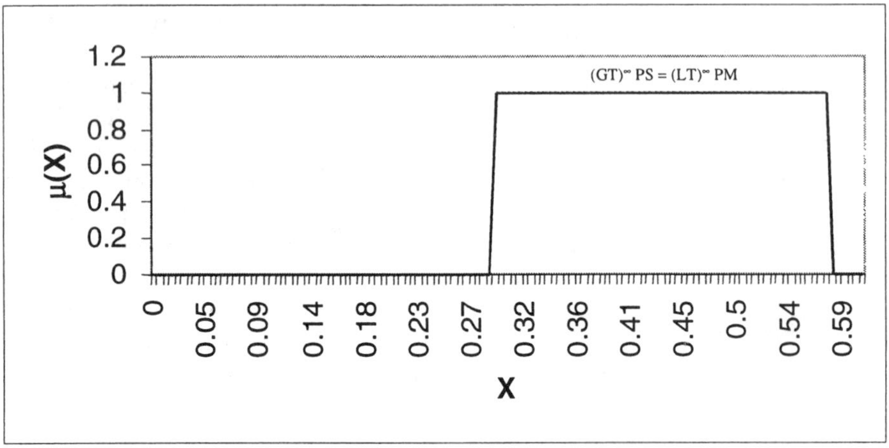

Figure 5: $\chi=\infty$

The second key concept in linguistic arithmetic is the idea of a difference between two linguistic terms. Linguistic differences have been studied in [61-63], amongst others. We include the following development for the sake of completeness. The point of departure for this discussion is contained in a paper by Stilman [51]. In this paper, Stilman defines a distance metric using a family of reachability sets. These sets are subsets of some universe of discourse, whose elements represent locations in a network. The set of all points y that can be reached from a point x in a single transition is a reachability set for x. Based on this idea, Stilman defines the distance between two points x and y as the minimum number of transitions required to reach y from x.

We adapt Stilman's idea of reachability to linguistic terms by noting that any two terms in the standard sequence σ will be separated by a known number of other entries in σ. For instance, there are exactly 14 terms that appear between AZ and PS in σ. We will henceforth refer to these "in-between" terms as *intervening* terms.

Definition 3: The distance $\lambda(t_1, t_2)$ between two terms t_1 and t_2 appearing in σ equals the number of intervening terms plus one. λ equals 0 iff $t_1 = t_2$.

Definition 4: The difference $D(t_1, t_2)$ between two terms t_1 and t_2 in σ is an integer whose absolute value equals $\lambda(t_1, t_2)$, and whose sign is positive if $t_1 < t_2$ and negative if $t_1 > t_2$.

Definitions (3) and (4) together define a difference for the term set τ, as ordered by the standard vector σ. While it would be possible to directly define linguistic addition and subtraction operators using these definitions, defining linguistic and scalar products would be more difficult. In the terminology of [52], the term set τ is currently an interval scale, in which the elements are totally ordered and there is a known distance between any two elements. Linear operations such as addition and subtraction can be defined for an interval scale; operations such as the product of two elements may *not* be defined, since there is no multiplicative zero. Thus, instead of using Definitions (3) and (4) to define operations directly, we will use them to generate a mapping that gives us a multiplicative zero as well as a common framework for defining all operations of linguistic arithmetic.

Definition 5: The function $\Gamma(t)$ is a mapping

$$\Gamma: \tau \rightarrow I$$
$$\Gamma(t) = D(AZ, t) \quad (9)$$

where $t \in \tau$, and I is the set of integers. When $t = AZ$, $\Gamma(t) = 0$, giving us a multiplicative zero. This mapping treats a linguistic value as a vector in a one-dimensional space of linguistic terms, which describes the location of that term relative to AZ. This corresponds to the classical interpretation of a number as a vector on the real line, describing the location of the number relative to 0. We will define the operations of linguistic arithmetic in terms of $\Gamma(t)$, which will allow us to use known properties of integer arithmetic to prove important properties of linguistic arithmetic. We will first show that τ is countable.

Theorem 1: The term set τ, as ordered by the standard vector σ, is countable.

Proof: We will prove Theorem (1) by showing that Γ is a bijection between τ and I. This will be accomplished by showing that Γ is (a) right unique, (b) left unique, and (c) a surjective mapping from τ to I [53].

(a) Γ is right unique iff $(\Gamma(t) = x) \wedge (\Gamma(t) = y) \Rightarrow x = y$. Since σ is fixed and a total ordering of τ, we can always determine the distance $\lambda(AZ, t)$, as well as which of the three mutually exclusive relations $AZ < t$, $AZ = t$, $AZ > t$ holds true. Thus, for all $t \in \tau$, the value of D(AZ, t) is known. Further, since σ is static, D(AZ, t) can map to exactly one value in I. Since $\Gamma(t) = D(AZ, t)$, Γ is right unique.

(b) Γ is left unique iff $(\Gamma(t_1) = x) \wedge (\Gamma(t_2) = x) \Rightarrow t_1 = t_2$. As observed in part (a), we can always determine the value of D(AZ, t) for any $t \in \tau$. In order for two different terms t_1 and t_2 to map to the same value of Γ, they must lie at the same distance λ from AZ, in the same direction. Thus, there must be two entries in σ that are equivalent to each other. An inspection of σ shows that this is not the case, and so Γ is left unique.

(c) $\Gamma: \tau \rightarrow I$ is a surjective mapping if the domain of Γ is τ and the range of Γ is I. Clearly, the domain of Γ is τ. To show that the range of Γ is I, observe first that $x \in \{-1, 0, 1\}$ is in the range of Γ, since $\Gamma(\text{LT AZ}) = -1$, $\Gamma(\text{AZ}) = 0$, and $\Gamma(\text{GT AZ}) = 1$. Next, observe that for every $x \in I \leq -2$, there exists some term $t < AZ$ for which there are $|x| - 1$ intervening terms between t and AZ. Thus D(AZ, t) must be x. Similarly, for every $x \in I \geq 2$, there will be some term $t > AZ$ for which there are $x-1$ intervening terms between AZ and t. Thus $D(AZ, t) = x$. Thus, the range of $\Gamma$is I.

Since (a), (b) and (c) are true, Γ is a bijection, and thus τ is a countable set, Q.E.D.

Definition 6: The inverse of the function Γ is a mapping
$\Gamma^{-1}: I \rightarrow \tau$
$\Gamma^{-1}(x)$ is the term t for which $\Gamma(t) = x$, $t \in \tau$, $x \in I$.

Since Γ is a bijection, its inverse Γ^{-1} is also a function. Since $\Gamma(AZ) = 0$ provides us with a multiplicative zero, τ is a ratio scale, on which we may define product operations as well as linear operations. We now use Γ and Γ^{-1} to define the operations of linguistic arithmetic.

Definition 7: The linguistic sum $\oplus$ of two terms t_1 and t_2 in τ is given by
$\oplus: \tau \times \tau \rightarrow \tau$

$$t_1 \oplus t_2 = \Gamma^{-1}(\Gamma(t_1) + \Gamma(t_2)) \tag{10}$$

Definition 8: The linguistic difference '-' of two terms t_1 and t_2 in τ is given by
$-: \tau \times \tau \rightarrow \tau$

$$t_1 - t_2 = \Gamma^{-1}(\Gamma(t_1) - \Gamma(t_2)) \tag{11}$$

Definition 9: Rounding a real value x to the nearest integer is defined as

$$round(x) = \begin{cases} \lfloor x+0.5 \rfloor, & if\ x > 0 \\ -\lfloor |x|+0.5 \rfloor, & if\ x < 0 \end{cases} \tag{12}$$

Definition 10: The scalar product $\otimes$ of a real value s and a term t in τ is given by
$\otimes: R \times \tau \rightarrow \tau$
$$s \otimes t = \Gamma^{-1}(\text{round}(s * \Gamma(t))) \tag{13}$$

Definition 11: The linguistic product $\otimes$ of two terms t_1 and t_2 in τ is given by
$\otimes: \tau \times \tau \rightarrow \tau$
$$t_1 \otimes t_2 = \Gamma^{-1}(\text{round}([\Gamma(t_1) * \Gamma(t_2)]/\chi)) \tag{14}$$

where χ is the crossover limit defined in Subsection 3.1. The rounding operation ensures that the argument of Γ^{-1} is always the integer that best represents the actual numeric value computed. The presence of this rounding operation, as well as the scaling factor χ in Definition (11), means that the linguistic arithmetic is not a simple morphism of integer arithmetic. Furthermore, those terms t for which $|\Gamma(t)| \leq \chi$ will act as "fractions"; the absolute value of the product of any other term with one of these "fractional" terms will be less than the absolute value of that other term. This is important for the dynamics of the LNN; the product of two small terms should be another small term, else the LNN could not converge to a stable result. In the next section, we shall offer a proof that the linguistic arithmetic forms a vector space over the field of integers.

3.3 A Vector Space of Linguistic Values

A vector space is an abstract means of representing a mathematical system that possesses addition and scalar multiplication operations (also known as a linear system) [54]. A vector space is a 4-tuple (F, V, +, *), where F is a field of scalars, V is a set of objects (known as vectors), + is the addition operator, and * is the scalar multiplication operator. The 4-tuple (F, V, +, *) qualifies as a vector space if it satisfies a collection of 10 axioms.

Theorem 2: The 4-tuple $(I, \tau, \oplus, \otimes)$ is a vector space, where I is the set of integers, τ is the set of values of ω ordered by the standard sequence σ, $\oplus$ is given by Definition (7), and $\otimes$ is given by Definition (10).
Proof: There are 10 axioms that $(I, \tau, \oplus, \otimes)$ must satisfy to qualify as a vector space [54]. These axioms are listed individually in parts (a) - (j) of this proof.

(a) For every $\alpha, \beta \in \tau, \alpha \oplus \beta \in \tau$
By definition, $\alpha \oplus \beta = \Gamma^{-1}(\Gamma(\alpha) + \Gamma(\beta))$. Since $\Gamma(\alpha), \Gamma(\beta) \in I$, therefore $\Gamma(\alpha) + \Gamma(\beta) \in I$. Since Γ is a bijection between τ and I, $\Gamma^{-1}(\Gamma(\alpha) + \Gamma(\beta)) \in \tau$. Hence, part (a) is proven.

(b) For every $\alpha, \beta \in \tau, \alpha \oplus \beta = \beta \oplus \alpha$
$$\Gamma^{-1}(\Gamma(\alpha) + \Gamma(\beta)) = \Gamma^{-1}(\Gamma(\beta) + \Gamma(\alpha)) \tag{15a}$$
$$\Gamma^{-1}(\Gamma(\alpha) + \Gamma(\beta)) = \Gamma^{-1}(\Gamma(\alpha) + \Gamma(\beta)) \tag{15b}$$
Since addition of integers is commutative, part (b) is proven.

(c) For all $\alpha, \beta, \gamma \in \tau$, $\alpha \oplus (\beta \oplus \gamma) = (\alpha \oplus \beta) \oplus \gamma$

$\Gamma^{-1}[\Gamma(\alpha) + \Gamma(\Gamma^{-1}(\Gamma(\beta) + \Gamma(\gamma)))] = \Gamma^{-1}[\Gamma(\Gamma^{-1}(\Gamma(\alpha) + \Gamma(\beta))) + \Gamma(\gamma)]$ (16a)

$\Gamma^{-1}[\Gamma(\alpha) + \Gamma(\beta) + \Gamma(\gamma)] = \Gamma^{-1}[\Gamma(\alpha) + \Gamma(\beta) + \Gamma(\gamma)]$ (16b)

Since Γ is bijective, $\Gamma(\Gamma^{-1}(x)) = x$. Thus, part © is proven.

(d) For all $\alpha \in \tau$, $AZ \oplus \alpha = \alpha$, where AZ is the unique additive identity

$\Gamma^{-1}(\Gamma(AZ) + \Gamma(\alpha)) = \alpha$ (17a)

$\Gamma^{-1}(0 + \Gamma(\alpha)) = \alpha$ (17b)

$\alpha = \alpha$ (17c)

Since 0 is the additive identity for integers, part (d) is proven.

(e) For every $\alpha \in \tau$, there is a unique vector $-\alpha$ such that $\alpha \oplus -\alpha = AZ$

We define $-\alpha$ as the element of τ for which $\Gamma(-\alpha) = -\Gamma(\alpha)$. Since Γ is a bijection between τ and I, $-\alpha$ exists and is unique.

$\Gamma^{-1}(\Gamma(\alpha) + \Gamma(-\alpha)) = AZ$ (18a)

$\Gamma^{-1}(\Gamma(\alpha) - \Gamma(\alpha)) = AZ$ (18b)

$AZ = AZ$ (18c)

(f) For all $c \in I$ and all $\alpha \in \tau$, $c \otimes \alpha \in \tau$

By Definition (10), $c \otimes \alpha = \Gamma^{-1}(\text{round}(c * \Gamma(\alpha)))$. Observe that for any $x \in I$, round(x) = x. Thus, for $c \in I$,

$c \otimes \alpha = \Gamma^{-1}(c * \Gamma(\alpha))$. Since $(c * \Gamma(\alpha)) \in I$ due to the closure of multiplication on integers, $\Gamma^{-1}(c * \Gamma(\alpha)) \in \tau$. This completes the proof for part (f).

(g) For all $\alpha \in \tau$, $1 \otimes \alpha = \alpha$

$\Gamma^{-1}(\text{round}(1 * \Gamma(\alpha))) = \alpha$ (19a)

$\Gamma^{-1}(\text{round}(\Gamma(\alpha))) = \alpha$ (19b)

$\alpha = \alpha$ (19c)

Since 1 is the multiplicative identity for integers, part (g) is proven.

(h) For all $c, d \in I$ and all $\alpha \in \tau$, $(c * d) \otimes \alpha = c \otimes (d \otimes \alpha)$

$\Gamma^{-1}(\text{round}((c * d) * \Gamma(\alpha))) = \Gamma^{-1}(\text{round}(c * \Gamma(\Gamma^{-1}(\text{round}(d * \Gamma(\alpha))))))$ (20a)

$\Gamma^{-1}(\text{round}(c * d * \Gamma(\alpha))) = \Gamma^{-1}(\text{round}(c * \Gamma(\Gamma^{-1}(d * \Gamma(\alpha)))))$ (20b)

$\Gamma^{-1}(\text{round}(c * d * \Gamma(\alpha))) = \Gamma^{-1}(\text{round}(c * (d * \Gamma(\alpha))))$ (20c)

$\Gamma^{-1}(\text{round}(c * d * \Gamma(\alpha))) = \Gamma^{-1}(\text{round}(c * d * \Gamma(\alpha)))$ (20d)

Using the associative law for integer multiplication, part (h) is proven.

(i) For all $c \in I$ and all $\alpha, \beta \in \tau$, $c \otimes (\alpha \oplus \beta) = c \otimes \alpha \oplus c \otimes \beta$

$\Gamma^{-1}(\text{round}(c * \Gamma(\Gamma^{-1}(\Gamma(\alpha) + \Gamma(\beta))))) = \Gamma^{-1}[\Gamma(\Gamma^{-1}(\text{round}(c * \Gamma(\alpha)))) + \Gamma(\Gamma^{-1}(\text{round}(c * \Gamma(\beta))))]$ (21a)

$\Gamma^{-1}(\text{round}(c * (\Gamma(\alpha) + \Gamma(\beta))) = \Gamma^{-1}[\Gamma(\Gamma^{-1}(c * \Gamma(\alpha)))+\Gamma(\Gamma^{-1}(c*\Gamma(\beta)))]$ (21b)

$\Gamma^{-1}(c * \Gamma(\alpha) + c * \Gamma(\beta)) = \Gamma^{-1}(c * \Gamma(\alpha) + c * \Gamma(\beta))$ (21c)

Using the distributive law for integers, part (i) is proven.

(j) For all c, d $\in$ I and all $\alpha \in \tau$, $(c + d) \otimes \alpha = c \otimes \alpha \oplus d \otimes \alpha$

$$\Gamma^{-1}(\text{round}((c + d) * \Gamma(\alpha))) = \Gamma^{-1}[\Gamma(\Gamma^{-1}(\text{round}(c * \Gamma(\alpha)))) + \Gamma(\Gamma^{-1}(\text{round}(d * \Gamma(\alpha))))] \quad (22a)$$

$$\Gamma^{-1}((c + d) * \Gamma(\alpha)) = \Gamma^{-1}((c * \Gamma(\alpha)) + (d * \Gamma(\alpha))) \quad (22b)$$

$$\Gamma^{-1}(c * \Gamma(\alpha) + d * \Gamma(\alpha)) = \Gamma^{-1}(c * \Gamma(\alpha) + d * \Gamma(\alpha)) \quad (22c)$$

Using the distributive law for integers, part (j) is proven. Since parts (a) - (j) are true, Theorem (2) is proven, Q.E.D.

4. The Linguistic Neural Network

In this section we present the architecture and learning rule of the LNN. The LNN is a feedforward network based on the multilayer perceptron architecture. Neurons in the network compute a weighted sum of their inputs and a bias term, and use the result as an argument to a nonlinear neuron activation function. Learning in the LNN is accomplished using backpropagation of errors from the network outputs. Each connection weight in the LNN is an independent copy of the weight variable ω introduced in Section 3. In both the computational and learning passes, the standard arithmetic operators have been replaced by their linguistic equivalents.

4.1 Network Architecture

The LNN network topology is identical to a multilayer perceptron. Neurons in the LNN are arranged in a layered feedforward neural network, having no shortcut links (see Figure 6). Each neuron is fully connected to the neurons in the next layer, and in the previous layer. Inputs are placed in the neurons of an input layer, and network outputs consist of the activation values of the neurons in the output layer. Layers in between these two are referred to as "hidden" layers; in a given network, the may be zero or more hidden layers. The input layer (layer 0) is composed of non-computational units, which each receive a single input, and distribute it to all the neurons in the next layer.

Each neuron in layer i of a network with L layers has m inputs plus a bias input, and 1 output, where m is the number of neurons in layer i-1, i=(1, 2,...,L) (see Figure 7). During the forward, computational pass of the network, each node computes a weighted sum of its inputs, and adds a bias term. In the terminology of [56], this quantity is the induced local field of the neuron. In the LNN, the connection weights and bias value are linguistic terms, drawn from the set τ, while the outputs of the previous layer are real numbers. Thus, the induced local field of neuron j in layer i, v_j is computed by

$$v_j = \left(\sum_k x_k \otimes w_{kj} \right) \oplus b_j \tag{23}$$

where x_k is the output of neuron k in layer i-1, w_{kj} is the connection weight between neuron k in layer i-1 and neuron j in layer i, b_j is the bias of neuron j in layer i, ⊕ represents the linguistic sum, ⊗ represents the scalar product, and the summation symbol represents linguistic summation. Clearly , v_j is a linguistic value. Once the induced local field has been calculated, we compute the activation value for the neuron. We use homogenous neurons for the hidden and output layers, so the output of a neuron must be a real value. This value is produced by a function

$$F : \tau \to [0,1]$$
$$F(t) = \frac{1}{1 + e^{-\Gamma(t)/(\chi/2)}} \tag{24}$$

where χ is the crossover limit for ω and $\Gamma(t)$ is given by Def. 5. Plainly, this is a modified sigmoid function, which is linear in roughly the neighborhood [NS, PS]. This mapping incorporates a change in granularity as a part of the neuron's activation function. We thus have a neural network which operates on numeric inputs and outputs, but whose knowledge is stored as a pattern of linguistic connection weights.

4.2 LNN Learning Algorithm

The learning algorithm of the LNN is a variant of the Backpropagation algorithm reported in [57]. In this variant, the operations of numeric arithmetic have been replaced by their linguistic equivalents. A note on the order of operations is required at this point; while the linguistic arithmetic forms a vector space over the field of integers, the same does not hold true for the field of real numbers. In particular, associativity does not hold in the scalar or linguistic products. Thus, the order of operations is critical. We shall make the order of operations explicit in our equations through parentheses.

We make two backward passes through the network. In the first pass, the error gradient of each node is computed. On the second pass, these gradients are used to compute an update for each weight in the network. At the output layer, we compute the function

$$e_j = d_j - y_j \tag{25}$$

where e_j is the error at output node j, d_j is the target value of node j, and y_j is the actual output of node j. These values are all numeric. The error gradient at this node γ_j is computed as

$$\gamma_f = [(d_j - y_j)\cdot(y_j\cdot(1-y_j))\cdot G]\otimes PL \qquad (26)$$

where the quantity $(y_j*(1-y_j))$ is the derivative of $y = F(t)$, G is a real-valued constant called the signal gain, and ⊗ denotes scalar multiplication. This equation differs from the normal Backpropagation expression in that a gain term is added, and the entire expression is multiplied by PL. The gain term was added after experimental evidence showed that the linguistic error signals decayed to AZ as

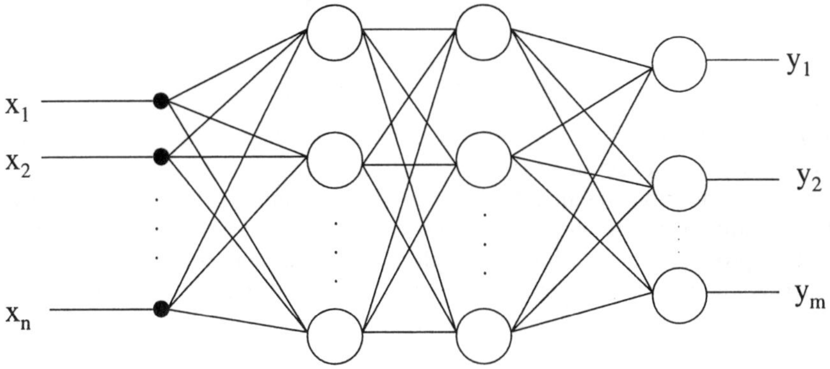

Figure 6: Network Architecture

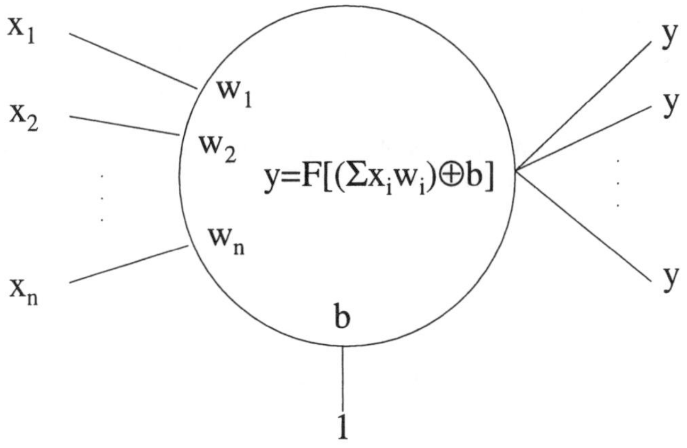

Figure 7: LNN Node

they were propagated backwards, preventing early layers from adapting. The error gradient needs to be linguistic in nature, and so the scalar multiplication operator was used. The choice of PL was motivated by the need to send sufficiently large error signals back through the network; other choices for this value can be made. Thus, a conversion from numeric to linguistic granularity and vice versa are key operations in the LNN, and fit quite naturally into the Backpropagation equations. To our knowledge, this is the first working system to exploit changes in granularity as basic computational steps.

For a hidden neuron j in layer l, we compute the local gradient γ_j in terms of the local gradients of layer $l + 1$:

$$\gamma_j = \left[(y_j \cdot (1 - y_j)) \cdot G\right] \otimes \left[\sum_k (\gamma_k \otimes w_{jk})\right] \tag{27}$$

where ⊗ represents either the scalar or linguistic product, the summation symbol represents linguistic summation, γ_k is the local gradient of node k in layer $l + 1$, w_{jk} is the connection weight between nodes j and k, and y_j is the output of node j. This equation again parallels the expression for the local gradient of a neuron in the Backpropagation algorithm. As before, the gain constant G appears in the numeric half of the expression. With the local gradient computed at each node, we execute a second pass through the network, updating the weight and bias terms as follows:

$$\Delta w_{ij} = (\eta * x_i) \otimes \gamma_j \tag{28a}$$
$$\Delta b_j = \eta \otimes \gamma_j \tag{28b}$$

where Δw_{ij} is the update to connection weight w_{ij} at node j, x_i is the i-th input to node j, γ_j is the local gradient of node j, Δb_j is the update to the bias of node j, η is a learning rate constant, and ⊗ represents the scalar product. Since the learning rate constant does not store knowledge, we define it as a numeric value. As discussed previously, the inputs and outputs of a neuron are also numeric, so the '*' operator represents the algebraic product.

4.3 Example

To demonstrate the learning process in the LNN, we will train an LNN to solve the Binary-OR problem for two inputs. The Binary-OR problem is to determine the disjunction of two bits, as in the following table:

X	Y	X or Y
0	0	0
0	1	1
1	0	1
1	1	1

Table 1: The Binary-OR problem

The network we will use is a two input, one output LNN, with two layers of two hidden nodes each. We will train the network on all four patterns, and test on all four patterns. The initial random weights of the network are given in Figure 8. After training, the network successfully classified all 4 input patterns, using the connection weights shown in Figure 9. The differences between the two figures represent the knowledge gained by the network during training.

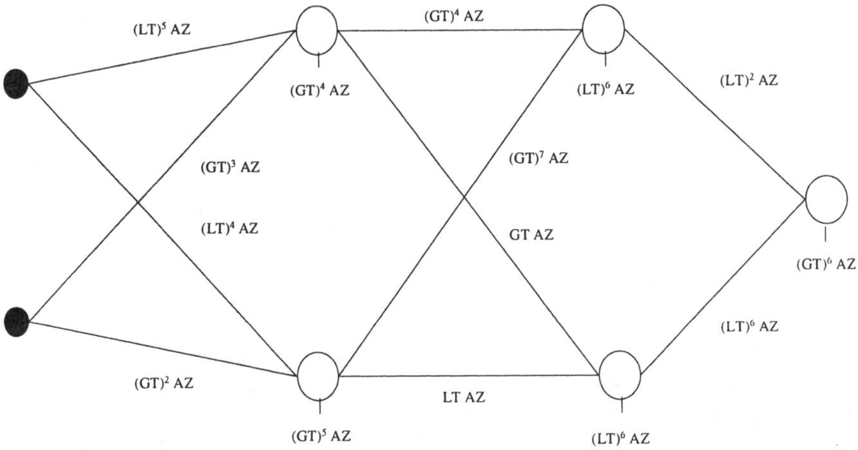

Figure 8: Initial Network Weights

5. Experiment

To test the granular computing premise, the LNN architecture and learning algorithm were implemented in C. We tested the LNN using the Iris data set [59,60]. We began by preprocessing the Iris data set, normalizing each input to the interval [0,1], and converting the class labels to a 1-of-n code. We then partitioned the data set into 10 disjoint subsets of equal size, with each example being assigned randomly to a subset. Using these 10 subsets, we performed a 10-fold cross-validation experiment, using a network with 2 hidden layers of 6 nodes each. The stopping criteria were a sum of squared error less than 18.0, or 100 iterations of training. The gain constant was set to 2.0, and the learning rate in each layer was 0.1.

The results of our experiments were intriguing. To begin with, the LNN demonstrated a habit of converging to poor local minima almost half of the time. In order to quantify this behavior, the 10-fold cross-validation experiment was repeated 30 times. On average, 5.6 of the ten partitions in one experiment would train successfully, with a standard deviation of 1.9. The remaining partitions would become trapped in one of several local minima. In order to measure the classification performance of the network, we looked at the test results from those

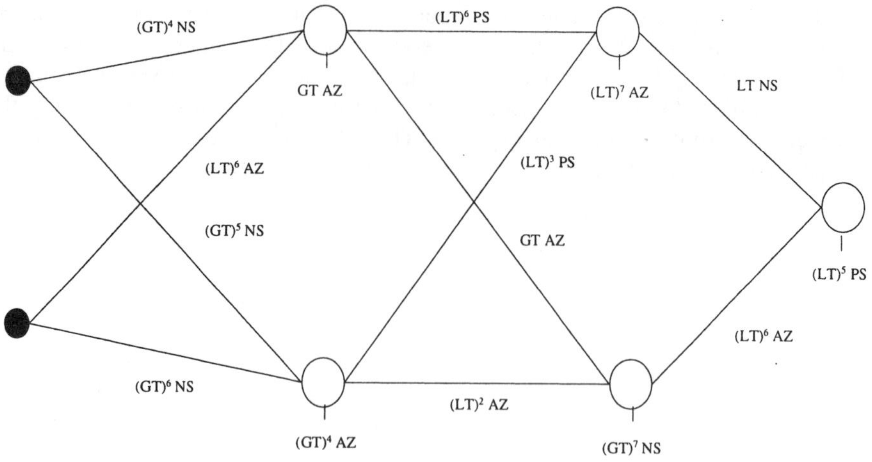

Figure 9: Trained Network Weights

same 30 runs. Partitions that did not train successfully were discarded. In the remaining partitions, the network achieved a classification rate of 90.2%. Broken down by classes, the network achieved classification rates of 99.3% for class 1, 85.2% for class 2, and 85.7% for class 3. In the Iris data set, class 1 is known to be linearly separable from classes 2 and 3, while classes 2 and 3 are not linearly separable.

Existing learning algorithms typically achieve a classification rate of about 95% on the Iris data set (see [2-4]), and neural networks that learn the Iris data set will also suffer from local minima. However, these algorithms will typically converge more often than 56% of the time. Thus, there is plainly a degradation in performance when the LNN is used. However, the degradation in performance is not overly large; when the LNN trains successfully, the testing accuracy is not much worse than in a standard neural network. Further, we believe that the LNN's increased vulnerability to local minima is actually a familiar tendency from multilayer perceptrons. When the learning rate in a multilayer perceptron is set too high, the network will not converge, and may entirely miss steep-sided valleys in weight space. It is our belief that a similar mechanism is at work in the LNN; the granulated weights are not precise enough to converge to a good local minimum from all initial positions in weight space. We thus conclude that the results of our experiment support our hypothesis that a granular neural network will learn a problem fairly well, but not as accurately as a numeric neural network.

6. Summary and Future Work

In this chapter, we proposed a new neuro-fuzzy architecture, which incorporates the ideas of granular computing. A multilayer perceptron is modified to use linguistic weights instead of numeric weights. The weights in this network were updated by adding linguistic hedges to the existing weights. The hedges used are novel, and lead to the development of a linguistic arithmetic. We proved that this linguistic arithmetic is a vector space over the field of integers. The computational and learning algorithms of our network were derived by replacing the numeric operations of a multilayer perceptron with their linguistic equivalents. We used this granular network to test the basic premise of granular computing: that by relaxing the requirement of precision in systems modeling, we suppress minor details while preserving the most significant relationships in the system. In an experiment on the Iris data set, this network achieved a classification rate of 90.2%. This is a small degradation in performance, and thus the network meets the goal of capturing the most significant relationships in a system, while ignoring fine details. Future work on this topic will include defining other linguistic network architectures (such as radial basis function networks), applying the LNN to data mining, and a rigorous analysis of why the LNN is more vulnerable to local minima than multilayer perceptrons.

Acknowledgements

This work was supported in part by the Natural Sciences and Engineering Research Council of Canada, under grant no. PGSB-222631-1999, and in part by the USF Center for Software Testing under grant no. 2108-004-00. The Iris data set was obtained from the Machine Learning Repository at the University of California, Irvine. Our thanks to Adam Schencker and Mark Last for their helpful comments.

References

[1] M. M. Gupta, "Fuzzy Neural Network Approach to Control Systems," in *Proceedings of the 1st International Symposium on Uncertainty Modeling and Analysis*, pp. 3019-3022, 1991.

[2] A. Gonzalez, R. Perez, "SLAVE: A Genetic Learning System Based on an Iterative Approach," *IEEE Transactions on Fuzzy Systems,* vol. 7 no. 2, pp. 176-191, April 1999.

[3] R. Setiono, H. Liu, "A Connectionist Approach to Generating Oblique Decision Trees," *IEEE Transactions on Systems, Man and Cybernetics, Part B: Cybernetics,* vol. 29 no. 3, pp. 440-444, June 1999.

[4] Y. Shi, R. Eberhart, Y. Chen, "Implementation of Evolutionary Fuzzy Systems," *IEEE Transactions on Fuzzy Systems,* vol. 7 no. 2, pp. 109-119, April 1999.

[5] L. A. Zadeh, "Fuzzy Sets," *Information and Control,* vol. 8, pp. 338-353, 1965

[6] G. J. Klir, B. Yuan, *Fuzzy Sets and Fuzzy Logic: Theory and Applications*, Upper Saddle River, NJ: Prentice Hall PTR, 1995.

[7] L. A. Zadeh, "Outline of a New Approach to the Analysis of Complex Systems and Decision Processes," *IEEE Transactions on Systems, Man and Cybernetics,* vol. 3 no. 1, pp. 28-44, Jan. 1973.

[8] L. A. Zadeh, "The Concept of a Linguistic Variable and its Application to Approximate Reasoning—Parts I, II, III," *Information Sciences,* vol. 8 pp. 199-249, 1975, vol. 8 pp. 301-357, 1975, vol. 9 pp. 43-80, 1975.

[9] L. A. Zadeh, "Quantitative Fuzzy Semantics," *Information Sciences,* vol. 3, pp. 159-176, 1971.

[10] L. A. Zadeh, "A Fuzzy-Set-Theoretic Interpretation of Linguistic Hedges," *Journal of Cybernetics,* vol. 2 no. 3, pp. 4-34, 1972.

[11] G. Lakoff, "Hedges: A Study in Meaning Criteria and the Logic of Fuzzy Concepts," *Journal of Philosophical Logic,* vol. 2, pp. 458-508, 1973.

[12] A. Kandel, *Fuzzy Mathematical Techniques with Applications,* Reading, MA: Addison-Wesley Pub. Co., 1986.

[13] R. R. Yager, "A Characterization of the Extension Principle," *Fuzzy Sets and Systems,* vol. 18, pp. 205-217, 1986.

[14] A. Markova-Stupnanova, "A Note to the Addition of Fuzzy Numbers Based on a Continuous Archimedean T-Norm," *Fuzzy Sets and Systems,* vol. 91, pp. 253-258, 1997.

[15] R. Fuller, "A Law of Large Numbers for Fuzzy Numbers," *Fuzzy Sets and Systems,* vol. 45, pp. 299-303, 1992.

[16] S. V. Krishna, K. K. M. Sarma, "Fuzzy Topological Vector Spaces—Topological Generation and Normability," *Fuzzy Sets and Systems,* vol. 41, pp. 89-99, 1991.

[17] D. P. Filev, R. R. Yager, "Operations on Fuzzy Numbers via Fuzzy Reasoning," *Fuzzy Sets and Systems,* vol. 91, pp. 137-142, 1997.

[18] R. Mesiar, "Triangular-Norm-Based Addition of Fuzzy Intervals," *Fuzzy Sets and Systems,* vol. 91, pp. 231-237, 1997.

[19] B. De Baets, A. Markova-Stupnanova, "Analytic Expression for the Addition of Fuzzy Intervals," *Fuzzy Sets and Systems,* vol. 91, pp. 203-213, 1997.

[20] B. Bouchon-Meunier, O. Kosheleva, V. Kreinovich, H. T. Nguyen, "Fuzzy Numbers are the Only Fuzzy Sets that Keep Invertible Operations Invertible," *Fuzzy Sets and Systems,* vol. 91, pp. 155-163, 1997.

[21] A. Kolesarova, "Similarity Preserving T-Norm-Based Additions of Fuzzy Numbers," *Fuzzy Sets and Systems,* vol. 91, pp. 215-229, 1997.

[22] D. H. Hong, C. Hwang, "A T-Sum Bound of LR Fuzzy Numbers," *Fuzzy Sets and Systems,* vol. 91, pp. 239-252, 1997.

[23] R. E. Giachetti, R. E. Young, "A Parametric Representation of Fuzzy Numbers and their Arithmetic Operators," *Fuzzy Sets and Systems,* vol. 91, pp. 185-202, 1997.

[24] D. Dubois, H. Prade, "Systems of Linear Fuzzy Constraints," *Fuzzy Sets and Systems,* vol. 3, pp. 37-48, 1980.

[25] D. Dubois, H. Prade, "On Several Definitions of the Differential of a Fuzzy Mapping," *Fuzzy Sets and Systems,* vol. 24, pp. 117-120, 1987.

[26] D. Dubois, H. Prade, "Additions of Interactive Fuzzy Numbers," *IEEE Transactions on Automatic Control,* vol. 26 no. 4, pp. 926-936, Aug. 1991.

[27] D. J. Dubois, H. M. Prade, "Various Kinds of Interactive Addition of Fuzzy Numbers. Application to Decision Analysis in Presence of Linguistic Probabilities," in *Proceedings of the 1979 IEEE Conference on Decision and Control,* pp. 783-787.

[28] K. Meier, "Methods for Decision Making with Cardinal Numbers and Additive Aggregation," *Fuzzy Sets and Systems,* vol. 88, pp. 135-159, 1997.

[29] M. Braae, D. A. Rutherford, "Theoretical and Linguistic Aspects of the Fuzzy Logic Controller," *Automatica,* vol. 15, pp. 553-577, 1979.

[30] M. Braae, D. A. Rutherford, "Selection of Parameters for a Fuzzy Logic Controller," *Fuzzy Sets and Systems,* vol. 2, pp. 185-199, 1979.

[31] D. Driankov, H. Hellendoorn, M. Reinfrank, *An Introduction to Fuzzy Control,* New York: Springer-Verlag, 1993.

[32] J. Aracil, A. Garcia-Cerezo, A. Ollero, "Stability Analysis of Fuzzy Control Systems: A Geometrical Approach," in *Artificial Intelligence, Expert Systems and Languages in Modeling and Simulation,* C. A. Kulikowski, R. M. Huber, G. A. Ferrate, Eds., New York: North-Holland, 1998.

[33] C. Liu, A. Shindhelm D. Li, K. Jin, "A Numerical Approach to Linguistic Variables and Linguistic Space," in *Proceedings of the 1996 IEEE International Conference on Fuzzy Systems,* pp. 954-959.

[34] J. -S. R. Jang, C. -T. Sun, E. Mizutani, *Neuro-Fuzzy and Soft Computing: A Computational Approach to Learning and Machine Intelligence,* Upper Saddle River, NJ: Prentice-Hall, Inc., 1997.

[35] J. -S. R. Jang, "ANFIS: Adaptive-Network-Based Fuzzy Inference System," *IEEE Transactions on Systems, Man and Cybernetics,* vol. 23 no. 3, pp. 665-685, May 1993.

[36] S. K. Pal, S. Mitra, "Multilayer Perceptron, Fuzzy Sets, and Classification," *IEEE Transactions on Neural Networks,* vol. 3 no. 5, pp. 683-697, Sept. 1992.

[37] J. J. Park, A. Kandel, "Processing of Linguistic Symbols in Multi-Level Neural Network," to appear.

[38] Y. Hayashi, J. J. Buckley, E. Czogala, "Fuzzy Neural Network with Fuzzy Signals and Weights," *International Journal of Intelligent Systems,* vol. 8, pp. 527-537, 1993.

[39] L. A. Zadeh, "Fuzzy Sets and Information Granularity," in *Advances in Fuzzy Set Theory and Applications,* M. M. Gupta, R. K. Ragade, R. R. Yager, Eds., New York: North-Holland, 1979, pp. 3-18. Reprinted in *Fuzzy Sets, Fuzzy Logic, and Fuzzy Systems: Selected Papers by Lotfi A. Zadeh,* G. J. Klir, B. Yuan, Eds., River Edge, NJ: World Scientific Pub. Co., 1996, pp. 433-448.

[40] L. A. Zadeh, "Fuzzy Logic = Computing with Words," *IEEE Transactions on Fuzzy Systems,* vol. 4 no. 2, pp. 103-111, May 1996.

[41] L. A. Zadeh, "Toward a Theory of Fuzzy Information Granulation and its Centrality in Human Reasoning and Fuzzy Logic," *Fuzzy Sets and Systems,* vol. 90, pp. 111-127, 1997.

[42] R. J. Hathaway, J. C. Bezdek, W. Pedrycz, "A Parametric Model for Fusing Heterogeneous Fuzzy Data," *IEEE Transactions on Fuzzy Systems,* vol. 4 no. 3, pp. 270-281, Aug. 1996.

[43] G. Bortolan, W. Pedrycz, "Reconstruction Problem and Information Granularity," *IEEE Transactions on Fuzzy Systems,* vol. 5 no. 2, pp. 234-248, May 1997.

[44] W. Pedrycz, J. C. Bezdek, R. J. Hathaway, G. W. Rogers, "Two Nonparametric Models for Fusing Heterogeneous Fuzzy Data," *IEEE Transactions on Fuzzy Systems,* vol. 6 no. 3, pp. 411-425, Aug. 1998.

[45] N. N. Karnik, J. M. Mendel, *An Introduction to Type-2 Fuzzy Logic Systems,* Signal and Image Processing Institute, Los Angeles, CA: Department of Electrical Engineering and Systems, University of Southern California, Technical Report, 1998.

[46] F. Wenstop, "Deductive Verbal Models of Organizations," *International Journal of Man-Machine Studies,* vol. 8, pp. 293-311, 1976.

[47] W. J. M. Kickert, "An Example of Linguistic Modeling: The Case of Mulder's Theory of Power," in *Advances in Fuzzy Set Theory and Applications,* M. M. Gupta, R. K. Ragade, R. R. Yager, Eds., New York: North-Holland Pub. Co., 1979, pp. 519-540.

[48] J. De Kleer, J. S. Brown, "A Qualitative Physics Based on Confluences," *Artificial Intelligence,* vol. 24 nos. 1-3, pp. 7-83, Dec. 1984.

[49] M. Sugeno, T. Yasukawa, "A Fuzzy-Logic-Based Approach to Qualitative Modeling," *IEEE Transactions on Fuzzy Systems,* vol. 1 no. 1, pp. 7-31, Feb. 1993.

[50] B. M. Seix, A. C. Mallofre, N. P. Carrete, "Qualitative Approach to Gradient Based Learning Algorithms," in *Lecture Notes in Computer Science No. 930,* New York: Springer-Verlag, 1995, pp. 478-485.

[51] B. Stilman, "Linguistic Geometry: Methodology and Techniques," *Cybernetics and Systems,* vol. 26, pp. 535-597, 1995.

[52] M. Grabisch, S. A. Orlovski, R. R. Yager, "Fuzzy Aggregation of Numerical Preferences," in *Fuzzy Sets in Decision Analysis, Operations Research, and Statistics,* R. Slowinski, Ed., Norwell, MA: Kluwer Academic Publishers, 1998, pp. 31-67.

[53] W. Gellert, H. Kustner, M. Hellwich, H. Kastner, Eds., *The VNR Concise Encyclopedia of Mathematics,* New York, Van Nostrand Reinhold Co., 1975.

[54] S. J. Leon, *Linear Algebra With Applications, 2nd Ed.,* New York: MacMillan Pub. Co., 1986.

[55] W. Pedrycz, A. V. Vasilakos, "Linguistic Models and Linguistic Modeling," *IEEE Transactions on Systems, Man and Cybernetics Part B: Cybernetics,* vol. 29 no. 6, 745-757, Dec. 1999.

[56] S. Haykin, *Neural Networks: A Comprehensive Foundation, 2nd Ed.,* Upper Saddle River, NJ: Prentice Hall, 1999.

[57] D. E. Rumelhart, G. E. Hinton, R. J. Williams, "Learning Internal Representations by Error Propagation," in *Parallel Distributed Processing: Explorations in the Microstructure of Cognition,* D. E. Rumelhart, J. L. McClelland, Eds., Cambridge, MA: MIT Press, 1986, pp. 318-362.

[58] L. T. Koczy, K. Hirota, "Ordering, Distance and Closeness of Fuzzy Sets," *Fuzzy Sets and Systems*, vol. 59, pp. 281-293, 1993.

[59] R. A. Fisher, "The Use of Multiple Measurements in Taxonomic Problems," *Annual Eugenics*, vol. 7 part II, pp. 179-188, 1936.

[60] R. A. Fisher, M. Marshall, "Iris Plants Database," July 1998. In C. Blake, E. Keogh, C. J. Merz, "UCI Repository of Machine Learning Databases, [http://www.ics.uci.edu/~mlearn/MLRepository.html]." Ivine, CA: Department of Information and Computer Science, University of California, September 1988.

[61] M. Delgado, J.L. Verdegay, M.A. Vila, "On aggregation operations of linguistic labels," *International Journal of Intelligent Systems,* vol. 8, pp. 351-370, 1993.

[62] T. Hasegawa, T. Furuhashi, "Stability analysis of control systems simplified as a discrete system," *Control and Cybernetics,* vol. 27 no. 4, pp. 565-577, 1998.

[63] T.J. Procyk, E.H. Mamdani, "A linguistic self-organizing process controller," *Automatica*, vol. 15, pp. 15-30, 1979.

Fuzzy Clustering for Multiple-Model Approaches in System Identification and Control

R. Babuška and M. Oosterom

Delft University of Technology, Department of Information Technology and Systems
Control Engineering Laboratory, P.O.Box 5031 2600 GA Delft, The Netherlands
e-mail: R.Babuska@its.tudelft.nl

Abstract. A review of fuzzy clustering and its use in the data-driven construction of nonlinear models and controllers is given. The focus is on algorithms of the fuzzy c-means type. Two application examples are presented: automated design of operating points for gain scheduling in flight control systems and nonlinear black-box identification. In the latter case, a comparison with an alternative technique is given. It is shown that fuzzy clustering is an effective technique for the decomposition of a complex nonlinear problem into a set of simpler local problems.

Keywords. Multiple-model control, gain scheduling, system identification, fuzzy clustering, fuzzy modeling, rue extraction, aircraft control, black-box modeling

1 Introduction

Clustering techniques can be used to organize data into groups based on similarities among the individual data items. The potential of clustering to group data and to reveal the underlying structures can be exploited in a wide variety of applications, including classification, image processing, pattern recognition, modeling and control.

Control engineering is a typical example of a field where granulation (local linearization, gain scheduling, decomposition) has been used as a standard technique to handle the inherent complexity of most real world problems. The construction of nonlinear models for control purposes requires modeling and identification techniques that provide simple, transparent and mathematically tractable models. A wide range of techniques are available ranging from mechanistic modeling to black-box models based on neural networks. In this variety, a main distinction can be made between global and local methods. Global methods describe the system under study using nonlinear functional relationships between the system's variables (such as neural networks). Local approaches cope with the nonlinearity by decomposing the operating range into a number of local operating regimes. In each regime, a linear model is constructed and an interpolating mechanism combines the outputs of the local models into a global model output. A major advantage is that local models can usually be more easily interpreted than complicated global models and this technique can directly be applied in control design [1].

A model that consists of multiple local submodels can be conveniently represented by means of fuzzy if-then rules [2]. The antecedent membership functions define fuzzy regions in the operating space in which the corresponding consequents (typically linear regression models) are valid. The overlap of the antecedent membership functions provides a smooth interpolation of the rules' consequents.

The decomposition of the operating range into the local regimes is clearly a critical step in the application of this methodology. The complexity of the system's behavior is typically not uniform – some operating regions can be well approximated by a single linear model, while other regions require rather fine partitioning. In order to obtain an efficient representation with as few local models as possible, the local regimes must be identified such that they correctly capture the nonuniform behavior of the system. Fuzzy clustering is an effective technique that can be used to obtain the partitioning into submodels based on data [3–5]. Clustering approaches do not require prior knowledge about the operating regimes and also an appropriate number of rules can be determined automatically.

This chapter first gives a brief overview of the basic notions and algorithms in fuzzy clustering. The use of clustering in the extraction of if-then rules from data is then addressed. Finally, two application examples from the control engineering domain are presented; automated design of operating points for gain scheduling in flight control systems and nonlinear black-box identification.

2 Fuzzy Clustering Techniques

An overview of fuzzy clustering based on the c-means functional is given. Readers interested in a deeper and more detailed treatment of fuzzy clustering may refer to classical literature [6–9].

2.1 The Data Set

Clustering techniques can be applied to data that are quantitative (numerical), qualitative (categorical), or a mixture of both. In this chapter, the clustering of quantitative data is considered. The data are typically observations of some physical process. Each observation consists of n measured variables, grouped into an n-dimensional column vector $\mathbf{z}_k = [z_{1k}, \ldots, z_{nk}]^T \in \mathbb{R}^n$. A set of N observations $\mathbf{Z} = \{\mathbf{z}_k | k = 1, 2, \ldots, N\}$, is represented as an $n \times N$ matrix:

$$\mathbf{Z} = \begin{bmatrix} z_{11} & z_{12} & \cdots & z_{1N} \\ z_{21} & z_{22} & \cdots & z_{2N} \\ \vdots & \vdots & \vdots & \vdots \\ z_{n1} & z_{n2} & \cdots & z_{nN} \end{bmatrix}. \tag{1}$$

In the pattern-recognition terminology, the columns of this matrix are called *patterns* or objects, the rows are called the *features* or attributes, and $\mathbf{Z}$ is called the *pattern* or *data matrix*. The meaning of the columns and rows of $\mathbf{Z}$ depends on the context. In medical diagnosis, for instance, the columns of $\mathbf{Z}$ may represent patients, and the rows are then symptoms, or laboratory measurements for these patients. When clustering is applied to the modeling and identification of dynamic systems, the columns of $\mathbf{Z}$ may contain samples of time signals, and the rows are, for instance, physical variables observed in the system (position, pressure, temperature, etc.). In order to represent the system's dynamics, past values or time derivatives of these variables are included in $\mathbf{Z}$ as well.

2.2 Clusters and Prototypes

Various definitions of a cluster can be formulated, depending on the objective of clustering. Generally, one may accept the view that a cluster is a group of objects that are more similar to one another than to members of other clusters [7,8]. The similarity is usually measured by means of a *distance* between the data vectors and some *prototype* of the cluster. The prototypes are typically not known beforehand, and are sought by the clustering algorithms simultaneously with the partitioning of the data. The prototypes may be vectors of the same dimension as the data objects, but they can also be defined as linear or nonlinear subspaces or functions.

2.3 Crisp and Fuzzy Clustering

One possible classification of clustering methods can be done according to whether the subsets are *fuzzy* or *crisp*. *Crisp clustering* methods are based on classical set theory, and require that an object either does or does not belong to a cluster. Crisp clustering thus means partitioning the data into a specified number of mutually exclusive subsets. *Fuzzy clustering* methods, however, allow the objects to belong to several clusters simultaneously, with different degrees of membership. In many situations, fuzzy clustering is more natural than crisp clustering. Objects on the boundaries between several classes are not forced to fully belong to one of the classes, but rather are assigned membership degrees between 0 and 1 indicating their partial membership.

In this chapter we focus on fuzzy clustering algorithms that use an *objective function* to measure the desirability of partitions. Nonlinear optimization is used to search for local optima of the objective function. These methods are relatively well understood, and mathematical results are available concerning the convergence properties and cluster validity assessment.

2.4 Fuzzy Partition

The objective of clustering is to partition the data set $\mathbf{Z}$ into c fuzzy subsets $\{A_i | 1 \leq i \leq c\}$. These subsets are defined by their *membership (characteristic) functions*, represented in the *partition matrix* $\mathbf{U} = [\mu_{ik}]_{c \times N}$. The ith row of this matrix contains values of the membership function of the ith fuzzy subset A_i of $\mathbf{Z}$. The membership degrees satisfy the following conditions:

$$\mu_{ik} \in [0,1], \quad 1 \leq i \leq c, \quad 1 \leq k \leq N, \tag{2a}$$

$$\sum_{i=1}^{c} \mu_{ik} = 1, \quad 1 \leq k \leq N, \tag{2b}$$

$$0 < \sum_{k=1}^{N} \mu_{ik} < N, \quad 1 \leq i \leq c. \tag{2c}$$

Equation (2a) states the well-known fact that the membership degrees are real numbers from the interval $[0, 1]$. Condition (2b) constrains the sum of each column to 1, and thus the total membership of each $\mathbf{z}_k$ in all the clusters equals one. Equation (2c) means that none of the fuzzy subsets is empty nor it contains all the data.

Let us illustrate this concept by a simple example. Consider a data set $\mathbf{Z} = [\mathbf{z}_1, \mathbf{z}_2, \ldots, \mathbf{z}_9]$, shown in Fig. 1.

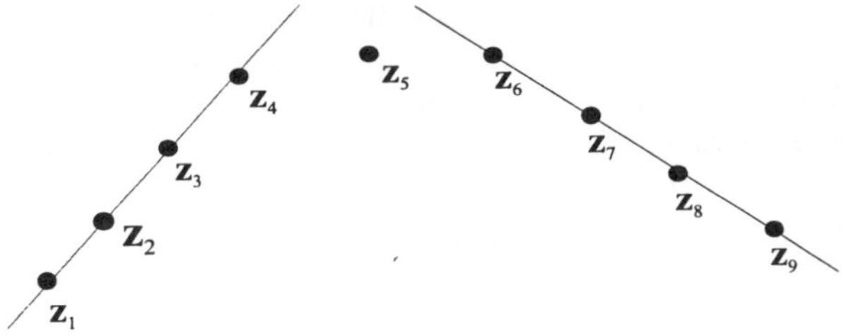

Fig. 1. A sample data set in $\mathbb{R}^2$.

By using multiple-model regression, this data set can be approximated by two lines (data points $\mathbf{z}_1$ to $\mathbf{z}_4$ and $\mathbf{z}_6$ to $\mathbf{z}_9$ respectively). Point $\mathbf{z}_5$ does not fully belong to either of the models. The corresponding fuzzy partitions of $\mathbf{Z}$ is:

$$\mathbf{U} = \begin{bmatrix} 1.0 & 1.0 & 1.0 & 1.0 & 0.5 & 0.0 & 0.0 & 0.0 & 0.0 \\ 0.0 & 0.0 & 0.0 & 0.0 & 0.5 & 1.0 & 1.0 & 1.0 & 1.0 \end{bmatrix}.$$

The first row of $\mathbf{U}$ defines the membership function for the first subset A_1, the second row defines the membership function of the second subset A_2. Membership degrees of $\mathbf{z}_5$ reflect the fuzziness of the partition. Note that the sum of the rows is one for all data points.

2.5 Clustering Algorithms

In most algorithms, the fuzzy partition matrix $\mathbf{U}$ is obtained by minimizing of the *fuzzy c-means* functional:

$$J(\mathbf{Z}; \mathbf{U}, \mathbf{V}) = \sum_{i=1}^{c} \sum_{k=1}^{N} (\mu_{ik})^m d_{ik\mathbf{A}_i}^2 \tag{3}$$

where $\mathbf{V} = [\mathbf{v}_1, \mathbf{v}_2, \ldots, \mathbf{v}_c] \in \mathbb{R}^n$ is a vector of *cluster prototypes* (centers), which have to be determined, and

$$d_{ik\mathbf{A}_i}^2 = (\mathbf{z}_k - \mathbf{v}_i)^T \mathbf{A}_i (\mathbf{z}_k - \mathbf{v}_i) \tag{4}$$

is the squared inner-product distance norm. The overlap of the clusters is controlled by the user-defined parameter $m \in [1, \infty)$. As m approaches one from above, the partition becomes crisp and as $m \to \infty$, the partition becomes completely fuzzy ($\mu_{ik} = 1/c$). The minimization of (3) represents a nonlinear optimization problem that is usually solved by using the Picard iteration through the first-order conditions for stationary points of (3).

Matrices $\mathbf{A}_i$ are either fixed *a priori* (fuzzy c-means (FCM) algorithm and its variants [7]) or are computed in the optimization algorithm based on local covariance of the data around each cluster center (Gustafson–Kessel (GK) algorithm [10], fuzzy maximum likelihood estimation algorithm [11]). This allows each cluster to adapt the distance norm to the local distribution of the data.

A limitation of the basic c-means algorithm based on fixed $\mathbf{A}_i$ is that the distance norm (4) forces the objective function to prefer clusters of a certain shape even if they are not present in the data. This is demonstrated in Fig. 2 which shows a synthetic data set in $\mathbb{R}^2$ with two well-separated clusters of different shapes. Note that the FCM algorithm (with $\mathbf{A}_i = \mathbf{I}$ for $i = 1, 2$) imposes a circular shape on both clusters, even though the lower cluster is rather elongated. The GK algorithm, however, adapts the distance norm to the underlying distribution of the data and detects clusters of different shape and orientation.

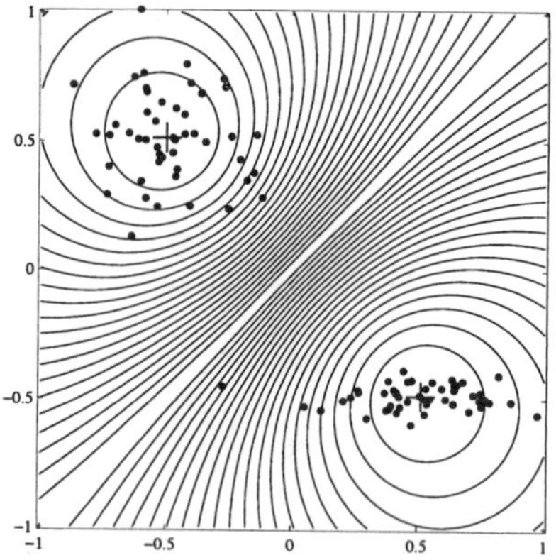

(a) Fuzzy c-means.

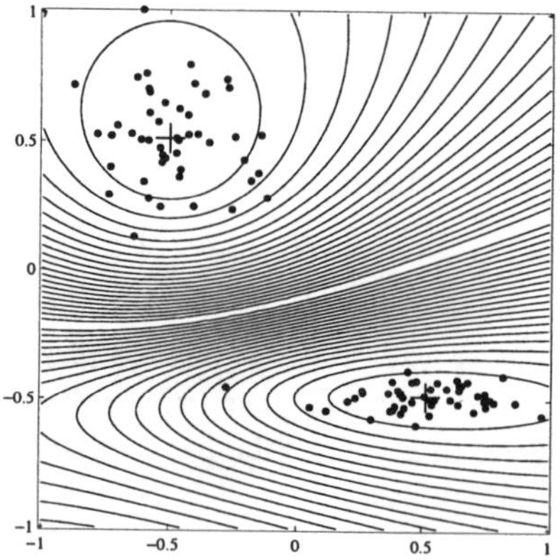

(b) Gustafson–Kessel.

Fig. 2. Comparison of the fuzzy c-means algorithm and the Gustafson–Kessel algorithm. The points represent the data, '+' are the cluster prototypes. Also shown are curves of constant membership.

Cluster prototypes can also be defined as linear and nonlinear hypersurfaces. The metric norm (4) is then redefined to account for the distance of the data from these hypersurfaces. Based on these considerations, different variants of the c-means algorithm have been proposed, including the fuzzy c-varieties [7], fuzzy c-elliptotypes [12], and fuzzy regression models [13].

2.6 Number of Clusters

In most algorithms based on the c-means functional, the number of clusters has to be specified beforehand. The algorithm then searches for these clusters, regardless of whether they are really present in the data or not. It is therefore important to check the validity of the obtained partition. Two main approaches can be distinguished:

Validity measures. Validity measures are scalar indices that assess the goodness of the obtained partition. Clustering algorithms generally aim at locating well-separated and compact clusters. When the number of clusters is chosen equal to the number of groups that actually exist in the data, it can be expected that the clustering algorithm will identify them correctly. When this is not the case, misclassifications appear, and the clusters are not likely to be well separated and compact. Hence, most cluster validity measures are designed to quantify the separation and the compactness of the clusters. However, as Bezdek [7] points out, the concept of cluster validity is open to interpretation and can be formulated in different ways. Consequently, many validity measures have been introduced in the literature, see [7,11,14] among others.

Iterative merging or insertion of clusters. The basic idea of cluster merging is to start with a sufficiently large number of clusters, and successively reduce this number by merging clusters that are similar (compatible) with respect to some predefined criteria [15,16]. One can also adopt an opposite approach, i.e., start with a small number of clusters and iteratively insert clusters in the regions where the data points have low degree of membership in the existing clusters [11].

3 Extraction of Fuzzy Rules from Data

Fuzzy clustering can be used to extract if–then rules from data and to build regression models. Fuzzy if–then rules are of the following general form:

If *antecedent proposition* **then** *consequent proposition.*

Depending on the antecedent and consequent propositions, two main types of rule-based fuzzy models are distinguished: *linguistic fuzzy models* [17], where both the antecedent and the consequent are fuzzy propositions, and *Takagi–Sugeno fuzzy models* [2], where the antecedent is a fuzzy proposition and the consequent is a crisp function.

3.1 Linguistic Rules

A linguistic model consists of a set of if–then rules:

$$R_i\text{: } \textbf{If } \mathbf{x} \text{ is } A_i \textbf{ then } y \text{ is } B_i, \qquad i = 1, 2, \ldots, K\,. \tag{5}$$

Here, $\mathbf{x}$ is the input and y is the output of the model. A_i and B_i are linguistic terms (like low, medium, high, etc.), represented by fuzzy sets. Models with multiple inputs are usually represented in the conjunctive form:

$$R_i\text{: } \mathbf{If}\ x_1 \text{ is } A_{i1} \ \mathbf{and}\ \ldots\ \mathbf{and}\ x_p \text{ is } A_{ip}\ \mathbf{then}\ y \text{ is } B_i, \quad i = 1, 2, \ldots, K\,. \tag{6}$$

The rules and the membership functions for A_{ij} and B_i can be extracted from input–output data by means of fuzzy clustering. Different approaches can be used:

- Cluster the data in the (one-dimensional) spaces of the individual antecedent and consequent variables.
- Cluster the data in one of the spaces (either antecedent or consequent) and induce the membership functions in the remaining space from the found partition.
- Cluster the data in the Cartesian product space of the antecedent and the consequent.

Figure 3 depicts the idea of rule extraction by using the last method – product space clustering. In this example, two clusters are found in the data. Each cluster induces one if–then rule. The membership functions for A_i are obtained by projecting the partition matrix on the x axis. Similarly, the membership functions for B_i are obtained by projecting the partition matrix on the y axis.

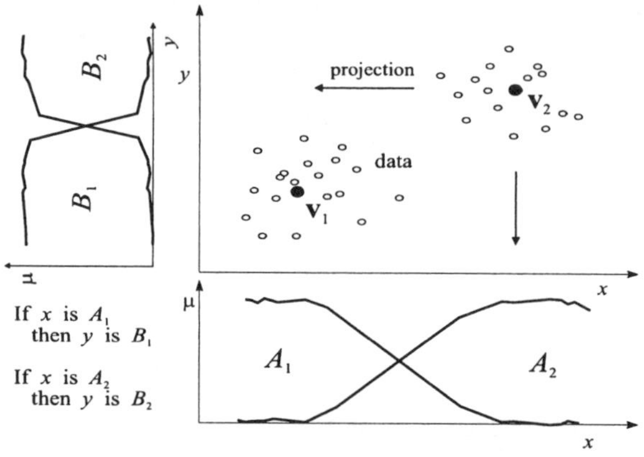

Fig. 3. Extraction of linguistic rules by product-space clustering.

Suitable linguistic labels can be attached to A_i and B_i after the projection. In our case, the obtained rules are:

If x is *Small* **then** y is *Small*

If x is *Large* **then** y is *Large*

This linguistic model is in fact a concatenation of locally constant models.

3.2 Takagi–Sugeno Rules

The Takagi–Sugeno (TS) model can be regarded as a combination of linguistic modeling and mathematical regression. Usually, affine linear local models are used:

$$R_i\text{: } \textbf{If } \mathbf{x} \text{ is } A_i \textbf{ then } y_i = \mathbf{a}_i^T \mathbf{x} + b_i, \qquad i = 1, 2, \ldots, K \tag{7}$$

where $\mathbf{a}_i$ is a parameter vector and b_i is a scalar offset. Antecedents of TS models with multiple inputs are typically represented in the conjunctive form (6).

TS rules can also be extracted from data by clustering in the product space of the inputs and outputs. By applying algorithms that are capable of detecting linear substructures in data (the Gustafson–Kessel algorithm, fuzzy c-elliptotypes, fuzzy regression models), a nonlinear regression problem can be automatically decomposed into several local linear subproblems. Each obtained cluster is transformed into one rule of the Takagi–Sugeno model. The antecedent membership functions are obtained by projection and the consequent parameters can be estimated by least-squares methods.

As an example, consider a nonlinear function $y = f(x)$ defined piece-wise by:

$$\begin{aligned} y &= 0.25x, && \text{for } x \leq 3, \\ y &= (x-3)^2 + 0.75, && \text{for } 3 < x \leq 6, \\ y &= 0.25x + 8.25, && \text{for } x > 6\,. \end{aligned} \tag{8}$$

Figure 4a shows a plot of this function evaluated in 50 samples uniformly distributed over $x \in [0, 10]$. Zero-mean, uniformly distributed noise with amplitude 0.1 was added to y.

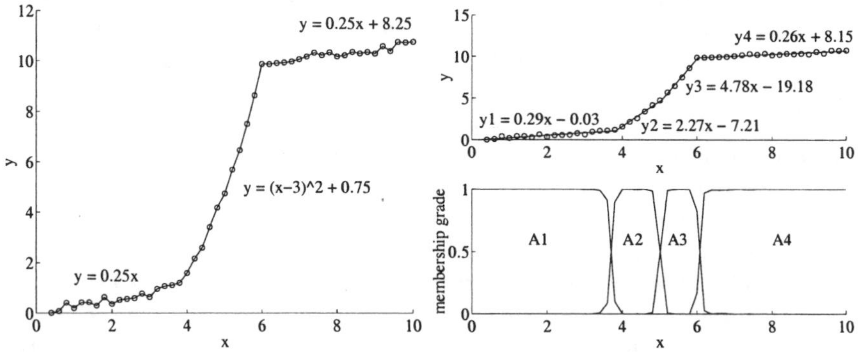

(a) *Nonlinear function (8).*

(b) *Local linear models and the corresponding membership functions.*

Fig. 4. Approximation of a static nonlinear function using a TS fuzzy model.

The data set $\mathbf{Z} = \{(x_i, y_i) | i = 1, 2, \ldots, 50\}$ was clustered into four hyper-ellipsoidal clusters, using the GK algorithm. The upper plot of Fig. 4b shows the local linear models obtained least-square estimation, the bottom plot gives the corresponding fuzzy partition. In terms of the TS rules, the fuzzy model is expressed as:

$$\begin{aligned}
R_1&: \textbf{If } x \text{ is } A_1 \textbf{ then } y = 0.29x - 0.03,\\
R_2&: \textbf{If } x \text{ is } A_2 \textbf{ then } y = 2.27x - 7.21,\\
R_3&: \textbf{If } x \text{ is } A_3 \textbf{ then } y = 4.78x - 19.18,\\
R_4&: \textbf{If } x \text{ is } A_4 \textbf{ then } y = 0.26x + 8.15\,.
\end{aligned}$$

Note that the consequents of R_1 and R_4 correspond almost exactly to the first and third equation (8). Consequents of R_2 and R_3 are approximate local linear models of the parabola defined by the second equation in (8).

This principle of clustering-based identification extends to input–output dynamic systems in a straightforward way. The product space is formed by the regressors (lagged input and output data) and the regressand (the output to be predicted). To illustrate this, assume a second-order NARX (nonlinear autoregressive with exogenous input) model $y(k+1) = f\left(y(k), y(k-1), u(k), u(k-1)\right)$. Given a set of data, $S = \{(u(j), y(j)) \mid j = 1, 2, \ldots, N\}$, the regressor matrix and the regressand vector are:

$$\mathbf{X} = \begin{bmatrix} y(2) & y(1) & u(2) & u(1) \\ y(3) & y(2) & u(3) & u(2) \\ \vdots & \vdots & \vdots & \vdots \\ y(N-1) & y(N-2) & u(N-1) & y(N-2) \end{bmatrix}, \quad \mathbf{y} = \begin{bmatrix} y(3) \\ y(4) \\ \vdots \\ y(N) \end{bmatrix}.$$

The unknown nonlinear function $y = f(\mathbf{x})$ represents a nonlinear (hyper)surface in the product space: $(X \times Y) \subset \mathbb{R}^{p+1}$, called the *regression surface*. By clustering the data $\mathbf{Z} = [\mathbf{X},\ \mathbf{y}]^T$, local linear models can be found that approximate the regression surface.

A major advantage of this approach is the possibility to interpret and validate the obtained models in terms of their local and global behavior. This contrasts with many purely black-box techniques, such as neural networks, where the validation mainly relies on numerical simulation. Prior knowledge also can be incorporated in the obtained model, in the form of additional rules, based on approximate mechanistic models, step-response identification or heuristics and experience [5].

In most cases, less complex models are obtained via product-space clustering than by optimizing antecedent membership functions for each variable separately, for instance, by adaptive spline methods [18], tree-construction approaches [19], table look-up schemes [20]. The reason is that the rules are created directly in the product space of the model variables and take into account the form of the system's nonlinearity and the coverage of the space by the identification data. Hence, more rules are obtained for regions characterized by complex nonlinear behavior, while (almost) linear regions are described by fewer rules. No rules are, however, generated in the regions that do not contain any data. With regard to generalization capabilities, this can be seen as a drawback, as the model gives no output for these regions. This can be turned into an advantage, since the analysis of the rule base completeness helps validate the model and design new identification experiments.

While clustering is very fast for small data sets, large data sets in combination with high-dimensional regression problems may drastically increase the computational demands of the algorithm. The technique has proven to be practically useful for medium-complexity identification problems, i.e., for data sets up to several thousands data points and low-order dynamical systems with few input variables (in our experience up to about 15 regressors). An advantage of the clustering approach is that is can be applied to well-behaved, smooth nonlinear systems, as well as to systems, whose behavior changes abruptly, as a result of 'hard' nonlinearities, such as dead-zone, hysteresis, stiction, etc. The character of the process nonlinearity is reflected in the form and overlap of the obtained membership functions.

A MATLAB toolbox has been developed by the authors (available free of charge at `http://lcewww.et.tudelft.nl/~babuska`). It contains functions that automatically generate dynamic TS fuzzy models from given input–output data by using a robust implementation of the GK clustering algorithm. Furthermore, tools are provided for the analysis, reduction and simulation of the obtained model. A detailed description of the implemented algorithms is given in [5].

4 Automated Design Procedure for Gain-Scheduling Control

This example demonstrates the application of fuzzy clustering to the selection of operating points for gain-scheduling in aircraft control. As the aircraft dynamics strongly vary with flight conditions, the design of the flight control laws is a non-linear control problem. Gain scheduling is typically used to cope with this nonlinearity [21]. The current design of gain-scheduling control laws in fly-by-wire aircraft systems is based on a time-consuming trial-and-error procedure which involves many manual iterations. In order to reduce the design effort, a more systematic method is sought which can automatically selected an optimal number of operating points.

In [22], a procedure has been proposed which uses fuzzy clustering to find in the flight envelope operating regimes where the aircraft dynamics are roughly linear. In each regime, conventional control design is used to tune the controller. The obtained membership functions provide an interpolation mechanism for the control parameters. The performance of this procedure is demonstrated using a realistic simulation model of a small commercial aircraft.

4.1 Application of Fuzzy Clustering

The operating range of an aircraft, called the flight envelope, is a subset in the space of possible velocities (Mach number) and dynamic pressures, see Fig. 5.

In order to determine the linear operating regimes, the flight envelope was discretized using a grid with steps of 0.01 in Mach number and $500\,\mathrm{kg{\cdot}m^{-1}{\cdot}s^{-2}}$ in dynamic pressure. The total number of grid points is equal to $N = 1586$. In each

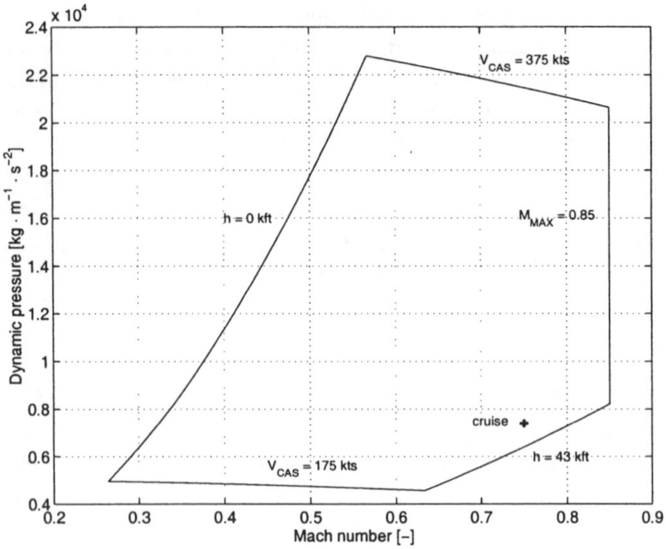

Fig. 5. The considered flight envelope in terms of the Mach number and altitude.

grid point, three aerodynamic derivatives M_α, M_q, and M_{δ_e} were computed using an available nonlinear model of the aircraft. These derivatives provide the required information about the (non)linearity of the aircraft dynamics of interest. If they are constant in a given region of the flight envelope, the dynamics are considered to be linear in that region. Hence, clustering was employed to search for subsets of similar aerodynamic derivatives, in this way identifying an appropriate number of operating points for tuning the controller. The data vectors to be clustered are:

$$z_k = [M_\alpha(k),\ M_q(k),\ M_{\delta_e}(k)]^T, \quad k = 1, ..., N$$

where $M_i(k)$ denotes the derivative M_i computed at the kth grid point. This three-dimensional data set was clustered using the GK algorithm with six clusters (selected by using the Xie-Beni validity index [23], among other validity measures).

The obtained partition was projected onto the scheduling variables which are the Mach number and the dynamic pressure (the values of these variables were stored for each grid point, but they were not clustered). In this way, membership functions are induced which partition the space of the scheduling variables into six operating regions (Fig. 6).

In each operating region, the parameters of the controller were tuned to obtain the desired dynamic response (a standard tuning procedure was used). These parameters are the pitch damper gain G_{Q}, the proportional and integral feedback gains K_{P} and K_{I}, respectively, the feedforward gain G_{FF}, and the lead-lag filter time constants τ_{LEAD} and τ_{LAG}. Through the interpolation mechanism, these parameters become nonlinear functions of the scheduling variables (Fig. 7).

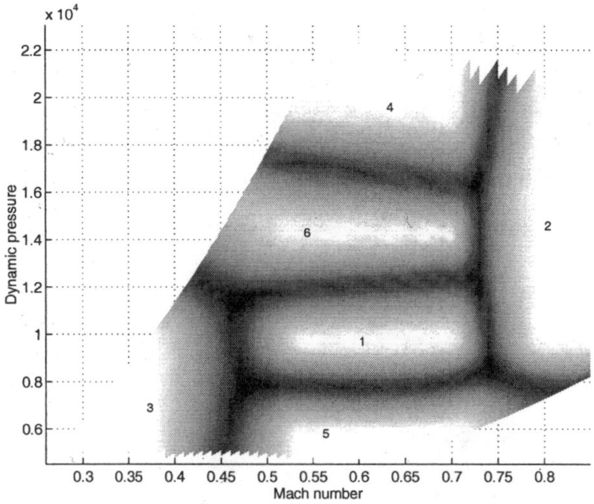

Fig. 6. Fuzzy clusters reconstructed in the space of the scheduling variables.

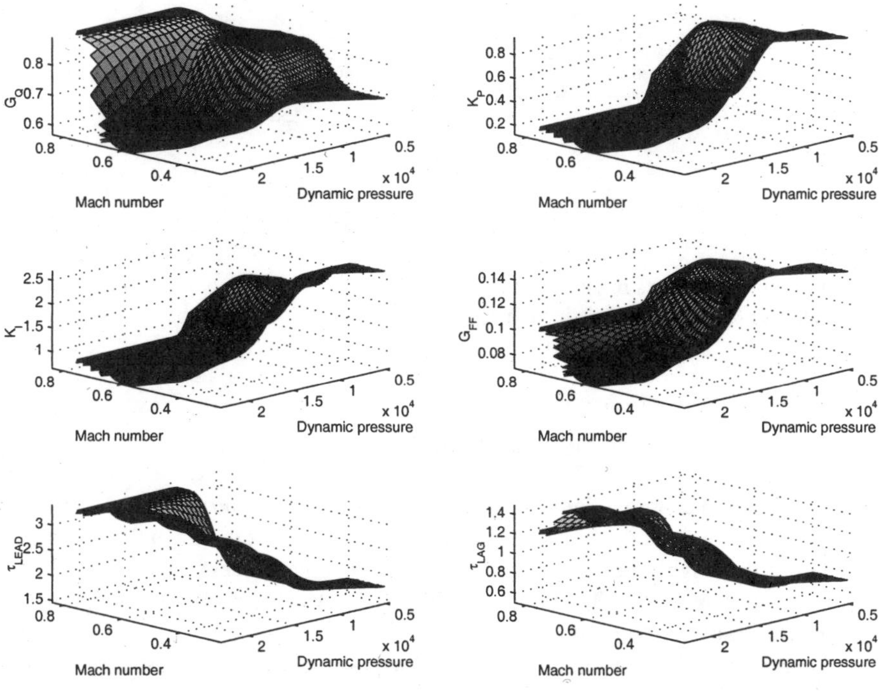

Fig. 7. The scheduled control parameters as a function of the scheduling variables.

Figure 8 shows a simulation result obtained with the aircraft model for the cruise flight condition, Mach $= 0.7$ and $h = 7400\,\text{kg}\cdot\text{m}^{-1}\cdot\text{s}^{-2}$ (see also Fig. 5). The pilot input is a block-shaped command initialized at $t = 1$ sec and terminated at $t = 7$ sec. One of the important design objectives is zero dropback in the pitch attitude. It can be seen from the time history of θ that this dropback is indeed negligible.

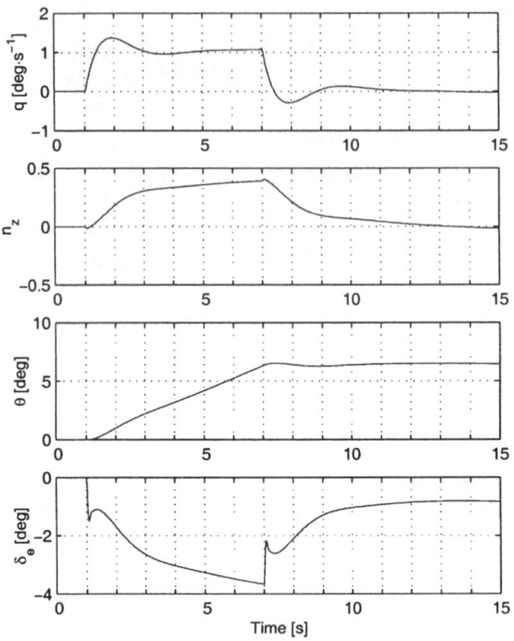

Fig. 8. Simulation results in the cruise flight condition: pitch rate q, normal acceleration n_z, pitch attitude θ, and elevator deflection δ_e as a function of time.

5 Nonlinear Black-Box Identification

This application example demonstrates the use of fuzzy clustering in the construction of a black-box model of a Diesel engine turbocharger. It is shown that via fuzzy clustering this highly nonlinear, multivariable dynamic process can accurately be approximated by a small TS model. A comparison with an alternative technique is given. The results are reproduced from [19].

Figure 9 depicts the charging process of a Diesel engine by an exhaust turbocharger. The charging process has a nonlinear input–output behavior as well as a strong dependency of the dynamic parameters on the operating point. This is known from physical insights.

The static behavior of the turbocharger may be sufficiently well described by characteristic maps (look-up tables) of the compressor and the turbine. However, if

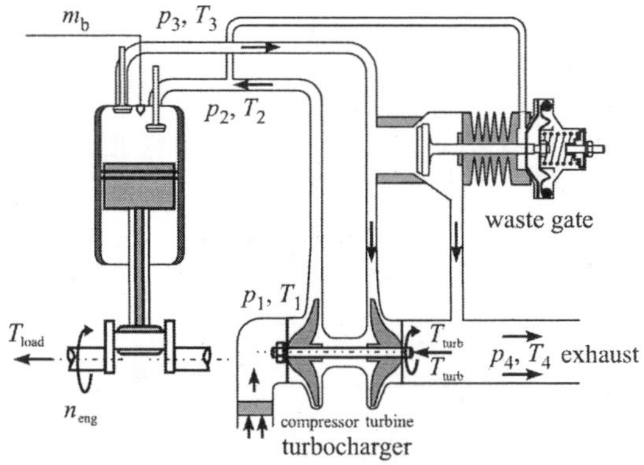

Fig. 9. Scheme of the combustion engine turbocharger.

the dynamics of the turbocharger need to be considered, mechanical and thermodynamical modeling is required. The quality of theoretical models essentially depends on the accurate knowledge of several process parameters, which have to be laboriously derived or estimated. Another disadvantage is the considerable computational effort due to the complexity of these models.

For these reasons, such methods are considered to be inconsistent with the requirement of typical control engineering applications such as controller design, fault diagnosis and hardware-in-the-loop simulations. Therefore, a nonlinear black-box model was developed from data. In this approach, no deep theoretical knowledge of the process is necessary. The rate of injection m_b and the engine speed n_{eng} are chosen as inputs while the charging pressure p_2 is the output. The sampling time is 0.2 sec.

The training data were generated by a special driving cycle in order to excite the system with the amplitudes and frequencies of interest, see Fig. 10. The measurements were recorded on a flat test track. Also, in order to operate the engine in high load ranges, the truck was driven with the biggest possible load. A total of 1300 data samples were used for identification. For validation, two special driving cycles were recorded, which represent realistic conditions in highway and urban traffic.

It was found that the turbocharger can be approximated with sufficient accuracy by assuming a second-order discrete-time model. Therefore, the charging pressure $p_2(k)$ at time instant k is modeled by the following nonlinear relationship

$$p_2(k) = f\big(m_b(k),\, m_b(k-1),\, m_b(k-2),\, n_{eng}(k),\, n_{eng}(k-1),\, n_{eng}(k-2),\\ p_2(k-1),\, p_2(k-2)\big)\,.$$

The above function f was parameterized by the TS fuzzy model (7) and identified by product-space fuzzy clustering. This approach was compared with a tree-construction method, called LOLIMOT [19]. The LOLIMOT algorithm arrived at a fuzzy model with 10 rules, achieving the RMS errors 0.021, 0.023 and 0.032 on the training and both validation data sets, respectively. In contrast, the product

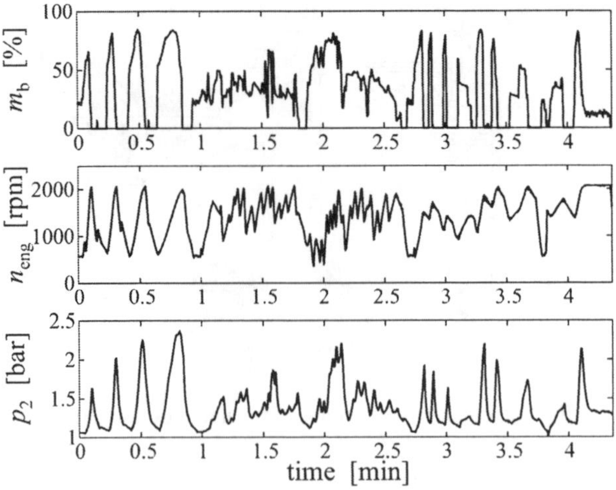

Fig. 10. Training data.

space clustering approach only required 3 rules to achieve a similar performance, 0.022, 0.023 and 0.032. In terms of computational demands, the identification took 6.68 seconds on a powerful Pentium computer and required 39 MFLOPS (millions of floating-point operations) in MATLAB. To compute a one step ahead prediction with the obtained model, only 238 FLOPS are needed.

Figure 11 shows a simulation of the clustering-based model on the training set and one of the validation sets. The nonlinear behavior can be seen from the steady-state characteristic shown in Figure 12.

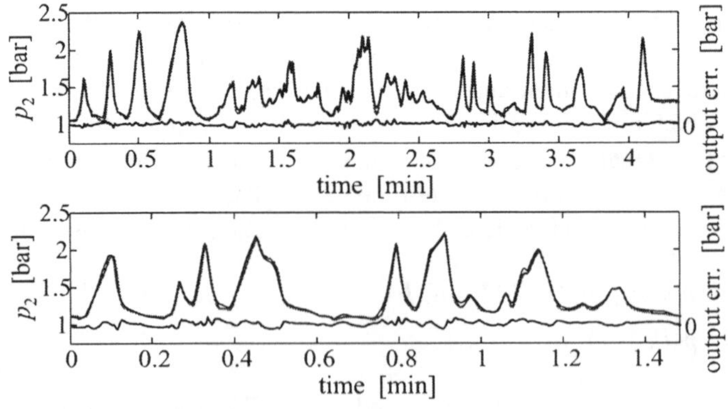

Fig. 11. Simulation of the fuzzy model on the training data (top) and the urban-traffic validation data (bottom).

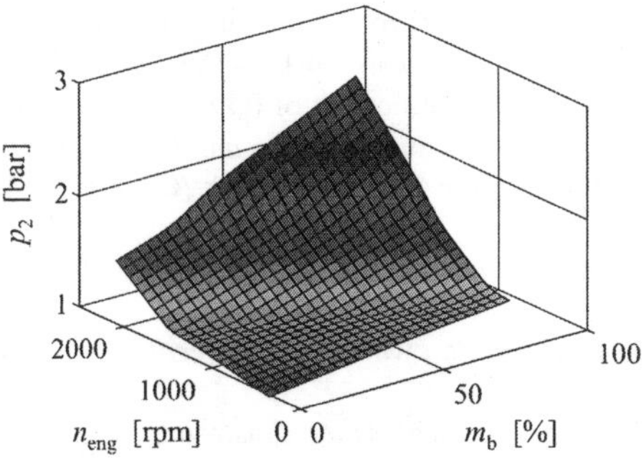

Fig. 12. Steady-state model characteristic.

A linear ARX (autoregressive with exogenous input) model of the same structure as the fuzzy models has the RMS errors 0.11, 0.088 and 0.11, which is about four times worse.

6 Concluding Remarks

Methods for generating nonlinear models from data by using clustering techniques were reviewed, including two application examples: automated design of operating points for gain scheduling in flight control systems and nonlinear black-box identification. Other applications can be found in the literature [24–28].

It should be noted that proper use application of clustering techniques requires certain skills and experience of the user. Rather than a black-box, fully automated procedure, the approach should be seen as an interactive method facilitating an active participation of the user in a computer-assisted modeling session.

Acknowledgement

The development of the automated design procedure for gain-scheduling control was supported by the Brite/EuRam project "Affordable Digital Fly-by-wire Flight Control Systems for Small Commercial Aircraft" (ADFCS).

The description of the turbocharger application and the data sets were provided by Alexander Fink of the Darmstadt University of Technology, Institute of Automatic Control, Darmstadt, Germany.

References

1. R. Murray-Smith and T. A. Johansen, editors. *Multiple Model Approaches to Nonlinear Modeling and Control.* Taylor & Francis, London, UK, 1997.
2. T. Takagi and M. Sugeno. Fuzzy identification of systems and its application to modeling and control. *IEEE Trans. Systems, Man and Cybernetics*, 15(1):116–132, 1985.

3. Y. Yoshinari, W. Pedrycz, and K. Hirota. Construction of fuzzy models through clustering techniques. *Fuzzy Sets and Systems*, 54:157–165, 1993.
4. Y. Nakamori and M. Ryoke. Identification of fuzzy prediction models through hyperellipsoidal clustering. *IEEE Trans. Systems, Man and Cybernetics*, 24(8):1153–73, 1994.
5. R. Babuška. *Fuzzy Modeling for Control*. Kluwer Academic Publishers, Boston, USA, 1998.
6. R.O. Duda and P.E. Hart. *Pattern Classification and Scene Analysis*. John Wiley & Sons, New York, 1973.
7. J.C. Bezdek. *Pattern Recognition with Fuzzy Objective Function*. Plenum Press, New York, 1981.
8. A.K. Jain and R.C. Dubes. *Algorithms for Clustering Data*. Prentice Hall, Englewood Cliffs, 1988.
9. J.C. Bezdek and S.K. Pal, editors. *Fuzzy Models for Pattern Recognition*. IEEE Press, New York, 1992.
10. D.E. Gustafson and W.C. Kessel. Fuzzy clustering with a fuzzy covariance matrix. In *Proc. IEEE CDC*, pages 761–766, San Diego, CA, USA, 1979.
11. I. Gath and A.B. Geva. Unsupervised optimal fuzzy clustering. *IEEE Trans. Pattern Analysis and Machine Intelligence*, 7:773–781, 1989.
12. J.C. Bezdek, C. Coray, R. Gunderson, and J. Watson. Detection and characterization of cluster substructure, I. Linear structure: Fuzzy c-lines. *SIAM J. Appl. Math.*, 40(2):339–357, 1981.
13. R.J. Hathaway and J.C. Bezdek. Switching regression models and fuzzy clustering. *IEEE Trans. Fuzzy Systems*, 1(3):195–204, 1993.
14. N.R. Pal and J.C. Bezdek. On cluster validity for the fuzzy c-means model. *IEEE Trans. Fuzzy Systems*, 3(3):370–379, 1995.
15. R. Krishnapuram and C.-P. Freg. Fitting an unknown number of lines and planes to image data through compatible cluster merging. *Pattern Recognition*, 25(4):385–400, 1992.
16. U. Kaymak and R. Babuška. Compatible cluster merging for fuzzy modeling. In *Proceedings FUZZ-IEEE/IFES'95*, pages 897–904, Yokohama, Japan, March 1995.
17. L.A. Zadeh. Outline of a new approach to the analysis of complex systems and decision processes. *IEEE Trans. Systems, Man, and Cybernetics*, 1:28–44, 1973.
18. M. Brown and C. Harris. *Neurofuzzy Adaptive Modelling and Control*. Prentice Hall, New York, 1994.
19. O. Nelles, A. Fink, R. Babuška, and M. Setnes. Comparison of two construction algorithms for takagi-sugeno fuzzy models. In *Proceedings Seventh European Congress on Intelligent Techniques and Soft Computing EUFIT'99*, pages 280–285, Aachen, Germany, September 1999.
20. L.-X. Wang. *Adaptive Fuzzy Systems and Control, Design and Stability Analysis*. Prentice Hall, New Jersey, 1994.
21. R.F. Stengel, J.R. Broussard, and P.W. Berry. Digital flight control design for a tandem-rotor helicopter. *Automatica*, 14:301–312, 1978.
22. M. Oosterom, G. Schram, R.Babuška, and H.B. Verbruggen. Automated procedure for gain scheduled flight control law design. In *Proceedings AIAA Guidance, Navigation and Control Conference*, pages AIAA–2000–4253, Denver, USA, 2000.
23. X. L. Xie and G. A. Beni. Validity measure for fuzzy clustering. *IEEE Trans. on Pattern Anal. and Machine Intell.*, 3(8):841–846, 1991.
24. M. Alvarez Grima and R. Babuška. Fuzzy model for the prediction of unconfined compressive strength of rock samples. *International Journal of Rock Mechanics and Mining Science*, 36(3):339–349, 1999.

25. R. Babuška, H.B. Verbruggen, and H.J.L. van Can. Fuzzy modeling of enzymatic penicillin–G conversion. *Engineering Applications of Artificial Intelligence*, 12(1):79–92, 1999.
26. H.L.H. van Ginneken, R. Babuška, J.Th. Groennou, and H.B. Verbruggen. Black-box modelling of a rapid sand filter. In *Proceedings IFAC/IMACS International Conference on Artificial Intelligence in Real-Time Control - AIRTC 98*, pages D–6, 1998.
27. J.M. Sousa, R. Babuška, and H.B. Verbruggen. Fuzzy predictive control applied to an air-conditioning system. *Control Engineering Practice*, 5(10):1395–1406, 1997.
28. R. Babuška, H.A.B. te Braake, A.J. Krijgsman, and H.B. Verbruggen. Comparison of intelligent control schemes for real-time pressure control. *Control Engineering Practice*, 4(11):1585–1592, 1996.

Information Granulation in Automated Modeling

Matthew Easley and Elizabeth Bradley

Department of Computer Science
University of Colorado
Boulder, CO 80309-0430 USA

Abstract. The goal of input-output modeling is to apply a test input to a system, analyze the results, and learn something useful from the cause-effect pair. Any automated modeling tool that takes this approach must be able to reason effectively about sensors and actuators and their interactions with the target system. The granulation level of the information involved in this process ranges from low-level data analysis techniques to abstract, qualitative observations about the system. This chapter describes a knowledge representation and reasoning framework that allows this process to be automated.

Keywords. Automated model building, system identification, input-output modeling, knowledge representation

1 Input-Output Modeling

One of the most powerful analysis and design tools in existence – and often one of the most difficult to create- is a good model. Modeling is an essential first step in a variety of engineering and scientific problems. Faced with the task of designing a controller for a robot arm, for instance, a mechanical engineer performs a few simple experiments on the system, observes the resulting behavior, makes some informed guesses about what model fragments could account for that behavior, and then combines those terms into a model and checks it against the physical system. The information involved in this process is heterogeneous both in type and in level. The observations of the target system, for instance, can range from detailed sensor data to abstract, qualitative information like " the temperature oscillates. "

In order to create a model that is both useful (i.e., that captures the desired behavior) and minimal (i.e., that does not contain extraneous modeling components), engineers must granulate that information in appropriate ways.

propriate ways. The topic of this chapter is a knowledge representation and reasoning framework that allows this process to be automated.

Modeling is an extremely broad research area, with roots—and applications—in fields ranging from artificial intelligence and cognitive psychology to control theory and engineering. We do not attempt a comprehensive survey of the uses of granulation across all of these fields; rather, we concentrate only upon the use of granular information in input-output modeling: that is, in the phase of the model-building process that is concerned with physical observations of the target system. We further restrict our attention to deterministic dynamical systems—in particular, those that can be modeled with ordinary differential equations (ODEs).

The goal of input-output modeling of dynamical systems is to apply a test input to the system, analyze the results, and learn something useful from the cause/effect pair. Reasoning about this procedure at a low level of granularity is tedious and difficult. Raw sensor data, for instance, is often both excessive and lacking: one has megabytes of noisy measurements, but of only one of the system's many state variables. Actuator interactions are even harder, since any reasoning about actuators must factor in the interface between the actuator and the system—a difficult modeling problem in its own right, and one that is all but impossible if one must solve it by manipulating voltage values and waveform timing.

An engineer's fundamental formalized knowledge lets him or her solve these problems by reasoning about sensors and actuators at a variety of levels, depending on the requirements of the problem at hand. Determining something as simple as "the voltage oscillates between 5 and 10 volts," for example, can be difficult if one attempts to scan an ASCII text file, but it is trivial to see when presented on an oscilloscope. This type of granular knowledge is critical in the model-building process, because qualitative observations play a much more wide-ranging role than a highly situation-specific sensor data set. Granulation is equally powerful in actuator-related reasoning. Any non-trivial dynamical system has multiple behavioral regimes, and identifying and characterizing these regimes is extremely useful (e.g., the *bifurcation diagram* that is commonly used to describe a dynamical system). The goal of this chapter is to show how automated tools can capture this kind of reasoning, generating and using high-level, granular knowledge that is useful to the model building process. The following two sections describe tools that use granular techniques to solve the sensor data analysis problem and the actuator control problem, respectively; we then close with a brief discussion of related work.

2 Granulating Sensor Data

Efficient model building requires the distillation of succinct, meaningful conclusions about a complicated system out of reams of sensor data. An effective way to do this is to apply geometric reasoning techniques to the data, as described in the following section. Many geometric reasoning techniques, however, require a full state-space trajectory, and fully *observable* systems are rare in engineering practice. Often, some of the state variables are either physically inaccessible or cannot be measured with available sensors. If the target system has 34 state variables, for example, and one can only measure one of those 34 signals, it would appear that the conclusions that one can draw from the sensor data are fundamentally limited. This is control theory's *observer problem*: the task of inferring the internal state of a system solely from observations of its outputs. Delay-coordinate embedding, a good solution to this problem, creates an m-dimensional *reconstruction-space* vector from m time-delayed samples of data from a single sensor; see Section 2.2. The central idea is that the reconstruction-space dynamics and the true (unobserved) state-space dynamics are topologically identical, which implies that a state-space portrait reconstructed from a single sensor time series is qualitatively identical to the true multidimensional dynamics of the system. Together, delay-coordinate embedding and geometric reasoning techniques allow effective granulation of sensor data for automated model building.

2.1 Distilling Qualitative Information from Quantitative Data

A variety of techniques have been developed for extracting qualitative properties from a numeric data set. The solution described here combines phase-portrait analysis, asymptote recognition, and computer-vision techniques. Dynamical systems practitioners typically reason about phase portraits, rather than time series, because the phase-portrait representation—which suppresses time and plots only the state variables—brings out the qualitative properties of the system under examination in a very natural way. For example, recognizing a damped oscillation in a time series from a linear system requires detailed examination of the amplitude decay rate of and the phase shift between two decaying sinusoidal time-domain signals. The same behavior manifests in a much more obvious form—a single symmetric spiral—on a phase portrait. Automated phase-portrait analysis techniques[3,24,25] are designed to capture this kind of information and generate the corresponding qualitative descriptions. This kind of granular information is perfectly suited for automated model building; its abstract, broadly applicable nature allows the automatic verification or rejection of large classes of candidate models. For example, if sensor measurements of a state variable indicate that it is undergoing a damped oscillation, one can immediately rule out all linear ODE models that are unstable, critically damped, or overdamped[1].

One of the earliest and most powerful phase-portrait analysis techniques is the cell dynamics formalism of Hsu[13,14], which discretizes a set of n-dimensional state vectors onto an n-dimensional mesh of uniform boxes or *cells*. In some implementations, cells are uniform in that each cell has the same size; the size and/or number of cells required for each dimension may vary. In others, the aspect ratio is held constant and the size varies. In general, cells need not be fixed in shape *or* size; the theorems involved require only that the cells form a *partition* of the space. Varying the cell size, shape, etc., might be useful—e.g., if the dynamics are smooth in one region and turbulent in another—but doing so vastly complicates the implementation. In Figure 1(a), for example, the circular trajectory—a sequence of two-vectors of floating-point numbers measured by a finite-precision sensor—can be represented as the *cell sequence*

$$[...(0,0)(1,0)(2,0)(3,0)(4,0)(4,1)(4,2)(4,3)(4,4)(3,4)...]$$

Because multiple trajectory points are mapped into each cell, this dis-

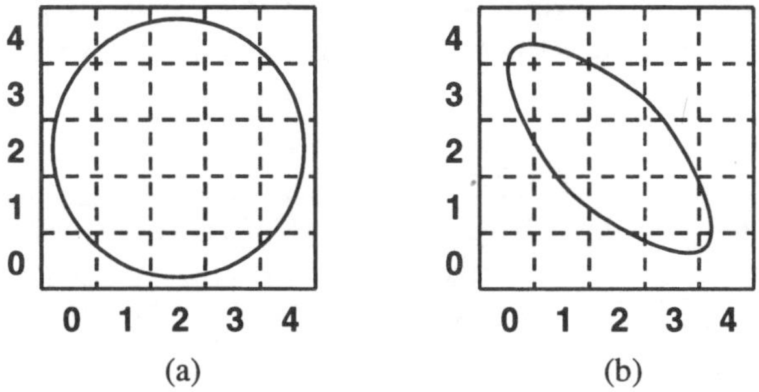

FIGURE 1. Identifying a limit cycle using simple cell mapping. After [4].

cretized representation of the dynamics is significantly more compact than the original series of floating-point numbers and therefore much easier to work with. This is particularly important when complex systems are involved, as the number of cells in the grid grows exponentially with the number of dimensions[2]. Although the approximate nature of the cell dynamics representation does abstract away much detailed information about the dynamics, it preserves many of its important invariant properties. This point

---

[1] Stability analysis of a linear ODE involves finding the roots of its *characteristic polynomial* (e.g., $as^2 + bs + c = 0$ for the ODE $a\ddot{x} + b\dot{x} + c = 0$). Only if those roots are imaginary can the system oscillate; only if their real parts are negative is the oscillation damped.

[2] The example of Figure 1 is two-dimensional, but the cell dynamics formalism generalizes smoothly to arbitrary dimension. The number of required cells is $O(m^n)$, where m is the maximum number of cells on a side and n is the dimension.

is crucial to the fidelity of this analysis method; it means that conclusions drawn from the discretized trajectory are also true of the real trajectory—e.g., a repeating sequence of cells implies that the true dynamics are also on a limit cycle. In this manner, low-level, finite-precision numerical data can be converted into a high-level qualitative classification. Its coarse-grained nature confers some important limitations upon this scheme—both subtle and obvious—many of which are described below; see [9] for more details. Another key concept of the cell dynamics formalism is that it allows a set of geometrically different and yet qualitatively similar trajectories—an "equivalence class" with respect to some important dynamical property—to be classified as a single coherent group of phase portraits. Part (b) of Figure 1 shows a different trajectory with identical topology; this, too, would be classified in the limit cycle equivalence class by the cell dynamics algorithm. See, e.g., Hao[11] or Lind & Marcus[17] for more details.

Given the cell dynamics formalism described in the previous paragraph, the dynamics of a discretized trajectory can be quickly and qualitatively classified using simple geometric heuristics. Some of these classification heuristics are trivial (e.g. determining if the trajectory exits the mesh), but detecting limit cycles or oscillations requires subtler pattern recognition techniques. Below are several of our geometric reasoner's dynamics classifications, the corresponding heuristics, and some associated implications for the ODE model:

- `fixed-cell`: when a trajectory relaxes to a single cell and remains within that cell for a fixed percentage of its total lifetime. This can, for instance, be used to recognize when a system is damped. Appropriate mesh geometry choices can extend this method to asymptote recognition.
- `limit-cycle`: when the trajectory contains a finite, repeating sequence of cells. These patterns are identified by discarding any transients and searching for periodic mapping sequences; they indicate that the system is either conservative or externally driven.
- `damped-oscillation`: when a trajectory enters a fixed cell via a decaying oscillation. This pattern is detected by recognition of an inward spiral; such dynamics can indicate, for instance, that a linear system is underdamped and thus that at least one pair of the model's natural frequencies must be complex.
- `constant`: when a state variable does not change over the duration of the trajectory. This computation involves a simple serial scan on each mesh axis; its results are particularly useful in the model-building process because they have wide-ranging implications about the order of the system.
- `sink-cell`: when a trajectory exits the mesh. This information is used to identify unstable trajectories.

Many other classifications are possible (e.g., that the system is chaotic); some are less useful than others—because their implications either are more limited in range or require processing at a less-abstract reasoning level.

The cell size, mesh boundary, and trajectory length affect the validity and efficiency of the cell dynamics classification. Among other things, a small limit cycle—one that is contained within a single cell—may be classified as a fixed point, and behavior outside the mesh will not be classified at all. All of these discretization and boundary effects are not, in fact, problems; rather, they actually allow one to explicitly represent and work with the abstraction levels implied by the finite range and resolution that are such fundamental features of a modeling hierarchy—e.g., to avoid including saturation and crossover distortion effects when asked to "model the *small-signal* behavior of the op amp *to within 10mV*." Specifically, the range and resolution information may be used to set up the mesh boundary and cell size, assuring that the behavior outside the range or below the resolution is not modeled.

2.2 Delay-Coordinate Methods for Observer Theory

If all of a system's state variables are identified and measured, the geometric reasoning techniques described in Section 2.1 can be applied directly to the sensor data. A fully *observable* system like this, however, is rare in engineering practice; as a rule, many—often, most—of the state variables either are physically inaccessible or cannot be measured with available sensors. Worse yet, the true state variables may not be known to the user; temperature, for instance, can play an important and often unanticipated role in the modeling of an electronic circuit. This is, as mentioned briefly before, part of control theory's observer problem: how to (1) identify the internal state variables of a system and (2) infer their values from the signals that *can* be observed. The arsenal of time-series analysis methods developed by the nonlinear dynamics community in the past decade[1] provides powerful solutions to both parts of this problem. This section describes two methods, Pineda-Sommerer (P-S)[21] and false near neighbor (FNN)[16], that may be used to infer the dimension of the internal system dynamics from a time series measured by a single output sensor[3].

Both P-S and FNN are based on *delay-coordinate embedding*, wherein one constructs m-dimensional *reconstruction-space* vectors from m time-delayed samples of the sensor data. For example, if the time series in Figure 2 is embedded in three dimensions ($m = 3$) with a delay of 0.2, the first two points in the reconstruction-space trajectory are (32.0 22.0 19.0)

[3] Techniques like divided differences can, in theory, be used to derive velocities from position data; in practice, however, these methods often fail because the associated arithmetic magnifies sensor error.

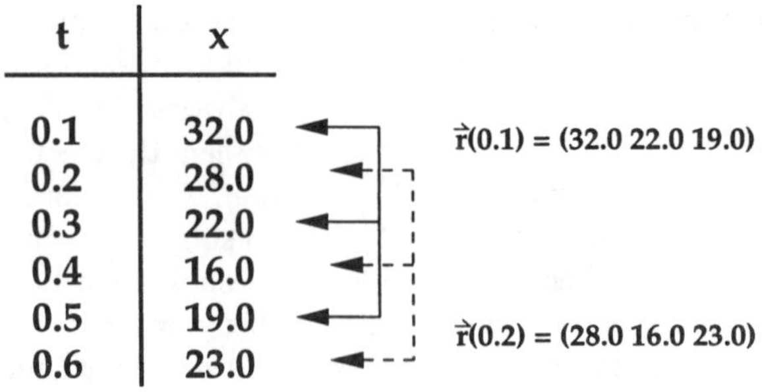
t
x
0.1
32.0
0.2
28.0
0.3
22.0
0.4
16.0
0.5
19.0
0.6
23.0
r(0.1) = (32.0 22.0 19.0)
r(0.2) = (28.0 16.0 23.0)

FIGURE 2. An example delay-coordinate embedding with an embedding dimension of three and a delay of 0.2. x is the measured state variable; $\boldsymbol{r}$ is the vector in reconstruction space.

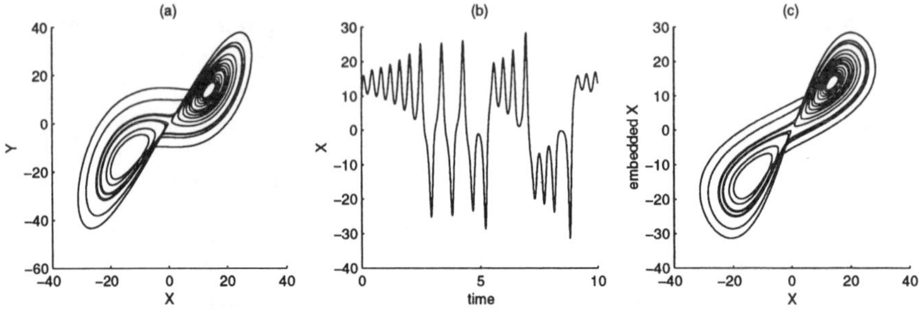

FIGURE 3. The Lorenz Attractor: (a) true version (x-y) (b) time series of x-axis (t-x) (c) embedded version (x-x').

and (28.0 16.0 23.0). Sampling a single system state variable is equivalent to projecting a d-dimensional state-space dynamics down onto one axis; embedding is akin to "unfolding" such a projection, albeit on different axes. Consider the classic Lorenz attractor, the first recognized instance of chaotic behavior[18], shown in Figure 3(a). Part (b) of the figure shows the results of sampling only the x coordinate of that three-dimensional trajectory and plotting it versus time—exactly the situation that would arise if one only had access to a single sensor. The embedded version of this one-dimensional time series, shown in part (c), is slightly distorted but qualitatively identical to part (a). The central theorem relating such embeddings to the true, underlying dynamics was suggested in [23] and proved in [20]; informally, it states that given enough dimensions (m) and the right delay (τ), the reconstruction-space dynamics and the true (unobserved) state-space dynamics are topologically identical[4]. This is an extremely powerful correspondence: it allows one to analyze the underlying dynamics using

[4] More formally, the reconstruction-space and state-space trajectories are diffeomorphic iff $m \geq 2d+1$, where d is the true dimension of the system.

only the output of a single sensor. In particular, many properties of the dynamics, such as dimension (i.e., whether the trajectory is a fixed point, limit cycle, chaotic attractor, etc.), are preserved by diffeomorphisms; if they are present in the embedding, they exist in the underlying dynamics as well. There are, of course, some important caveats, and the difficulties that they pose are the source of most of the effort and subtlety in these types of methods. Specifically, in order to embed a data set, one needs m and τ, and neither of these parameters can be measured or derived from the data set, either directly or indirectly, so algorithms typically rely on numeric and geometric heuristics to estimate them.

The Pineda-Sommerer algorithm creates such estimates; it takes a time series and returns the delay τ and a variety of different estimates of the dimension m. The procedure has three major steps: it estimates τ using the mutual information function, uses that estimated value τ_0 to compute a temporary embedding dimension E, and uses E and τ_0 to compute the *generalized dimensions* D_q, also known as "fractal dimensions." The standard algorithm for computing the fractal dimension of a trajectory, loosely described, is to discretize state space into ϵ-boxes, count the number of boxes occupied by the trajectory, and let $\epsilon \to 0$. Generalized dimensions are defined as

$$D_q = \frac{1}{q-1} \limsup_{\epsilon \to 0} \frac{\log \sum_i p_i^q}{\log \epsilon} \tag{1}$$

where p_i is some measure of the trajectory on box i. D_0, D_1, and D_2 are known, respectively, as the capacity, information, and correlation dimensions; all three are useful as estimates of the number of state variables in the system. The actual details of the P-S algorithm are quite involved; we will only present a qualitative description:

- Construct 1- and 2-embeddings of the data for a range of τs and compute the saturation dimension D_* of each; the first minimum in this function is τ_0. The D_* computation entails:
  - Computing the information dimension D_1 for a range of embedding dimensions E and identifying the saturation point of this curve, which occurs at D_*. The D_1 computation entails:
    - Embedding the data in E-dimensional space, dividing that space into E-cubes that are ϵ on a side, and computing D_1 using equation (1) with $q = 1$.

Ideally, of course, one lets $\epsilon \to 0$ in the third step, but floating-point arithmetic and computational complexity place obvious limits on this; instead, one repeats the calculation for a range of ϵs and finds the power-law asymptote in the middle of the log-log plot of dimension versus ϵ. P-S incorporates an ingenious complexity-reduction technique: the ϵs are chosen to be of the form 2^{-k} for integers k and the data are integerized; this allows most of the mathematical operations to proceed at the bit level and vastly accelerates the algorithm. To increase the precision of this computation, we have implemented an arbitrary-length virtual integer package that facilitates the integerization.

The false near neighbor algorithm is far simpler than P-S. It takes a τ and a time series[5] and returns m. FNN is based on the observation that neighboring points may in reality be projections of points that are very far apart, as shown in Figure 4. The algorithm starts with $m = 1$, finds each

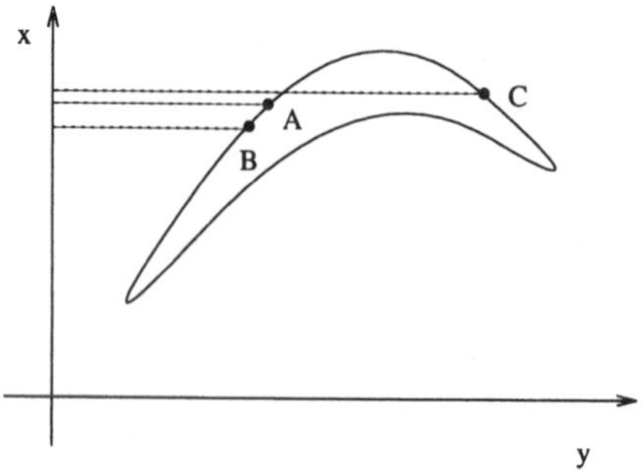

FIGURE 4. The geometric basis of the FNN algorithm: the points labeled A and B are true near neighbors in the x-projection, while A and C are false near neighbors. After [4].

point's nearest neighbor, and then re-embeds the data with $m = 2$. If the point separations change abruptly between the 1- and 2-embeddings, then the points were *false* neighbors (like A and C in the x-projection of Figure 4). The FNN algorithm continues adding dimensions until an acceptably small[6] number of false near neighbors remain, and returns the last m-value as the estimated dimension. A K-D tree implementation[10] reduces the complexity of the nearest-neighbor step from $O(N^2)$ to $O(N \log N)$, where N is the length of the time series.

As both FNN and P-S are based on heuristics, their estimates of the embedding dimension m are not necessarily the same. Since both algorithms provide conservative estimates, choosing the minimum of the two results gives a reasonable lower bound for the dimension of the model. This knowledge facilitates granulation of the reasoning involved in the model-building process, as it allows the modeler to rule out entire branches of the search space of possible models (e.g., all models whose dimension is below that lower bound).

---

[5] P-S may be used to generate τ for use in FNN. Other methods, such as autocorrelation[1] can also be used to estimate τ.

[6] An algorithm that removes *all* false near neighbors can be unduly sensitive to noise.

3 Granulating Actuator Control

Analysis of sensor data is only a small part of understanding the entire system, since dynamic systems—by their very nature—are not passive objects. Rather, they have inputs and outputs, and the relationship between the two is a critical feature of the system's behavior, and thus an important part of its model. Moreover, any non-trivial dynamical system has multiple behavioral regimes, and any successful model builder must be able to reason about this property. This can be a daunting task, even for human experts; selecting and exploring appropriate ranges of state variables, parameter values, etc., is a subtle and difficult problem that has received much attention in the qualitative reasoning community[2,3,24,25]. Manipulation of the available actuators and sensors so as to actually carry out a specific experiment is another difficult problem. Fortunately, granular information can help engineers and scientists manipulate a system's inputs *and* observe its outputs, exploring its different operating regimes without generating overwhelming amounts of data.

This section describes a knowledge representation and reasoning framework called *qualitative bifurcation analysis* or QBA which allows a computer to mimic the kind of analysis an engineer would perform in the input-output analysis of an unknown system. This framework, which is designed to support reasoning about the effects of control parameters and the existence of multiple behavioral regimes, is based on ideas from hybrid systems, nonlinear dynamics, and computer vision. Its representation is a hybrid construct termed the *qualitative state/parameter (QS/P)* space, which combines information about the behavior of a complex system and the effects of its control parameters (inputs) upon its behavior. QBA's reasoning procedures emulate a classic technique in the dynamical systems literature known as *bifurcation analysis*, wherein a human expert changes a control parameter, classifies the resulting behavior, determines the regime boundaries, and groups similar behaviors into equivalence classes. Putting QBA's ideas into physical practice requires yet another layer of granulation, which translates abstract concepts about experiments, such as "measure the step response," into low-level commands that manipulate actuators and sensors in appropriate ways.

Working together, QBA's representation and reasoning processes allow the automatic generation of the kind of observations that a human engineer would make about the system, such as

> "in the temperature range from 0 to $50°C$, the system undergoes a damped oscillation to a fixed point at $(x, y) = (1.4, -8)$; when $T > 50°C$, it follows a period-two limit cycle located at..."

As described before, this kind of granulated information is useful in that it raises the abstraction level of the model-building process.

3.1 Qualitative Bifurcation Analysis: the Representation

As described in Section 2.1, a discretized version of the state-space representation can effectively abstract away many of the low-level details about the dynamics of a system while preserving its important qualitative and invariant properties. Using the cell dynamics representation, in particular, the dynamics of a trajectory can be quickly and qualitatively classified using simple geometric heuristics. This type of discretized geometric reasoning "distills" out the qualitative features of a given state-space portrait, allowing the representation of and reasoning about these features to proceed at a much higher (and cheaper) abstraction level[4].

Raising the granularity level of the analysis of individual sensor data sets, however, is only a very small part of the power of qualitative reasoning about phase portraits. Dynamical systems can be extremely complicated. Attempting to understand one by analyzing a single behavior instance—e.g., system evolution from *one* initial condition at *one* parameter value, like the limit cycle shown in Figure 1(a)—is generally inadequate. Rather, one must vary a system's inputs and control parameters and study the change in the response. Even in one-parameter systems, however, this procedure can be difficult[2]; as the parameter is varied, the behavior may vary smoothly in some ranges and then change abruptly ("bifurcate") at critical parameter values. A thorough representation of this behavior, then, requires a "stack" of state-space portraits: at least one for each interesting and distinct range of parameter values. Constructing such a stack requires automatic recognition of the boundaries between behavioral regimes, and the cell dynamics representation makes this very easy, as described in conjunction with Figure 1. Specifically, it allows a set of geometrically different and yet qualitatively similar trajectories—an "equivalence class" with respect to some important dynamical property—to be classified as a single coherent group of state-space portraits.

Similar kinds of problems arise in the hybrid systems literature[12]. Hybrid modeling techniques describe continuous nonlinear behavior using an ontology of piecewise-continuous regimes and discrete inter-regime transitions[19]. In this representation, if a control parameter is changed or a state variable moves into a prescribed state-space region, a *transition function* moves or "jumps" the hybrid model into that new operating regime and simultaneously invokes the appropriate governing equations. If one attempts to use this representation to capture the behavior of a nonlinear dynamical system, however, the requirement that different operating regimes occupy physically distinct state-space regions poses some serious problems. The same state-space region may exhibit radically different behaviors for different control parameter values, and the simple hybrid system representation cannot handle this.

Consider, for example, a driven pendulum model described by the ODE

$$\ddot{\theta}(t) + \frac{\beta}{m}\dot{\theta}(t) + \frac{g}{l}\sin\theta(t) = \frac{\gamma}{ml}\sin\alpha t$$

with mass (m), arm length (l), gravity constant (g), damping factor (β), drive amplitude (γ) and drive frequency (α). m, l, g and β are constants; the state variables of this system are θ and $\omega = \dot{\theta}$. In many experimental setups, the drive amplitude and/or frequency are controllable: these are the "control parameter" inputs of the system. The behavior of this apparently simple device is really quite complicated and interesting. For low drive frequencies, it has a single stable fixed point; as the drive frequency is raised, the attractor undergoes a series of bifurcations between chaotic and periodic behavior[7]. These bifurcations do not, however, necessarily cause the attractor to *move*. That is, the qualitative behavior of the system changes and the operating regime (in state space) does not. Traditional bifurcation analysis of this system would involve constructing phase portraits of the system, like the ones shown in Figure 1, at closely spaced control parameter values across some interesting range. Traditional hybrid representations do not handle this smoothly, as the operating regimes involved are not distinct. If, however, one adds a parameter axis to the state space, most of

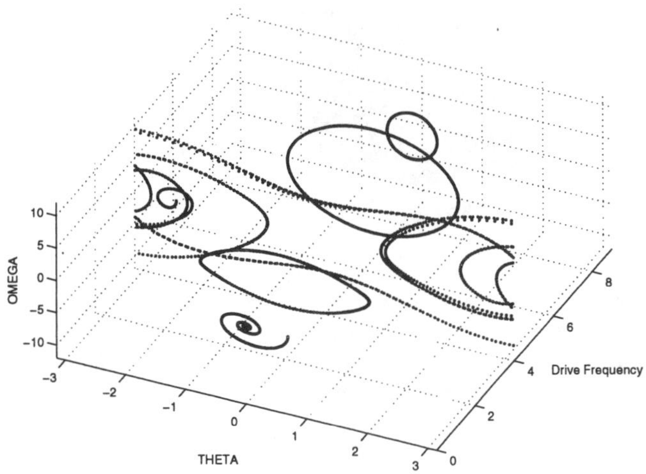

FIGURE 5. A state/parameter (S/P) space portrait of the driven pendulum: a parameterized collection of state-space portraits of the device at various Drive Frequencies. Each (θ, ω) slice of this *S/P-space portrait* is a standard phase portrait at one parameter value. For example, when the drive frequency is zero, the pendulum's attractor is a fixed point; as the frequency is raised, the device goes through various chaotic and periodic regimes and finally settles into a family of limit cycles (the ellipses at drive frequency values of 6 and above). After [9].

these problems vanish. Figure 5 describes the behavior of the driven pendulum in this new *state/parameter space* (S/P space). Each θ, ω slice of this plot is a state-space portrait, and the control parameter varies along the Drive Frequency axis. This idea is similar to the hybrid systems community's idea of forming a cross product of the input space of a system with its output space.

The final step in our development of a good representation for the qualitative bifurcation analysis framework was to combine the state/parameter space idea pictured in Figure 5 with the qualitative abstraction of Hsu's cell dynamics (Section 2.1), to produce the *qualitative state/parameter space* (QS/P-space) representation. A QS/P-space portrait of the driven pendulum is shown in Figure 6. This representation is similar to the state/parameter-space portrait shown in Figure 5, but it groups similar behaviors into equivalence classes, and then uses those groupings to define the boundaries of qualitatively distinct regions—an effective, useful, application-specific granulation of the behavioral description.

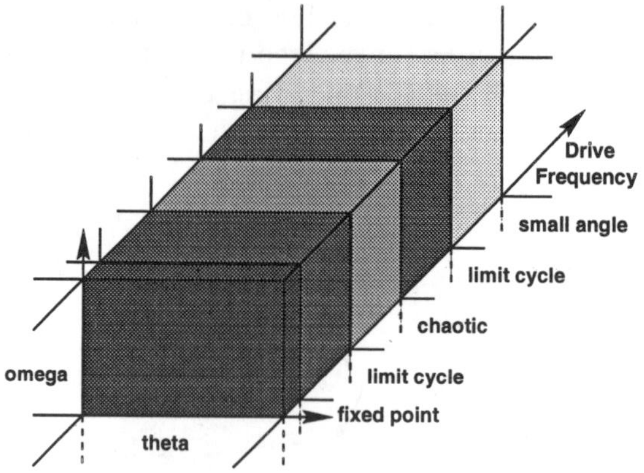

FIGURE 6. A *qualitative state/parameter-space* (QS/P-space) portrait of the driven pendulum. This is a granularization of the state/parameter space portrait in the previous figure; it groups qualitatively similar behaviors into equivalence classes and uses those groupings to define the boundaries of qualitatively distinct regions of state/parameter space. After [9]

This qualitative state/parameter-space representation is an extremely powerful modeling tool. One can use it to identify the individual operating regimes, then create a separate model in each, and perhaps use a finite-state machine to model transitions between them. More importantly, however, the QS/P-space representation lets the model builder leverage the knowledge that its regions—e.g., the five slabs in Figure 6—all describe the behavior of the *same system*, at different parameter values. This is exactly the type of high-level, granular knowledge needed to plan how to learn more about a system by changing its inputs and observing the results. This representation is also widely applicable, as it works on all classes of one-parameter, continuous systems of ordinary differential equations. Some specific examples of these types of systems are:

- A non-linear spring, $\dot{x} = \alpha x - x^3$, where α is the control parameter.
- Duffing's equation, $\ddot{x} + k\dot{x} + x^3 = p\cos t$, where p is the control parameter.

Although the QS/P-space representation is not limited to lower order systems, most of the systems from the dynamic systems literature are typically order four or less as higher order systems quickly become too complex to reason about effectively. The remainder of this section expands upon these ideas, describing how the QS/P-space representation assists in the automation of input/output modeling of dynamical systems.

3.2 QBA:Reasoning about Granular Knowledge

Reasoning about actuators is much more difficult than reasoning about sensors. The problem lies in the inherent difference between passive and active modeling. It is easy to recognize damped oscillations in sensor data without knowing anything about the system or the sensor, but using an actuator requires a lot of knowledge about both. Different actuators have different characteristics (range, resolution, response time, etc.); consider the difference between the time constants involved in turning off a burner on a gas versus an electric stove, and what happens if the cook is unaware of this difference. The effect of an actuator on a system also depends intimately on how the two are connected. For example, a DC motor may be viewed as a voltage-to-RPM converter with a linear range and a saturation limit. How that motor affects a driven pendulum depends not only on those characteristics, but also on the linkage between the two devices, (e.g., a direct rotary drive configuration versus a slot/cam-follower setup). Planning and executing experiments successfully requires modeling these kinds of properties and effects. One way to do so is to use another granular computing abstraction such as bond graphs[15]. Exploiting the bond graph's two-port nature allows the effects of an actuator to be incorporated into the model quite naturally, via additional modeling components. Since actuators themselves can be complex nonlinear systems, actuators should be modeled initially off-line; that knowledge can then be used as an invariant (that is, regime-independent) part of the actuator/system model.

Qualitative bifurcation analysis requires interleaved input and output reasoning, in which sensors and actuators are used to probe the system at a variety of control parameter values to find interesting behaviors and identify boundaries between different regimes. The QBA process simply scans the range of an actuator at predefined intervals, classifying the results as described above; it then zeroes in on the bifurcations using a simple bisecting search. These bifurcations are the dividing lines between regimes in the QS/P-space portrait of the system, and the qualitative classifications are the labels for the regimes. The results of the QBA process are twofold: a QS/P-space representation of the system dynamics—like the one shown in Figure 6—and a set of qualitative observations similar to those a human engineer would make about the system, such as "When the control parameter ρ is in the range [1.2, 5.6], the state variable x_1 oscillates." The granular information captured in the QS/P-space portrait is useful well beyond the input-output phase of the model-building process; it can also streamline the generate phase, for instance, since a model that is valid in one regime is often also valid in other regimes that have the same qualitative behavior. And even when the qualitative behavior is different, continuity suggests that a neighboring regime's model is a very good starting point.

4 Related Work

A number of researchers have combined numerical techniques with ideas from symbolic computation and artificial intelligence to create granular computational tools for scientists and engineers. One such class of tools autonomously plans, executes, and interprets simulation experiments from high-level specifications of physical models. Abelson's *Bifurcation Interpreter*[2], for instance, autonomously explores the steady-state orbits of one-parameter families of periodically driven oscillators; it automatically generates a bifurcation diagram and a text description of the findings. A related class of tools addresses the problem of automated phase-portrait analysis, combining ideas from dynamical systems, discrete mathematics, and artificial intelligence to generate qualitative descriptions of different kinds of systems. Bradley's *Perfect Moment* explores a system's state space with particular attention to chaotic features, then uses that information to design nonlinear controllers[3,6]. Yip's KAM extracts useful, high-level information about the phase portraits of Hamiltonian systems by combining computer vision techniques with sophisticated mathematical invariants[24]. Zhao's *Phase Space Navigator* analyzes phase portraits, producing a detailed description of the system dynamics—equilibrium points, boundaries of stability regions, etc.—using a granular analysis tool called the *flow pipe*[25]. The work described in this chapter, which is part of the PRET project[4,5,8,22] is similar to these tools in that it combines traditional numerical computation techniques with symbolic artificial intelligence; it even uses some similar analysis tools (e.g., phase-portrait analysis).

5 Summary

The goal of input-output modeling is to apply a test input to a system, analyze the results, and learn something useful from the cause/effect pair.

Automating this analysis procedure is not only important from an engineering standpoint, but also hard and interesting from an artificial intelligence standpoint. In particular, planning, executing, and interpreting experiments requires some fairly difficult reasoning about what experiments are useful and possible, and information granulation plays an extremely important role in this process. The approach described in this chapter uses a hybrid representation termed the *state/parameter (S/P)* space, which granulates information about the behavior of a complex system and the effects of the control parameter upon the behavior. This information then undergoes a second level of granulation—termed qualitative bifurcation analysis—wherein the S/P space is decomposed into discrete regions, each associated with an equivalence class of dynamical behaviors, derived qualitatively using geometric reasoning. In this representation, each trajectory is effectively equivalent, in a well-known sense, to all the other trajectories in the same region, which allows the model builder to describe the behavior of a multiregime system in a significantly simpler way, which results in ease of analysis—and great computational savings.

Acknowledgments. Joe Iwanski and Reinhard Stolle contributed code, and/or ideas to this paper.

6 References

[1] H. Abarbanel. *Analysis of Observed Chaotic Data.* Springer, 1995.

[2] H. Abelson. The Bifurcation Interpreter: A step towards the automatic analysis of dynamical systems. *International Journal of Computers and Mathematics with Applications*, 20:13, 1990.

[3] E. Bradley. Autonomous exploration and control of chaotic systems. *Cybernetics and Systems*, 26:299–319, 1995.

[4] E. Bradley and M. Easley. Reasoning about sensor data for automated system identification. *Intelligent Data Analysis*, 2(2):123–138, 1998.

[5] E. Bradley, M. Easley, and R. Stolle. Reasoning about nonlinear system identification. Technical Report CU-CS-894-99, University of Colorado at Boulder, 2000. In review for *Artificial Intelligence.*

[6] E. Bradley and F. Zhao. Phase space control system design. *IEEE Control Systems Magazine*, 13:39–46, 1993.

[7] D. D'Humieres, M. Beasley, B. Huberman, and A. Libchaber. Chaotic states and routes to chaos in the forced pendulum. *Physical Review A*, 26:3483–3496, 1982.

[8] M. Easley and E. Bradley. Generalized physical networks for model building. In *Proceedings of the International Joint Conference on Artificial Intelligence (IJCAI-99)*, 1999.

[9] M. Easley and E. Bradley. Reasoning about input-output modeling of dynamical systems. In *Proceedings of the Third International Sympo-*

sium on Intelligent Data Analysis (IDA-99), LNCS 1642, pages 343–355. Springer, 1999. Amsterdam, The Netherlands.

[10] J. H. Friedman, J. L. Bentley, and R. A. Finkel. An algorithm for finding best matches in logarithmic expected time. *ACM Transactions on Mathematical Software*, 3:209–226, 1977.

[11] B.-L. Hao. Symbolic dynamics and characterization of complexity. *Physica D*, 51:161–176, 1991.

[12] T. Henzinger and S. Sastry, editors. *Hybrid Systems: Computation and Control, Proceedings of the First International Workshop, HSCC'98*, LNCS 1386. Springer, 1998.

[13] C. Hsu. A theory of cell-to-cell mapping dynamical systems. *Journal of Applied Mechanics*, 47:931–939, 1980.

[14] C. Hsu. *Cell-to-Cell Mapping*. Springer-Verlag, New York, 1987.

[15] D. Karnopp, D. Margolis, and R. Rosenberg. *System Dynamics: A Unified Approach*. Wiley, New York, second edition, 1990.

[16] M. B. Kennel, R. Brown, and H. D. I. Abarbanel. Determining minimum embedding dimension using a geometrical construction. *Physical Review A*, 45:3403–3411, 1992.

[17] D. Lind and B. Marcus. *Symbolic Dynamics and Coding*. Cambridge University Press, Cambridge, 1995.

[18] E. N. Lorenz. Deterministic nonperiodic flow. *Journal of the Atmospheric Sciences*, 20:130–141, 1963.

[19] P. J. Mosterman, F. Zhao, and G. Biswas. An ontology for transitions in physical dynamic systems. In *Proceedings of the AAAI-98*, pages 219–223, 1998.

[20] N. Packard, J. Crutchfield, J. Farmer, and R. Shaw. Geometry from a time series. *Physical Review Letters*, 45:712, 1980.

[21] F. J. Pineda and J. C. Sommerer. Estimating generalized dimensions and choosing time delays: A fast algorithm. In *Time Series Prediction: Forecasting the Future and Understanding the Past*. Santa Fe Institute, Santa Fe, NM, 1993.

[22] R. Stolle and E. Bradley. Multimodal reasoning for automatic model construction. In *Proceedings of the AAAI-98*, pages 181–188, 1998.

[23] F. Takens. Detecting strange attractors in fluid turbulence. In D. Rand and L.-S. Young, editors, *Dynamical Systems and Turbulence*, pages 366–381. Springer, Berlin, 1981.

[24] K. Yip. *KAM: A System for Intelligently Guiding Numerical Experimentation by Computer*. MIT Press, 1991.

[25] F. Zhao. Computational dynamics: Modeling and visualizing trajectory flows in phase space. *Annals of Mathematics and Artificial Intelligence*, 8:285–300, 1993.

Optical Music Recognition: the Case of Granular Computing

Władysław Homenda

Faculty of Mathematics and Information Science
Warsaw University of Technology, 00-661 Warsaw, Poland

Abstract. The paper deals with optical music recognition (OMR) as a process of structured data processing applied to music notation. Granularity of OMR in both its aspects: data representation and data processing is especially emphasised in the paper. OMR is a challenge in intelligent computing technologies, especially in such fields as pattern recognition and knowledge representation and processing. Music notation is a language allowing for communication in music, one of most sophisticated field of human activity, and has a high level of complexity itself. On the one hand, music notation symbols vary in size and have complex shapes; they often touch and overlap each other. This feature makes the recognition of music symbols a very difficult and complicated task. On the other hand, music notation is a two dimensional language in which importance of geometrical and logical relations between its symbols may be compared to the importance of the symbols alone. Due to complexity of music nature and music notation, music representation, necessary to store and reuse recognised information, is also the key issue in music notation recognition and music processing. Both: the data representation and the data processing used in OMR is highly structured, granular rather than numeric. OMR technology fits paradigm of granular computing

Keywords: *information granules, information granulation, pyramid architectures, encoding and decoding, utility of granular information, knowledge representation, music notation, music recognition, music representation.*

1. Introduction

Music notation may be interpreted as a language allowing documenting musical information in a legible, archival form. Recognition of music notation can of course be modelled as mappings between the printed notation and the information it represents [3], cf. Figure 1. However, this general formulation of the task of music notation recognition does not reflect the conception and technical problems that developers of recognition systems are faced with.

First of all, music notation does not have a universal definition. Although attempts to codify printing standard for music notation have been undertaken, c.f.

[20,21], in practice composers and publishers feel free to adopt different rules and invent new ones. Though most of scores keep the standard, they still can vary in details. Moreover, music notation, as a subject of human creative activity, constantly develops and probably will be unrestricted by any codified set of rules. Thus, it may not be possible to build a universal recognition system accepting all dialects of printed music notation. This paper deals only with problems that are common for most of dialects of music notation. Furthermore, the nature and structure of music, even the printed one, is much more complicated than, for instance, the structure of a text, so representation of music is comparably much more difficult than representation of printed text. As the result, comparing music notation recognition with text recognition, there are several applicable computer systems for automated text recognition with considerably high rate of recognition (which can be calculated as ratio of recognised characters to all characters in the text). As to music notation recognition, commercial systems are still very rare despite the fact that several research systems of music notation recognition have already been developed c.f. [3,6,13,14]. This paper is based on an experience gained during development of Musitek's Smart Score and derivatives.

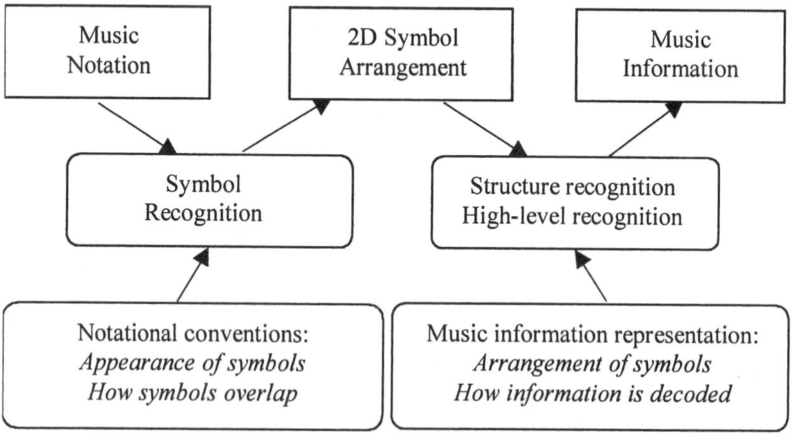

Figure 1. Music notation recognition

2. Music knowledge structure

Making one more analogy between computer processing of a text and a printed music it is worth underlining that it is considerably easy to define and implement data representation, which keeps all features of the text, allows for processing it, etc. Rich Text Format, RTF format is an example of such representation. Because recognition confidence of scanning program usually is not satisfactory enough for further use, it is necessary to correct acquired data. These tasks need storing of acquired data in some form suitable for both editing and further processing and use.

Music data representation is far more difficult and, up to now, no universal representation widely used and commonly accepted has been defined. Music data formats used in computer systems are intended more for particular task rather than for common use. Even if a particular format is widely spread and commonly applied, it is used for special tasks rather than for any purpose. For example, "MIDI (Music Instrument Digital Interface) data format was established as a hardware and software specification which would make it possible to exchange information between different musical instruments or other devices such as sequencers, computers, lighting controllers, mixers, etc. This ability to transmit and receive data was originally conceived for live performances, although subsequent developments have had enormous impact in recording studios, audio and video production, and composition environments" (c.f. [16]). Nevertheless, MIDI format, as performance oriented, is not a universal one. For instance, it is very difficult or even impossible to graphical features related to music notation represent in MIDI format.

On the other hand, the format used by notation programs is notation oriented and cannot be easily used as a universal format of music representation.

Several attempts have been made lately in order to define a universal format of music notation, cf. [4,5,8]. A special format, Notation Interchange File Format, NIFF, was defined in 1995, cf. [18]. This format has a potential ability to become the commonly accepted universal format of music notation. Despite the achievement mentioned above, there is still a number of difficulties in music data representation which make many recognition systems concentrated on locating staves and isolating and recognizing symbols. The interpretation resulting from the 2-D arrangement of symbols and precise representation of acquired information in suitable format are not done in such systems.

2.1. Terminology

Though music notation does not have a universal definition, it has common features for all its dialects. There are some terms commonly found in discussions of music notation that most musicians intuitively understand. Music notation, a complicated product of human productivity, may be interpreted from different perspectives: logical, geometrical, temporal, etc. In this paper the logical and geometrical aspects of music notation are considered. The geometrical point of view obviously affects the recognition process. The logical perspective is important from the knowledge representation angle and also affects recognition methods. This section defines the usage of these terms in context of logical and geometrical perspectives. The examples of terms defined here are given in Figures 2 and 3.

Figure 2. Score, its first page

- *Score.* Each music representation file usually contains one score. The score usually holds notation describing a piece of music like a song, a piano sonata, a quartet, a symphony.
- *Part.* A stream of musical events and associated symbols, text and graphics which can be extracted and printed on a part score for an individual performer. A part may contain music to be played sequentially on different related instruments by the same performer (e.g. oboe/English horn).
- *Voice.* A rhythmically independent stream of musical events within a part, and its associated symbols, text and graphics. Voices can appear and disappear at any time within a part.

  Voice. A musical line. A voice may correspond to a single instrument; a piano part is usually notated as two and sometimes more voices. Several voices may be printed together on one staff: In an orchestra score, the Flute 1 and Flute 2 voices may be printed together on one staff (with opposite stem direction), but they are printed separately to make the individual instrumental part.

- *Page.* A sheet of paper with music notation printed on it, or a graphical file being its computer representation.
- *System. (logical)* The visual framework on a page, where symbols representing simultaneous events in the various parts are more or less vertically aligned. All parts of the score are logically present in every system although some may be "hidden".
  System. (geometrical) A set of staves that are played in parallel; in printed music all of these staves are connected by barline drawn through from one stave to next on their left end. A braces and/or brackets may be drown in front of all or some of them. Symbol of a system are performed according to their horizontal placement in system, independently on the stave they belong to.
- *Staff line.* A long thin horizontal line. Typically five parallel staff lines are drawn to form a stave (a staff).
- *Stave (or Staff). (logical)* An arrangement of parallel horizontal lines and its neighbourhood serving as the locale for displaying musical symbols. It is a sort of vessel within a system into which musical symbols can be "poured". Displayed on it are music symbols, text and graphics belonging to one or more parts.
  Stave (or Staff). (geometrical) Five staff lines that are horizontal, parallel, equally spaced, unbroken and of constant thickness. Staff lines play central role in music notation. They define the vertical coordinate system for pitches and provide a horizontal direction for temporal coordinate system.
- *Barline.* A vertical line (or two or three vertical lines close to each other) splitting notation on units called measure. Barlines are drawn on all staves of a system and may be extended going from one stave to next one.

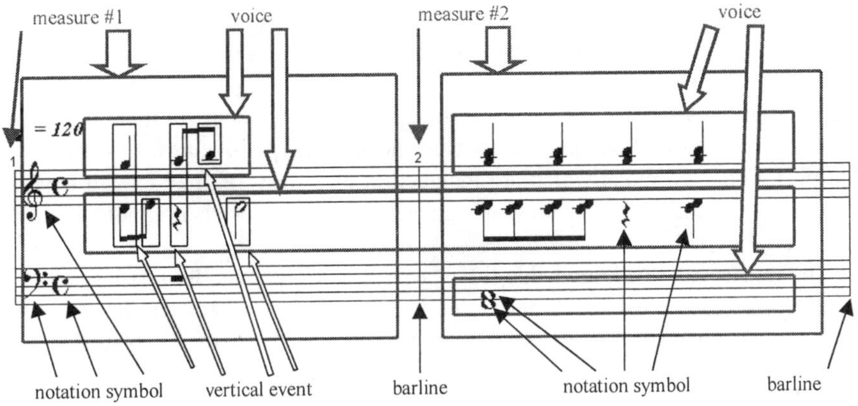

Figure 3 Examples of music items and terminology

- *Vertical Event.* The notion by which a specific point in time is identified in the score. Musical symbols representing simultaneous events are logically grouped within the same vertical event. The music symbols in a music representation file are physically grouped by page and staff, so symbols belonging to a common logical vertical event may be physically separated in the file.
- *Slur.* Curved line used to show that two or more notes are to be sung to one syllable or performed legato.
- *Tie.* Curved line joining two notesof the same pitch that are to be played or sung as one.
- *Tuplet.* Curved line with numerical character centered within the arc or at the apex of the arc to show that the shown number of notes are to be played or sung in time of other number of notes. Triplet is a case of tuplet used to show that three notes of the same duration are to be pleyed in time of two notes of the same duration.
- *Notation symbols.* Other symbols used in music notation, e.g. clefs, notes, rests, signatures, etc.

2.2. Geometrical structure

The geometrical structure of music notation is displayed in Figure 4. This feature of data structure, or better to say: knowledge structure, can be symbolically well expressed as the granular structure of this knowledge and, following the idea expressed in [19], can be outlined as a hierarchical system of granules. Every unit in this hierarchy consists of a collection of units of lower levels. For instance, the score is a collection of pages and – indirectly – includes granules of lower levels; the page is a collection of systems and – indirectly - includes granules of lower levels. This simple inclusion is broken at lower levels: the measure, as a vertical part of the system, is not included in any stave. On the other hand, apart from one-stave systems and measures occupying the whole system, the measure does not include any whole stave. A similar irregularity of granules pyramid is observed in case of music notation symbols. Most of symbols belong to one stave and are included in one measure, e.g. notes, rests, clefs. However, some of them belong to one stave, but extend beyond the measure horizontally, e.g. slures, ties, dynamic hairpins. Other symbols, which are included in one measure, may extend beyond one stave vertically, e.g. barlines, arpeggios, notes of voices visiting neighbouring stave, cf. Figure 3.

In this structure of knowledge the page is a unit separated by restrictions of a sheet of paper. In most cases there is no important relation between the page and its systems, so page can be seen as a simple collection of its systems. The exception of the first page, where the composer's name, the title of the piece, etc., are printed, does not break this rule because these items of information are logically connected to the level of the whole score rather than particular page. Despite this, it is convenient to keep this level in the knowledge hierarchy.

Although the page can be seen as the simple union of its system, other relations in knowledge structure are not so straightforward and simple. Neither the stave is a simple collection of its symbols nor the system is only a set of staves. Unlike page, the whole score is not only a collection of its systems.

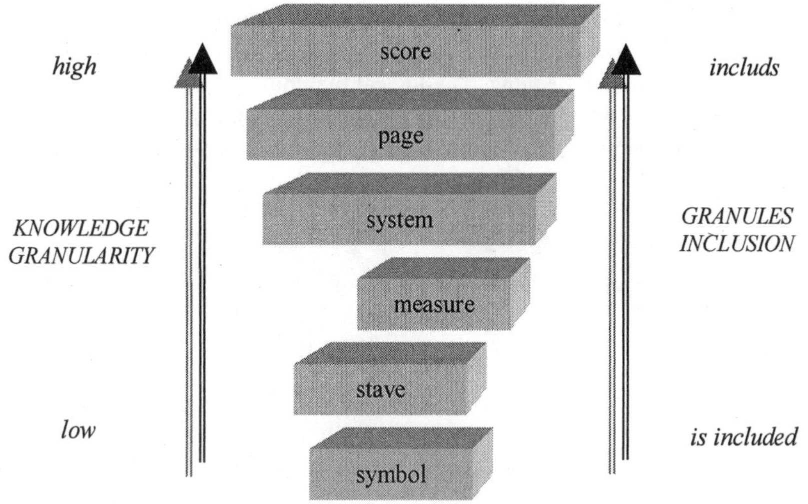

Figure 4. Music notation, the geometrical knowledge granularity

Apart from the parted score, i.e. a score with all parts separated one from another rather than joined into systems, there are relations between systems and between staves, e.g. barlines, repetitions, ties, slurs, dynamic hairpins, etc. Although these relations are examples of a logical structure of a score, they obviously have to be considered as geometrical features. The importance of relations between symbols of music notation from geometrical point of view comes from their influence on recognition of music symbols. Recognition of music symbols is merely understood as a process that is geometrically based unlike in classification of symbols and their context identification which are mostly seen as a process based on logical structure of the score.

2.3. Logical structure

The geometrical structure does not include another important entities of knowledge, let us say granules: parts, voices, vertical event, etc. And, what is more important, the interpretation of granules is significantly different than in the geometrical case. The entities of logical structure of music notation are understood as units of data describing symbols and units representing the relations between symbol. In this duality of data relations between symbols are at least as important as symbols themselfs. So, the levels of geometrical and logical pyramids of granules are not the same even if they have the same labels. For instance, the score is a simple collection of pages without attention put to relations between other granules in the meaning of geometrical structures.

And vice versa, the score is more a net of relations between systems, parts and voices etc. when the logical meaning of this term is understood. In this case the page structure has little meaning. The logical pyramid contains granules that have no geometrical meaning. Examples of such granules are: parts, voices, vertical events. This logical pyramid of knowledge granularity is shown in Figure 5.

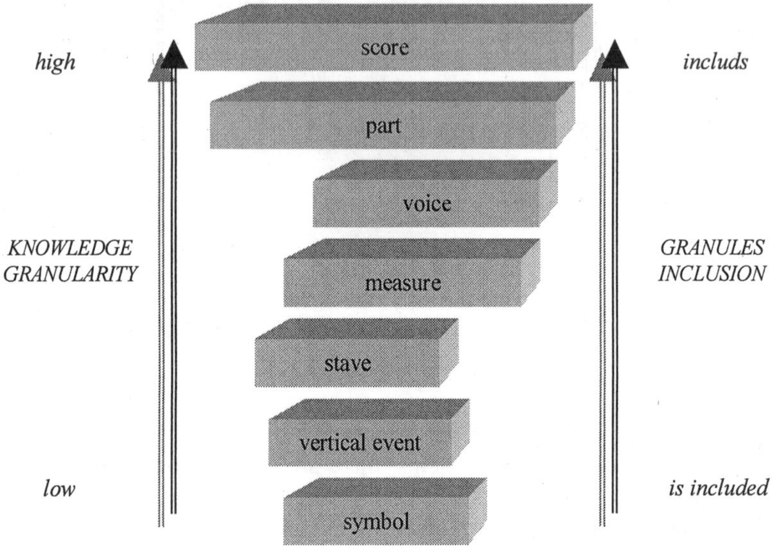

Figure 5. Music notation, the logical knowledge granularity

3. Music notation recognition

It is not quite clear, what *music information* - the goal of recognition task means, cf. Figure 1. In fact, the meaning of this term may depend on the application of recognition task. For some applications, e.g. from the point of view of a musician, a complete solution to the music recognition problem is the specification of: which notes are present, what order they are played in, their time values or durations, and volume, tempo, and interpretation. Thus, in this case the scanning units are generally quite small, for example, a clef, one or more key signatures, accidentals, a time signature, note heads in a chord, and the following music symbols are not included in scanning units: staff lines, beams, slurs, ties, brackets, text, and crescendo or decrescendo signs. However, this set of information is not sufficient for producing parts from a score – a wider spectrum of symbols and their features must be recognized.

The recognition of printed music notation may be undertaken for different reasons: extracting the pitch, duration and timing of notes to create a basic MIDI file; identification of logical structure of music; recognition of geometrical data of a score; pouring in music representation format; editing music notation; etc. Cf. [3] for details. For the purpose of this paper the aspects of logical and geometrical structures of information are considered.

3.1. Score structure identification

The recognition of staff lines is one of the first step in music notation recognition. Although staff lines are characteristic elements of music notation and are obviously noticable, their localization on the page of real notation is not easy. Having image of music notation not distorted, cf. Figure 6a, the staff lines are easily localized. The obvious method is to find five lines that are longest, straight and equidistant.

The staff lines found on a piece of sheet music are not exactly parallel, horizontal, equidistant, of constant thickness, or even straight; scanning noise and quantization noise escalate these problems. So, the staff-line analysis techniques must be able to cope with staff-line inclination and curvature, as well as with the interfering effects of beams and other linear elements in the score. In some cases, staff lines may be obscured to a significant extent by multiple beams, particularly when these are horizontal, cf. Figure 6b. Thus, the standard image-processing techniques of line-finding do not often suffice for locating of staff lines. Two main strategies are applied in recognition systems regarding staff lines handling. The first one is based on staff line removal, the second one leaves staff lines unchanged applying only information about staff lines placement to avoid their influence in symbol recognition.

Thus, proceessing the whole length of staff line is highly inconvenient. Instead, processing a part of staff lines, a vertical strip in the middle of an image, results in finding horizontal lines' handles. This technique allows for avoiding bowing and skewing in finding the vertical location of staff lines. However, symbols other than staff lines can easily influence recognition process. Assuming that the recognition method accepts staff lines that are fragmented, not equidistant and of varying thickness, different symbols may be classified as possible staff line handle. In Figure 6c six or more handles will be classified as potential staff lines' handles at the bottom of the system. Then the iterative process allows for extending handles of horizontal lines and dropping false staff line handles, cf. Figure 6b to see the whole staves.

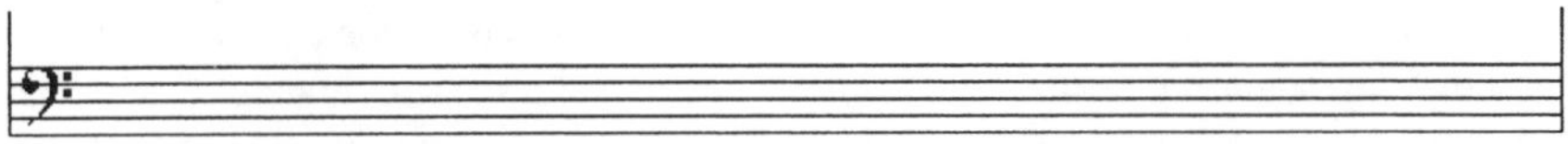

Figure 6a. Empty, not distorted staff

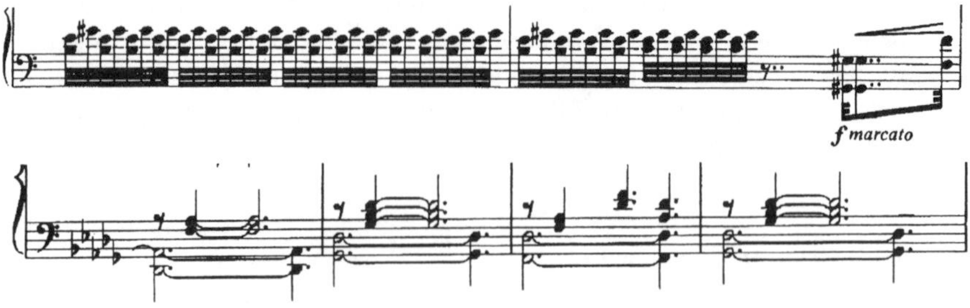

Figure 6b. Staves of real score superimposed by other symbols

Figure 6c. Handles of horizontal lines supposedly detected

Figure 7 presents the interpretation of the process of score structure identification as a granular structure. The meaning of granules of this hierarchy is understood as a unity of data/knowledge representation and algorithms applied to data/knowledge processing. The lower levels are data representation and numeric computing oriented. Transforming of raster bitmap into run lengths of black and while pixels is obviously a kind of numeric computing. It transfers data from its basic form to more compressed data. However, the next levels of the hierarchy, i.e. finding handles of horizontal lines, begins the process of data concentration that becomes to be embryos of knowledge units rather than more compressed data entity. Thus, data representation and numeric computing turns to knowledge representation and granular computing, cf. [19]. The process of data to knowledge transformation does not have the clear boundaries. It can be rather compared with incipient crystal of knowledge emerging from an ocean of data which is going to be a mature entity knowledge, i.e. a knowledge of granule at the end of the data to knowledge transformation path. In the presented example the stave can be interpreted as a knowledge entity. The process of data to knowledge transformation does not have strictly separated stages. The numerical processing of data affects the whole process of transformation until the highest level of the granules pyramid. For instance, the identification of systems is a task of knowledge processing, i.e. knowledge about staff lines and barlines. So, this can be called granular computing rather than numerical computing. Nevertheless, data processing and numerical computing is also involved, e.g. tuning of system localization still refers to image data encoded as runs of pixels. Nevertheless, this is only a side effect of data to knowledge transformation rather than numeric computing: crystallisation of knowledge still absorbs material from the ocean of data.

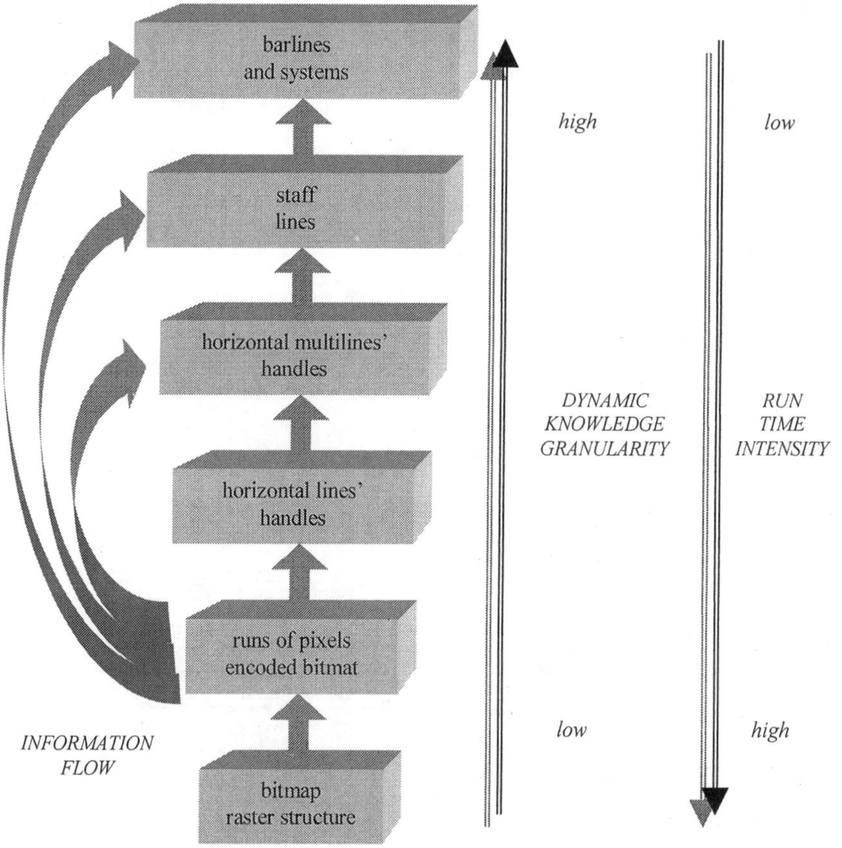

Figure 7. Granularity of the process of score structure identification

3.2. Symbols' recognition – the geometrical case of granular computing

Once the staff lines have been identified and/or removed from the image, the next major processing step is to classify music symbols according to a predefined spectrum of geometrical figures. Pre-classification usually done in OMR is basically interpreted as the identification of geometrical features of notation symbols without deriving contextual information associated with these symbols. In fact, there is no strict separation between recognition and classification. In this paper the term *recognition* is applied to the symbol identification process involving geometrical features rather than logical relations of the symbol. And vice versa, the term *classification* described identification of the symbol that basically involves logical relations between symbols. Thus, the title of this passage uses the term *recognition* rather than *classification* although the term *classification* in the meaning explained above will also be used in the passage.

A great variety of techniques have been applied to this problem. In this chapter some recognition methods are mentioned. A more detailed survey was done in [3]. Brief description of the most important recognition methods is done in the following paragraphs:

- *height/width space exploration.* The bounding-box dimensions (height/width space) of every symbol are used to find possible matches. There are several possible matches for each symbol being analyzed, so further processing must be done to classify a given symbol.
- *extensive use of projections.* The intensive use of projections is applied, for instance, in [6]. For simple music notation and restricted subset of notational symbols projections can be successfully applied. However, for complex notation and an extended set of symbols this method is ineffective.

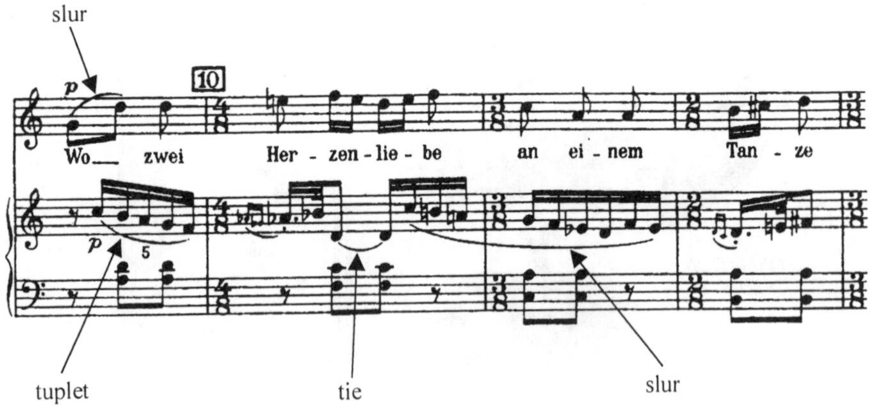

Figure 8 Slur, Tie, Tuplet – the cases of arc

- *examining the pixels in a few particular rows and columns of the symbol-image.* For example, when going along the sections of a vertical line of a given symbol, two columns of pixels are used to differentiate an accidental as a sharp or a natural. The vertical line sections of theses symbols may be located by examining its projection on OX axis.
- *syntactical methods locally applied.* Music symbols, such as notes, chords, accidentals, etc. are composed of pattern primitives, such as line sections, note heads, stems, beams and flags. Recognition of pattern primitives may be simpler that recognition of such complex structures as a beamed group of notes or a chord containing notes of different time value. It is simpler to give a valuable description for a beam and notes recognition separately than to formulate such a description for a beamed note sequence; and it is easier to design flexible recognition procedures around simple descriptions.

3.3. Symbols classification – the logical case of granular computing

Classification of musical symbols is a very important process which involves the semantics of music notation. It significantly extends the recognition process that is seen as finding locations of symbols from given repertoire on a page. However, the semantic of many music symbols can be identified only by exploring contextual information of the symbols. Referring to Figure 5., symbols of music notation are interpreted as a low-level granule of knowledge. The same or similar symbols may play a different role depending on the context of their placement and relations with other symbols. For instance, numbers found on a sheet of music notation can stand for time signature, fingering, barlines numbers, tuplet labels and for other goals. This low-level granule, perhaps obtained from the recognition process, is being added new information based on its context in music notation – it becomes an element of a wider structure, a higher level knowledge granule. It is worth to underline that the higher level of the pyramid of knowledge structure is not only a simple union of lower level granules, it also includes new quality built on relationships between lower level granules.

In Figure 8 the example of contextual knowledge is presented. An arc can either be a slur, or a tie, or a tuplet. As for many symbols, classification of an arc can be done only on the basis of symbols that are in relation to the arc. The features of an arc: its length, rotation, curvature, direction, etc. is not sufficient for final classification due to deep immerse of their meaning into context of music notation. The following reasoning rule, taken out of wider reasoning system, which has a form of an expert system, illustrates an attempt to classify the arc, cf. [17]. Though this example brings a case of applicable supervised learning in the case, the learning scheme is of little interest of the discussion.

Assumptions: Arcs are either ties, or slurs or tuplets.
Apply following classification rules

Do: Find arcs

If *numerical character centered within the arc* **or** *at the apex of the arc is found* **then**

classify arc as tuplet

If *the end points of found arc are within +/- 5 degrees of level* **then**

add w_{t2+} amount of credit to **Tie**

remove w_{s2+} amount of credit from **Slur**

else

add w_{s2-} amount of credit to **Slur**

remove w_{t2-} amount of credit from **Tie**

If *noteheads horizontally aligned are found or they share the same pitch* **then**

add w_{t3+} amount of credit to **Tie**

remove w_{s3+} amount of credit from **Slur**

else

add w_{s3-} amount of credit to **Slur**

remove w_{t3-} amount of credit from **Tie**

If *there are no other noteheads between the two found noteheads* **then**

add w_{t4+} amount of credit to **Tie**

remove w_{s4+} amount of credit from **Slur**

else

add w_{s4-} amount of credit to **Slur**

remove w_{t4-} amount of credit from **Tie**

Final classification rule: If the arc is not classified as a tuplet then include it to a respective class according to a greater amount of credit.

The values of credits are given by an expert and, then, are tuned at the basis of tests done on example scores. Since the premises of *If Then Else* rules come from the recognition system, they may vary between different recognition systems and, what is really important, may be changed any time the recognition process is changed or tuned. So, the expert's knowledge must be updated basing on the observation of recognition results. Then tuning of credits of *If Then Else* rules is necessary as a supplement of primary values given by the expert.

This simple classification rule is potentially powerfull in all cases when assumptions of *If Then Else* rules are certain. However, *If Then Else* assumptions obtained from recognition of a real score are not certain. For instance, the condition: *the end points of found arc are within +/- 5 degrees of level*, that defines rotation of the arc, cannot be based on the absolute placement of endpoints of the arc on the page. The rotation angle of the arc should be calculated on the basis of a respective stave skew angle. The crisp condition *the end points of found arc are within +/- 5 degrees of level* does not reflects the fuzzy nature of the subject. In real notations the difference between 4.9 degree and 5.1 degree rotations is practically of less importance while 1.0 degree and 4.9 degree rotations significantly differ one from another. Thus, this condition should rather be formulated as a more flexible condition, e.g. as the linguistic label *arc is flat*. Similarly, the crisp condition *numerical character centered within the arc or at the apex of the arc is found* should rather be expressed in the form: *numerical character close to center of the arc or to the apex of the arc is found* allowing for soft interpretation of the linguistic label *close*.

The results computed as a simple arithmetic sum can easily yield values greater than 1. Furthemore, the composition of results is solely based on subtraction of credits, which easily gives a negative number. This is the reason why the classical theory of fuzzy sets, which can implement the flexibility of assumption, is not applicable. The solution to the problem is given in the next chapter, where algebraic structures of fuzzy sets are introduced and applied to reasoning in OMR.

Let us consider a fuzzy reasoning rule of the form:

$$
\begin{array}{llll}
\textit{If} & X \textit{ is } A_1 & \textit{then} & Y \textit{ is } B_1 \\
\textit{If} & X \textit{ is } A_2 & \textit{then} & Y \textit{ is } B_2 \\
 & & \cdots\cdots\cdots & \\
\textit{If} & X \textit{ is } A_k & \textit{then} & Y \textit{ is } B_k \\
 & X \textit{ is } A_0 & & \\
\hline
 & & & Y \textit{ is } B_0
\end{array}
$$

where X and Y are names of variables, A_0 , A_1 , ... A_k , and B_0 , B_1 , ... B_k are fuzzy sets in a finite universe of discourse representing values of these variables, i.e.

A_0 , A_1 , ... A_k are the elements of the space $F(\{x_0 , x_1 , \ldots x_k \})$,

B_0 , B_1 , ... B_k are the elements of the space $F(\{y_0 , y_1 , \ldots y_k \})$.

This form of fuzzy reasoning rule is similar to that given by an expert in a reasoning model in the process of knowledge acquisition. Such a model is of particular interest since it copes with uncertainty existing in many real-life cases concerning medical diagnosis, diagnosis of technical devices, reasoning in knowledge bases, and is successfully applicable in OMR. This model of fuzzy reasoning will bring consideration to the formula that describes reasoning in the neural network applied in the arc classification task.

Let us assume that fuzzy concepts $A_1, \ldots A_k$ can be considered as coming from the space of assumptions and fuzzy concepts $B_1, \ldots B_k$ from the space of results. Descriptions obtained from the expert are of the following form: if symptoms x_1, ..., x_k are observed with certainty factors $r_{i1}, \ldots, r_{ik}$, respectively then results $y_1, \ldots y_k$ should be considered with certainty factors $b_{i1}, \ldots, b_{ik}$.Having a given space of symptoms $x_1, \ldots, x_k$ with certainty factors $a_1, \ldots, a_k$, we are faced with the problem of concluding on the relevant space of results $y_1, \ldots y_k$ and on their certainty factors $b_1, \ldots b_k$.

This form of fuzzy reasoning may also be expressed as a fuzzy relational equation

$$\begin{cases} r_{11} * x_1 \circ r_{12} * x_2 \circ \quad \cdots \quad * r_{1n} * x_n = y_1 \\ r_{21} * x_1 \circ r_{22} * x_2 \circ \quad \cdots \quad * r_{2n} * x_n = y_2 \\ \qquad\qquad \cdots \\ r_{m1} * x_1 \circ r_{m2} * x_2 \circ \quad \cdots \quad * r_{mn} * x_n = y_m \end{cases}$$

or in the symbolic form:

$$R \sigma X = Y$$

where $x_1, \cdots, x_n, y_1, \cdots, y_m, r_{11}, \cdots, r_{mn}$ are fuzzy sets of assumptions, results and dependencies respectively, while $\circ$ and $*$ are the operators defined respectively to a reasoning system. For instance, operators may be interpreted as classical fuzzy set max-min operators, triangular norms, etc.

For extended discussion on this subject see [9, 11, 12]

4. Algebraic system of fuzzy reasoning

A brief look at an alternative approach to fuzzy sets is given in this chapter. This new approach: algebraic structures of fuzzy sets is applied in modelling classical neuron and neural network. The formula describing classical reasoning models such as the classical neural network, the expert's reasoning rule formulated in the previous passage, the fuzzy neural network are considered from the point of view of a general formula being a template for formulas describing a given reasoning model. This template is also applicable in algebraic structures of fuzzy sets. This result illustrates the analogy between classical approaches to reasoning systems and the new one introduced here. Finally, two fuzzy reasoning models are applied to arc classification task to outline the suitability of applied fuzzy reasoning models.

It is worth to say that this chapter is aimed in presentation of quite new approach to reasoning – the approach applying algebraic structures of fuzzy sets as an alternative to so called classical approaches in the meaning of fuzzy sets, expert systems, standard neural networks. The implementations of reasoning systems based on algebraic structures of fuzzy sets compared with respective implementations of such systems in classical models of such systems, i.e. in fuzzy sets theory, standard neural networks, expert systems, is expected to bring interesting research subjects. But, on the other hand, discussion on these subjects is out of the scope of this work. Considerations from now on are firmly based on an example of reasoning system applied to given music notation symbols. In its full implementation, the system is created on the case of supervised learning, though details of this aspect are dropped as of less importance for the subject.

4.1. Introductory remarks - algebraic structures of fuzzy sets

Membership functions are usually defined as mappings from any universe of discourse into the unit interval. This fact comes from tradition rather than is a result of theoretical requirements. Membership functions may refer to values from [-1, 1]. This symmetrical bipolar representation was used in dealing with uncertainty in historic MYCIN system, it is also applied in fuzzy neural networks, cf. [7], in cognitive maps, cf. [15] etc. Representation of negative information is also more obvious in the case of bipolarity: positive value defines the grade of inclusion of the element into fuzzy set, while negative - the grade of exclusion. The greater the absolute value of membership, the more certain inclusion/exclusion information. Treating intersection and union as implemented by minimum and maximum respectively and viewing complement as changing the sign of the value of the membership function, the spaces of grades of membership becomes isomorphic for both cases: unipolar and bipolar.

Take for instance mappings:

- $i:\{0,1\}^X \to \{-1,1\}^X$ such that $i(x)=2x-1$ for crisp sets
- $i:[0,1]^X \to [-1,1]^X$ and as before $i(x)=2x-1$ for fuzzy sets

These mappings are isomorphic functions, where a straightforward proof of satisfaction of this property can be easily derived, cf. [11].
In general let us introduce a transforming function h from the open interval (-1, 1) onto the real line R, such that it is

- continuous,
- symmetrical, i.e.: $h(-x)=h(x)$
- strictly increasing.

Obviously, we have:

- $h(0)=0$

- $\lim h(x) = +\infty$ for $x \to \pm 1$
- h^{-1} exists.

Examples of the transforming function are: hyperbolic arc tangent, x/(1-abs(x)), tan(π/2*x), etc. C.f. [12] for discussion on the problem of selection of transforming function.

Now let a binary operator (additive operator) be such that:

$$\oplus : G(X) \times G(X) \to G(X) \quad \text{and for all } x \text{ within the domain } X:$$

$$\oplus(A, B)(x) = (A \oplus B)(x) = h^{-1}(h(A(x)) + h(B(x)))$$

where A, B are fuzzy sets, + is the addition operator on real numbers and $\times$ is Cartesian product and $G(X)$ is the space of membership functions over universe X and with values in interval [-1,1], i.e. the space of bipolar fuzzy sets.

The following is held:

Proposition 1. The pair $(G(X), \oplus)$ is a commutative group.

Let us define new operators on the space $G(X)$ of fuzzy sets based on the transforming function h introduced above.

The definition of the (inner) multiplication operator $\otimes$ is as follows:

$$\otimes : G(X) \times G(X) \to G(X) \quad \text{and for all } x \text{ within the domain } X:$$

$$\otimes(A, B)(x) = (A \otimes B)(x) = h^{-1}(h(A(x)) * h(B(x)))$$

where A, B are fuzzy sets, $*$ is the multiplication of real numbers and $\times$ is Cartesian product.

Proposition 2. The pair $(G(X), \otimes)$ is a commutative group.

Proposition 3. $\otimes$ operator is distributive in respect to $\oplus$ operator:

$$A \otimes (B \oplus C) = (A \otimes B) \oplus (A \otimes C)$$

or if we assume $\otimes$ is of higher precedence than $\oplus$ this yields

$$A \otimes (B \oplus C) = A \otimes B \oplus A \otimes C$$

Corollary. A structure $R = (G(X), \oplus, \otimes)$ is a ring.

Cf. [9,12] for detailed discussion on algebraic structures of fuzzy sets.

4.2. Modelling an artificial neuron with algebraic operators

In this chapter, the fuzzy neural network is explored from the perspective of the task performed by a single neuron in one network pulse. Since the computational model of a fuzzy neuron involves classical fuzzy operations such as triangular norms, it still suffers from the drawbacks outlined in the previous chapter that are related to negative and repetitive character of input signals being processed by the neuron. To avoid these disadvantages, the model of ***algebraic neuron*** is proposed on the basis of more general operators s and t.

The architecture of a (classical and fuzzy) neuron was extensively developed and numerous of references can be recalled here. The model of the classical neuron is shown in Figure 3a (see also Bezdek [1,2]).

From mathematical point of view processing of information within a neuron involves two distinct mathematical operations, namely:

- an integration function (the synapsic operation) *f* first integrates the (synapsic) weights $W = (w_1, \ldots, w_n)$ with the pulse input $X = (x_1, \ldots, x_n)$, and is followed by
- a transfer (activation) function (the somatic operation) *F* applied to the value $y = f(x)$.

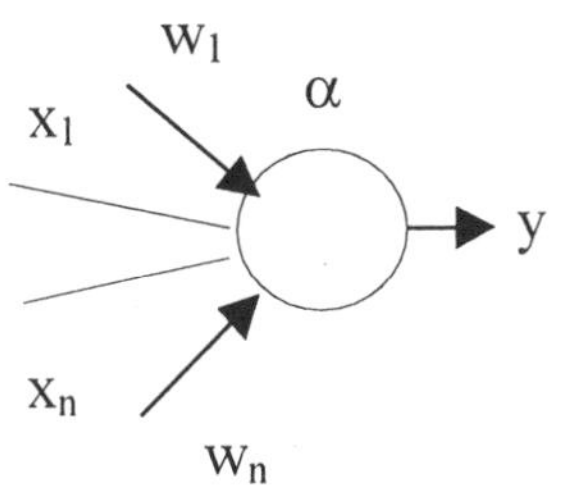

Figure 9a

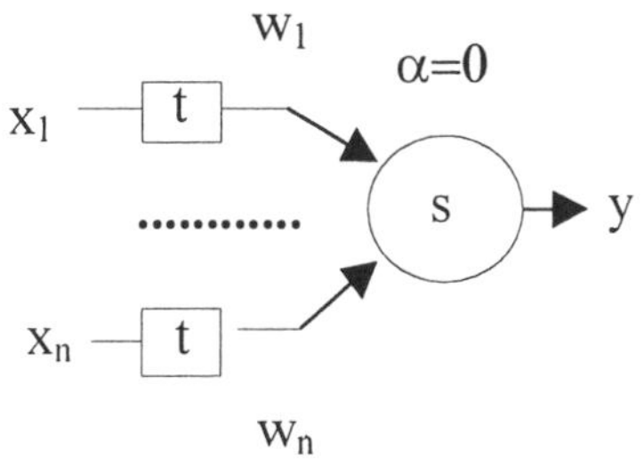

Figure 9b

The integration function *f* is usually an inner product:

$$f(X) = w_1x_1 + \ldots w_nx_n + \alpha$$

where α is the bias or offset from the origin of $\mathbf{R}^n$ to the hyperplane normal to W defined by equation:

$$f(X) = 0 \text{ i.e.} \qquad w_1x_1 + \ldots w_nx_n + \alpha = 0$$

This hyperplane is a simple separation of input space into two symmetrical subspaces.

This model of artificial neuron is called first order neuron because *f* is an affine (linear when $\alpha = 0$) function of its input, cf. [2]. Although more complicated models of neurons have been developed, including higher order neurons, e.g. neurons with quadratic integration function, the attention is focused here on this simple, first order neuron because this model is the basis of the artificial fuzzy neuron.

Activation function *F* is used to decide, whether the node should fire and if so, how much charge, and of what sign, should be broadcast to the network in response to the node inputs. *F* is typically the logistic (sigmoidal) function, for example $F(z) = F(f(X)) = 1/(1+\exp(-z))$.

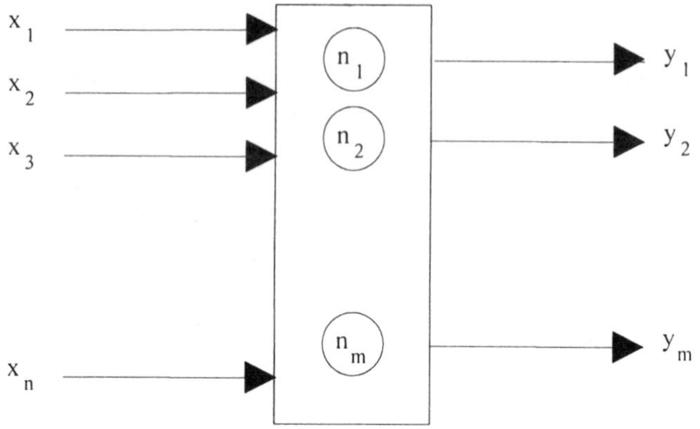

Figure 10

One of the models of fuzzy neuron is defined by replacing the integration function f with triangular norms:

$$u(t) = s(\, w_1 \, t \, x_1, \ldots, w_n \, t \, x_n \,)$$

where t and s are the triangular norm and co-norm, respectively.

This model of neuron is obviously a higher order (non-linear) one since its integration function is non-linear. The separation of its input space by the equation $u(t) = 0$ is more complex than in case of a first order neuron and allows for more flexible separation of input signals into different classes.

With respect to one layer of fuzzy neural network (see Figure 9), this model of fuzzy neuron leads to the fuzzy relational equation as a mathematical model of the layer. It may be assumed that every input of the layer is copied the number of times equal to the number of neurons in this layer. So, every input of the layer comes to every neuron of this layer. If this assumption is not true, the network may be extended by adding new connections going to respective neurons of the layer with weights equal to 0. One can label all input connections by putting two indices for every weight: the first index is equal to the number of neurons receiving the connections, the second - to the number of input connections. In this way, all weights of the layer may be arranged in an array and this neural network can be modelled as a universal equation:

$$R \sigma X = Y$$

or in an expanded form:

$$\begin{cases} w_{11} * x_1 \circ w_{12} * x_2 \circ \cdots * w_{1n} * x_n \circ \alpha_1 = y_1 \\ w_{21} * x_1 \circ w_{22} * x_2 \circ \cdots * w_{2n} * x_n \circ \alpha_2 = y_2 \\ \cdots \\ w_{m1} * x_1 \circ w_{m2} * x_2 \circ \cdots * w_{mn} * x_n \circ \alpha_m = y_m \end{cases} \quad (1)$$

where $*$ and $\circ$ denotes operators respective to the applied reasoning system. This equation can be seen as a template of a reasoning form that can be filled in depending on a given formal system. For instance, in classical fuzzy set theory the equation will be changed into the relational equation with max-min operators replacing $*$ and $\circ$. Or, more generally, triangular norms can be used instead of max-min operators. On the other hand, the algebraic structures of fuzzy sets introduced in Chapter 4.1 will turn this equation into an algebraic equation with operators $\otimes$ and $\oplus$ replacing $*$ and $\circ$ respectively.

4.3. Algebraic system of fuzzy reasoning - an OMR example

In this chapter two different reasoning systems are applied to the example of arc classification given in Chapter 3.3: fuzzy neural network based on triangular norms and fuzzy neural network based on algebraic operators.

In Figure 11 the architecture of neural network is shown. This architecture is used for arc classification. Four input data of this network takes following features of an arc:

- P1: the arc is flat,
- P2: the numerical character centered within the arc or at the apex of the arc is found,
- P3: the horizontally aligned noteheads are found or they share the same pitch,
- P4: other noteheads are found between the two outer noteheads of this arc,

The network yields arc classification to three classes as the results of its computation:

- R1: the arc is classified as aslur,
- R2: the arc is classified as a tie,
- R3: the arc is classified as a tuplet.

Of course, the input, the results and also the weights of above neurons are represented as fuzzy sets.

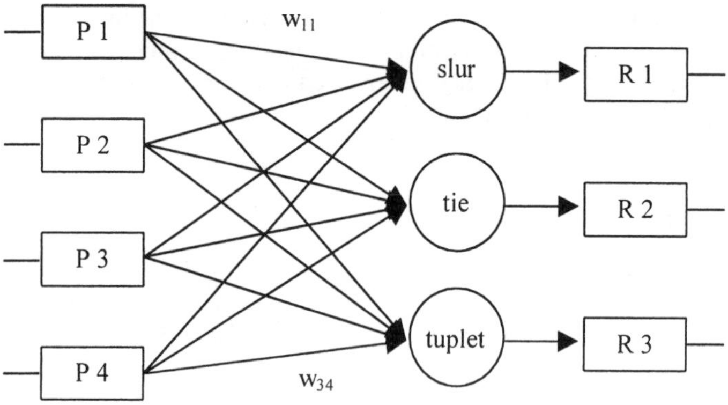

Figure 11 Architecture of neural network

Two different systems of fuzzy sets were used that resulted in:

- The fuzzy neural network (using unipolar fuzzy sets) based on triangular norms t and s replacing $*$ and $\circ$ operators. Triangular norms are generated by the function. The equations (1) is turned to the relational equation:

$$\begin{cases} w_{11}\ t\ p_1\ \ s\ w_{12}\ t\ p_2\ \ s\ w_{13}\ t\ p_3\ \ s\ w_{14}\ t\ p_4 = r_1 \\ w_{21}\ t\ p_1\ s\ w_{22}\ t\ p_2\ \ s\ w_{23}\ t\ p_3\ s\ w_{24}\ t\ p_4 = r_2 \\ w_{31}\ t\ p_1\ s\ w_{32}\ t\ p_2\ \ s\ w_{33}\ t\ p_3\ s\ w_{34}\ t\ p_4 = r_3 \end{cases}$$

- The fuzzy neural network (using bipolar fuzzy sets) based on algebraic operators described in section 4.1. The function $f(x) = sign(x) * x^4 /(1 - x^4)$ was used to generate $\oplus$ and $\otimes$ operators.

This operators were applied as $\circ$ and $*$ operators. The equation (1) is immersed in the ring structure:

$$\begin{cases} w_{11} \otimes p_1 \oplus w_{12} \otimes p_2 \oplus w_{13} \otimes p_3 \oplus w_{14} \otimes p_4 = r_1 \\ w_{21} \otimes p_1 \oplus w_{22} \otimes p_2 \oplus w_{23} \otimes p_3 \oplus w_{24} \otimes p_4 = r_2 \\ w_{31} \otimes p_1 \oplus w_{32} \otimes p_2 \oplus w_{33} \otimes p_3 \oplus w_{34} \otimes p_4 = r_3 \end{cases}$$

Tables 1 and 2 outline classification examples of the network shown in Figure 11. The examples were taken from a broader recognition system, c.f. [10]. Two examples of every class: slurs, ties and tuplets were considered. The premises, results and weights of the first example are isomorphic with respective values of the second example. For instance, in Table 1 premisses of the arc being classified are equal to (0.25, 0.95, 0.50, 0.95), the classification results computed by the fuzzy neural network based on triangular norms is equal to (0.95, 0.99, 0.96). In other words, if the grade of the feature *"the arc is flat"* is equal to 0.25, the grade of the feature "the *numerical character centered within the arc or at the apex of the arc is found"*, is equal to 0.95 etc. Then the result was yielded: *"the arc is a slur"* with the grade equal to 0.95, *"the arc is a tie"* with the grade equal to 0.99, *"the arc is a tuplet"* with the grade equal to 0.96.

					slur		tie		tuplet		
					0.45	0.25	0.25	0.25	0.95	0.95	P 1
					0.45	0.95	0.95	0.85	0.65	0.95	P 2
					0.25	0.50	0.95	0.95	0.65	0.95	P 3
w_{ij}	1	2	3	4	0.50	0.95	0.10	0.10	0.75	0.90	P 4
1	0.05	0.30	0.15	0.85	0.51	0.95	0.44	0.32	0.76	0.91	R 1
2	0.05	0.95	0.95	0.05	0.64	0.99	0.99	0.99	0.90	0.99	R 2
3	0.98	0.70	0.75	0.85	0.78	0.96	0.94	0.92	1.00	1.00	R 3

Table 1. The example of computation of fuzzy neural network based on triangular norms

					slur		tie		tuplet		
					-0.10	-0.50	-0.50	-0.50	0.90	0.90	P 1
					-0.10	0.90	0.90	0.70	0.30	0.90	P 2
					-0.50	0.00	0.90	0.90	0.30	0.90	P 3
w_{ij}	1	2	3	4	0.00	0.90	-0.80	-0.80	0.50	0.80	P 4
1	-0.90	-0.40	-0.70	0.70	0.38	0.80	-0.81	-0.80	-0.94	-0.95	R 1
2	-0.90	0.90	0.90	-0.90	-0.58	0.58	0.97	0.96	-0.94	0.91	R 2
3	0.97	0.40	0.50	0.70	-0.27	0.59	-0.77	-0.78	0.98	0.98	R 3

Table 2. The example of computation of fuzzy neural network based on algebraic operators

In the Table 2 respective features of the arc are equal to (-0.10, -0.10, -0.50, 0.00) and the classification results of fuzzy neural network based on algebraic operators are equal to (0.38, -0.58, -0.27).

The premises, results and weights of the first example are isomorphic with respective values of the second example. This choice gives the opportunity to compare behaviour of the network for the two systems of fuzzy sets implementing the network.

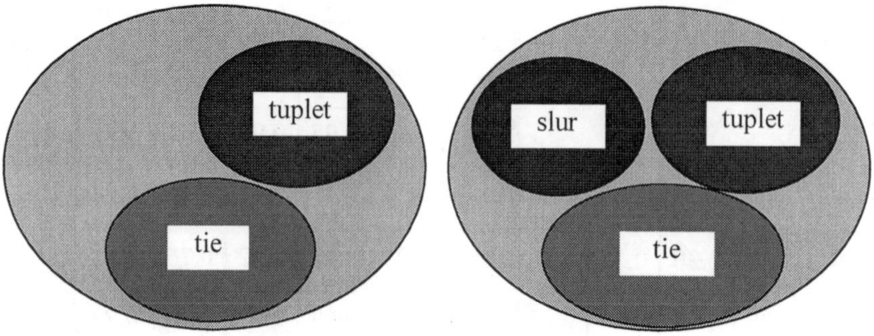

fuzzy neural network based on triangular norms

fuzzy neural network based on algebraic operators

Figure 12 Separation of input space

Both reasoning systems were tested with regard to separation of input space. The input datum was classified in a given class only if the result of its grade exceeded the grades of two other classes by at least: 0.1 for the fuzzy neural network based on triangular norms and 0.2 for the fuzzy neural network based on algebraic operators. The fuzzy neural network based on triangular norms yielded 0% of the input space ascribed to the slur area, 21,3% - as ascribed to the tie area, 21.1% - as ascribed to the tuplet area. The fuzzy neural network based on algebraic operators yielded respectively: 16.2% part of input space - as ascribed to the slur area, 19.0% - as ascribed to the tie area, 24.2% - as ascribed to the tuplet area. Figure 12 outlines results of input data separation.

5. Conclusions

Music notation symbols, though well characterised by their features, are arranged in an elaborated way in real music notation, which makes the recognition task very difficult and still open for new ideas. On the other hand, the aim of the system, i.e. the application of acquired printed music into further processing requires special representation of music data. Due to complexity of music nature and music notation, music representation is one of the key issue in music notation recognition and music processing.

The idea of granular computing seems to be a perfect tool to express the problems of optical music recognition. It gives a view at OMR from a new, fresh and interesting perspective. This new perspective could allow for better understanding of the structure of music knowledge representation and processing, of the structures of music notation from the point of its recognition, as well as the structure of the recognition process. Thus, it could be fruitful in new ideas in the subject of optical music recognition including learning patterns utilised in reasoning system training.

Several points of view on granularity of the field of optical music recognition were given in this work. Granularity of the geometrical and logical static pyramids of music information is supplemented by the dynamic pyramid of data that is processed at the stage of score structure identification. This last topic shows a granule as a unity of data being processed and algorithms that are applied to data processing. An example of arc symbol classification presents detailed description of such a granule. This example employs the system of fuzzy reasoning, which is based on algebraic structures of fuzzy sets. This system presents interesting structuring of fuzzy sets – algebraic structuring – that gives better results in comparison to classical systems based on max-min operators and triangular norms.

Acknowledgements

Professor Witold Pedrycz encouraged me to prepare this paper and suggested the final shape of it. Without his help this paper could not have been prepared. I wish to thank him a lot. The collaboration with Christopher Newell has been fruitful in interesting ideas and has enriched my experience in OMR. I would like to express my thanks to him

References

1 J.C. Bezdek , *Computing with uncertainty*, IEEE Communication Magazine, September 1992, pp. 24-36.

2 J.C. Bezdek, *On the Relationship Between Neural Networks, Pattern Recognition and Intelligence*, Int. J. of Approximate Reasoning, 6(1992)85-107.

3 D. Blostein, H.S.Baird, *A Critical Survey of Music Image Analysis*, pp. 405-434 in: Structured Document Analysis, H.S.Baird, H.Bunke, K.Yamamoto (Eds), Springer-Verlag, 1992.

4 R. Dannenberg, *Music Representation Issues, Techniques, and Systems*, Computer Music Journal, 17:3 (1993) 20-30.

5 H.S. Field-Richards, *Cadenza: A Music Description Language*, Computer Music Journal, 17:4 (1993) 60-72.

6 I. Fujinaga, *Optical music recognition using projections*, Master's thesis, McGill University, Montreal, Canada, September 1988.

7 M.M. Gupta, D.H. Rao, *On the principles of Fuzzy Neural Networks*, Fuzzy Sets and Systems, 61 (1994) 1-18.

8 L. Haken, D. Blostein, *The Tilia Music Representation: Extensibility, Abstraction, and Notation Contexts for the Lime Music Editor*, Computer Music Journal, 17:3 (1993) 43-58.

9 W. Homenda, *Algebraic Operators: an Alternative Approach to Fuzzy Sets*, Appl. Math. and Comp. Sci., vol. 6, No. 3, pp. 505-527, 1996.

10 W. Homenda, *Smart Score v.2.0 – Technical Specification*, Intelligent Computing Technologies' Internal Document, Warsaw, Poland, December 1999.

11 W. Homenda, W. Pedrycz, *Processing of uncertain information in linear space of fuzzy sets*, Fuzzy Sets & Systems, 44 (1991) 187-198.

12 W. Homenda, W. Pedrycz, *Fuzzy neuron modelling based on algebraic approach to fuzzy sets*, Proc. of the SPIE's International Symposium on Aerospace/Defense Sensing & Control and Dual-Use Photonics, Orlando, Florida, April 17-21, 1995, pp. 71-82,

13 T. Itagaki et al., *Automatic Recognition of Several Types of Music Notation*, pp. 466-476 in: Struct. Document Analysis.

14 H. Kato, S Inokuchi, *A Recognition System for Printed Piano Music Using Musical Knowledge and Constraines*, pp. 435-455 in: Struct. Document Analysis.

15 B. Kosko, *Fuzzy cognitive maps*, Int, J. Man-Machine Studies, vol. 24, pp. 65-75, Jan. 1986.

16 *MIDI 1.0, Detailed Specification*, Document version 4.1.1, February 1990.

17. C. Newell, *Smart Score v.1.3 - technical report,* Musitek's Internal Document, Ojai, CA, November 1999.

18 *NIFF, Notation Interchange File Format*, Document version 6a.3, October 1998.

19 W. Pedrycz, *Neural Networks in the Framework of Granular Computing*

20 T. Ross, *The Art of Music Engraving and Processing*, Hansen Books, Miami, 1970.

21 K. Stone, *Music Notation in Twentieth Century: A Practical Guidebook*, W.W.Norton & Co.,New York, 1980.

Modeling MPEG VBR Video Traffic Using Type-2 Fuzzy Logic Systems

Qilian Liang and Jerry M. Mendel

Signal and Image Processing Institute, Department of Electrical Engineering-Systems, University of Southern California, Los Angeles, CA 90089-2564 USA

Abstract. In this chapter, we present a new approach for MPEG variable bit rate (VBR) video modeling using a type-2 fuzzy logic system (FLS). We demonstrate that a type-2 fuzzy membership function, i.e., a Gaussian MF with uncertain variance, is most appropriate to model the log-value of I/P/B frame sizes in MPEG VBR video. We treat the video traffic as a dynamic system, and use a type-2 FLS to model this system. Simulation results show that a type-2 FLS performs much better than a type-1 FLS in video traffic modeling.

1 Introduction

Multimedia technologies will profoundly change the way we access information, conduct business, communicate, educate, learn, and entertain. Among the various kinds of multimedia services, video service is becoming an important component. Video service refers to the transmission of moving images together with sound [27]. Research on video transfers for multimedia services has been quite active in recent years, and video applications are expected to be the major source of traffic in future broad-band networks [32]. In this chapter, we treat the video traffic as a dynamic system, and apply a type-2 FLS to model it.

Dawood and Ghanbari [4] [5] used linguistic labels to model MPEG video traffic, and classified them into 9 classes based on texture and motion complexity. They used crisp values obtained from the mean values of training prototype video sequences to define *low*, *medium*, and *high* texture and motion. Chang and Hu [3] investigated the applications of pipelined recurrent neural networks to MPEG video traffic prediction and modeling. I/P/B pictures were characterized by a general nonlinear ARMA process. Pancha et al. [29] observed that a gamma distribution fits the statistical distribution of the packetized bits/frame of video traffic with low bit rates. Heyman et al. [10] showed that the number of bits/frame distribution of I-frames has a lognormal distribution and its autocorrelation follows a geometrical function, and they concluded that there is no specific distribution that can fit P and B frames. Krunz et al. [16], however, found that the lognormal distribution is the best match for all three types. All these methods belong to the statistical signal processing-based approaches, which match the mean and variance to a known statistical distribution. Recently, Krunz and Makowski [17] observed that $M/G/\infty$ input models are good candidates for modeling many types of correlated traffic (such as video traffic) in computer networks.

As noted in [25], a shortcoming to model-based statistical signal processing is "$\cdots$ the assumed probability model, for which model-based statistical signal processing results will be good if the data agrees with the model, but may not be so good if the data does not." In real variable bit rate (VBR) video traffic, the traffic is highly bursty, and we believe that no statistical model can really demonstrate the uncertain nature of the I/P/B frames. Fuzzy logic systems (FLS) are model free. Their membership functions are not based on statistical distributions. In this chapter, we, therefore, apply fuzzy techniques to MPEG VBR video traffic modeling.

A survey of recent advances in fuzzy logic (FL) applied to telecommunications networks is discussed in [8]; it shows that FL is very promising for every aspect of communication networks. Recently, Tsang, Bensaou, and Lam [32] proposed a fuzzy-based real-time MPEG video rate control scheme to avoid a long delay or excessive loss at the user-network interface (UNI) in an ATM network. The success of fuzzy logic applied to communication networks motivates us to apply FL to video traffic modeling.

In Section 2, we briefly introduce MPEG video traffic. In Section 3, we introduce type-2 fuzzy sets. In Section 4, we model I/P/B frame sizes using supervised clustering. In Section 5, we present an interval type-2 TSK FLS where antecedents are type-2 and consequents are type-1. A design method for the type-2 TSK FLS is provided in Section 6. In Section 7, we apply the type-2 TSK FLS to MPEG video traffic modeling, and compare it against a type-1 TSK FLS. Conclusions are presented in Section 8.

In this chapter, A denotes a type-1 fuzzy set; $\mu_{\mathrm{A}}(x)$ denotes the membership grade of x in the type-1 fuzzy set A; $\tilde{\mathrm{A}}$ denotes a type-2 fuzzy set; $\mu_{\tilde{\mathrm{A}}}(x)$ denotes the membership grade of x in the type-2 fuzzy set $\tilde{\mathrm{A}}$, i.e., $\mu_{\tilde{\mathrm{A}}}(x) = \int_u f_x(u)/u\,,\ \ u \in J_x \subseteq [0,1]$; $\sqcap$ denotes *meet* operation; and, $\sqcup$ denotes *join* operation. Meet and join are defined and explained in great detail in [12] [15].

2 Introduction to MPEG Video Traffic

MPEG (Moving Picture Expert Group) is an ISO/IEC standard for digital video compression coding, and has been extensively used to overcome the problem of storage of prerecorded video on digital storage media, because of the high compression ratios it achieves. MPEG video traffic is composed of a Group of Pictures (GoP) including some of the encoded frames: I (intra-coded), P (predicted) and B (bidirectional). I frames are coded with respect to the current frame using a two-dimensional discrete cosine transform, and they have a relatively low compression ratio; P frames are coded with reference to previous I or P frames using interframe coding, and they can achieve a better compression ratio than I frames; and, B frames are coded with reference to the next and previous I or P frame, and, B frames can achieve the highest compression ratio of the three frame types. The sequence of frames is specified by two parameters: M, the distance between I and P frames; and, N, the distance between I frames. The use of these three types of frames allows MPEG to be both robust (I frames permit error recovery) and efficient (B and P frames have high compression ratio). Variable bit-rate (VBR) MPEG video is used in ATM networks, and constant bit-rate (CBR) MPEG video is often used in narrowband ISDN. We focus on MPEG VBR video traffic. In Fig. 2, we plot the I/P/B frame sizes for 3000 frames of an MPEG coded video of *The Silence of the Lambs.*

3 Type-2 Fuzzy Sets

The concept of type-2 fuzzy sets was introduced by Zadeh [34] as an extension of the concept of an ordinary fuzzy set, i.e., a type-1 fuzzy set [23]. Type-2 fuzzy sets have grades of membership that are themselves fuzzy [9]. A type-2 membership grade can be any subset in $[0, 1]$ – the *primary membership*; and, corresponding to each primary membership, there is a *secondary membership* (which can also be in $[0, 1]$) that defines the possibilities for the primary membership. A type-1 fuzzy set is a special case of a type-2 fuzzy set; its secondary membership function is a subset with only one element, unity. Type-2 fuzzy sets allow us to handle linguistic uncertainties, as typified by the adage "words can mean different things to different people." A Fuzzy relation of higher type (e.g., type-2) has been regarded as one way to increase the fuzziness of a relation, and, according to Hisdal, "increased fuzziness in a description means increased ability to handle inexact information in a logically correct manner [11]".

A general type-2 fuzzy set has lots of parameters to be determined [12] [15], but things simplify a lot when its secondary memberships are interval sets. Interval sets are very useful when we have no other a priori knowledge about membership function uncertainties. In our video traffic modeling, we will focus on using interval type-2 fuzzy sets. An interval type-2 fuzzy set can be represented by its *upper* and *lower* membership functions (MFs) [19]. An upper MF and a lower MF are two type-1 MFs which are bounds for the footprint of uncertainty (the union of all primary membership grades) of an interval type-2 MF. The upper MF is a subset which has the maximum membership grade of the footprint of uncertainty; and, the lower MF is a subset which has the minimum membership grade of the footprint of uncertainty. Determing the footprint of uncertainty is crucial for the use of type-2 fuzzy sets, and is application dependent. We establish the footprint of uncertainty for MPEG coded videos in Section 4.

We use an overbar (underbar) to denote the upper (lower) MF. For example, let $\tilde{\mathrm{F}}_k^l(x_k)$ denote the type-2 MF for the kth antecedent of the lth rule, then the upper and lower MFs of $\mu_{\tilde{F}_k^l}(x_k)$ are $\overline{\mu}_{\tilde{F}_k^l}(x_k)$ and $\underline{\mu}_{\tilde{F}_k^l}(x_k)$, respectively, so that

$$\mu_{\tilde{F}_k^l}(x_k) = \int_{q^l \in [\underline{\mu}_{\tilde{F}_k^l}(x_k), \overline{\mu}_{\tilde{F}_k^l}(x_k)]} 1/q^l \tag{1}$$

where $\int$ denotes the union of individual points of each set in the continuum.

Example 1: Gaussian Primary MF with Uncertain Standard Deviation

Consider the case of a Gaussian primary MF having a fixed mean, m_k^l, and an uncertain variance that takes on values in $[\sigma_{k1}^l, \sigma_{k2}^l]$, i.e.,

$$\mu_k^l(x_k) = \exp\left[-\frac{1}{2}(\frac{x_k - m_k^l}{\sigma_k^l})^2\right], \quad \sigma_k^l \in [\sigma_{k1}^l, \sigma_{k2}^l] \tag{2}$$

where: $k = 1, \ldots, p$; p is the number of antecedents; $l = 1, \ldots, M$; and, M is the number of rules. The upper MF, $\overline{\mu}_k^l(x_k)$, is (see Fig. 1)

$$\overline{\mu}_k^l(x_k) = \mathcal{N}(m_k^l, \sigma_{k2}^l; x_k), \tag{3}$$

and the lower MF, $\underline{\mu}_k^l(x_k)$, is (see Fig. 1)

$$\underline{\mu}_k^l(x_k) = \mathcal{N}(m_k^l, \sigma_{k1}^l; x_k) \tag{4}$$

□

This examples illustrate how to define $\overline{\mu}$ and $\underline{\mu}$, so that it is clear how to define these membership functions for other situations (e.g., triangular, trapezoidal, bell MFs).

Sometimes an upper (or a lower) MF cannot be represented by one mathematical function over its entire domain (e.g., Gaussian primary MF with uncertain mean [19]). It may consist of several branches each defined over a different segment of the entire domain. When the input is located in one domain-segment, we call its corresponding MF branch an *active branch*,

4 Modeling I/P/B Frame Sizes Using Supervised Clustering

Clustering of numerical data forms the basis of many classification and modeling algorithms. The purpose of clustering is to distill natural groupings of data from a large data set, producing a concise representation of a system's behavior.

To the best of our knowledge, all current approaches (e.g., [29], [10], [16]) for modeling I/P/B frame sizes belong to supervised clustering, i.e., they

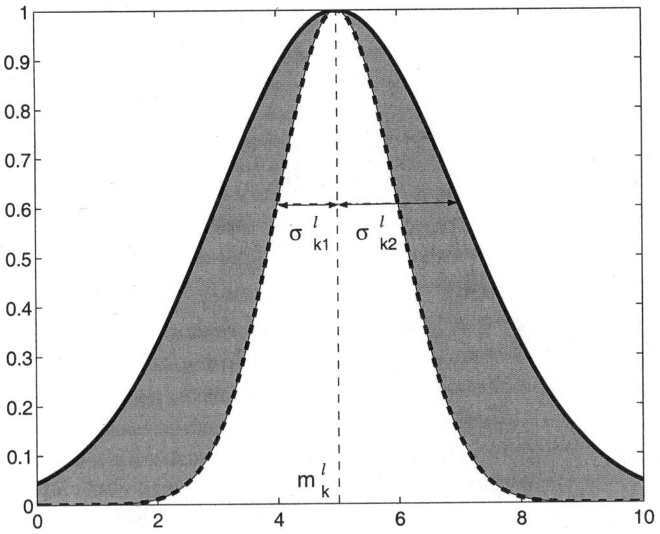

Fig. 1. The type-2 MFs for Example 1. The thick solid lines denote upper MFs, and the thick dashed lines denote lower MFs. The shaded regions are the footprints of uncertainty for interval secondaries. The center of the Gaussian MFs is 5, and the variance varies from 1.0 to 2.0.

assume the I/P/B frame categories are known ahead of time; consequently, we shall use a supervised clustering algorithm, assuming there are three clusters, so as to analyze the statistical nature of these clusters.

We will show that the statistical knowledge (mean and std) about the size (bits/fame) of I, P, or B clusters in different groups of frames is distinct, even in the same video product; hence, this will motivate us to use type-2 fuzzy sets to model the number of bits/frame in I/P/B traffic, and is consistent with our belief that the frame sizes of video traffic are not really wide-sense stationary (WSS), and that their distribution varies with respect to the frame index.

Thanks to Oliver Rose [30] who made 20 MPEG-1 video traces on-line (at FTP://ftp-info3.informatik.uni-wuerzburg.de/pub/MPEG/), we were able to perform our experiments. Lots of works by others have been done based on these resources, e.g., Rose [30] analyzed their statistical properties and observed that the frame and GoP sizes can be approximated by Gamma or Lognormal distributions, Manzoni et al. [22] studied the workload models of VBR video traffic based on these video traces, and, Adas [1] used adaptive linear prediction to forecast the VBR video for dynamic bandwidth allocation.

The videos were compressed by Rose using an MPEG-1 encoder. The quantization values were: $I = 10$, $P = 14$, and $B = 18$ using the pattern

$IBBPBBPBBPBB$, which means GoP size 12. Each MPEG video stream consisted of 40,000 video frames, which at 25 frames/s represented about 30 minutes of real-time full motion video. Fig. 2 shows portions of the I/P/B frame size sequences of the *the Silence of the Lambs* video.

Krunz et al. [16] found that the lognormal distribution is the best match for all I/P/B frames, i.e., if the I, P, or B frame size at time j is s_j, then

$$\log_{10} s_j \sim \mathcal{N}(\cdot; m, \sigma^2) \quad (5)$$

We, therefore, tried to model the logarithm of the frame size, to see if a Gaussian MF can match its nature. We chose *German Talk Show (talk2)* and *MTV video clips (mtv2)* as examples. For each video traffic, we decomposed the I/P/B frames into 8 segments, and computed the mean m_i and std σ_i of the logarithm of the frame size of the ith segment, $i = 1, 2, \cdots, 8$. We also computed the mean m and std σ of the entire video traffic in a video product. To see which value $-m_i$ or σ_i- varies more, we normalized the mean and std of each segment using m_i/m, and σ_i/σ, and we then computed the std of their normalized values, σ_m and σ_{std}. As we see from the last row of Tables 1 and 2, $\sigma_m \ll \sigma_{std}$. We conclude, therefore, that if the I/P/B frames of each segment (short range) of the video traffic are lognormally distributed, then the logarithm of the I, P, or B frame sizes in an entire video traffic (long range) is more appropriately modeled as a Gaussian with uncertain variance. This justifies the use of the Gaussian MFs, given in Example 1, to model the video traffic.

Table 1. Mean and std values for 8 segments and the entire *talk2* video traffic, and their normalized std.

Video data	I Frame		P Frame		B Frame	
	mean	std	mean	std	mean	std
Segment 1	4.8883	0.0394	4.2928	0.1114	4.0883	0.0758
Segment 2	4.8605	0.0738	4.2450	0.1133	4.0414	0.0841
Segment 3	4.8176	0.0749	4.1383	0.1901	3.9662	0.1244
Segment 4	4.8848	0.0825	4.2360	0.1880	4.0335	0.1349
Segment 5	4.8533	0.0507	4.1961	0.1375	3.9970	0.0964
Segment 6	4.8539	0.0605	4.1845	0.1435	3.9907	0.0971
Segment 7	4.8438	0.0789	4.1600	0.1425	3.9646	0.1000
Segment 8	4.8897	0.0901	4.2792	0.2452	4.0817	0.1898
Entire Traffic	4.8614	0.0745	4.2165	0.1726	4.0204	0.1263
Normalized std	0.0052	0.2307	0.0132	0.2643	0.0121	0.2904

The approach in this Section has been based on the assumption that the I/P/B category of a frame is known, which is a form of *supervised clustering*. This kind of modeling is very useful in network workload analysis (for

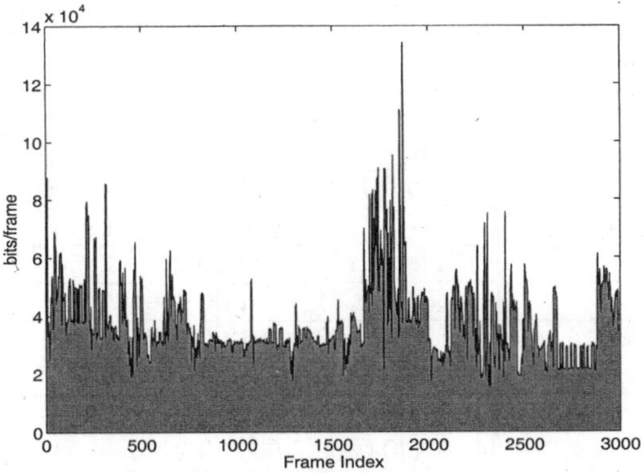

(a)

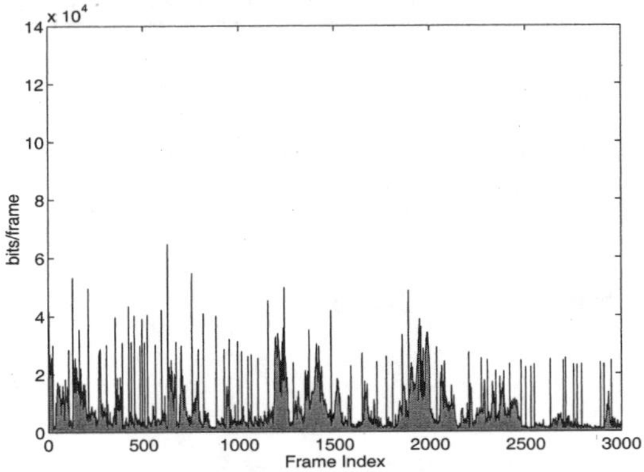

(b)

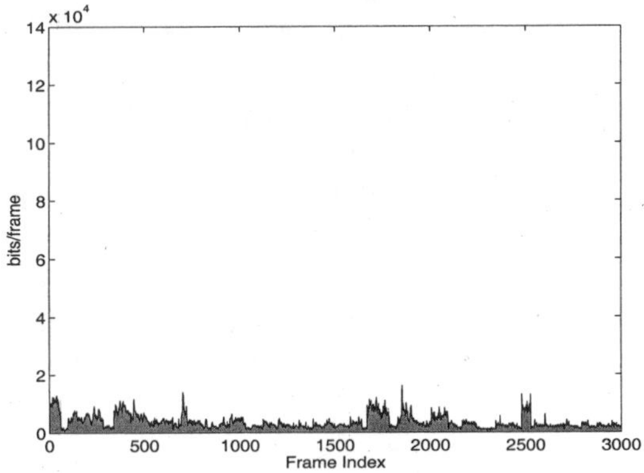

(c)

Fig. 2. Portions of I/P/B frame sizes of *The Silence of the Lambs* video. (a) I frame, (b) P frame, (c) B frame.

Table 2. Mean and std values for 8 segments and the entire *mtv2* video traffic, and their normalized std.

Video data	I Frame		P Frame		B Frame	
	mean	std	mean	std	mean	std
Segment 1	4.7623	0.1125	4.3718	0.2230	4.0820	0.2097
Segment 2	4.7517	0.1294	4.2415	0.3197	3.9157	0.3187
Segment 3	4.8428	0.1218	4.3807	0.2462	4.0547	0.2315
Segment 4	4.6683	0.1281	4.2511	0.2666	3.9438	0.2532
Segment 5	4.7828	0.1497	4.3012	0.3448	4.0038	0.3442
Segment 6	4.7201	0.1304	4.1269	0.3098	3.8307	0.2528
Segment 7	4.8909	0.1960	4.5381	0.3724	4.2473	0.3813
Segment 8	4.6516	0.0809	3.9668	0.2186	3.7039	0.1839
Entire Traffic	4.7590	0.1545	4.2723	0.3342	3.9727	0.3194
Normalized std	0.0171	0.2122	0.0404	0.1718	0.0416	0.2161

dynamic bandwidth allocation), and network control (such as connection admission control).

5 Type-2 TSK FLS A2-C0

5.1 Rules

In a first-order type-2 TSK FLS A2-C0, with a rule base of M rules, in which each rule has p antecedents, the ith rule, R^i, is denoted as

$$R^i\text{: IF } x_1 \text{ is } \tilde{\mathrm{F}}_1^i \text{ and } x_2 \text{ is } \tilde{\mathrm{F}}_2^i \text{ and } \ldots \text{ and } x_p \text{ is } \tilde{\mathrm{F}}_p^i \text{ THEN}$$
$$y^i = c_0^i + c_1^i x_1 + c_2^i x_2 + \cdots + c_p^i x_p$$

where $i = 1, 2, \ldots, M$; the c_j^i $(j = 0, 1, \ldots, p)$ are the consequent parameters that are crisp numbers; y^i is an output from the ith IF-THEN rule, which is a crisp number; and, the $\tilde{\mathrm{F}}_k^i$ $(k = 1, 2, \ldots, p)$ are type-2 fuzzy sets. These rules are for the case when all the uncertainties occur only in their antecedents, a situation that occurs, for example, in Bayesian equalization of a time-varying channel [21].

5.2 Inference

In type-2 TSK A2-C0, given an input $\mathbf{x} = (x_1, x_2, \cdots, x_p)$, the firing strength of the ith rule is

$$F^i = \mu_{\tilde{\mathrm{F}}_1^i}(x_1) \sqcap \mu_{\tilde{\mathrm{F}}_2^i}(x_2) \sqcap \cdots \sqcap \mu_{\tilde{\mathrm{F}}_p^i}(x_p) \qquad (6)$$

where $\sqcap$ is the meet operation [12]. The final output of this type-2 TSK FLS is

$$Y(F^1, \cdots, F^M) = \int_{f^1} \cdots \int_{f^M} \mathcal{T}_{i=1}^{M} \mu_{\mathrm{F}^i}(f^i) \Big/ \frac{\sum_{i=1}^{M} f^i y^i}{\sum_{i=1}^{M} f^i} \tag{7}$$

where M is the number of rules fired, $f^i \in F^i$, and, $\mathcal{T}$ and $\star$ indicate the chosen t-norm. We refer to Y as the *extended output* of a TSK FLS (although it resembles the so called *type-reduced set* for a type-2 Mamdani FLS [12] [15], there is no type-reduction needed in a type-2 TSK FLS). It reveals the uncertainty at the output of a type-2 TSK FLS due to antecedent uncertainties.

General discussions on how to compute (7) are given in [14], [12]. Because this computation can be very complicated, except for the case of interval set MFs, we focus on this case. More specifically, we focus on the case where interval type-2 sets are used in the antecedents, which means $\mu_{\tilde{\mathrm{F}}_k^i}(x_k)$ $(k = 1, \ldots, p)$ are interval sets, i.e.,

$$\mu_{\tilde{\mathrm{F}}_k^i}(x_k) = [\underline{\mu}_{\tilde{\mathrm{F}}_k^i}(x_k), \overline{\mu}_{\tilde{\mathrm{F}}_k^i}(x_k)] \triangleq [\underline{f}_k^i, \overline{f}_k^i]; \tag{8}$$

For this case, we are able to easily compute the firing strength of type-2 TSK FLS A2-C0, as summarized in the following:

Theorem 1 *In interval type-2 TSK FLS A2-C0, with meet under minimum or product t-norm, the firing strength in (6) for rule R^i is an interval set, i.e., $F^i = [\underline{f}^i, \overline{f}^i]$, where $(i = 1, \ldots, M)$*

$$\underline{f}^i = \underline{\mu}_{\tilde{F}_1^i}(x_1) \star \ldots \star \underline{\mu}_{\tilde{F}_p^i}(x_p) \triangleq \mathcal{T}_{k=1}^{p} \underline{f}_k^i, \tag{9}$$

and

$$\overline{f}^i = \overline{\mu}_{\tilde{F}_1^i}(x_1) \star \ldots \star \overline{\mu}_{\tilde{F}_p^i}(x_p) \triangleq \mathcal{T}_{k=1}^{p} \overline{f}_k^i; \tag{10}$$

The proof of this Theorem is given in [20].

When interval type-2 sets are used in the antecedents, (7) simplifies to

$$Y(F^1, \cdots, F^M) = [y_l, y_r] = \int_{f^1} \cdots \int_{f^M} 1 \Big/ \frac{\sum_{i=1}^{M} f^i y^i}{\sum_{i=1}^{M} f^i} \tag{11}$$

where $f^i \in [\underline{f}^i, \overline{f}^i]$. It is proven in [14] [12] that y_r and y_l depend only on a mixture of $\overline{f}^i$ or $\underline{f}^i$ values [and not on the values in $(\overline{f}^i, \underline{f}^i)$]. Consequently, y_l can be represented as

$$y_l = \frac{\sum_{i=1}^{M} f_l^i y^i}{\sum_{i=1}^{M} f_l^i} \triangleq \sum_{i=1}^{M} y^i p_l^i \tag{12}$$

where f_l^i denotes the firing strength membership grade (either $\underline{f}^i$ or $\overline{f}^i$) contributing to y_l, and $p_l^i \triangleq f_l^i \Big/ \sum_{i=1}^M f_l^i$; and,

$$y_r = \frac{\sum_{i=1}^M f_r^i y^i}{\sum_{i=1}^M f_r^i} \triangleq \sum_{i=1}^M y^i p_r^i \tag{13}$$

where f_r^i denotes the firing strength membership grade (either $\underline{f}^i$ or $\overline{f}^i$) contributing to y_r, and $p_r^i \triangleq f_r^i \Big/ \sum_{i=1}^M f_r^i$.

In order to complete the computation of y_l and y_r, we need to establish $\{f_l^i, i = 1, 2, \ldots, M\}$ and $\{f_r^i, i = 1, 2, \ldots, M\}$. This can be done using the exact computational procedure given in [15]. Here we briefly provide the computation procedure for y_r. Without loss of generality, assume the y^i are arranged in ascending order, i.e., $y^1 \leq y^2 \leq \ldots \leq y^M$.

1. Compute y_r in (13) by initially setting $f_r^i = (\overline{f}^i + \underline{f}^i)\big/2$ for $i = 1, \ldots, M$, where $\overline{f}^i$ and $\underline{f}^i$ have been previously computed using (9) and (10); and, let $y_r' \triangleq y_r$.
2. Find R $(1 \leq R \leq M-1)$ such that $y^R \leq y_r' \leq y^{R+1}$.
3. Compute y_r in (13), where $f_r^i = \underline{f}^i$ for $i \leq R$, and $f_r^i = \overline{f}^i$ for $i > R$; and, let $y_r'' \triangleq y_r$.
4. If $y_r'' \neq y_r'$, then go to step 5. If $y_r'' = y_r'$, then stop, and set $y_r \triangleq y_r''$.
5. Set y_r' equal to y_r'', and return to step 2.

This 4 step computation procedure (step 1 is an initialization step) has been proven to converge to the exact solution in no more than M iterations [14] [12]. Observe that in this procedure, the number R is very important. For $i \leq R$, $f_r^i = \underline{f}^i$; and, for $i > R$, $f_r^i = \overline{f}^i$; so, y_r can be represented as

$$y_r = y_r(\underline{f}^1, \ldots, \underline{f}^R, \overline{f}^{R+1}, \ldots, \overline{f}^M, y^1, \ldots, y^M) \tag{14}$$

The procedure for computing y_l is very similar. In step 2, find L $(1 \leq L \leq M-1)$ such that $y^L \leq y_l' \leq y^{L+1}$; and, in step 3, let $f_l^i = \overline{f}^i$ for $i \leq L$, and $f_l^i = \underline{f}^i$ for $i > L$. Then y_l can be represented as

$$y_l = y_l(\overline{f}^1, \ldots, \overline{f}^L, \underline{f}^{L+1}, \ldots, \underline{f}^M, y^1, \ldots, y^M) \tag{15}$$

In a type-2 TSK FLS, the extended output Y is a type-1 fuzzy set. If a crisp value is needed at its output, then defuzzification is necessary. In an interval type-2 TSK FLS, Y is an interval type-1 fuzzy set, so we defuzzify it using the average of y_l and y_r; hence, the defuzzified output of any interval type-2 TSK FLS is

$$y = f(\mathbf{x}) = \frac{y_l + y_r}{2} \tag{16}$$

6 Designing an Interval Type-2 TSK FLS A2-C0 Based on Tuning

Given an input-output training pair $(\mathbf{x}, d)$, $\mathbf{x} \in R^p$ and $d \in R$, we wish to design an interval type-2 TSK FLS with output (16) so that the error function

$$e = \frac{1}{2}\left[f(\mathbf{x}) - d\right]^2 \tag{17}$$

is minimized. Based on the analysis in Section 5, we know that only the upper and lower MFs, and the consequent y^i (with parameter c_k^i) determine $f(\mathbf{x})$. So we want to tune the upper and lower MFs, and the consequent parameters c_k^i.

Given T input-output training samples $(\mathbf{x}^t, d^t)$ $(t = 1, \ldots, T)$, we wish to update the design parameters so that (17) is minimized for E training epochs (updating the parameters using all the T training samples one time is called one epoch). A general method for doing this is described next for interval type-2 TSK FLS A2-C0. Our method is to:

1. Initialize all the parameters, including the parameters in antecedent and consequent MFs; and, choose appropriate step size α.
2. Set the counter of training epoch $e \triangleq 0$.
3. Set the counter of training data sample $t \triangleq 0$.
4. Apply input $\mathbf{x}^t$ to the type-2 TSK FLS, and compute the total firing degree and consequent for each rule, i.e., compute $\overline{f}^i$ and $\underline{f}^i$ using Theorem 1.
5. Compute y_l and y_r, as described in Section 5 (which leads to a re-ordering of the M rules, but they are then renumbered 1, 2, ..., M). This will establish L and R, so that y_l and y_r can be expressed as:

$$\begin{aligned} y_l &= y_l(\overline{f}^1, \ldots, \overline{f}^L, \underline{f}^{L+1}, \ldots, \underline{f}^M, y^1, \ldots, y^M) \\ &= \frac{\sum_{i=1}^{L} \overline{f}^i y^i + \sum_{j=L+1}^{M} \underline{f}^j y^j}{\sum_{i=1}^{L} \overline{f}^i + \sum_{j=L+1}^{M} \underline{f}^j} \end{aligned} \tag{18}$$

$$\begin{aligned} y_r &= y_r(\underline{f}^1, \ldots, \underline{f}^R, \overline{f}^{R+1}, \ldots, \overline{f}^M, y^1, \ldots, y^M) \\ &= \frac{\sum_{i=1}^{R} \underline{f}^i y^i + \sum_{j=R+1}^{M} \overline{f}^j y^j}{\sum_{i=1}^{R} \underline{f}^i + \sum_{j=R+1}^{M} \overline{f}^j} \end{aligned} \tag{19}$$

6. Compute $f(\mathbf{x}^t) = (y_l + y_r)/2$, which is the defuzzified output of the type-2 TSK FLS.
7. Determine the explicit dependence of y_l and y_r on membership functions. To do this, first determine the dependence of $\underline{f}^i$ and $\overline{f}^i$ on membership

functions, using (9) and (10). Consequently,

$$y_l = y_l\Big(\overline{\mu}_{\tilde{F}_1^1}(x_1), \ldots, \overline{\mu}_{\tilde{F}_p^1}(x_p), \cdots, \overline{\mu}_{\tilde{F}_1^L}(x_1), \ldots, \overline{\mu}_{\tilde{F}_p^L}(x_p),$$
$$\underline{\mu}_{\tilde{F}_1^{L+1}}(x_1), \ldots, \underline{\mu}_{\tilde{F}_p^{L+1}}(x_p), \cdots, \underline{\mu}_{\tilde{F}_1^M}(x_1), \ldots, \underline{\mu}_{\tilde{F}_p^M}(x_p),$$
$$c_0^1, \ldots, c_p^1, \cdots, c_0^M, \ldots, c_p^M\Big) \qquad (20)$$

Similarly,

$$y_r = y_r\Big(\underline{\mu}_{\tilde{F}_1^1}(x_1), \ldots, \underline{\mu}_{\tilde{F}_p^1}(x_p), \cdots, \underline{\mu}_{\tilde{F}_1^R}(x_1), \ldots, \underline{\mu}_{\tilde{F}_p^R}(x_p),$$
$$\overline{\mu}_{\tilde{F}_1^{R+1}}(x_1), \ldots, \overline{\mu}_{\tilde{F}_p^{R+1}}(x_p), \cdots, \overline{\mu}_{\tilde{F}_1^M}(x_1), \ldots, \overline{\mu}_{\tilde{F}_p^M}(x_p),$$
$$c_0^1, \ldots, c_p^1, \cdots, c_0^M, \ldots, c_p^M\Big) \qquad (21)$$

Although the dependence of y_l and y_r on their activated upper and lower MFs and consequent parameters have been shown, the exact formulas for y_l and y_r need to be worked out so that partial derivatives of y_l and y_r with respect to parameters can be determined.

8. Test each component of $\mathbf{x}$ to determine the active branches in $\underline{\mu}_{\tilde{F}_k^l}(x_k)$, and $\overline{\mu}_{\tilde{F}_k^l}(x_k)$, $k = 1, 2, \ldots, p$.
9. Represent the active branches as functions of their associated parameters. This step depends on the kinds of MFs and the locations of x_k, $(k = 1, \ldots, p)$ in relation to these MFs.
10. Train the parameters of the active branches and the parameters in the consequent using a steepest-descent (or any other optimization) method. The error function is $e^t = [f(\mathbf{x}^t) - d^t]^2/2$. Observe that the parameters in the consequents are shared by both y_l and y_r, so their update need to be connected with both y_l and y_r using chain rules.
11. Set $t \triangleq t + 1$. If $t = T$, go to step 12; otherwise go to step 4.
12. Set $e \triangleq e + 1$. If $e = E$, stop; otherwise go to step 3.

7 MPEG VBR Video Traffic Modeling

We used MPEG video traffic of the sports, *formula 1 race: GP Hockenheim 1994*, for our experiment.

Our simulations were based on the logarithm of the first 1000 I frame sizes of the sports video, $s(1)$, $s(2)$, ..., $s(1000)$. We used four antecedents for forecasting, i.e., $s(k-3)$, $s(k-2)$, $s(k-1)$, and $s(k)$ were used to predict $s(k+1)$. The first 504 data, $s(1)$, $s(2)$, ..., $s(504)$, are for training, and the remaining 496 data, $s(505), s(506), \ldots, s(1000)$ are for testing. In Fig. 3, we plot the 1000 I-frames, $s(1)$, $s(2)$, ..., $s(1000)$.

We designed a type-1 TSK FLS and an interval type-2 TSK FLS A2-C0, and we used 5 rules for each FLS.

For the type-1 TSK FLS, the ith rule R^i is:

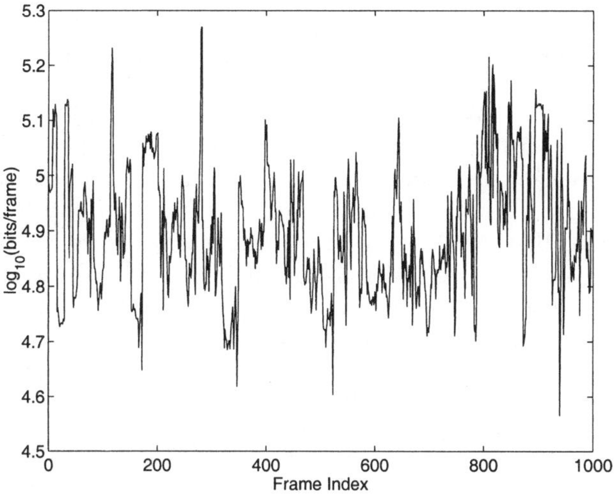

Fig. 3. The first 1000 I frames of *formula 1 race: GP Hockenheim 1994* MPEG-1 video traces.

R^i: IF $s(k-3)$ is F_1^i and $s(k-2)$ is F_2^i and ... and $s(k)$ is F_4^i THEN
$\hat{s}^i(k+1) = c_0^i + c_1^i s(k-3) + c_2^i s(k-2) + c_3^i s(k-1) + c_4^i s(k)$.

where $i = 1, 2, \ldots, 5$. We initially chose F_j^i $(i = 1, 2, \ldots, 5;\ j = 1, \cdots, 4)$ to be the same for all i and j, and used a Gaussian MF for them, one whose initial mean and std were chosen from the first 500 I frames as $m = 4.8851$ and $\sigma = 0.1080$ (see Table 3).

Table 3. Mean and std values for 5 segments (frames 1–1200, 1201–2400, 2401–3600, 3601–4800, 4801–6000) and the entire segment (frames 1–6000) of *formula 1 race: GP Hockenheim 1994* MPEG-1 traffic, and their normalized std.

Video Data	I Frame		P Frame		B Frame	
	mean	std	mean	std	mean	std
Segment 1	4.9028	0.1263	4.4363	0.2946	4.2267	0.2122
Segment 2	4.9366	0.1289	4.5972	0.2224	4.3579	0.1715
Segment 3	4.8975	0.0943	4.6048	0.1964	4.3672	0.1962
Segment 4	4.8471	0.1001	4.4402	0.2093	4.2317	0.1673
Segment 5	4.8939	0.0783	4.5384	0.1660	4.3053	0.1371
Entire Segment	4.8851	0.1080	4.5313	0.2167	4.3013	0.1813
Normalized std	0.0065	0.2003	0.0181	0.2207	0.0156	0.1591

For the type-2 TSK FLS A2-C0, the ith rule R^i is:

R^i: IF $s(k-3)$ is $\tilde{\mathrm{F}}_1^i$ and $s(k-2)$ is $\tilde{\mathrm{F}}_2^i$ and ... and $s(k)$ is $\tilde{\mathrm{F}}_4^i$ THEN
$\hat{s}^i(k+1) = c_0^i + c_1^i s(k-3) + c_2^i s(k-2) + c_3^i s(k-1) + c_4^i s(k)$.

where $i = 1, 2, \ldots, 5$. We initially chose $\tilde{\mathrm{F}}_j^i$ ($i = 1, 2, \ldots, 5$; $j = 1, \cdots, 4$) to be the same for all i and j, and used a Gaussian MF for them (with fixed mean and uncertain std), one whose initial mean and std were chosen from the first 500 I frames (contained in the first 6000 frames of *race* traffic because the GoP is IBBPBBPBBPBB, and the I frame only occurs one time in each GoP) as $m = 4.8851$ (see Table 3) and $\sigma \in [\sigma_1, \sigma_2]$. We divided the 500 frames into 5 equal-length (100 frames) segments, and computed the initial std σ^j of the jth segment (see Table 3), i.e.,

$$\sigma_1 = \min_{j=1,\cdots,5} \sigma^j = 0.0783 \tag{22}$$

$$\sigma_2 = \max_{j=1,\cdots,5} \sigma^j = 0.1289 \tag{23}$$

The number of parameters in the type-1 TSK FLS is $(3p+1)M = (3 \times 4 + 1)5 = 65$, and that in the type-2 TSK FLS A2-C0 is $(4p+1)M = (4 \times 4 + 1)5 = 85$.

We randomly chose the consequents of the two FLSs, i.e., c_j^i ($i = 1, 2, \ldots, 5$; $j = 0, 1, \cdots, 4$) were chosen randomly in $[0, 0.2]$ with uniform distribution. Then we used a steepest descent algorithm to tune all the parameters in the antecedents and consequents, using a step size $\alpha = 0.001$. After training, the rules were fixed, and we tested the FL forecaster based on the remaining 496 frames, $s(1505), s(1506), \ldots, s(2000)$.

Each FLS was tuned using a simple steepest-descent algorithm for 10 epochs. We used the testing data to see how each FLS performed by evaluating the RMSE between the defuzzified output of each FLS and the actual frame sizes, i.e.,

$$\mathrm{RMSE} = \sqrt{\frac{1}{496} \sum_{k=504}^{999} [s(k+1) - f(\mathbf{s}^k)]^2} \tag{24}$$

where $\mathbf{s}^k = [s(k-3), s(k-2), s(k-1), s(k)]^T$, and T denotes transpose.

For each of the 2 designs, we ran 50 Monte-Carlo realizations. Since there are $50 \times 10 = 500$ RMSE values for each design, we summarize the average RMSEs for each epoch and for each design in Fig. 4. Observe, from the figures:

1. The type-2 TSK FLS outperforms the type-1 TSK FLS. The minimum average RMSE of the type-1 TSK FLS was 0.0882; and, the minimum average RMSE of the type-2 TSK FLS was 0.0798, which shows that the type-2 TSK FLS A2-C0 outperforms the type-1 TSK FLS by $\frac{0.0882-0.0798}{0.0882} = 9.52\%$ in RMSE.
2. After the first epoch, the type-2 TSK FLS outperforms the type-1 TSK FLS by $\frac{0.1736-0.1267}{0.1736} = 27.02\%$ in RMSE. This shows that a type-2 TSK

FLS will be very useful in real-time signal processing where more than one epoch of tuning is impossible.

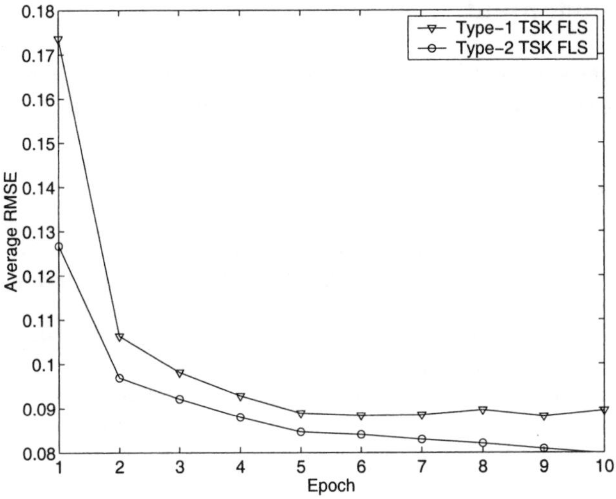

Fig. 4. The average RMSE (for the test traffic) for the 2 FLS designs, averaged over 50 Monte-Carlo realizations. Tuning was performed in each realization for 10 epochs.

8 Conclusions and Future Works

We have demonstrated that I/P/B frame sizes of video traffic can be represented using type-2 fuzzy sets. We observed that for long-range video traffic, a Gaussian MF with uncertain std is appropriate for modeling the frame sizes.

We treated the video traffic as a dynamic system, and applied a type-2 TSK FLS to model the video traffic, and showed that a type-2 TSK FLS performs better than a type-1 TSK FLS.

The emergence of the ATM technique, with its flexibility to accommodate a diverse mixture of traffic that possesses different traffic characteristics and quality of service (QoS) requirements, has made the MPEG video compression standard appropriate for the transmission of real-time digital video signals over high-speed communication networks. Future research includes developing a dynamic bandwidth allocation scheme, and making connection admission control decision based on the modeled video traffic.

Finally, MATLAB files for performing the training procedure of type-2 TSK FLS discussed in this chapter are available at URL :
http://sipi.usc.edu/ mendel/software.

Acknowledgement

This work was supported by the Center for Research on Applied Signal Processing (CRASP), at the University of Southern California.

References

1. A. M. Adas, "Using adaptive linear prediction to support real-time VBR video under RCBR network service model," *IEEE Trans. on Networking*, vol. 6, no. 5, Oct. 1998.
2. J. C. Bezdek, *Pattern Recognition with Fuzzy Objective Function Algorithms*, Plenum Press, New York, 1981.
3. P.-R. Chang and J.-T. Hu, " Optimal nonlinear adaptive prediction and modeling of MPEG video in ATM networks using pipelined recurrent neural networks," *IEEE J. of Selected Areas in Communications*, vol. 15, no. 6, pp. 1087-1100, August 1997.
4. A. M. Dawood and M. Ghanbari, "MPEG video modelling based on scene description," *IEEE Int'l. Conf. Image Processing*, vol. 2, pp. 351-355, Chicago, IL, Oct. 1998.
5. A. M. Dawood and M. Ghanbari, "Content-based MPEG video traffic modeling," *IEEE Trans. on Multimedia*, vol. 1, no. 1, pp. 77-87, March 1999.
6. N. Dimitrova and F. Golshani, "Motion recovery for video content classification," *ACM Trans. Information Systems*, vol. 13, no. 4, Oct. 1995, pp. 408-439.
7. R. O. Duda and P. E. Hart, *Pattern Classification and Scene Analysis*, John Wiley & Sons, Inc, USA, 1973.
8. S. Ghosh, Q. Razouqi, H. J. Schumacher, and A. Celmins, " A survey of recent advances in fuzzy logic in telecommunications networks and new challenges ," *IEEE Trans. Fuzzy Systems*, vol. 6, no. 3, pp. 443-447, August 1998.
9. D. Dubois and H. Prade, *Fuzzy Sets and Systems: Theory and Applications*, Academic Press, New York, USA, 1980.
10. D. P. Heyman, A. Tabatabi, and T. V. Lakshman, " Statistical analysis of MPEG-2 coded VBR video traffic," *6th Int'l Workshop on Packet Video*, Portland, OR, Sept 1994.
11. E. Hisdal, "The IF THEN ELSE statement and interval-valued fuzzy sets of higher type," *Int'l. J. Man-Machine Studies*, vol. 15, pp. 385-455, 1981.
12. N. N. Karnik and J. M. Mendel, *An Introduction to Type-2 Fuzzy Logic Systems*, October 1998, USC Report, http://sipi.usc.edu/~mendel/report.
13. N. N. Karnik and J. M. Mendel, "Operations on type-2 fuzzy sets," submitted to *Fuzzy Sets and Systems*.
14. N. N. Karnik and J. M. Mendel, "Centroid of a type-2 fuzzy set," submitted to *Information Sciences*.
15. N. N. Karnik, J. M. Mendel, and Q. Liang, "Type-2 fuzzy logic systems", *IEEE Trans. Fuzzy Systems*, vol. 7, no. 6, pp. 643-658, Dec. 1999.

16. M. Krunz, R. Sass, and H. Hughes, " Statistical characteristics and multiplexing of MPEG streams," *Proc. IEEE Int'l Conf. Computer Communications, INFOCOM'95*, Boston, MA, Apr. 1995, vol. 2, pp. 455-462.
17. M. Krunz and A. M. Makowski, " Modeling video traffic using $M/G/\infty$ input processes: a compromise between Markovian and LRD models," *IEEE J. of Selected Areas in Communications*, vol. 16, no. 5, pp. 733-748, June 1998.
18. S.-Y. Kung and J.-N. Hwang, " Neural networks for intelligent multimedia processing ," *Proc. of the IEEE* , vol. 86, no. 6, pp. 1244-1272, June 1998.
19. Q. Liang and J. M. Mendel, " Interval type-2 fuzzy logic systems: theory and design," submitted to *IEEE Trans. Fuzzy Systems.*
20. Q. Liang and J. M. Mendel, "Interval type-2 TSK fuzzy logic systems," submitted to *IEEE Trans. Systems, Man, and Cybernetics.*
21. Q. Liang and J. M. Mendel, "Equalization of nonlinear time-varying channels using type-2 fuzzy adaptive filters," submitted to *IEEE Trans. Fuzzy Systems.*
22. P. Manzoni, P. Cremonesi, and G. Serazzi, " Workload models of VBR video traffic and their use in resource allocation policies," *IEEE Trans. on Networking*, vol. 7, no. 3, pp. 387-397, June 1999.
23. J. M. Mendel, " Fuzzy logic systems for engineering: a tutorial ," *Proc. of the IEEE*, vol. 83, no. 3, pp. 345-377, March 1995.
24. J. M. Mendel, " Computing with words when words can mean different things to different people ," presented at *Int'l. ICSC Congress on Computational Intelligence: Methods & Applications, Third Annual Symposium on Fuzzy Logic and Applications*, Rochester, NY, June 22-25, 1999.
25. J. M. Mendel, " Uncertainty, fuzzy logic, and signal processing," *Signal Processing*, to appear in 2000.
26. G. C. Mouzouris and J. M. Mendel, "Nonsingleton fuzzy logic systems: theory and application," *IEEE Trans. Fuzzy Systems*, vol. 5, no. 1, pp. 56-71, Feb 1997.
27. G. Pacifici, G. Karlsson, M. Garrett, and N. Ohta, " Guest editorial real-time video services in multimedia networks," *IEEE J. of Selected Areas in Communications*, vol. 15, no. 6, pp. 961-964, August 1997.
28. N. Patel and I. K. Sethi, " Video shot detection and characterization for video databases ," *Pattern Recognition*, vol. 30, no. 4, pp. 583-592, 1997.
29. P. Pancha and M. El-Zarki, " A look at the MPEG video coding standard for variable bit rate video transmission," *IEEE INFOCOM'92*, Florence, Italy, 1992.
30. O. Rose, "Satistical properties of MPEG video traffic and their impact on traffic modeling in ATM systems," University of Wurzburg, Institute of Computer Science, Research Report 101, Feb 1995.
31. K. Tanaka, M. Sano, and H. Watanabe, "Modeling and control of carbon monoxide concentration using a neuro-fuzzy technique" *IEEE Trans Fuzzy Systems*, vol. 3, no. 3, pp. 271-279, Aug 1995.
32. D. Tsang, B. Bensaou, and S. Lam, " Fuzzy-based rate control for real-time MPEG video ," *IEEE Trans. Fuzzy Systems*, vol. 6, no. 4, pp. 504-516, Nov. 1998.
33. R. Zabih, J. Miller, and K. Mai, " A feature-based algorithm for detecting and classifying production effects ," *Multimedia Systems*, vol. 7, pp. 119-128, 1999.
34. L. A. Zadeh, " The concept of a linguistic variable and its application to approximate reasoning - I," *Information Sciences*, vol. 8, pp. 199-249, 1975.

Induction of Rules about Complications with the use of Rough Sets

Shusaku Tsumoto

Department of Medical Informatics,
Shimane Medical University, School of Medicine,
89-1 Enya-cho, Izumo 693-8501 Japan
E-mail: tsumoto@computer.org

Abstract. One of the most important problems on rule induction methods is that they cannot extract the rules that plausibly represent experts' decision processes: the induced rules are too short to represent the reasoning of domain experts. In this paper, the characteristics of experts' rules are closely examined and a new approach to extract plausible rules is introduced, which consists of the following three procedures. First, the characterization of decision attributes (given classes) is extracted from databases and the classes are classified into several groups with respect to the characterization. Then, two kinds of sub-rules, characterization rules for each group and discrimination rules for each class in the group are induced. Finally, those two parts are integrated into one rule for each decision attribute. The proposed method was evaluated on medical databases, the experimental results of which show that induced rules correctly represent experts' decision processes.
Keywords. rough sets, rough inclusion, rule induction, hierarchical rules, data mining, knowledge discovery, medical decision support, knowledge acquisition, focusing mechanism, diagnostic reasoning.

1 Introduction

One of the most important problems in developing expert systems is knowledge acquisition from experts[3]. In order to automate this problem, many inductive learning methods, such as induction of decision trees[2,13], rule induction methods[6,8,10,13,14] and rough set theory[11,16,20], are introduced and applied to extract knowledge from databases, and the results show that these methods are appropriate.

However, it has been pointed out that conventional rule induction methods cannot extract rules, which plausibly represent experts' decision processes[16,17]: the description length of induced rules is too short, compared with the experts' rules (Those results are shown in Appendix B). For example, rule induction methods, including AQ15[10] and PRIMEROSE[16], induce the following common rule for muscle contraction headache from databases on differential diagnosis of headache[17]:

$$[location = whole] \wedge [\text{Jolt Headache} = no] \wedge [\text{Tenderness of M1} = yes] \\ \rightarrow \text{muscle contraction headache.}$$

This rule is shorter than the following rule given by medical experts.

$$[\text{Jolt Headache} = no] \wedge [\text{Tenderness of M1} = yes] \wedge [\text{Tenderness of B1} = no] \wedge [\text{Tenderness of C1} = no] \rightarrow \text{muscle contraction headache},$$

where [Tenderness of B1 $= no$] and [Tenderness of C1 $= no$] are added.

These results suggest that conventional rule induction methods do not reflect a mechanism of knowledge acquisition of medical experts.

In this paper, the characteristics of experts' rules are closely examined and a new approach to extract plausible rules is introduced, which consists of the following three procedures. First, the characterization of each decision attribute (a given class), a list of attribute-value pairs the supporting set of which covers all the samples of the class, is extracted from databases and the classes are classified into several groups with respect to the characterization. Then, two kinds of sub-rules, rules discriminating between each group and rules classifying each class in the group are induced. Finally, those two parts are integrated into one rule for each decision attribute. The proposed method was evaluated on medical databases, the experimental results of which show that induced rules correctly represent experts' decision processes. The paper is organized as follows: in Section 2, we make a brief description about rough set theory and the definition of probabilistic rules based on this theory. Section 3 discusses interpretation of medical experts' rules. Then, Section 4 presents an induction algorithm for incremental learning. Section 5 gives experimental results. Section 6 discusses the problems of our work and related work, and finally, Section 7 concludes our paper.

2 Rough Set Theory and Probabilistic Rules

Table 1. An Example of Database

	age	loc	nat	prod	nau	M1	class
1	50-59	occ	per	no	no	yes	m.c.h.
2	40-49	who	per	no	no	yes	m.c.h.
3	40-49	lat	thr	yes	yes	no	migra
4	40-49	who	thr	yes	yes	no	migra
5	40-49	who	rad	no	no	yes	m.c.h.
6	50-59	who	per	no	yes	yes	psycho

DEFINITIONS: loc: location, nat: nature, prod: prodrome, nau: nausea, M1: tenderness of M1, who: whole, occ: occular, lat: lateral, per: persistent, thr: throbbing, rad: radiating, m.c.h.: muscle contraction headache, migra: migraine, psycho: psychological pain,

In this section, a probabilistic rule is defined by the use of the following three notations of rough set theory[11]. The main ideas of these rules are illustrated by a small database shown in Table 1.

First, a combination of attribute-value pairs, corresponding to a complex in AQ terminology[9], is denoted by a formula R. For example, $[age = 50-59] \wedge [loc = occular]$ will be one formula, denoted by $R = [age = 50-59] \wedge [loc = occular]$.

Secondly, a set of samples which satisfy R is denoted by $[x]_R$, corresponding to a star in AQ terminology. For example, when $\{2,3,4,5\}$ is a set of samples which satisfy $[age = 40-49]$, $[x]_{[age=40-49]}$ is equal to $\{2,3,4,5\}$. [1]

Finally, U, which stands for "Universe", denotes all training samples.

According to these notations, a rule is defined as follows:

Definition 1 (Representation of Rule). Let R be a formula (conjunction of attribute-value pairs), D denote a set whose elements belong to a class d, or positive examples in all training samples (the universe), U. Finally, let $|D|$ denote the cardinality of D. A rule of D is defined as a tripule, $< R \overset{\alpha,\kappa}{\rightarrow} d, \alpha_R(D), \kappa_R(D) >$, where $R \overset{\alpha,\kappa}{\rightarrow} d$ satisfies the following conditions:

$$\begin{aligned} &(1) \quad [x]_R \cap D \neq \phi, \\ &(2) \quad \alpha_R(D) = \frac{|[x]_R \cap D)|}{|[x]_R|}, \\ &(3) \quad \kappa_R(D) = \frac{|[x]_R \cap D)|}{|D|}. \end{aligned}$$

In the above definition, α corresponds to the accuracy measure: if α of a rule is equal to 0.9, then the accuracy is also equal to 0.9. On the other hand, κ is a statistical measure of what proportion of D is covered by this rule, that is, a coverage or a true positive rate: when κ is equal to 0.5, half of the members of a class belong to the set whose members satisfy that formula.

For example, let us consider a case of a rule $[age = 40-49] \rightarrow m.c.h.$ Since $[x]_{[age=40-49]} = \{2,3,4,5\}$ and $D = \{1,2,5\}$, the values of accuracy and coverage are obtained as: $\alpha_{[age=40-49]}(D) = |\{2,5\}|/|\{2,3,4,5\}| = 0.5$ and $\kappa_{[age=40-49]}(D) = |\{2,5\}|/|\{1,2,5\}| = 0.67$. Thus, if a patient, who complains of a headache, is 40 to 49 years old, then m.c.h. is suspected, whose accuracy and coverage are equal to 0.5 and 0.67, respectively .

Probabilistic Rules The simplest probabilistic model is that which only uses classification rules which have high accuracy and high coverage.

This model is applicable when rules of high accuracy can be derived. Such rules can be defined as:

$$\begin{aligned} R \overset{\alpha,\kappa}{\rightarrow} d \text{ s.t.} \quad & R = \vee_i R_i = \vee \wedge_j [a_j = v_k], \\ & \alpha_{R_i}(D) \geq \delta_\alpha \text{ and } \kappa_{R_i}(D) \geq \delta_\kappa, \end{aligned}$$

[1] In this notation, "n" denotes the nth sample in a dataset (Table 1).

where δ_α and δ_κ denote given thresholds for accuracy and coverage, respectively. For the above example shown in Table 1, probabilistic rules for m.c.h. are given as follows:

$$[M1 = yes] \rightarrow m.c.h.\ \alpha = 3/4 = 0.75,\ \kappa = 1.0,$$
$$[nau = no] \rightarrow m.c.h.\ \alpha = 3/3 = 1.0,\ \kappa = 1.0,$$

where δ_α and δ_κ are set to 0.75 and 0.5, respectively.

It is notable that this rule is a kind of probabilistic proposition with two statistical measures, which is an extension of Ziarko's variable precision model (VPRS) [20]. [2]

2.1 Rough Inclusion

In order to measure the similarity between classes with respect to characterization, we introduce a rough inclusion measure μ, which is defined as follows.

$$\mu(S,T) = \frac{|S \bigcap T|}{|S|}.$$

It is notable that if $S \subseteq T$, then $\mu(S,T) = 1.0$, which shows that this relation extends subset and superset relations. This measure is introduced by Polkowski and Skowron in their study on rough mereology[12]. Whereas rough mereology firstly applies to distributed information systems, its essential idea is rough inclusion: Rough inclusion focuses on set-inclusion to characterize a hierarchical structure based on a relation between a subset and superset. Thus, application of rough inclusion to capturing the relations between classes is equivalent to constructing rough hierarchical structure between classes, which is also closely related with information granulation proposed by Zadeh[18].

3 Interpretation of Medical Experts' Rules

As shown in Section 1, rules acquired from medical experts are much longer than those induced from databases the decision attributes of which are given by the same experts.[3]

[2] In VPRS model, the two kinds of precision of accuracy is given, and the probabilistic proposition with accuracy and two precision conserves the characteristics of the ordinary proposition. Thus, our model is to introduce the probabilistic proposition not only with accuracy, but also with coverage.

[3] This is because rule induction methods generally search for shorter rules, compared with decision tree induction. In the latter cases, the induced trees are sometimes too deep and in order for the trees to be learningful, pruning and examination by experts are required. One of the main reasons why rules are short and decision trees are sometimes long is that these patterns are generated only by one criteria, such as high accuracy or high information gain. The comparative study in this section suggests that experts should acquire rules not only by one criteria but by the usage of several measures.

Those characteristics of medical experts' rules are fully examined not by comparing between those rules for the same class, but by comparing experts' rules with those for another class. For example, a classification rule for muscle contraction headache is given by:

$$
\begin{aligned}
[\text{Jolt Headache} = no] \ \wedge (& [\text{Tenderness of M0} = yes] \\
& \vee [\text{Tenderness of M1} = yes] \vee [\text{Tenderness of M2} = yes]) \\
& \wedge [\text{Tenderness of B1} = no] \wedge [\text{Tenderness of B2} = no] \\
& \wedge [\text{Tenderness of B3} = no] \\
& \wedge [\text{Tenderness of C1} = no] \wedge [\text{Tenderness of C2} = no] \\
& \wedge [\text{Tenderness of C3} = no] \wedge [\text{Tenderness of C4} = no] \\
& \rightarrow \text{muscle contraction headache}
\end{aligned}
$$

This rule is very similar to the following classification rule for disease of cervical spine:

$$
\begin{aligned}
[\text{Jolt Headache} = no] \ \wedge (& [\text{Tenderness of M0} = yes] \\
& \vee [\text{Tenderness of M1} = yes] \vee [\text{Tenderness of M2} = yes]) \\
& \wedge ([\text{Tenderness of B1} = yes] \vee [\text{Tenderness of B2} = yes] \\
& \quad \vee [\text{Tenderness of B3} = yes] \\
& \quad \vee [\text{Tenderness of C1} = yes] \vee [\text{Tenderness of C2} = yes] \\
& \quad \vee [\text{Tenderness of C3} = yes] \vee [\text{Tenderness of C4} = yes]) \\
& \rightarrow \text{disease of cervical spine}
\end{aligned}
$$

The differences between these two rules are attribute-value pairs, from tenderness of B1 to C4. Thus, these two rules can be simplified into the following form:

$$
\begin{aligned}
a_1 \wedge A_2 \wedge \neg A_3 &\rightarrow \textit{muscle contraction headache} \\
a_1 \wedge A_2 \wedge A_3 &\rightarrow \textit{disease of cervical spine}
\end{aligned}
$$

The first two terms and the third one represent different reasoning. The first and second term a_1 and A_2 are used to differentiate muscle contraction headache and disease of cervical spine from other diseases. The third term A_3 is used to make a differential diagnosis between these two diseases. Thus, medical experts firstly selects several diagnostic candidates, which are very similar to each other, from many diseases and then make a final diagnosis from those candidates.

In the next section, a new approach for inducing the above rules is introduced.

4 Rule Induction

Rule induction consists of the following three procedures. First, the characterization of each decision attribute (a given class), a list of attribute-value pairs the supporting set of which covers all the samples of the class, is extracted from databases and the classes are classified into several groups with respect to the characterization. Then, two kinds of sub-rules, rules discriminating between each group and rules classifying each class in the group are induced. Finally, those two parts are integrated into one rule for each decision attribute.

4.1 An Algorithm for Rule Induction

An algorithm for rule induction is given in Fig.1.

procedure *Rule Induction* (*Total Process*);
  var
    $i : integer$; $M, L, R : List$;
    $L_D : List$; /* A list of all classes */
  begin
    Calculate $\alpha_R(D_i)$ and $\kappa_R(D_i)$ for each elementary relation R and each class D_i;
    Make a list $L(D_i) = \{R|\kappa_R(D) = 1.0\})$ for each class D_i;
    while $(L_D \neq \phi)$ **do**
      begin
        $D_i := first(L_D)$; $M := L_D - D_i$;
        while $(M \neq \phi)$ **do**
          begin
            $D_j := first(M)$;
              if $(\mu(L(D_j), L(D_i)) \leq \delta_\mu)$ **then** $L_2(D_i) := L_2(D_i) + \{D_j\}$;
            $M := M - D_j$;
          end
        Make a new decision attribute D'_i for $L_2(D_i)$;
        $L_D := L_D - D_i$;
      end
      Construct a new table $(T_2(D_i))$for $L_2(D_i)$.
      Construct a new table$(T(D'_i))$ for each decision attribute D'_i;
        Induce classification rules R_2 for each $L_2(D)$; /* Fig.2 */
        Store Rules into a List $R(D)$
        Induce classification rules R_d for each D' in $T(D')$; /* Fig.2 */
        Store Rules into a List $R(D')(= R(L_2(D_i)))$
      Integrate R_2 and R_d into a rule R_D; /* Fig.3 */
  end {*Rule Induction* };

Fig. 1. An Algorithm for Rule Induction

Induction of Classification Rules For induction of rules, the algorithm introduced in PRIMEROSE[16] is applied, which is shown in Fig. 2.

Integration of Rules An algorithm for integration is given as shown in Fig. 3.

4.2 Example

Let us illustrate how the introduced algorithm works by using a small database in Table 1. For simplicity, two thresholds δ_α and δ_μ are set to 1.0, which means that

```
procedure Induction of Classification Rules;
  var
    i : integer;   M, L_i : List;
  begin
    L_1 := L_er; /* L_er: List of Elementary Relations */
    i := 1;   M := {};
    for i := 1 to n do        /* n: Total number of attributes */
      begin
        while ( L_i ≠ {} ) do
          begin
            Select one pair R = ∧[a_i = v_j] from L_i;
            L_i := L_i − {R};
            if  (α_R(D) ≥ δ_α)   and   (κ_R(D) ≥ δ_κ)
              then  do S_ir := S_ir + {R}; /* Include R as Inclusive Rule */
            else M := M + {R};
          end
        L_{i+1} := (A list of the whole combination of the conjunction formulae in M);
      end
  end {Induction of Classification Rules };
```

Fig. 2. An Algorithm for Classification Rules

```
procedure Rule Integration;
  var
    i : integer;   M, L_2 : List;
    R(D_i) : List; /* A list of rules for D_i */
    L_D : List; /* A list of all classes */
  begin
    while(L_D ≠ φ) do
      begin
        D_i := first(L_D); M := L_2(D_i);
        Select one rule R' → D'_i from R(L_2(D_i)).
        while (M ≠ φ) do
          begin
            D_j := first(M);
              Select one rule R → d_j for D_j;
              Integrate two rules: R ∧ R' → d_j.
            M := M − {D_j};
          end
        L_D := L_D − D_i;
      end
  end {Rule Combination}
```

Fig. 3. An Algorithm for Rule Integration

only deterministic rules should be induced and that only subset and superset relations should be considered for grouping classes.

After the first and second step, the following three $L(D_i)$ will be obtained: $L(m.c.h.) = \{[prod = no], [M1 = yes]\}$, $L(migra) = \{[age = 40-49], [nat = who], [prod = yes], [nau = yes], [M1 = no]\}$, and $L(psycho) = \{[age = 50-59], [loc = who], [nat = per], [prod = no], [nau = no], [M1 = yes]\}$.

Thus, since a relation $L(psycho) \subset L(m.c.h.)$ holds (i.e., $\mu(L(m.c.h.), L(psycho)) = 1.0$), a new decision attribute is $D_1 = \{m.c.h., psycho\}$ and $D_2 = \{migra\}$, and a partition $P = \{D_1, D_2\}$ is obtained. From this partition, two decision tables will be generated, as shown in Table 2 and Table 3 in the fifth step.

Table 2. A Table for a New Partition P

	age	loc	nat	prod	nau	M1	class
1	50-59	occ	per	0	0	1	D_1
2	40-49	who	per	0	0	1	D_1
3	40-49	lat	thr	1	1	0	D_2
4	40-49	who	thr	1	1	0	D_2
5	40-49	who	rad	0	0	1	D_1
6	50-59	who	per	0	1	1	D_1

Table 3. A Table for D_1

	age	loc	nat	prod	nau	M1	class
1	50-59	occ	per	0	0	1	m.c.h.
2	40-49	who	per	0	0	1	m.c.h.
5	40-49	who	rad	0	0	1	m.c.h.
6	50-59	who	per	0	1	1	psycho

In the sixth step, classification rules for D_1 and D_2 are induced from Table 2. For example, the following rules are obtained for D_1.

$$
\begin{array}{lll}
[M1 = yes] & \rightarrow D_1 \ \alpha = 1.0, \kappa = 1.0, & \text{supported by } \{1,2,5,6\} \\
[prod = no] & \rightarrow D_1 \ \alpha = 1.0, \kappa = 1.0, & \text{supported by } \{1,2,5,6\} \\
[nau = no] & \rightarrow D_1 \ \alpha = 1.0, \kappa = 0.75, & \text{supported by } \{1,2,5\} \\
[nat = per] & \rightarrow D_1 \ \alpha = 1.0, \kappa = 0.75, & \text{supported by } \{1,2,6\} \\
[loc = who] & \rightarrow D_1 \ \alpha = 1.0, \kappa = 0.75, & \text{supported by } \{2,5,6\} \\
[age = 50-59] & \rightarrow D_1 \ \alpha = 1.0, \kappa = 0.5, & \text{supported by } \{2,6\}
\end{array}
$$

In the seventh step, classification rules for *m.c.h.* and *psycho* are induced from Table 3. For example, the following rules are obtained from *m.c.h.*.

$$[nau = no] \rightarrow m.c.h.\ \alpha = 1.0,\ \kappa = 1.0,\ \text{supported by } \{1,2,5\}$$
$$[age = 40-49] \rightarrow m.c.h.\ \alpha = 1.0,\ \kappa = 0.67,\ \text{supported by } \{2,5\}$$

In the eighth step, these two kinds of rules are integrated in the following way. For a rule $[M1 = yes] \rightarrow D_1$, $[nau = no] \rightarrow m.c.h.$ and $[age = 40-49] \rightarrow m.c.h.$ have a supporting set which is a subset of $\{1,2,5,6\}$. Thus, the following rules are obtained:

$$[M1 = yes]\ \&\ [\text{nau=no}] \rightarrow m.c.h.\ \alpha = 1.0,\ \kappa = 1.0,\ \text{supported by } \{1,2,5\}$$
$$[M1 = yes]\ \&\ [\text{age=40-49}] \rightarrow m.c.h.\ \alpha = 1.0,\ \kappa = 0.67,\ \text{supported by } \{2,5\}$$

5 Experimental Results

The above rule induction algorithm is implemented in PRIMEROSE4 (Probabilistic Rule Induction Method based on Rough Sets Ver 4.0), [4] and was applied to databases on differential diagnosis of headache, meningitis and cerebrovascular diseases (CVD), whose precise information is given in Table 4. In these experiments, δ_α, δ_κ and δ_μ were set to 0.75, 0.5 and 0.9, respectively. [5]

Table 4. Information about Databases

Domain	Samples	Classes	Attributes
headache	1477	20	20
meningitis	198	3	25
CVD	261	6	27

This system was compared with PRIMEROSE [16], C4.5[13], CN2[4], AQ15 and *k*-NN [6] with respect to the following points: length of rules, similarities between induced rules and expert's rules and performance of rules.

In this experiment, length was measured by the number of attribute-value pairs used in an induced rule and Jaccard's coefficient was adopted as a similarity measure, the definition of which is shown in the Appendix. Concerning the performance of rules, ten-fold cross-validation was applied to estimate classification accuracy.

Table 5 shows the experimental results, which suggest that PRIMEROSE4 outperforms the other four rule induction methods and induces rules very similar to medical experts' ones.

---

[4] The program is implemented by using SWI-prolog [15] on Sparc Station 20.

[5] These values are given by medical experts as good thresholds for rules in these three domains.

[6] The most optimal *k* for each domain is attached to Table 5.

Table 5. Experimental Results

Method	Length	Similarity	Accuracy
		Headache	
PRIMEROSE4	8.6 ± 0.27	0.93 ± 0.08	93.3 ± 2.7%
Experts	9.1 ± 0.33	1.00 ± 0.00	98.0 ± 1.9%
PRIMEROSE	5.3 ± 0.35	0.54 ± 0.05	88.3 ± 3.6%
C4.5	4.9 ± 0.39	0.53 ± 0.10	85.8 ± 1.9%
CN2	4.8 ± 0.34	0.51 ± 0.08	87.0 ± 3.1%
AQ15	4.7 ± 0.35	0.51 ± 0.09	86.2 ± 2.9%
k-NN (7)	6.7 ± 0.25	0.61 ± 0.09	88.2 ± 1.5%
		Meningitis	
PRIMEROSE4	2.6 ± 0.19	0.92 ± 0.08	92.0 ± 3.7%
Experts	3.1 ± 0.32	1.00 ± 0.00	98.0 ± 1.9%
PRIMEROSE	1.8 ± 0.45	0.64 ± 0.25	82.1 ± 2.5%
C4.5	1.9 ± 0.47	0.63 ± 0.20	83.8 ± 2.3%
CN2	1.8 ± 0.54	0.62 ± 0.36	85.0 ± 3.5%
AQ15	1.7 ± 0.44	0.65 ± 0.19	84.7 ± 3.3%
k-NN (5)	2.3 ± 0.41	0.71 ± 0.33	83.5 ± 2.3%
		CVD	
PRIMEROSE4	7.6 ± 0.37	0.89 ± 0.05	91.3 ± 3.2%
Experts	8.5 ± 0.43	1.00 ± 0.00	92.9 ± 2.8%
PRIMEROSE	4.3 ± 0.35	0.69 ± 0.05	84.3 ± 3.1%
C4.5	4.0 ± 0.49	0.65 ± 0.09	79.7 ± 2.9%
CN2	4.1 ± 0.44	0.64 ± 0.10	78.7 ± 3.4%
AQ15	4.2 ± 0.47	0.68 ± 0.08	78.9 ± 2.3%
k-NN (6)	6.2 ± 0.37	0.78 ± 0.18	83.9 ± 2.1%

k-NN (i) shows the value of i which gives the highest performance in k ($1 \leq k \leq 20$).

6 Discussion

6.1 Focusing Mechanism

One of the most interesting features in medical reasoning is that medical experts make a differential diagnosis based on focusing mechanisms: with several inputs, they eliminate some candidates and proceed into further steps. In this elimination, our empirical results suggest that grouping of diseases are very important to realize automated acquisition of medical knowledge from clinical databases. Readers may say that conceptual clustering or nearest neighborhood methods(*k*-NN)[1,14] will be useful for grouping. However, those two methods are based on classification accuracy, that is, they induce grouping of diseases, whose rules are of high accuracy. Their weak point is that they do not reflect medical reasoning: focusing mechanisms of medical experts are chiefly based not on classification accuracy, but on coverage.

Thus, we focus on the role of coverage in focusing mechanisms and propose an algorithm on grouping of diseases by using this measure. The above experiments show that rule induction with this grouping generates rules, which are similar to medical experts' rules and they suggest that our proposed method should capture medical experts' reasoning.

6.2 Granular Fuzzy Partition

Coverage is also closely related with granular fuzzy partition, which is introduced by Lin[7] in the context of granular computing.

Since coverage $\kappa_R(D)$ is equivalent to a conditional probability, $P(R|D)$,this measure will satisfy the condition on partition of unity, called BH-partition (If we select a suitable partition of universe, then this partition will satisfy $\sum_\kappa \kappa_R(D) = 1.0$.) Also, from the definition of coverage,it is also equivalent to the counting measure for $|[x]_R \bigcap D|$, since $|D|$ is constant in a given universe U. Thus, this measure satisfies a "nice context", which holds:

$$|[x]_{R_1} \bigcap D| + |[x]_{R_2} \bigcap D| \leq |D|.$$

Hence, all these features show that a partition generated by coverage is a kind of granular fuzzy partition[7]. This result also shows that the characterization by coverage is closely related with information granulation.

From this point of view, the usage of coverage for characterization and grouping of classes means that we focus on some specific partition generated by attribute-value pairs, the coverage of which are equal to 1.0 and that we consider the second-order relations between these pairs. It is also notable that if the second-order relation makes partition, as shown in the example above, then this structure can also be viewed as granular fuzzy partition.

However, rough inclusion and accuracy do not always hold the nice context. It would be our future work to examine the formal characteristics of coverage (and also accuracy) and rough inclusion from the viewpoint of granular fuzzy sets.

7 Conclusion

In this paper, the characteristics of experts' rules are closely examined, whose empirical results suggest that grouping of diseases are very important to realize automated acquisition of medical knowledge from clinical databases. Thus, we focus on the role of coverage in focusing mechanisms and propose an algorithm on grouping of diseases by using this measure. The above experiments show that rule induction with this grouping generates rules, which are similar to medical experts' rules and they suggest that our proposed method should capture medical experts' reasoning. Interestingly, the idea of this proposed procedure is very similar to rough mereology. The proposed method was evaluated on three medical databases, the experimental results of which show that induced rules correctly represent experts' decision processes and also suggests that rough mereology may be useful to capture medical experts' decision process.

References

1. Aha, D. W., Kibler, D., and Albert, M. K., Instance-based learning algorithm. *Machine Learning*, 6, 37-66, 1991.
2. Breiman, L., Freidman, J., Olshen, R., and Stone, C., *Classification And Regression Trees*, Wadsworth International Group, Belmont, 1984.
3. Buchnan, B. G. and Shortliffe, E. H., *Rule-Based Expert Systems*, Addison-Wesley, New York, 1984.
4. Clark, P. and Niblett, T., The CN2 Induction Algorithm. *Machine Learning*, 3, 261-283, 1989.
5. Everitt, B. S., *Cluster Analysis*, 3rd Edition, John Wiley & Son, London, 1996.
6. Langley, P. *Elements of Machine Learning*, Morgan Kaufmann, CA, 1996.
7. Lin, T.Y. Fuzzy Partitions: Rough Set Theory, in *Proceedings of Seventh International Conference on Information Processing and Management of Uncertainty in Knowledge-based Systems(IPMU'98)*, Paris, pp. 1167-1174, 1998.
8. Mannila, H., Toivonen, H., Verkamo, A.I., Efficient Algorithms for Discovering Association Rules, in *Proceedings of the AAAI Workshop on Knowledge Discovery in Databases (KDD-94)*, pp.181-192, AAAI press, Menlo Park, 1994.
9. Michalski, R. S., A Theory and Methodology of Machine Learning. *Machine Learning - An Artificial Intelligence Approach.* (Michalski, R.S., Carbonell, J.G. and Mitchell, T.M., eds.) Morgan Kaufmann, Palo Alto, 1983.
10. Michalski, R. S., Mozetic, I., Hong, J., and Lavrac, N., The Multi-Purpose Incremental Learning System AQ15 and its Testing Application to Three Medical Domains, in *Proceedings of the fifth National Conference on Artificial Intelligence*, 1041-1045, AAAI Press, Menlo Park, 1986.
11. Pawlak, Z., *Rough Sets*. Kluwer Academic Publishers, Dordrecht, 1991.
12. Polkowski, L. and Skowron, A.: Rough mereology: a new paradigm for approximate reasoning. Intern. J. Approx. Reasoning **15**, 333–365, 1996.
13. Quinlan, J.R., *C4.5 - Programs for Machine Learning*, Morgan Kaufmann, Palo Alto, 1993.
14. *Readings in Machine Learning*, (Shavlik, J. W. and Dietterich, T.G., eds.) Morgan Kaufmann, Palo Alto, 1990.
15. SWI-Prolog Version 2.0.9 Manual, University of Amsterdam, 1995.
16. Tsumoto, S. and Tanaka, H., PRIMEROSE: Probabilistic Rule Induction Method based on Rough Sets and Resampling Methods. *Computational Intelligence*, **11**, 389-405, 1995.
17. Tsumoto, S., Empirical Induction of Medical Expert System Rules based on Rough Set Model. PhD dissertation, 1997(in Japanese).
18. Zadeh, L.A., Toward a theory of fuzzy information granulation and its certainty in human reasoning and fuzzy logic. *Fuzzy Sets and Systems* **90**, 111-127, 1997.
19. Ziarko, W., The Discovery, Analysis, and Representation of Data Dependencies in Databases. in: *Knowledge Discovery in Databases*, (Shapiro, G. P. and Frawley, W. J., eds.), AAAI press, Menlo Park, pp.195-209, 1991.
20. Ziarko, W., Variable Precision Rough Set Model. *Journal of Computer and System Sciences*. 46, 39-59, 1993.

A Comparison of Rule Length

Table 6 shows comparison between induced rules and medical experts' rules with respect to the number of attribute-value pairs used to describe. The most important difference is that medical experts' rules are longer than induced rules for diseases of high prevalence.

Table 6. Comparision of Rule Length between Induced Rules and Medical Experts' Rules

Disease	Samples	PRIMEROSE	RHINOS
Muscle Contraction Headache	923	3.00	9.00
Disease of Cervical Spine	163	5.50	3.50
Common Migraine	112	4.00	7.50
Psychological Headache	79	6.67	3.67
Tension Vascular Headache	79	11.00	10.50
Classical Migraine	49	4.50	9.00
Teeth Disease	21	3.25	6.00
Costen Syndrome	19	4.00	3.00
Sinusitus	11	4.50	5.00
Neuritis of Occipital Nerves	5	10.00	14.00
Ear Disease	5	8.50	7.00
Intracranial Mass Lesion	2	2.75	3.75
Intracranial Aneurysm	2	4.00	2.00
Autonomic Disturbance	1	5.25	3.50
Trigeminus Neuralgia	1	5.25	3.50
Inflammation of Eyes	1	6.00	8.00
Arteriosclerotic Headache	1	9.50	11.00
Herpes Zoster	1	3.00	1.00
Tolosa-Hunt syndrome	1	6.00	4.00
Ramsey-Hunt syndrome	1	3.00	7.00
Total	1477		

B Similarity Measure

PRIMEROSE4 calculates the following similarity measure from all the inputs. Although there are many kinds of similarities[5], a family of similarity measures based on a contingency table is adopted. Let us consider a contingency table for a rule of a certain disease (Table 6). The first and second column denote the positive and negative information of an experts' rule. The first and second row denote the positive and negative information of an induced rule. Then, for example, a denotes the number of attributes in an induced rule which matches an experts' rule. From this table, several kinds of similarity measures can be defined. The best similarity measures in the statistical literature are four measures shown in Table 7. In PRIMEROSE4, users can choose a similarity measure from these four. As a default, Jaccard's coefficient, is used for defining similaritites, because it satisfies not only the low computational complexity, but also a good performance.

Table 7. Contigency Table for Similarity

		Rule		
		1	0	Total
Sample	1	a	b	a+b
	0	c	d	c+d

Table 8. Definition of Similarity Measures

(1) Matching Number	a
(2) Jaccard's coefficient	$a/(a+b+c)$
(3) χ^2-statistics	$N(ad-bc)^2/M$
(4) point correlation coefficient	$(ad-bc)/\sqrt{M}$

$N=a+b+c+d$, $M=(a+b)(b+c)(c+d)(d+a)$

Druck: Strauss Offsetdruck, Mörlenbach
Verarbeitung: Schäffer, Grünstadt